Birkhäuser

Pseudo-Differential Operators

Theory and Applications

Vol. 14

Pseudo-Differential Operators: Theory and Applications is a series of moderately priced graduate-level textbooks and monographs appealing to students and experts alike. Pseudo-differential operators are understood in a very broad sense and include such topics as harmonic analysis, PDE, geometry, mathematical physics, microlocal analysis, time-frequency analysis, imaging and computations. Modern trends and novel applications in mathematics, natural sciences, medicine, scientific computing, and engineering are highlighted.

More information about this series at http://www.springer.com/series/7390

Johannes Sjöstrand

Non-Self-Adjoint Differential Operators, Spectral Asymptotics and Random Perturbations

Birkhäuser

Johannes Sjöstrand
Université de Bourgogne Franche-Comté
Dijon, France

ISSN 2297-0355 ISSN 2297-0363 (electronic)
Pseudo-Differential Operators
ISBN 978-3-030-10818-2 ISBN 978-3-030-10819-9 (eBook)
https://doi.org/10.1007/978-3-030-10819-9

Mathematics Subject Classification (2010): 30C15, 35P20, 81Q12

This book is published under the imprint Birkhäuser, www.birkhauser-science.com by the registered company Springer Nature Switzerland AG.
The registered company address is: Gewerbestrasse 11, 6330 Cham, Switzerland

Abstract

The main purpose of this monograph is to describe some recent developments around the Weyl asymptotics for the eigenvalues of non-self-adjoint differential operators with random perturbations. It was natural to add some other results about non-self-adjoint operators, such as spectral stability and instability, growth bounds for semi-groups, and to review some of the classical theory.[1]

[1]Chapter 20 has been prepared in collaboration with Agnès E. Sjöstrand.

Contents

1 Introduction

Non-self-adjoint operators is an old, sophisticated and highly developed subject. See for instance Carleman [28] for an early result on Weyl type asymptotics for the real parts of the large eigenvalues of operators that are close to self-adjoint ones, with later results by Markus and Matseev [96, 97] in the same direction. (See the classical works Weyl [159], Avakumović [8], Levitan [94], Hörmander [80] for the asymptotics of large eigenvalues of elliptic self-adjoint operators and Robert [119] and Dimassi and Sjöstrand [41] for corresponding results in the semi-classical case, not to mention numerous deep and sophisticated results by Ivrii [86] and others.) Abstract theory with the machinery of s-numbers can be found in the book of Gohberg and Krein [48]. Other quite classical results concern upper bounds on the number of eigenvalues in various regions of the complex plane and questions about completeness of the set of all generalized eigenvectors. (Cf. Agmon [1, 2].)

In my own work I have often encountered non-self-adjoint operators. Thus for instance problems about analytic regularity for PDE's with pointwise degeneracy or in domains with conic singularities turned out to boil down to problems about non-self-adjoint operators; estimates on their resolvents and completeness of the set of generalized eigenvectors. Cf. [9, 10] with Baouendi. Non-self-adjoint quadratic operators turned up naturally in the study of certain classes of hypoelliptic operators with double characteristics, [24, 131] and in a number of recent works, this theme has been taken up again and is now quite an active field, see e.g., [34, 68–70, 75–77, 109, 112–114, 153, 154]. The study of resonances (scattering poles) of operators of Kramers–Fokker–Planck type with related problems of return to equilibrium are very rich domains where non-self-adjointness is an important ingredient; these, however, are beyond the scope of this book.

A major difficulty in the non-self-adjoint theory is that the norm of the resolvent may be very large even when the spectral parameter is far from the spectrum. This causes the

J. Sjöstrand, *Non-Self-Adjoint Differential Operators, Spectral Asymptotics and Random Perturbations*, Pseudo-Differential Operators 14,
https://doi.org/10.1007/978-3-030-10819-9_1

spectrum (and in this book we will only consider the discrete spectrum) to be unstable under small perturbations of the operator and this can be a source of numerical errors. Starting in the 90'ies the perspectives have changed thanks to works by numerical analysts, (see for instance the article Trefethen, [151], and the monograph [152] by Trefethen and Embree) who emphasized that the pseudospectrum—roughly the zone in the complex plane where the resolvent norm is large—can be of independent interest, for instance to understand the onset of turbulence from the (non-self-adjoint) linearizations of stationary flows. This spurred new interest among analysts like Davies [33, 34], Zworski and others [40, 161, 162].

Given the spectral instability it has been natural to study the effect on the eigenvalues of small random perturbations, and beautiful numerical results can be found for instance in [152]. In her thesis [53–55], M. Hager made a mathematical study of the spectrum of the sum of the simple operator $hD_x + g(x)$ on S^1 and a small random term. Here, $D_x = (1/i)d/dx = (1/i)\partial_x$ and as quite often in this text, we work in the semi-classical limit $h \searrow 0$. The spectrum of the unperturbated operator is just an arithmetic progression of eigenvalues on the line $\Im z = \langle \Im g \rangle = (2\pi)^{-1} \int_0^{2\pi} \Im g(x)dx$ and the effect of the random perturbation is to spread out the eigenvalues in the band $\{z \in \mathbf{C};\ \inf \Im g \le \Im z \le \sup \Im g\} = p(T^*S^1)$, where $p(x,\xi) = \xi + g(x)$ is the semi-classical principal symbol. The main result is that if Ω is a fixed bounded domain with smooth boundary in the interior of the band, then with probability very close to 1, the number of eigenvalues $N(\Omega)$ of the perturbed operator in Ω is given by

$$N(\Omega) = (2\pi h)^{-1}(\mathrm{vol}_{T^*S^1}(p^{-1}(\Omega)) + o(1)), \tag{1.0.1}$$

in the semi-classical limit, $h \to 0$, with an explicit estimate on the remainder $o(1)$.[1] We here recognize the natural non-self-adjoint version of Weyl asymptotics in the semi-classical limit, well established for large eigenvalues of self-adjoint differential operators since more than a century and later in the semi-classical self-adjoint case.

This came as a big surprise since in the cases known to me, one has to assume analyticity to get eigenvalue asymptotics via complex Bohr–Sommerfeld conditions. In one dimension such eigenvalues typically sit on curves which is incompatible with (1.0.1). (We here discuss genuinely non-self-adjoint operators with complex valued principal symbol.)

Thus with hindsight one can say that the random perturbation will typically destroy (uniform) analyticity and hence destroy all possible asymptotic formulae in terms of complex phase space (like complex Bohr–Sommerfeld conditions). Among the possible remaining formulae in terms of real phase space, the Weyl asymptotics seems to be the only possible one.

[1] At first, the formula appeared more complicated, depending on the method of proof, and the simpler form was pointed out to me by E. Amar-Servat.

After this first result, there have been several works, by Hager [55], Bordeaux Montrieux [16–18] as well as [19, 56, 137, 139] that treated more general situations and obtained more precise results. During this process the methods were improved and the main purpose of this research monograph is to give a unified account,[2] leaving out many other recent developments for non-self-adjoint operators. We also leave out some related results for resonances (see [144] and references given there). A related, very promising approach for studying directly the expected eigenvalue density and correlations can be found in work of Christiansen and Zworski [32] and Vogel [147, 155, 156]. Since this approach is currently less developed for spectral problems (especially in higher dimensions), we will not treat it here.

Quite naturally, the classical theory of non-self-adjoint operators [48] is a fundamental ingredient of the methods. We will also use some results of complex analysis, in particular a result on counting the zeros of holomorphic functions with exponential growth. Microlocal analysis in its classical C^∞ version will be important also. Except for the complex WKB method in Chap. 7, we do not touch upon the subject of analytic operators and methods of analytic microlocal analysis, a subject that would require a much longer and deeper text.

We will assume that the reader has some familiarity with standard microlocal analysis, roughly corresponding to the books [51] with A. Grigis and [41], but we think that even without formal prerequisites in that area, much of the book should remain accessible. Applications have not been our main motivation, but rather to present a coherent piece of mathematical analysis where Weyl asymptotic appeared first as a surprise.

The text is split into three parts.

- **Part I** is devoted to some functional analysis and to spectral theory in dimension 1.
  - In Chap. 2 we review some notions including spectrum and the pseudospectrum in the spirit of Davies [35] and Trefethen [151].
  - In Chap. 3 we discuss the original result of Hager on Weyl asymptotics for random perturbations of the model operator $hD + g$, where many ideas appear in a non-technical context.
  - In Chap. 4 we use a classical WKB-approach to construct quasimodes for differential operators, generalizing a result of Davies [33] for the non-self-adjoint Schrödinger operators. (The construction in any dimension is given in Chap. 9, and goes back to Hörmander [78, 79].)
  - Chapter 5 is devoted to Weyl asymptotics for more general differential operators in one dimension, in spirit close to [55].
  - In Chap. 6 we establish a result of Bordeaux Montrieux [18] about the norm of the resolvent near the boundary of the range of the symbol. What is remarkable is that

[2] Including several articles of the author, often with minor changes.

we get not only a precise upper bound, but even the asymptotics of the norm. Later, in Chap. 10 we review upper bounds in any dimension.
  - In Chap. 7 we explain the complex WKB method for ordinary differential equations.
  - In Chap. 8 we review abstract theory of non-self-adjoint operators, following essentially [48].
- **Part II** deals with various general facts.
  - In Chap. 9 we generalize the construction of quasimodes to any dimension, cf. [40, 78, 79, 161].
  - In Chap. 10 we give resolvent bounds near the boundary of the range of the symbol for semi-classical operators in any dimension. We follow closely [140].
  - In Chap. 11 we discuss some abstract questions around the Gearhardt–Prüss–Hwang–Greiner theorem, getting estimates on semigroups from estimates on resolvents. This is mainly joint work with B. Helffer, originally published in [138] and later also in [57].
  - In Chap. 12 we give a result about the number of zeros of a holomorphic function with exponential growth. A simpler version of this result was originally proved by Hager, see Proposition 3.4.6. The improvements here are essential for the more precise and general results in Chap. 13 and in Part III.
  - In Chap. 13, we study the distribution of eigenvalues of small random perturbations of large Jordan blocks. We show that with probability close to 1, the eigenvalues concentrate to a certain circle and have an approximately uniform angular distribution there. Originally I planned a last section, based on a joint work with Vogel [147] about the expected density of eigenvalues inside that circle, but we now refer to that work instead, as well as to related more recent works.
- **Part III.** This part deals with spectral asymptotics for differential operators in arbitrary dimension.
  - In Chap. 14 we review a result of Markus and Matseev [97] (see also [96]) about Weyl distribution of the real parts of the eigenfrequencies of the damped wave equation. We choose to use Chap. 12.
  - In Chaps. 15–17 we present and prove a general result on Weyl asymptotics for semi-classical (pseudo-)differential operators on a compact manifold which basically improves the main result of [139]. The proof follows the strategy of [137, 139], with the difference that we can now use the improved results of Chap. 12.
  - Chapter 18 gives almost sure Weyl asymptotics for the large eigenvalues of differential operators, no longer in the semi-classical limit. We here follow [19].
  - In Chap. 19 we apply the results to $\mathcal{PT}$-symmetric operators. Such operators are generally non-self-adjoint, but with an additional symmetry that forces the spectrum to be symmetric around the real line. They have been proposed by physicists as building blocks in new versions of quantum mechanics and the reality of the spectrum is then important. We first show (following [143]) that most $\mathcal{PT}$-symmetric operators have most of their eigenvalues away from $\mathbf{R}$ in the semi-classical case as well as

in that of large eigenvalues. Then we describe without detailed proofs some results of [23] about the reality of eigenvalues for semi-classical $\mathcal{PT}$-symmetric *analytic* Schrödinger operators with a simple potential well in dimension one and a result of [101] about the non-reality of the eigenvalues for semi-classical Schrödinger operators with a double-well potential.
- In Chap. 20, prepared jointly with A. Sjöstrand, we give numerical illustrations to Weyl asymptotics and spectral instability, as well as double-well Schrödinger operator. The examples are one-dimensional, but often illustrate results that hold in all dimensions.

As already mentioned, this text is a coherent description of some developments for non-self-adjoint operators, but far from all. Even within our special horizons, some aspects are missing and should be the subject of further research:

- Give more general and intrinsic classes of random perturbations: a look at the results clearly indicates that this should be possible.
- Treat non-elliptic operators like the Kramers–Fokker–Planck operator.
- Study boundary value problems.
- Discuss applications to resonances (scattering poles). Results for generic perturbations have been obtained by Christiansen et al. [29–31] and in probability in [144]
- From the proofs for Weyl asymptotics for operators with random perturbations, it is quite clear that the resolvent bounds improve in the pseudospectral region (cf. [18]). We think that this is an important phenomenon, to be studied further.

This book is a research monograph on topics in non-self-adjoint operator theory, and an attempt to give a systematic presentation of some recent developments. It is not self-contained, and for the full understanding of some chapters, the reader is supposed to have some familiarity with microlocal analysis (or to revise that topic in parallel with the study of selected parts of this book). We will also use some complex analysis and some basic functional analysis, as well as some very basic probability theory, covered for instance by the first two chapters in the book [87] by Kallenberg. The following chapters are more elementary (not excluding various amounts of general mathematical analysis):

Chapters 2–4, 7–9, assuming some notions of symplectic geometry, Chaps. 11–13, 15 (the proofs in Chaps. 16 and 17 do use some microlocal analysis), Chaps. 18–20.

Acknowledgements Discussions with colleagues and coworkers have been a very important basis for this work. I am grateful to W. Bordeaux Montrieux, M. Hager, M. Hitrik, B. Helffer, K. Pravda-Starov, J. Viola, M. Vogel, X. P. Wang, M. Zworski and many others. We also thank the two referees for a very careful work and most useful comments.

Part I

Basic Notions, Differential Operators in One Dimension

2 Spectrum and Pseudo-Spectrum

2.1 Operators in Hilbert Spaces, a Quick Review

In this book all Hilbert spaces will be assumed to separable for simplicity. In this section we review some basic definitions and properties; we refer to [88, 116, 117] for more substantial presentations.

Let $\mathcal{H}$ be a complex Hilbert space and denote by $\|\cdot\|$, $(\cdot|\cdot)$ the norm and the scalar product, respectively.[1] A unbounded (or rather not necessarily bounded) operator $S : \mathcal{H} \to \mathcal{H}$ is given by a linear subspace $\mathcal{D}(S) \subset \mathcal{H}$, the domain of S, and a linear mapping $S : \mathcal{D}(S) \to \mathcal{H}$. We say that S is bounded if $\mathcal{D}(S) = \mathcal{H}$ and

$$\|S\| = \sup_{0 \neq x \in \mathcal{H}} \frac{\|Sx\|}{\|x\|} < \infty. \tag{2.1.1}$$

Let $\mathcal{L}(\mathcal{H}, \mathcal{H})$ denote the corresponding normed space of bounded operators. These definitions and many of the facts below have straightforward extensions to the case of operators $S : \mathcal{H}_1 \to \mathcal{H}_2$, where $\mathcal{H}_j$ are two Hilbert spaces.

If $S : \mathcal{H} \to \mathcal{H}$ is an unbounded operator, we introduce its graph

$$\text{graph}\,(S) = \{(x, Sx);\ x \in \mathcal{D}(S)\}, \tag{2.1.2}$$

which is a linear subspace of $\mathcal{H} \times \mathcal{H}$. We say that S is closed if graph (S) is closed. Every bounded operator $S \in \mathcal{L}(\mathcal{H}, \mathcal{H})$ is closed; conversely, the closed graph theorem tells us

[1] Linear in the first argument and anti-linear in the second.

J. Sjöstrand, *Non-Self-Adjoint Differential Operators, Spectral Asymptotics and Random Perturbations*, Pseudo-Differential Operators 14,
https://doi.org/10.1007/978-3-030-10819-9_2

that if $S : \mathcal{H} \to \mathcal{H}$ is closed and $\mathcal{D}(S) = \mathcal{H}$, then S is bounded, i.e., $S \in \mathcal{L}(\mathcal{H}, \mathcal{H})$. In the next proposition, we introduce the adjoint S^* of a densely defined operator:

Proposition 2.1.1 *Let $S : \mathcal{H} \to \mathcal{H}$ be an unbounded operator whose domain $\mathcal{D}(S)$ is dense in $\mathcal{H}$. Then there exists an unbounded operator $S^* : \mathcal{H} \to \mathcal{H}$ characterized by:*

$$\mathcal{D}(S^*) = \{v \in \mathcal{H};\ \exists C(v) \in]0, +\infty[,\ |(Su|v)| \le C(v)\|u\|,\ \forall u \in \mathcal{D}(S)\}, \tag{2.1.3}$$

$$(Su|v) = (u|S^*v);\ \forall u \in \mathcal{D}(S),\ v \in \mathcal{D}(S^*). \tag{2.1.4}$$

Proof (2.1.3) is just a definition and the space $\mathcal{D}(S^*)$ so defined is linear. For $v \in \mathcal{D}(S^*)$, the map $\mathcal{H} \ni u \mapsto (Su|v)$ is a bounded linear functional, so there is a unique element $w \in \mathcal{H}$ such that $(Su|v) = (u|w)$ for all $u \in \mathcal{D}(S)$. By definition, $S^*v = w$ and it is straightforward to see that S^*v depends linearly on v. □

We have the following easily verified properties:

- S^* is closed.
- Let $\overline{S}$ denote the closure of S, so that in general $\overline{S}$ is the relation $\mathcal{H} \to \mathcal{H}$ whose graph is the closure of the graph of S; graph $(\overline{S}) = \overline{\text{graph}(S)}$. We say that S is closable if $\overline{S}$ is an operator.

  If $\mathcal{D}(S^*)$ is dense, then $S^{**} = \overline{S}$. In particular, S is closable.

Definition 2.1.2 Let $A, B : \mathcal{H} \to \mathcal{H}$ be unbounded operators. We say that $A \subset B$ if graph $(A) \subset$ graph (B), or equivalently if $\mathcal{D}(A) \subset \mathcal{D}(B)$ and $Ax = Bx$ for all $x \in \mathcal{D}(A)$.

If $\mathcal{D}(A)$ is dense and $A \subset B$, then $B^* \subset A^*$.

Definition 2.1.3 A densely defined operator $A : \mathcal{H} \to \mathcal{H}$ is symmetric if $A \subset A^*$ and self-adjoint if $A = A^*$.

Notice that a densely defined operator $A : \mathcal{H} \to \mathcal{H}$ is symmetric iff

$$(Ax|y) = (x|Ay),\ \forall x, y \in \mathcal{D}(A).$$

An important general problem is to determine when a symmetric operator $A : \mathcal{H} \to \mathcal{H}$ has a self-adjoint extension $\widetilde{A} : \mathcal{H} \to \mathcal{H}$ (in the sense that $A \subset \widetilde{A}$). Notice that a self-adjoint operator is always closed, so every self-adjoint extension $\widetilde{A}$ of a symmetric operator A has to contain the closure $\overline{A}$. The "best" case is when A is essentially self-adjoint in the sense that A has a unique self-adjoint extension. It can be shown that when a symmetric operator A is essentially self-adjoint, then the unique self-adjoint extension of A is the closure $\overline{A}$. Equivalently, a symmetric operator A is essentially self-adjoint precisely when

$\overline{A}$ is self-adjoint. Recall that many of these statements are easy to understand if we make the observation that

$$(J(\text{graph}\,(A))^{\perp} = \text{graph}\,(A^*),$$

where the superscript $\perp$ indicates the orthogonal space in $\mathcal{H} \times \mathcal{H}$ and

$$J = \begin{pmatrix} 0 & 1 \\ -1 & 0 \end{pmatrix} : \mathcal{H} \times \mathcal{H} \to \mathcal{H} \times \mathcal{H},$$

so that $J^*J = 1$ and $JJ^* = 1$ on $\mathcal{H} \times \mathcal{H}$, where 1 denotes the identity operator.

We have the following theorem of von Neumann, see [88, p. 275]:

Theorem 2.1.4 *Let $T : \mathcal{H} \to \mathcal{H}$ be closed and densely defined. Then the operator T^*T with its natural domain $\mathcal{D}(T^*T) = \{x \in \mathcal{D}(T);\ Tx \in \mathcal{D}(T^*)\}$ is self-adjoint. Moreover, $\mathcal{D}(T^*T)$ is a core for T in the sense that $\{(u, Tu);\ u \in \mathcal{D}(T^*T)\}$ is dense in $\mathcal{D}(T)$.*

Definition 2.1.5 A closed densely defined operator $T : \mathcal{H} \to \mathcal{H}$ is normal if $T^*T = TT^*$ (where the second operator TT^* is equipped with its natural domain).

Every self-adjoint operator is normal. Many properties of self-adjoint operators extend to normal ones.

Spectrum, Resolvent Let $T : \mathcal{H} \to \mathcal{H}$ be closed and densely defined. It is often practical to equip $\mathcal{D}(T)$ with then norm $\|u\|_{\mathcal{D}(T)} = (\|u\|^2 + \|Tu\|^2)^{\frac{1}{2}} := \|(u, Tu)\|_{\mathcal{H}\times\mathcal{H}}$. We say that $z_0 \in \mathbf{C}$ belongs to the resolvent set $\rho(T)$ of T if $T - z_0 : \mathcal{D}(T) \to \mathcal{H}$ is bijective and the inverse $(z_0 - T)^{-1} : \mathcal{H} \to \mathcal{H}$ is bounded. Here we write $(T - z_0)u = Tu - z_0u$. (Notice that $(T - z_0)^{-1}$ is bounded $\mathcal{H} \to \mathcal{H}$ precisely when it is bounded $\mathcal{H} \to \mathcal{D}(T)$ and that the closed graph theorem tells us that it is indeed bounded, once it is well defined.)

When $z_0 \in \rho(T)$ we define the resolvent $R(z_0) = (z_0 - T)^{-1}$. For $z \in \mathbf{C}$, we have

$$(z - T)R(z_0) = 1 + (z - z_0)R(z_0), \tag{2.1.5}$$

whereas above "1" denotes the identity operator. Here

$$\|(z - z_0)R(z_0)\| \le |z - z_0|\|R(z_0)\|$$

is < 1 when z belongs to the open disc $D(z_0, 1/\|R(z_0)\|)$ of center z_0 and radius $1/\|R(z_0)\|$, and the operator $1 + (z - z_0)R(z_0)$ then has the bounded inverse given by the Neumann series

$$(1 + (z - z_0)R(z_0))^{-1} = 1 - (z - z_0)R(z_0) + ((z - z_0)R(z_0))^2 - \cdots$$

We also see that

$$\|(1+(z-z_0)R(z_0))^{-1}\| \le \frac{1}{1-|z-z_0|\|R(z_0)\|}. \tag{2.1.6}$$

(Here we use that the normed linear space $\mathcal{L}(\mathcal{H},\mathcal{H})$ is complete.) From (2.1.5) it now follows that $z-T:\mathcal{D}(T)\to\mathcal{H}$ has the right inverse

$$\widetilde{R}(z)=R(z_0)(1+(z-z_0)R(z_0))^{-1} \tag{2.1.7}$$

with norm

$$\|\widetilde{R}(z)\| \le \frac{\|R(z_0)\|}{1-|z-z_0|\|R(z_0)\|}.$$

Since $1+(z-z_0)R(z_0)$ maps $\mathcal{D}(T)$ into itself, we see that the inverse has the same property. Using also that

$$R(z_0)(z-T)=1+(z-z_0)R(z_0), \tag{2.1.8}$$

we see that $\widetilde{R}(z)$, defined in (2.1.7), is also a left inverse, so z belongs to the resolvent set and $R(z)=\widetilde{R}(z)$.

Proposition 2.1.6 *Let $z_0\in\rho(T)$, $z\in D(z_0,1/\|R(z_0)\|)$. Then $z\in\rho(T)$ and*

$$\|R(z)\| \le \frac{\|R(z_0)\|}{1-|z-z_0|\|R(z_0)\|}.$$

It follows in particular that the resolvent set $\rho(T)$ is open.

Definition 2.1.7 Let $T:\mathcal{H}\to\mathcal{H}$ be a closed densely defined operator. The spectrum $\sigma(T)$ is the closed subset of $\mathbf{C}$ defined as

$$\sigma(T)=\mathbf{C}\setminus\rho(T). \tag{2.1.9}$$

From Proposition 2.1.6 we see that

$$\|R(z)\| \ge \frac{1}{\mathrm{dist}\,(z,\sigma(T))},\quad z\in\rho(T), \tag{2.1.10}$$

where $\mathrm{dist}\,(z,\sigma(T))=\inf_{w\in\sigma(T)}|z-w|$ denotes the distance from z to the set $\sigma(T)$. From the spectral resolution theorem for self-adjoint operators, we have

Theorem 2.1.8 *Let $T : \mathcal{H} \to \mathcal{H}$ be self-adjoint. Then $\sigma(T) \subset \mathbf{R}$ and*

$$\|R(z)\| = \frac{1}{\operatorname{dist}(z, \sigma(T))}, \quad z \in \rho(T). \tag{2.1.11}$$

2.2 Pseudospectrum

Let $P : \mathcal{H} \to \mathcal{H}$ be closed and densely defined.

Definition 2.2.1 Let $\epsilon > 0$. The ϵ-pseudospectrum of P is the set

$$\sigma_\epsilon(P) = \sigma(P) \cup \{z \in \rho(P);\ \|(z - P)^{-1}\| > 1/\epsilon\}. \tag{2.2.1}$$

Some authors put "$\geq$" rather than "$>$" in this definition. We follow [35, 151] and the book [152]. With this choice, $\sigma_\epsilon(P)$ becomes an open subset of $\mathbf{C}$.

From (2.1.10) it follows that

$$\sigma(P) + D(0, \epsilon) \subset \sigma_\epsilon(P). \tag{2.2.2}$$

It is a standard fact for non-self-adjoint operators that the second set in (2.2.2) may be much larger than the first one. For self-adjoint operators we have equality by (2.1.11). Thus for general non-self-adjoint operators, the existence of quasimodes for $P - z$ does not necessarily imply that z is close to the spectrum of P. The following proposition shows though that there is a close link between the existence of quasimodes and the ϵ-pseudospectrum. (The notion of "quasimode" is implicitly defined in (2.2.3) below.)

Proposition 2.2.2 *Let $\epsilon > 0$, $z \in \mathbf{C}$. The following two statements are equivalent:*

1) $z \in \sigma_\epsilon(P)$.
2) $z \in \sigma(P)$, *or*

$$\exists u \in \mathcal{D}(P), \text{ such that } \|u\| = 1 \text{ and } \|(P - z)u\| < \epsilon. \tag{2.2.3}$$

Proof It suffices to show that for $z \in \rho(P)$ the statement

$$z \in \sigma_\epsilon(P) \tag{2.2.4}$$

is equivalent to (2.2.3). Now (2.2.3) is equivalent to:

$$\exists 0 \neq v \in \mathcal{H}, \text{ such that } \|v\| < \epsilon,\ \|(z - P)^{-1}v\| = 1,$$

which in turn is equivalent to (2.2.4) with the choice $u = (z - P)^{-1}v$. □

In the above situation we call u a *quasimode* and z the corresponding *quasi-eigenvalue*. The ϵ-pseudospectrum is a set of spectral instability: a small perturbation of P may change the spectrum a lot. This is formalized in the following easy result:

Theorem 2.2.3 *For every* $\epsilon > 0$,

$$\sigma_\epsilon(P) = \bigcup_{\substack{A\in\mathcal{L}(\mathcal{H},\mathcal{H})\\ \|A\|<\epsilon}} \sigma(P+A). \tag{2.2.5}$$

Proof Denote the right-hand side of (2.2.5) by $\widetilde{\sigma}_\epsilon(P)$. Clearly $\widetilde{\sigma}_\epsilon(P)$ contains $\sigma(P)$, so we only have to identify $\sigma_\epsilon(P)\setminus\sigma(P)$ and $\widetilde{\sigma}_\epsilon(P)\setminus\sigma(P)$.

Let $z\in\sigma_\epsilon(P)\setminus\sigma(P)$ so that $(z-P)^{-1}$ exists and is of norm $> 1/\epsilon$. Then, by Proposition 2.2.2, there exist a vector u as in (2.2.3). Let $v=(P-z)u$, so that $\|v\|<\epsilon$. Now we can find $A\in\mathcal{L}(\mathcal{H},\mathcal{H})$ with $\|A\|<\epsilon$, $Au=v$. For instance, we can define A by $Ax=(x|u)v$. Then $(P+A-z)u=0$, so $z\in\sigma(P+A)$, and hence $z\in\widetilde{\sigma}_\epsilon(P)\setminus\sigma(P)$.

Next, let $z\in\mathbf{C}\setminus\sigma_\epsilon(P)$, so that $\|(P-z)^{-1}\|\le 1/\epsilon$. Let $A\in\mathcal{L}(\mathcal{H},\mathcal{H})$, $\|A\|<\epsilon$. Then

$$(P+A-z)(P-z)^{-1}=1+A(P-z)^{-1},\quad \|A(P-z)^{-1}\|<1.$$

Thus $1+A(P-z)^{-1}:\mathcal{H}\to\mathcal{H}$ has a bounded inverse and we see that $P+A-z:\mathcal{D}(P)\to\mathcal{H}$ has the bounded right inverse $(P-z)^{-1}(1+A(P-z)^{-1})^{-1}$.

Similarly, $(P-z)^{-1}(P+A-z)=1+(P-z)^{-1}A$ is bijective $\mathcal{H}\to\mathcal{H}$ and $\mathcal{D}(P)\to\mathcal{D}(P)$ and $P+A-z$ has the bounded left inverse $(1+(P-z)^{-1}A)^{-1}(P-z)^{-1}$. We conclude that $z\notin\sigma(P+A)$ and varying A it follows that $z\notin\widetilde{\sigma}_\epsilon(P)$. □

Subharmonicity This property allows us to use the maximum principle in order to establish some general properties of pseudospectra. We start by recalling some general properties, see [84].

Definition 2.2.4 Let $\Omega\subset\mathbf{C}$ be open and let $u:\Omega\to[-\infty,+\infty[$. We say that u is subharmonic if

(a) u is upper semi-continuous, i.e., $u^{-1}([-\infty,s[)$ is open for every $s\in\mathbf{R}$,
(b) If $K\subset\Omega$ is compact, $h\in C(K;\mathbf{R})$ is harmonic on the interior of K (in the sense that $\Delta h=0$ there) and $h\ge u$ everywhere on ∂K, then $h\ge u$ in K.

In this definition we can restrict K to the set of closed discs contained in $\mathcal{O}$. (Theorem 1.6.3 in [84].) Another important property (Theorem 1.6.2 in [84]) is

Theorem 2.2.5 *Let* u_α, $\alpha\in\mathcal{A}$, *be family of subharmonic functions such that* $u:=\sup_{\alpha\in\mathcal{A}}u_\alpha$ *is pointwise* $<+\infty$ *and upper semi-continuous. Then* u *is subharmonic.*

As the name indicates, every harmonic function is subharmonic.

We recall the characterization of subharmonic functions as those for which $\Delta u \geq 0$ in the sense of distributions [84, Theorem 1.6.9–1.6.11].

Theorem 2.2.6 *Let Ω be open and connected and let $L(dz)$ denote the Lebesgue measure on* $\mathbf{C}$.

(a) *If u is a subharmonic function on Ω, not identically $-\infty$, then $u \in L^1_{\mathrm{loc}}(\Omega)$ and Δu is a positive distribution:*

$$\int u \Delta v L(dz) \geq 0, \text{ for all } 0 \leq v \in C_0^2(\Omega). \tag{2.2.6}$$

(b) *Conversely, if $u \in L^1_{\mathrm{loc}}(\Omega)$ and* (2.2.6) *holds, then there exists a unique subharmonic function U on Ω that is equal to u almost everywhere.*

We now return to our general closed and densely defined operator $P : \mathcal{H} \to \mathcal{H}$.

Proposition 2.2.7 *The function $z \mapsto \|(z-P)^{-1}\|$ is subharmonic on $\rho(P)$.*

Proof We use the variational formula

$$\|u\| = \sup_{\substack{v \in \mathcal{H} \\ \|v\|=1}} \Re(u|v), \;\; u \in \mathcal{H},$$

to deduce that for $z \in \rho(P)$,

$$\|(z-P)^{-1}\| = \sup_{\substack{u,v \in \mathcal{H} \\ \|u\|,\|v\| \leq 1}} \Re((z-P)^{-1} u | v).$$

Here $\rho(P) \ni z \mapsto \Re((z-P)^{-1}u|v)$ is harmonic for every fixed $(u, v) \in \mathcal{H} \times \mathcal{H}$. Moreover, $\rho(P) \ni z \mapsto \|(z-P)^{-1}\|$ is continuous and pointwise finite, so Theorem 2.2.5 gives the desired conclusion. □

Applying the maximum principle, we get

Theorem 2.2.8 *Every bounded connected component of $\sigma_\epsilon(P)$ contains some point of $\sigma(P)$.*

Proof We first remark that if z_0 is a point in the spectrum $\sigma(P)$ of P but not in the interior of $\sigma(P)$, then $\|(z-P)^{-1}\| \to +\infty$, $\rho(P) \ni z \to z_0$.

Let $V \subset \mathbf{C}$ be a bounded connected component of $\sigma_\epsilon(P)$. We notice that $\|(z-P)^{-1}\| = 1/\epsilon$ everywhere on the boundary of V.

If V does not intersect the spectrum of P, then the function $f(z) := \|(z-P)^{-1}\|$ is continuous and subharmonic in a small open neighborhood Ω of $\overline{V}$. Thus the maximum principle for subharmonic functions (apply Proposition 2.2.4 with $K = \overline{V}$, $h = 1/\epsilon$) implies that $f(z) \le 1/\epsilon$ in V, which is in contradiction with the fact that $\|(z-P)^{-1}\| > 1/\epsilon$ in V. □

2.3 Numerical Range

Another set which has interesting connections with the spectrum and the pseudospectrum is the numerical range. As above, let $P : \mathcal{H} \to \mathcal{H}$ be a closed densely defined operator.

Definition 2.3.1 We define the numerical range $W(P) \subset \mathbf{C}$ of P by

$$W(P) = \left\{ \frac{(Pu|u)}{\|u\|^2};\ 0 \neq u \in \mathcal{D}(P) \right\}. \tag{2.3.1}$$

Notice that we get the same set if we restrict u to the set of all $u \in \mathcal{D}(P)$ with $\|u\| = 1$. The following theorem is due to Hausdorff and Toeplitz (see [85]).

Theorem 2.3.2 *$W(P)$ is convex.*

Proof Let $z_0, z_1 \in W(P)$ be two distinct points, so that $z_j = (Pe_j|e_j)$, $e_j \in \mathcal{D}(P)$, $\|e_j\| = 1$. Then e_0, e_1 are linearly independent and it suffices to show that $\{(Pu|u)/\|u\|^2;\ u \in \Sigma\}$ is convex, where $\Sigma := \mathbf{C}e_0 \oplus \mathbf{C}e_1$. For $u \in \Sigma$, we have $(Pu|u) = (\Pi_\Sigma Pu|u)$, where $\Pi_\Sigma : \mathcal{H} \to \Sigma$ is the orthogonal projection onto Σ. Thus we may replace P by $\Pi_\Sigma P : \Sigma \to \Sigma$ and thus reduce the proof of the theorem to the case when $\mathcal{H}$ is a two-dimensional Hilbert space. After choosing some orthonormal basis in $\mathcal{H}$ we can identify P with a square matrix and $\mathcal{H}$ with $\mathbf{C}^2$.

Write $P = \Re P + i\Im P$, where $\Re P = (P + P^*)/2$ and $\Im P = (P - P^*)/(2i)$ are Hermitian matrices. After conjugation with a unitary matrix, we can assume that

$$\Re P = \begin{pmatrix} \lambda_1 & 0 \\ 0 & \lambda_2 \end{pmatrix}$$

where λ_1 and λ_2 are real. Thus for $u = (u_1, u_2) \in \mathbf{C}^2 \setminus 0$, we have

$$\frac{(Pu|u)}{\|u\|^2} = \frac{\lambda_1|u_1|^2 + \lambda_2|u_2|^2}{|u_1|^2 + |u_2|^2} + i\frac{(\Im Pu|u)}{|u_1|^2 + |u_2|^2}. \tag{2.3.2}$$

Now the Hermitian matrix $\Im P$ takes the form

$$\Im P = \begin{pmatrix} \mu_1 & a \\ \overline{a} & \mu_2 \end{pmatrix}, \quad \mu_1, \mu_2 \in \mathbf{R},$$

so finally

$$\frac{(Pu|u)}{\|u\|^2} = \frac{(\lambda_1 + i\mu_1)|u_1|^2 + (\lambda_2 + i\mu_2)|u_2|^2}{|u_1|^2 + |u_2|^2} + i\frac{2\Re(\overline{a}u_1\overline{u}_2)}{|u_1|^2 + |u_2|^2}. \tag{2.3.3}$$

When investigating the set of these values, we may assume that $\|u\|^2 = 1$, so that

$$|u_1|^2 = 1 - t, \ |u_2|^2 = t, \quad 0 \le t \le 1. \tag{2.3.4}$$

Then

$$|2\Re(\overline{a}u_1\overline{u}_2)| \le 2|a|\sqrt{t(1-t)}.$$

More precisely, for every $\theta \in [-2|a|\sqrt{t(1-t)}, 2|a|\sqrt{t(1-t)}]$ there exists a u satisfying (2.3.4) such that $2\Re(\overline{a}u_1\overline{u}_2) = \theta$, and we conclude that the set of values in (2.3.2) is equal to

$$\{(1-t)(\lambda_1 + i\mu_1) + t(\lambda_2 + i\mu_2) + i\theta; \ |\theta| \le 2|a|\sqrt{t(1-t)}, \ 0 \le t \le 1\}. \tag{2.3.5}$$

The function $\sqrt{t(1-t)}$ is concave, so the set (2.3.5) is convex. More precisely, it is an ellipsoid. □

If $\lambda \in \mathbf{C}$ is an eigenvalue of P, so that $Pu = \lambda u$ for some $0 \ne u \in \mathcal{D}(P)$, then it is immediate that $\lambda \in W(P)$. We have

Theorem 2.3.3 *Let $\Omega \subset$ be a connected component of the open set $\mathbf{C} \setminus \overline{W(P)}$, so that Ω is open. If Ω contains a point z_0 which is not in $\sigma(P)$, then $\Omega \cap \sigma(P) = \emptyset$, and moreover*

$$\|(z-P)^{-1}\| \le \frac{1}{\operatorname{dist}(z, W(P))}, \quad z \in \Omega. \tag{2.3.6}$$

For the proof we will use the following proposition of independent interest:

Proposition 2.3.4 *For every $z \in \mathbf{C}$, we have*

$$\operatorname{dist}(z, W(P))\|u\| \le \|(P-z)u\|, \quad u \in \mathcal{D}(P). \tag{2.3.7}$$

Proof For every non-vanishing $u \in \mathcal{D}(P)$, we have

$$|((z-P)u|u)| = \left| z\|u\|^2 - (Pu|u) \right| = |z-w|\|u\|^2,$$

where $w = (Pu|u)/\|u\|^2$ belongs to $W(P)$. Thus,

$$\mathrm{dist}\,(z, W)\|u\|^2 \le |z-w|\|u\|^2 \le |((P-z)u|u)| \le \|(P-z)u\|\|u\|.$$

Dividing by the norm of u, we get (2.3.7) for non-zero u. When $u = 0$, (2.3.7) holds trivially. □

Proof of Theorem 2.3.3 Let $z_0 \in \Omega \setminus \sigma(P)$ and let $z_1 \in \Omega$. Knowing that the resolvent $(z_0 - P)^{-1}$ exists, we can apply (2.3.7) to see that

$$\|(z_0 - P)^{-1}\| \le \frac{1}{\mathrm{dist}\,(z_0, W(P))},$$

and hence that $D(z_0, \mathrm{dist}\,(z_0, W(P))) \subset \rho(P)$. Now let γ be a C^1 curve in Ω that connects z_0 to z_1. Then we can find finitely many points $w_0, w_1, \dots, w_N$ with $w_0 = z_0$, $w_N = z_1$ in (the image of) γ such that $w_{k+1} \in D(w_k, \mathrm{dist}\,(w_k, W(P)))$. Iteratively, we see that $D(w_k, \mathrm{dist}\,(w_k, W(P))) \subset \rho(P)$ for $k = 1, 2, \dots, N$ and in particular for $k = N$ we get $z_1 \in \rho(P)$ and again by (2.3.7) that (2.3.6) holds when $z = z_1$. Now z_1 is an arbitrary point in Ω and the theorem follows. □

2.4 A Simple Example of a Large Matrix

Consider a Jordan block $A_0 : \mathbf{C}^N \to \mathbf{C}^N$ that we identify with its matrix

$$A_0 = \begin{pmatrix} 0\,1\,0\,0 \dots 0 \\ 0\,0\,1\,0 \dots 0 \\ 0\,0\,0\,1 \dots 0 \\ \dots\dots\dots \\ 0\,0\,0\,0 \dots 1 \\ 0\,0\,0\,0 \dots 0 \end{pmatrix}. \tag{2.4.1}$$

It was observed by Zworski [162] that the open unit disc is a region of spectral instability for A_0, so we expect the eigenvalues to move in a vicinity of that disc when we add a small perturbation to A_0. Here we shall just look at the simple case of

$$A_\delta^0 = \begin{pmatrix} 0\,1\,0\,0 \ldots 0 \\ 0\,0\,1\,0 \ldots 0 \\ 0\,0\,0\,1 \ldots 0 \\ \cdot\;\cdot\;\cdot\;\cdot\;\cdot\cdot\cdot\;\cdot \\ 0\,0\,0\,0 \ldots 1 \\ \delta\,0\,0\,0 \ldots 0 \end{pmatrix}, \ \delta > 0, \tag{2.4.2}$$

leaving the study of more general random perturbations for Chap. 13. If we identify $\mathbf{C}^N$ with $\ell^2(\{1, 2, \ldots, N\})$ in the natural way, then $A_\delta^0 u(j) = u(j+1)$, $j = 1, \ldots, N-1$, $A_\delta^0 u(N) = \delta u(1)$.

Since A_0 is a Jordan block, we already know that the spectrum of A_0 is reduced to the eigenvalue $\lambda = 0$ which has the algebraic multiplicity N.

We look for the eigenvalues of A_δ^0, $\delta > 0$. Now $\lambda \in \mathbf{C}$ is such an eigenvalue iff there exists $0 \neq u \in \ell^2(\{1, \ldots, N\})$ such that

$$u(j+1) = \lambda u(j), \ 1 \leq j \leq N-1,$$
$$\delta u(1) = \lambda u(N).$$

The spectrum of A_δ^0 consists of the N distinct simple eigenvalues

$$\lambda_k = \delta^{\frac{1}{N}} e^{\frac{2\pi i k}{N}}, \ k = 0, \ldots, N-1,$$

which are equidistributed on the circle of radius $\delta^{1/N}$. If $\delta > 0$ is fixed and $N \to +\infty$, then "the spectrum converges to the circle S^1".

We next look at the ϵ-pseudospectrum of A_0. Using that $A_0^N = 0$, we find for $z \neq 0$:

$$(z - A_0)^{-1} = \frac{1}{z}(1 - \frac{1}{z}A_0)^{-1} = \frac{1}{z}(1 + \frac{1}{z}A_0 + \cdots + \frac{1}{z^{N-1}}A_0^{N-1}),$$

which has the matrix

$$\begin{pmatrix} \frac{1}{z} & \frac{1}{z^2} & \frac{1}{z^3} & \frac{1}{z^4} & \cdots & \frac{1}{z^N} \\ 0 & \frac{1}{z} & \frac{1}{z^2} & \frac{1}{z^3} & \cdots & \frac{1}{z^{N-1}} \\ 0 & 0 & \frac{1}{z} & \frac{1}{z^2} & \cdots & \frac{1}{z^{N-2}} \\ \cdot & \cdot & \cdot & \cdot & \cdots & \cdot \\ 0 & 0 & 0 & 0 & \cdots & \frac{1}{z^2} \\ 0 & 0 & 0 & 0 & \cdots & \frac{1}{z} \end{pmatrix}$$

Applying this to the vector u with $u(j) = 0$ for $1 \leq j \leq N-1$, $u(N) = 1$, we see that

$$\|(z - A_0)^{-1}\| \geq \frac{1}{|z|^N}. \tag{2.4.3}$$

Moreover,

$$\begin{aligned}\|(z - A_0)^{-1}\| &\leq \frac{1}{|z|} + \frac{\|A_0\|}{|z|^2} + \cdots + \frac{\|A_0^{N-1}\|}{|z|^N} \\ &= \frac{1}{|z|} + \frac{1}{|z|^2} + \cdots + \frac{1}{|z|^N}.\end{aligned} \tag{2.4.4}$$

Inequality (2.4.3) shows that the resolvent is large inside $D(0, 1)$ when N is large, while (2.4.4) implies

$$\|(z - A_0)^{-1}\| \leq \frac{1}{|z| - 1}, \ |z| > 1. \tag{2.4.5}$$

Combining (2.4.3) and (2.4.5) we see that, for $0 < \epsilon < 1$,

$$D(0, \epsilon^{\frac{1}{N}}) \subset \sigma_\epsilon(A_0) \subset D(0, 1 + \epsilon). \tag{2.4.6}$$

Let us next study the numerical range of A_δ^0. Notice that A_1^0 describes translation by -1 on $\ell^2(\mathbf{Z}/N\mathbf{Z})$ with the natural identification $\mathbf{Z}/N\mathbf{Z} \equiv \{1, \dots, N\}$, so A_1^0 is unitary and hence normal. It has the eigenvalues $\lambda_k = e^{2\pi ik/N}$, $0 \leq k \leq N-1$ (already computed) and the eigenvectors e_k given by $e_k(j) = N^{-1/2}e^{2\pi ijk/N}$. Since A_1^0 is normal, we can use the spectral resolution theorem to see that

$$W(A_1^0) = \mathrm{ch}\,\{e^{\frac{2\pi ik}{N}};\ 0 \leq k \leq N-1\} \tag{2.4.7}$$

which is contained in the unit disc and contains a disc $D(0, 1 - \mathcal{O}(1/N^2))$. Here ch B denotes the convex null of a set $B \subset C$, i.e., the smallest convex set containing B.

We have similar estimates for $W(A_\delta^0)$, $0 \leq \delta \leq 1$:

1. $W(A_\delta^0) \subset \overline{D(0, \|A_\delta^0\|)} = \overline{D(0, 1)}$ (except in the trivial case $N = 1$).
2. $(A_\delta^0 e_k|e_k) = \lambda_k - ((A_1^0 - A_\delta^0)e_k|e_k) = \lambda_k + (1-\delta)\epsilon_{k,N}$, where $|\epsilon_{k,N}| := |e_k(1)\overline{e_k(N)}| \leq 1/N$, so $W(A_\delta^0)$ contains the points $\lambda_k + (1-\delta)\epsilon_{k,N} = \lambda_k + \mathcal{O}(1/N)$.

We conclude that

$$D\left(0, 1 - \mathcal{O}\left(\frac{1}{N}\right)\right) \subset W(A_\delta^0) \subset \overline{D(0, 1)}. \tag{2.4.8}$$

We pause to fix some conventions. If $\mathcal{B}$ is a normed space, we write "$a = \mathcal{O}(b)$" , if "there is a constant C such that the norm of a is $\leq C|b|$" for a family of $(a, b) \in \mathcal{B} \times \mathbf{C}$, implicitly defined from the context. We adopt the less standard convention that $1/\mathcal{O}(b)$ with $b > 0$ refers to a positive quantity which is $\geq 1/(Cb)$ for some constant $C > 0$.

We shall return to A_0 and its perturbations in Chap. 13 and as a preparation we establish some slightly more refined bounds on the resolvents.

When $|z| > 1$ we can estimate the sum in (2.4.4) by that of the corresponding infinite series. When $|z| < 1$ we can write it as $|z|^{-N} \sum_{j=0}^{N-1} |z|^j$ and estimate the sum by that of the corresponding infinite series. However, (which is of interest when $|z| - 1 = \mathcal{O}(1/N)$) the finite sums can also be estimated by N and we get:

$$\|(z - A_0)^{-1}\| \leq F(|z|),\ F(R) := \begin{cases} \frac{1}{R} \min(N, \frac{R}{R-1}),\ R \geq 1, \\ \frac{1}{R^N} \min(N, \frac{1}{1-R}),\ R \leq 1. \end{cases} \tag{2.4.9}$$

A straightforward calculation gives the more explicit expression

$$F(R) = \begin{cases} \dfrac{1}{R^N(1-R)}, & R \leq \frac{N-1}{N}, \\ N/R^N, & \frac{N-1}{N} \leq R \leq 1, \\ N/R, & 1 \leq R \leq \frac{N}{N-1}, \\ 1/(R-1), & R \geq \frac{N}{N-1}. \end{cases} \tag{2.4.10}$$

We see that this is a continuous strictly decreasing function on $]0, +\infty[$ with range $]0, +\infty[$.

Consider the perturbation

$$A_\delta = A_0 + \delta Q, \tag{2.4.11}$$

where Q is a general complex $N \times N$-matrix, so that

$$z - A_\delta = (z - A_0)(1 - (z - A_0)^{-1}\delta Q).$$

If

$$F(|z|)\delta\|Q\| < 1, \tag{2.4.12}$$

(having in mind also the case $|z| < 1$ below), we can expand the last factor in a Neumann series and get

$$\|(z - A_\delta)^{-1}\| \leq F(|z|)\frac{1}{1 - F(|z|)\delta\|Q\|}. \tag{2.4.13}$$

When $|z| > 1$, we have $F(|z|) \leq 1/(|z|-1)$ and (2.4.13) gives

$$\|(z-A_\delta)^{-1}\| \leq \frac{1}{|z|-1} \cdot \frac{1}{1-\frac{\delta\|Q\|}{|z|-1}} = \frac{1}{|z|-1-\delta\|Q\|}, \tag{2.4.14}$$

provided that we replace (2.4.12) by the stronger assumption

$$\frac{\delta\|Q\|}{|z|-1} < 1. \tag{2.4.15}$$

The spectrum of A_δ is confined to the disc $\overline{D(0,R)}$, if $\delta\|Q\|F(|z|) < 1$ for every z outside that disc. Since F is strictly decreasing, we conclude that $\sigma(A_\delta) \subset \overline{D(0,R)}$ if

$$F(R) = \frac{1}{\delta\|Q\|}. \tag{2.4.16}$$

In view of (2.4.10), this relation splits into four different cases,

$$\begin{aligned}
&R^N(1-R) = \delta\|Q\|, \text{ if } R \leq 1-\frac{1}{N}, \text{ implying } \delta\|Q\| \leq \frac{\left(1+\mathcal{O}(\frac{1}{N})\right)}{eN}, \\
&\frac{R^N}{N} = \delta\|Q\|, \text{ if } 1-\frac{1}{N} \leq R \leq 1, \text{ implying } \frac{1+\mathcal{O}(\frac{1}{N})}{eN} \leq \delta\|Q\| \leq \frac{1}{N}, \\
&R = N\delta\|Q\|, \text{ if } 1 \leq R \leq \frac{N}{N-1}, \text{ implying } \frac{1}{N} \leq \delta\|Q\| \leq \frac{1}{N-1}, \\
&R = 1+\delta\|Q\|, \text{ if } R \geq \frac{N}{N-1}, \text{ implying } \frac{1}{N-1} \leq \delta\|Q\|.
\end{aligned} \tag{2.4.17}$$

In these four cases we get, respectively,

$$\begin{aligned}
&(\delta\|Q\|)^{1/N} \leq R \leq (\delta\|Q\|)^{1/N} N^{1/N}, \\
&R = (\delta\|Q\|)^{1/N} N^{1/N}, \\
&R = N\delta\|Q\|, \\
&R = 1+\delta\|Q\|.
\end{aligned} \tag{2.4.18}$$

2.5 The Non-self-adjoint Harmonic Oscillator

Let $c \in \mathbf{C}\setminus\,]-\infty, 0\,]$. We consider the non-self-adjoint harmonic oscillator on $\mathbf{R}$:

$$P_c = D_x^2 + cx^2 : L^2(\mathbf{R}) \rightarrow L^2(\mathbf{R}). \tag{2.5.1}$$

P_c is a closed operator with the dense domain $B^2 \subset L^2(\mathbf{R})$, where we define, for general $k \in \mathbf{N} := \{0, 1, 2, \dots\}$,

$$B^k = \{u \in L^2(\mathbf{R});\ x^\nu D^\mu u \in L^2,\ \nu + \mu \le k\}. \tag{2.5.2}$$

(For $0 > k \in \mathbf{Z}$, we define $B^k \subset \mathcal{S}'(\mathbf{R})$ to be the dual space of B^{-k}.) Here we use the notation $D = D_x = \frac{1}{i}\frac{d}{dx}$.

We will use the following facts, see for example [131]:

1. B^k are Hilbert spaces when equipped with the natural norms and we have the *compact* inclusion maps: $B^k \hookrightarrow B^j$ when $k > j$.
2. $P_c - z : B^2 \to L^2$ are Fredholm operators of index 0 (cf. Sect. 8.1), depending holomorphically on $z \in \mathbf{C}$. (Here the assumption that $c \notin]-\infty, 0]$ guarantees a basic ellipticity property.)
3. The spectrum of $P_c : L^2 \to L^2$ is discrete:
   - It is a discrete subset $\sigma(P_c)$ of $\mathbf{C}$.
   - Each element λ_0 of $\sigma(P_c)$ is an eigenvalue of finite algebraic multiplicity, in the sense that the spectral projection

     $$\Pi_{\lambda_0} = \frac{1}{2\pi i}\int_{\partial D(\lambda_0,\epsilon)} (z - P)^{-1} dz$$

     is of finite rank. Here $\epsilon > 0$ is small enough so that $\sigma(P_c) \cap D(\lambda_0, \epsilon) = \{\lambda_0\}$.
4. When $c = 1$ we get a self-adjoint operator with spectrum $\{\lambda_k = 2k + 1;\ k = 0, 1, 2, \dots\}$, and a corresponding orthonormal basis of eigenfunctions is given by

   $$e_0 = \frac{1}{\sqrt{2\pi}} e^{-x^2/2},\ e_k = C_k \left(x - \frac{d}{dx}\right)^k (e_0) \text{ for } k \ge 1,$$

   where $x - \frac{d}{dx}$ is the creation operator and $C_k > 0$ are normalization constants. Thus,

   $$e_k =: p_k(x) e^{-x^2/2} \tag{2.5.3}$$

   are the Hermite functions and p_k are the Hermite polynomials. Notice that p_k is of order k of the form $p_{k,k}x^k + p_{k,k-2}x^{k-2} + \cdots$, $p_{k,k} > 0$ and that p_k is even/odd when k is even/odd, respectively.

Calculation of the Spectrum of P_c When $c > 0$ this is easy by means of a dilation: Put $x = c^{-1/4}y$, so that $D_x = c^{1/4}D_y$. Then we get

$$P_c = c^{1/2}(D_y^2 + y^2), \tag{2.5.4}$$

and we conclude that the eigenvalues and eigenfunctions of P_c are given by

$$\lambda(k, c) = c^{\frac{1}{2}}\lambda_k =: c^{\frac{1}{2}}(2k + 1), \quad e_{k,c}(x) = c^{\frac{1}{8}}e_k(c^{\frac{1}{4}}x). \tag{2.5.5}$$

Here $c^{1/8}$ is the normalization constant, ensuring that $\{e_{k,c}\}_{k\in\mathbf{N}}$ is an orthonormal basis in L^2.

A more formal proof of this is to introduce the unitary operator

$$Uu(x) = c^{\frac{1}{8}}u(c^{\frac{1}{4}}x) \tag{2.5.6}$$

and to check that

$$P_cU = c^{\frac{1}{2}}UP_1, \quad P_c = c^{\frac{1}{2}}UP_1U^{-1}. \tag{2.5.7}$$

In the general case when c is no longer real, we can no longer define Uu as in (2.5.6) for general L^2 functions u, but (2.5.5) still makes sense because e_k is an entire function. When defining the fractional powers of c we here use the convention that $\arg c \in]-\pi, \pi[$, so that the argument of $c^{1/4}$ is in $[-\pi/4, \pi/4[$. We still have that $e_{k,c} \in B^2$ and

$$(P_c - \lambda(k, c))e_{k,c} = 0, \tag{2.5.8}$$

so $\lambda(k, c)$ are eigenvalues of P_c. We shall prove (cf. [24, 34, 131])

Proposition 2.5.1

$$\sigma(P_c) = \{\lambda(k, c);\ k \in \mathbf{N}\}.$$

Proof It remains to show that there are no other eigenvalues. (The fact 3 above tells us that every element of $\sigma(P_c)$ is an eigenvalue.) For that, we shall use the (formal) adjoint of P_c which is given by $P_c^* = P_{\overline{c}}$ and more precisely that

$$(P_cu|v) = (u|P_{\overline{c}}v),\ \forall u, v \in B^2. \tag{2.5.9}$$

We have

Lemma 2.5.2 $\{e_{k,\overline{c}};\ k \in \mathbf{N}\}$ *spans a dense subspace of* $L^2(\mathbf{R})$.

Proof Let $L \subset L^2(\mathbf{R})$ be the space of all finite linear combinations of the $e_{k,\overline{c}}$, $k \in \mathbf{N}$. Then u belongs to the orthogonal space $L^\perp$ iff

$$(u|e_{k,\overline{c}}) = 0, \ \forall k \in \mathbf{N}, \tag{2.5.10}$$

and it suffices to show that this fact implies that $u = 0$. Rewrite (2.5.10) as

$$0 = \int u(x)e^{-c^{1/2}x^2/2} p_{k,c}(x)dx, \tag{2.5.11}$$

where we used that

$$\overline{e_{k,\overline{c}}} = e_{k,c} =: p_{k,c}(x)e^{-c^{1/2}x^2/2}, \quad p_{k,c}(x) = c^{1/8}p_k(c^{1/4}x).$$

It is easy to see that the Fourier transform $\widehat{v}(\xi)$ of the function $v(x) = u(x)e^{-c^{1/2}x^2/2}$ can be extended to all of $\mathbf{C}_\xi$ as an entire function and (2.5.11) means that

$$p_{k,c}(-D_\xi)\widehat{v}(0) = 0, \ \forall k. \tag{2.5.12}$$

Now, every polynomial can be expressed as a finite linear combination of the $p_{k,c}$, $k \in \mathbf{N}$, so (2.5.12) means that

$$(D_\xi^\alpha \widehat{v})(0) = 0, \ \forall \alpha \in \mathbf{N}. \tag{2.5.13}$$

In other words, the power series expansion of the entire function $\widehat{v}(\xi)$ vanishes identically, hence $\widehat{v} = 0$. Therefore, $v = 0$ and finally $u = 0$. □

Now we can finish the proof of the proposition. Let $\lambda \in \sigma(P_c) \setminus \{\lambda(k, c);\ k \in \mathbf{N}\}$ and assume that

$$(P_c - \lambda)u = 0, \tag{2.5.14}$$

for some $0 \neq u \in B^2$. Then, for every $k \in \mathbf{N}$,

$$(u|e_{k,\overline{c}}) = \frac{1}{\lambda(k,c) - \lambda}(u|(P_{\overline{c}} - \overline{\lambda})e_{k,\overline{c}}) = \frac{1}{\lambda(k,c) - \lambda}((P_c - \lambda)u|e_{k,\overline{c}}) = 0,$$

so Lemma 2.5.2 shows that $u = 0$. □

Remark 2.5.3 The arguments in the proof can be pushed further to show that the generalized eigenspaces (i.e., the ranges of the spectral projections) corresponding to $\lambda(k, c)$ are one-dimensional:

Let $\lambda_0 = \lambda(k_0, c) \in \sigma(P_c)$ and let u be a corresponding eigenfunction, so that $(P_c - \lambda_0)u = 0$. As in the proof above, we see that $(u|e_{k,\bar{c}}) = 0$, $\forall k \neq k_0$. By unique holomorphic extension with respect to c from $]0, +\infty[$, we have also that

$$(e_{k,c}|e_{\ell,\bar{c}}) = \begin{cases} 0, & \text{if } \ell \neq k, \\ 1, & \text{if } \ell = k. \end{cases}$$

Hence we can find $d \in \mathbf{C}$ such that $(u - de_{k_0,c}|e_{k,\bar{c}}) = 0$ for all $k \in \mathbf{N}$, and as above we conclude that $u - de_{k_0,c} = 0$.

Furthermore, there can be no Jordan blocks: if $(P_c - \lambda_0)^2 u = 0$ for some $u \in B^2$ with $P_c u \in B^2$, then

$$(u|e_{k,\bar{c}}) = \frac{1}{(\lambda(k,c) - \lambda_0)^2}(u|(P_{\bar{c}} - \overline{\lambda}_0)^2 e_{k,\bar{c}}) = 0,$$

for $k \neq k_0$, and as above we see that $u = de_{k,c}$.

A different way to reach the same conclusion would be to choose a continuous deformation $[0, 1] \ni t \mapsto c_t \in \mathbf{C} \setminus]-\infty, 0]$ with $c_0 = 1$, $c_1 = c$ and to notice that the corresponding spectral projections $\Pi_{\lambda(k,c_t)}$ vary continuously with t for every fixed $k \in \mathbf{N}$. Then the rank of $\Pi_{\lambda(k,c_t)}$ is independent of t and equal to 1 for $t = 0$. Thus $\Pi_{\lambda(k,c)} = \Pi_{\lambda(k,c_1)}$ is also of rank 1.

The Numerical Range We have the following result of Boulton [22]:

Proposition 2.5.4

$$W(P_c) = \{t + sc;\ t, s > 0,\ ts \geq \frac{1}{4}\}.$$

We have already recalled that the lowest eigenvalue of P_1 is 1 and this implies (by the spectral resolution theorem) that $(P_1 u|u) \geq \|u\|^2$ for all $u \in B^2$. By integration by parts, we also have $(P_1 u|u) = \|Du\|^2 + \|xu\|^2$. We now recall an additional inequality, the uncertainty relation:

Lemma 2.5.5 *We have*

$$\|u\|^2 \leq 2\|xu\|\|Du\| \leq \|Du\|^2 + \|xu\|^2,\ \forall u \in B^1. \tag{2.5.15}$$

Proof The second inequality follows from "$2ab \leq a^2 + b^2$" and since $\mathcal{S}$ is dense in B^1 it suffices to show the first inequality for every u in $\mathcal{S}$. For such a function, we have by

Cauchy-Schwarz,

$$|(xDu|u)| = |(Du|xu)| \leq \|xu\|\|Du\|$$
$$|(Dxu|u)| = |(xu|Du)| \leq \|xu\|\|Du\|.$$

Thus,

$$|(Dx - xD)u|u)| \leq 2\|xu\|\|Du\|.$$

But $Dx - xD = [D, x] = 1/i$, so $|((Dx - xD)u|u)| = \|u\|^2$, and the lemma follows. □

Proof of the Proposition For $u \in B^2$, $\|u\| = 1$, we have

$$(P_c u|u) = \|Du\|^2 + c\|xu\|^2 = t + sc,$$

with $t = \|Du\|^2$, $s = \|xu\|^2$, and (2.5.15) implies that $1 \leq 4ts$. It is then clear that

$$W(P_c) \subset \{t + sc;\ t, s > 0,\ ts \geq \frac{1}{4}\}.$$

To show the opposite inclusion, we first notice that if $u = e_0$, we have equalities in (2.5.15). Indeed, the first expression is equal to 1 since e_0 is normalized in L^2, and the last expression is equal to $((D^2 + x^2)e_0|e_0) = (e_0|e_0) = 1$, so the inequalities have to be equalities for this choice of u. Since $\|De_0\| = \|xe_0\|$, we also have

$$\|De_0\| = \|xe_0\| = \frac{1}{\sqrt{2}}. \tag{2.5.16}$$

Now dilate and consider the functions

$$f_\lambda(x) := \lambda^{\frac{1}{2}} e_0(\lambda x),\ \lambda > 0. \tag{2.5.17}$$

Using the same change of variables as in the calculation of the L^2 norms, we see that

$$\|xf_\lambda\| = \lambda^{-1}\|xe_0\| = \lambda^{-1}\frac{1}{\sqrt{2}},$$
$$\|Df_\lambda\| = \lambda\|De_0\| = \lambda\frac{1}{\sqrt{2}}.$$

Hence we still have equality in the first part of (2.5.15),

$$\|f_\lambda\|^2 = 2\|xf_\lambda\|\|Df_\lambda\|, \tag{2.5.18}$$

and

$$(P_c u|u) = \|Df_\lambda\|^2 + c\|xf_\lambda\|^2 = t + sc,$$

where now $t = \|Df_\lambda\|^2 = \lambda^2/2$ and $s = \|xf_\lambda\|^2 = 1/2\lambda^2$ can take arbitrary positive values with $ts = 1/4$.

We conclude that $\{t + sc;\ t, s > 0,\ ts = 1/4\}$ is contained in $W(P_c)$. The convex hull of (i.e., the smallest convex set containing) this set is precisely $\{t+sc;\ t, s > 0,\ ts \geq 1/4\}$ and the latter set is therefore contained in $W(P_c)$. □

Notes

Sect. 2.1 is a review of standard material in Hilbert space theory. See e.g., [88, 116, 117].

The discussion of pseudospectra in Sect. 2.2 has been inspired by the influential papers [35, 151] and the monograph [152].

The discussion of the numerical range in relation to pseudospectra in Sect. 2.3 follows [35] and [152].

The discussion of large Jordan blocks in Sect. 2.4 owes a lot to the work [162], where the pseudospectral aspects where highlighted. Some of the results in this section will be used in Chap. 13, devoted to random perturbations of a large Jordan block.

A discussion of non-self-adjoint harmonic oscillators from the pseudospectral point of view was given in [34], while the computation of the spectrum is more classical [24, 131] appearing in the study of PDE with double characteristics. See also [22, 152, 162].

Quadratic non self-adjoint operators have been the subject of rather intense investigations in recent years, in connection with resolvent estimates and evolution equations; see [5, 6, 68, 70, 75–77, 109, 112–114, 153, 154].

3 Weyl Asymptotics and Random Perturbations in a One-Dimensional Semi-classical Case

We consider a simple model operator P in dimension 1 and show how random perturbations give rise to Weyl asymptotics in the interior of the range of P. We follow rather closely the work of Hager [55] with some input also from Bordeaux Montrieux [16] and Hager–Sjöstrand [56]. Some of the general ideas appear perhaps more clearly in this special situation.

Let $P = hD_x + g(x)$, $g \in C^\infty(S^1)$, $S^1 \simeq \mathbf{R}/2\pi\mathbf{Z}$, with symbol $p(x,\xi) = \xi + g(x)$, $(x,\xi) \in T^*S^1 \simeq S^1 \times \mathbf{R}$, and assume that $\Im g$ has precisely two critical points, a unique maximum and a unique minimum. Here and at many other places of this book we work in the semi-classical limit, i.e., for $h > 0$ sufficiently small, even though we may sometimes omit the wording "then for $h > 0$ small enough". We notice that P is a closed operator $L^2(S^1) \to L^2(S^1)$ with domain equal to the Sobolev space $H^1(S^1)$. The spectrum is discrete and confined to the line

$$\Im z = \Im\langle g\rangle, \quad \langle g\rangle := \frac{1}{2\pi}\int_0^{2\pi} g(x)dx.$$

Indeed, a direct computation shows that the eigenvalues are simple and given by

$$z_k = \langle g\rangle + kh, \ k \in \mathbf{Z}$$

Let $\Omega \Subset \{z \in \mathbf{C};\ \min \Im g < \Im z < \max \Im g\}$ be open. Put

$$\begin{aligned} P_\delta &= P_{\delta,\omega} = hD_x + g(x) + \delta Q_\omega, \\ Q_\omega u(x) &= \sum_{|k|,|\ell| \le \frac{C_1}{h}} \alpha_{k,\ell}(\omega)(u|e^k)e^\ell(x), \end{aligned} \tag{3.0.1}$$

J. Sjöstrand, *Non-Self-Adjoint Differential Operators, Spectral Asymptotics and Random Perturbations*, Pseudo-Differential Operators 14,
https://doi.org/10.1007/978-3-030-10819-9_3

where $C_1 > 0$ is sufficiently large, $e^k(x) = (2\pi)^{-1/2}e^{ikx}$, $k \in \mathbf{Z}$, and $\alpha_{j,k} \sim \mathcal{N}_{\mathbf{C}}(0, 1)$ are independent complex Gaussian random variables, centered with variance 1.[1] In (3.4.1) we will review the definition. In this book we only use basic probability theory, see for instance the first two chapters in [87]. Q_ω is compact, so P_δ has discrete spectrum. Let $\Gamma \Subset \Omega$ have smooth boundary.

Theorem 3.0.6 *Let $\kappa > 5/2$ and let $\epsilon_0 > 0$ be sufficiently small. Let $\delta = \delta(h)$ satisfy $e^{-\epsilon_0/h} \ll \delta \ll h^\kappa$ and put $\epsilon = \epsilon(h) = h\ln(1/\delta)$. Then there exists a constant $C > 0$ such that for $h > 0$ small enough, we have with probability $\geq 1 - \mathcal{O}(\frac{\delta^2}{\sqrt{\epsilon}h^5})$ that the number of eigenvalues of P_δ in Γ satisfies*

$$|\#(\sigma(P_\delta) \cap \Gamma) - \frac{1}{2\pi h}\mathrm{vol}(p^{-1}(\Gamma))| \leq C\frac{\sqrt{\epsilon}}{h}. \tag{3.0.2}$$

Here, we recall that $p : T^*S^1 \to \mathbf{C}$ and we take the volume with respect to the natural symplectic volume element, $dxd\xi$.

The conclusion in the theorem is of interest when

$$\frac{\delta^2}{\sqrt{\epsilon}h^5} \ll 1, \ \sqrt{\epsilon} \ll 1,$$

which is satisfied when

$$\kappa > \frac{11}{4}.$$

If instead we let Γ vary in a collection of subsets that satisfy the assumptions uniformly, then with probability $\geq 1 - \mathcal{O}(\frac{\delta^2}{\epsilon h^5})$ we have (3.0.2) uniformly for all Γ in that collection. The remainder of the chapter is devoted to the proof of this result.

Remark 3.0.7 The estimate on the probability in Theorem 3.0.6 is rather crude rough and can be improved by adapting the arguments in [56], as we will do in Sects. 5.5 and 13.7.

[1] The choice of Gaussian random variables is convenient in order to formulate a less technical result. In Chap. 15 more general classes of perturbations will be treated, also in higher dimensions.

3.1 Preparations for the Unperturbed Operator

For $z \in \Omega$, let $x_+(z), x_-(z) \in S^1$ be the solutions of the equation $\Im g(x) = \Im z$, with $\pm \Im g'(x_\pm) < 0$. We sometimes view $x_\pm$ as elements in $\mathbf{R}$ (unique mod $2\pi\mathbf{Z}$) chosen so that $x_- - 2\pi < x_+ < x_-$. Define $\xi_\pm(z)$ by $\xi_\pm + \Re g(x_\pm) = \Re z$. Putting $\rho_\pm = (x_\pm, \xi_\pm)$, we have

$$p(\rho_\pm) = z, \quad \pm \frac{1}{i}\{p, \overline{p}\}(\rho_\pm) = \mp \Im g'(x_\pm) > 0.$$

Here, $\{a, b\} = \partial_\xi a\, \partial_x b - \partial_x a\, \partial_\xi b$ denotes the Poisson bracket of two sufficiently smooth functions $a(x, \xi)$, $b(x, \xi)$.

In the following we us the notation "neigh (A, B)" for "some neighborhood of A in B", when A is an element or a subset of the topological space B.

Let $\chi \in C_0^\infty(\text{neigh}(0, \mathbf{R}))$ be equal to 1 in a neighborhood of 0 and consider the function

$$u(x) = u(x, z; h) = \chi(x - x_+(z)) e^{\frac{i}{h}\phi_+(x)},$$

where

$$\phi_+(x) = \phi_+(x, z) = \int_{x_+(z)}^{x} (z - g(y))dy.$$

Then

$$\Im \phi_+(x) \asymp |x - x_+(z)|^2, x \in \text{neigh}(x_+(z), \mathbf{R})$$

and we choose the support of χ small enough, so that this holds on supp $(\chi(\cdot - x_+(z))$.[2] Then

$$\int |u(x)|^2 dx = \int |\chi(x - x_+(z))|^2 e^{-\frac{2}{h}\Im \phi_+(x)} dx.$$

Applying the Morse lemma (here in a simple one-dimensional situation) as explained for instance in Chapter 2 of [51] and Exercise 2.4 of that book, we see that

$$\int |u(x)|^2 dx = h^{\frac{1}{2}} b(z; h),$$

[2]We write $a \asymp b$, when a, b are real numbers with the same sign, such that $|a| = \mathcal{O}(|b|)$ and $|b| = \mathcal{O}(|a|)$. This notion has a natural extension to the case when a, b are positive Radon measures on the same set in $\mathbf{R}^n$.

where the symbol b satisfies[3]

$$b(z;h) \sim b_0(z) + hb_1(z) + \cdots \text{ in } C^\infty(\Omega),$$

and in particular

$$\partial_z^\alpha \partial_{\overline{z}}^\beta b = \mathcal{O}_{\alpha,\beta}(1), \text{ for all } \alpha, \beta \in \mathbf{N}.$$

Here the subscripts α, β indicate that the estimate is uniform for each fixed (α, β), but not uniformly with respect to these parameters. Moreover,

$$b_0(z) = \frac{\sqrt{2\pi}}{\sqrt{2\Im \partial_x^2 \phi_+(x_+(z))}} > 0.$$

In particular, $b(z;h) > 0$ on Ω for $h > 0$ small enough (since the above holds in a relatively compact neighborhood of $\overline{\Omega}$) and we can form $a(z;h) = (b(z;h))^{-1/2}$, which satisfies

$$a(z;h) \sim a_0(z) + ha_1(z) + \cdots \text{ in } C^\infty(\Omega), \quad a_0(z) = b_0(z)^{-1/2}.$$

Put

$$e_{\mathrm{wkb}}(x) = h^{-1/4} a(z;h) \chi(x - x_+(z)) e^{\frac{i}{h}\phi_+(x)}.$$

Then $\|e_{\mathrm{wkb}}\| = 1$ where we take the L^2 norm over $]x_-(z) - 2\pi, x_+(z)[$. Moreover,

$$(P - z)e_{\mathrm{wkb}} = \mathcal{O}(e^{-\frac{1}{Ch}}),$$

where the remainder to the right comes from the cutoff function $\chi(x - x_+)$.

Define z-dependent elliptic self-adjoint operators

$$Q = (P-z)^*(P-z),\ \widetilde{Q} = (P-z)(P-z)^* : L^2(S^1) \to L^2(S^1),$$

with domain $\mathcal{D}(Q)$, $\mathcal{D}(\widetilde{Q}) = H^2(S^1)$ (the usual Sobolev space of order 2). They have discrete spectrum $\subset [0, +\infty[$ and smooth eigenfunctions. (It is a standard fact that elliptic formally self-adjoint differential operators on a compact manifold M with smooth coefficients are essentially self-adjoint with domain $H^m(M)$, where m is the order of the operator. Furthermore the spectrum is discrete.) Using that $P - z : H^1 \to L^2$ is Fredholm

[3]If $\mathcal{F}$ is a seminormed space, we write $b(h) \sim b_0 + hb_1 + \cdots$ in $\mathcal{F}$, if for every seminorm q and every $N \in \mathbf{N}$, we have $q(b - (b_0 + hb_1 + \cdots + b_N h^N)) = \mathcal{O}(h^{N+1})$.

of index zero (exercise!), we see that $\dim \mathcal{N}(Q) = \dim \mathcal{N}(\widetilde{Q}) \leq 1$. Here $\mathcal{N}(A)$ denotes the kernel of the linear operator A and we shall use $\mathcal{R}(A)$ to denote the range. By elliptic regularity we know that the kernel of $P - z$ in H^1 agrees with that of Q in H^2. If $\mu \neq 0$ is an eigenvalue of Q, with the corresponding eigenfunction $e \in C^\infty$, then $f := (P - z)e$ is an eigenfunction for $\widetilde{Q}$ with the same eigenvalue μ. Pursuing this observation, we see that

$$\sigma(Q) = \sigma(\widetilde{Q}) = \{t_0^2, t_1^2, \ldots\},\ 0 \leq t_j \nearrow +\infty.$$

Proposition 3.1.1 *There exists a constant $C > 0$ such that $t_0^2 = \mathcal{O}(e^{-1/(Ch)})$, $t_1^2 - t_0^2 \geq h/C$ for $h > 0$ small enough.*

Proof We have $Qe_{\mathrm{wkb}} = r$, $\|r\| = \mathcal{O}(e^{-1/Ch})$, and since Q is self-adjoint, we deduce that t_0^2 is exponentially small. (Cf. (2.1.11).) If e_0 denotes the corresponding normalized eigenfunction uniquely determined up to a factor of modulus 1, we see that $(P - z)e_0 =: v$ with $\|v\|$ exponentially small. Considering this ODE on $]x_-(z) - 2\pi, x_-(z)[$, we get

$$e_0(x) = Ch^{-\frac{1}{4}}a(h)e^{\frac{i}{h}\phi_+(x)} + Fv(x),\ C = C(h), \tag{3.1.1}$$

$$Fv(x) = \frac{i}{h}\int_{x_+}^{x} e^{\frac{i}{h}(\phi_+(x)-\phi_+(y))}v(y)dy, \tag{3.1.2}$$

where $\phi_+(x) = \int_{x_+}^{x}(z - g(y))dy$. We observe that $\Im(\phi_+(x) - \phi_+(y)) \geq 0$ on the domain of integration.

Lemma 3.1.2 *We have*

$$\|F\|_{\mathcal{L}(L^2,L^2)} = \mathcal{O}(h^{-1/2}).$$

Proof Indeed,

$$Fw(x) = \int K(x, y)w(y)dy,$$

where

$$K(x, y) = \frac{i}{h}\left(1_{\{x_+ \leq y \leq x \leq x_-\}}(x, y) - \frac{i}{h}1_{\{x_- - 2\pi \leq x \leq y \leq x_+\}}(x, y)\right)e^{\frac{i}{h}(\phi_+(x)-\phi_+(y))}.$$

By treating the three cases

- $x, y = x_+ + \mathcal{O}(h^{1/2})$,
- $x, y = x_- + \mathcal{O}(h^{1/2})$,
- $x, y = x_- - 2\pi + \mathcal{O}(h^{1/2})$

separately, we check that for some $C > 0$,

$$\Im\left(\phi_+(x) - \phi_+(y)\right) \geq \frac{h^{\frac{1}{2}}}{C}|x-y| - Ch, \text{ on } \operatorname{supp} K.$$

It follows that

$$\sup_x \int |K(x,y)|dy = \mathcal{O}(h^{-\frac{1}{2}}), \quad \sup_y \int |K(x,y)|dx = \mathcal{O}(h^{-\frac{1}{2}})$$

and the Schur lemma tells us that $\|F\|_{L^2}$ is bounded by the geometric mean of these two quantities. □

For our particular v, we see that Fv decays exponentially decaying in L^2 and hence, since e_0 is normalized, that C in (3.1.1) satisfies $|C| = 1 + \mathcal{O}(e^{-1/\mathcal{O}(h)})$. Replacing e_0 by $e^{i\theta}e_0$ for a suitable $\theta \in \mathbf{R}$ and recalling the form of $e_{\mathrm{wkb}}(x)$, we conclude that $\|e_0 - e_{\mathrm{wkb}}\|$ is exponentially small.

To show that $t_1^2 - t_0^2 \geq h/C$, it suffices to show that $(Qu|u) \geq \frac{h}{C}\|u\|^2$ when $u \perp e_0$, or, in other words, that

$$\|u\| \leq \sqrt{\frac{C}{h}}\|(P-z)u\|. \tag{3.1.3}$$

If $v := (P-z)u$, we again have on $]x_-(z) - 2\pi, x_-(z)[$ that

$$u = Ch^{-\frac{1}{4}}a(h)e^{\frac{i}{h}\phi_+(x)} + Fv$$

for some constant C, and the orthogonality requirement on u implies that

$$0 = (1 + \mathcal{O}(h^\infty))C + (Fv|e_0),$$

where $(Fv|e_0) = \mathcal{O}(h^{-\frac{1}{2}})\|v\|$, so $C = \mathcal{O}(h^{-1/2})\|v\|$ and we get the desired estimate on $\|u\|$. Here we write $\mathcal{O}(h^\infty)$ for a quantity which is $\mathcal{O}(h^N)$ for every $N \in \mathbf{N}$. □

3.2 Grushin (Schur, Feschbach, Bifurcation) Approach

Let $\{e_0, e_1, \ldots\}$ and $\{f_0, f_1, \ldots\}$ be orthonormal bases of eigenfunctions of $Q = (P-z)^*(P-z)$ and $\widetilde{Q} = (P-z)(P-z)^*$, respectively, so that $e_j, f_j \in H^2$, $Qe_j = t_j^2 e_j$, $\widetilde{Q}f_j = t_j^2 f_j$. As observed prior to Proposition 3.1.1, we have

$$(P-z)e_j = \alpha_j f_j, \; (P-z)^* f_j = \beta_j e_j, \quad \alpha_j\beta_j = t_j^2,$$

and combining this with $((P-z)e_j|f_j) = (e_j|(P-z)^* f_j)$, we see that $\alpha_j = \overline{\beta}_j$. Replacing f_j by $e^{i\theta_j} f_j$ for suitable real values of θ_j, we can arrange so that $\alpha_j = \beta_j = t_j$, which will somewhat simplify the notations.

Proposition 3.2.1 *Define $R_+ : H^1(S^1) \to \mathbf{C}$, $R_- : \mathbf{C} \to L^2(S^1)$ by*

$$R_+ u = (u|e_0), \quad R_- u_- = u_- f_0.$$

Then

$$\mathcal{P}(z) := \begin{pmatrix} P - z & R_- \\ R_+ & 0 \end{pmatrix} : H^1 \times \mathbf{C} \to L^2 \times \mathbf{C}$$

is bijective with the bounded inverse

$$\mathcal{E}(z) = \begin{pmatrix} E & E_+ \\ E_- & E_{-+} \end{pmatrix} = \begin{pmatrix} \mathcal{O}(h^{-\frac{1}{2}}) & \mathcal{O}(1) \\ \mathcal{O}(1) & \mathcal{O}(e^{-1/Ch}) \end{pmatrix},$$

where the estimates refer to the norms in $\mathcal{L}(L^2, H^1)$, $\mathcal{L}(\mathbf{C}, H^1)$, $\mathcal{L}(L^2, \mathbf{C})$, $\mathcal{L}(\mathbf{C}, \mathbf{C})$, respectively. (Here when B_1, B_2 are normed spaces.) We let $\mathcal{L}(B_1, B_2)$ denote the space of bounded operators $B_1 \to B_2$ Moreover,

$$E_+ v_+ = v_+ e_0, \quad E_- v = (v|f_0).$$

Here we use the semi-classical norm on H^1:

$$\|u\|_{H^1_h} = \left(\|u\|^2 + \|hDu\|^2\right)^{\frac{1}{2}}.$$

It is a general feature of such auxiliary (Grushin) operators that

$$z \in \sigma(P) \iff E_{-+}(z) = 0.$$

Indeed, it is easy to show the formulae

$$(P - z)^{-1} = E - E_+ E_{-+}^{-1} E_-, \quad E_{-+}^{-1} = -R_+(P - z)^{-1} R_-,$$

under the assumptions that E_{-+} is bijective and that $P - z$ is bijective, respectively.

Proof of the Proposition We shall show that for every $(v, v_+) \in L^2 \times \mathbf{C}$, there is a unique solution $(u, u_-) \in H^1 \times \mathbf{C}$ of the system

$$\begin{cases} (P - z)u + R_- u_- &= v, \\ R_+ u &= v_+, \end{cases} \tag{3.2.1}$$

and give estimates and explicit formulae for the solution. Actually, it suffices to find a unique solution $(u, u_-) \in L^2 \times \mathbf{C}$, since we can then deduce that $hDu \in L^2$ from the first equation in (3.2.1).

Express u and v in the bases introduced above:

$$u = \sum_0^\infty u_j e_j = u_0 e_0 + u^\perp, \quad v = \sum_0^\infty v_j e_j = v_0 e_0 + v^\perp,$$

and recall that $\|v\|^2 = \sum_0^\infty |v_j|^2 = |v_0|^2 + \|v^\perp\|^2$, $v_j = (v|e_j)$, and similarly for u. Then (3.2.1) becomes

$$\begin{cases} \sum_0^\infty t_j u_j f_j + u_- f_0 = \sum_0^\infty v_j f_j, \\ u_0 = v_+, \end{cases}$$

i.e.,

$$t_0 v_+ + u_- = v_0\ , \quad t_j u_j = v_j \text{ for } j \geq 1.$$

Thus, we get the unique solution

$$u_0 = v_+, \quad u^\perp = \sum_1^\infty \frac{v_j}{t_j} e_j, \ u_- = v_0 - t_0 v_+,$$

from which we deduce the expressions for E, $E_\pm$, E_{-+} where $v = \sum_0^\infty v_j f_j$ and $v_+ \in \mathbf{C}$:

$$E v = \sum_1^\infty (v_j/t_j) e_j, \quad E_+ v_+ = v_+ e_0,$$

$$E_- v = v_0, \quad E_{-+} v_+ = -t_0 v_+.$$

It then suffices to recall that $t_j \geq \sqrt{h}/C$ for $j \geq 1$. □

3.3 d-Bar Equation for E_{-+}

We will use the following notations for the holomorphic and anti-holomorphic derivatives in the complex variable z:

$$\partial_z = \frac{\partial}{\partial z} = \frac{1}{2}\left(\frac{\partial}{\partial \Re z} + \frac{1}{i}\frac{\partial}{\partial \Im z}\right),$$
$$\partial_{\overline{z}} = \frac{\partial}{\partial \overline{z}} = \frac{1}{2}\left(\frac{\partial}{\partial \Re z} - \frac{1}{i}\frac{\partial}{\partial \Im z}\right).$$

Also recall the expressions for the Laplace operator on $\mathbf{C} \simeq \mathbf{R}^2$,

$$\Delta = \left(\frac{\partial}{\partial \Re z}\right)^2 + \left(\frac{\partial}{\partial \Im z}\right)^2 = 4\partial z \partial \overline{z}.$$

Proposition 3.3.1 *We have*

$$\partial_{\overline{z}} E_{-+}(z) + f(z)E_{-+}(z) = 0,\ z \in \Omega, \tag{3.3.1}$$

where

$$f(z) = f_+(z) + f_-(z), \quad f_+(z) = (\partial_{\overline{z}} R_+)E_+, \quad f_-(z) = E_-\partial_{\overline{z}} R_-. \tag{3.3.2}$$

Thus,

$$\partial_{\overline{z}}(e^{F(z)}E_{-+}(z)) = 0 \text{ if } \partial_{\overline{z}} F(z) = f(z). \tag{3.3.3}$$

Moreover,

$$\Re\Delta F(z) = 4\Re\partial_z f = \frac{2}{h}\left(\frac{1}{\frac{1}{i}\{p,\overline{p}\}(\rho_+)} - \frac{1}{\frac{1}{i}\{p,\overline{p}\}(\rho_-)}\right) + \mathcal{O}(1). \tag{3.3.4}$$

Proof Formulae (3.3.1)–(3.3.3) follow from the general formula for the differentiation of the inverse of an operator, here

$$\partial_{\overline{z}}\mathcal{E} + \mathcal{E}(\partial_{\overline{z}}\mathcal{P})\mathcal{E} = 0.$$

□

Let $\Pi(z) : L^2 \to \mathbf{C}e_0$ be the spectral projection of Q corresponding to the exponentially small eigenvalue t_0^2, and choose e_0 to be the normalization of $\Pi(z)e_{wkb}$.

It is easy to see that the various z and $\overline{z}$ derivatives of e_{wkb} have at most temperate growth in $1/h$. The same fact holds for $\Pi(z)$:

Lemma 3.3.2 *For every $(\alpha, \beta) \in \mathbf{N} \times \mathbf{N}$, there exists a constant $N_{\alpha,\beta} \geq 0$ such that*

$$\|\partial_z^\alpha \partial_{\overline{z}}^\beta \Pi(z)\|_{\mathcal{L}(L^2,L^2)} = \mathcal{O}(h^{-N_{\alpha,\beta}}),\ z \in \Omega.$$

Proof By the Cauchy-Riesz functional calculus,

$$\Pi(z) = \frac{1}{2\pi i} \int_\gamma (w - Q(z))^{-1} dw, \tag{3.3.5}$$

where γ is the oriented boundary of $D(0, h/C)$ for a fixed sufficiently large constant $C > 0$. For $w \in \gamma$, we have

$$\|(w - Q(z))^{-1}\|_{\mathcal{L}(L^2,H_h^1)} = \mathcal{O}\left(\frac{1}{h}\right), \tag{3.3.6}$$

where we let $H^1 = H_h^1$ be the Sobolev space of order 1 in the semi-classical (h dependent) variant, equipped with the semi-classical norm

$$\|u\|_{H_h^1} = \left(\|u\|^2 + \|hDu\|^2\right)^{\frac{1}{2}}$$

with $\|\cdot\|$ denoting the L^2-norm on S^1 . (Relation (3.3.6) even holds with H_h^1 replaced by $H_h^2 = \mathcal{D}(Q)$, equipped with its natural semi-classical norm, but this will not be needed.) In fact, we first notice that the resolvent $(w - Q(z))^{-1}$ is $\mathcal{O}(1/h)$ as a bounded operator in L^2 when $w \in \gamma$, so it suffices to show that $hD \circ (w - Q)^{-1}$ is $\mathcal{O}(1/h)$ as a bounded operator in L^2. Now $Q = (P - z)^*(P - z)$ is of the form $(hD)^2 + a(x)hD + hD \circ \overline{a}(x) + b(x)$, where a, b are smooth and b is real-valued. It follows that for $u \in H^1(S^1)$

$$\begin{aligned}(Qu|u) &\geq \|hDu\|^2 - \mathcal{O}(1)\|u\|\|hDu\| - \mathcal{O}(1)\|u\|^2 \\ &\geq (1/2)\|hDu\|^2 - \mathcal{O}(1)\|u\|^2.\end{aligned}$$

On the other hand,

$$|(Qu|u)| \leq \frac{1}{2}\left(\|Qu\|^2 + \|u\|^2\right).$$

Hence,

$$\|hDu\|^2 \leq \|Qu\|^2 + \mathcal{O}(1)\|u\|^2,$$

and we get $\|hDu\| \le \mathcal{O}(1)(\|Qu\|+\|u\|)$ which in turn implies that $\|hDu\| \le \mathcal{O}(1)(\|(Q-w)u\| + \|u\|)$ uniformly for $w \in \gamma$. Since the $\mathcal{L}(L^2, L^2)$-norm of the resolvent is $\mathcal{O}(1/h)$, we have $\|u\| \le \mathcal{O}(1/h)\|(Q-w)u\|$, so $\|hDu\| \le \mathcal{O}(1/h)\|(Q-w)u\|$, which gives the desired estimate on the $L^2 \to L^2$ norm of $hD(w-Q)^{-1}$, and we have verified (3.3.6).

On the other hand, $\partial_z^\alpha \partial_{\overline{z}}^\beta Q(z)$ vanishes when $\max(|\alpha|, |\beta|) \ge 2$, and for $(\alpha, \beta) \in \{(1,0), (0,1), (1,1)\}$ we see that

$$\partial_z^\alpha \partial_{\overline{z}}^\beta Q(z) = \mathcal{O}(1) : H_h^1 \to L^2.$$

For $(\alpha, \beta) \neq (0,0)$, $\partial_z^\alpha \partial_{\overline{z}}^\beta (w - Q(z))^{-1}$ is a linear combination of terms,

$$(w - Q(z))^{-1}(\partial_z^{\alpha_1} \partial_{\overline{z}}^{\beta_1} Q(z))(w - Q(z))^{-1} \cdots (\partial_z^{\alpha_N} \partial_{\overline{z}}^{\beta_N} Q(z))(w - Q(z))^{-1},$$

with $(\alpha_j, \beta_j) \neq (0,0)$, $\alpha_1 + \cdots + \alpha_N = \alpha$, $\beta_1 + \cdots + \beta_N = \beta$. It now suffices to apply $\partial_z^\alpha \partial_{\overline{z}}^\beta$ to (3.3.5) and use (3.3.6). □

Since e_0 is the normalization of $\Pi(z)e_{\mathrm{wkb}}$ we see that the various $(z, \overline{z})$-derivatives of e_0, and hence also of $e_0 - e_{\mathrm{wkb}}$, are of temperate growth. The last quantity is exponentially small in L^2, and by elementary interpolation estimates for the successive derivatives in $z, \overline{z}$ we get the same conclusion for the higher derivatives of $e_0 - e_{\mathrm{wkb}}$. (Cf. [51, Proposition 1.6].)

It follows that

$$f_+(z) = (e_0(z)|\partial_z e_0(z)) = (e_{\mathrm{wkb}}(z)|\partial_z e_{\mathrm{wkb}}(z)) + \mathcal{O}(e^{-\frac{1}{Ch}}), \tag{3.3.7}$$

and the various $z, \overline{z}$-derivatives of the remainder are also exponentially decaying.

Using the same simple variant of the method of stationary phase as in Sect. 3.1, we get

$$\begin{aligned}(e_{\mathrm{wkb}}|\partial_z e_{\mathrm{wkb}}) \\ &= \mathcal{O}(1) - \frac{i}{h} h^{-\frac{1}{2}} |a(z;h)|^2 \int |\chi(x - x_+(z))|^2 \overline{\partial_z \phi_+(x,z)} e^{-\frac{2}{h}\Im \phi_+(x,z)} dx \\ &= -\frac{i}{h} \overline{(\partial_z \phi_+)(x_+(z), z)} + \mathcal{O}(1),\end{aligned} \tag{3.3.8}$$

where the remainder has a complete asymptotic expansion in powers of h, namely, $\sim r_0(z) + hr_1(z) + \cdots$ in the space of smooth functions and, in particular, it remains bounded after taking $z, \overline{z}$ derivatives.

Using that $\phi_+(x_+(z), z) = 0$, $(\phi_+)'_x(x_+(z), z) = \xi_+(z)$, we get after applying ∂_z to the first of these relations, that

$$(\partial_z \phi_+)(x_+(z), z) = -\xi_+(z) \partial_z x_+(z). \tag{3.3.9}$$

On the other hand, if we apply ∂_z and $\partial_{\overline{z}}$ to the equation $p(x_+(z), \xi_+(z)) = z$, we get

$$\begin{cases} p'_x \partial_z x_+ + p'_\xi \partial_z \xi_+ = 1, \\ p'_x \partial_{\overline{z}} x_+ + p'_\xi \partial_{\overline{z}} \xi_+ = 0, \end{cases} \tag{3.3.10}$$

and using that $x_+(z)$ and $\xi_+(z)$ are real valued,

$$\begin{cases} \overline{p}'_x \partial_{\overline{z}} x_+ + \overline{p}'_\xi \partial_{\overline{z}} \xi_+ = 1, \\ p'_x \partial_{\overline{z}} x_+ + p'_\xi \partial_{\overline{z}} \xi_+ = 0, \end{cases} \tag{3.3.11}$$

which has the solution

$$\partial_{\overline{z}} x_+ = \frac{p'_\xi}{\{p, \overline{p}\}}(\rho_+), \quad \partial_{\overline{z}} \xi_+ = \frac{-p'_x}{\{p, \overline{p}\}}(\rho_+). \tag{3.3.12}$$

Combining (3.3.7)–(3.3.9), we get $f_+ = \mathcal{O}(1) + \frac{i}{h}\xi_+ \partial_{\overline{z}} x_+$, where the $\mathcal{O}(1)$ is stable under differentiation. To this we apply ∂_z, then take real parts and notice that $\partial_z \partial_{\overline{z}} x_+$ is real:

$$\Re \partial_z f_+ = \mathcal{O}(1) + \Re \frac{i}{h} \partial_z \xi_+ \partial_{\overline{z}} x_+ .$$

It now suffices to apply (3.3.12) to get the second (non-trivial) identity in (3.3.4) for the contribution from f_+. The one from f_- can be treated similarly. □

Using the expressions for the z-derivatives of x_+, ξ_+ and the analogous ones for x_-, ξ_-, we have the following easy result relating (3.3.4) to the symplectic volume:

Proposition 3.3.3 *Writing $z = x + iy$, we have*

$$d\xi_+(z) \wedge dx_+(z) = \frac{2}{\frac{1}{i}\{p, \overline{p}\}(\rho_+)} dy \wedge dx,$$

$$-d\xi_-(z) \wedge dx_-(z) = -\frac{2}{\frac{1}{i}\{p, \overline{p}\}(\rho_-)} dy \wedge dx,$$

so by (3.3.4),

$$\Re \Delta F(z) dy \wedge dx = \frac{1}{h}(d\xi_+ \wedge dx_+ - d\xi_- \wedge dx_-) + \mathcal{O}(1). \tag{3.3.13}$$

3.4 Adding the Random Perturbation

Let $X \sim \mathcal{N}_{\mathbf{C}}(0, \sigma^2)$ be a complex Gaussian random variable, meaning that X has the probability distribution

$$X_*(\mathbf{P}(d\omega)) = \frac{1}{\pi\sigma^2} e^{-\frac{|X|^2}{\sigma^2}} d(\Re X) d(\Im X). \tag{3.4.1}$$

Here $\sigma > 0$. For $t < 1/\sigma^2$, we have the expectation value

$$E(e^{t|X|^2}) = \frac{1}{1 - \sigma^2 t}. \tag{3.4.2}$$

Bordeaux Montrieux [16] observed that we have the following possibly classical result (improving a similar statement in [56]).

Proposition 3.4.1 *There exists $C_0 > 0$ such that the following holds: Let $X_j \sim \mathcal{N}_{\mathbf{C}}(0, \sigma_j^2)$, $1 \le j \le N < \infty$ be independent complex Gaussian random variables. Put $s_1 = \max \sigma_j^2$. Then for every $x > 0$, the probability that $\sum_1^N |X_j|^2 \ge x$ statistics*

$$\mathbf{P}(\sum_1^N |X_j|^2 \ge x) \le \exp\left(\frac{C_0}{2s_1} \sum_1^N \sigma_j^2 - \frac{x}{2s_1}\right).$$

Proof For $t \le 1/(2s_1)$, we have

$$\mathbf{P}(\sum |X_j|^2 \ge x) \le E(e^{t(\sum |X_j|^2 - x)}) = e^{-tx} \prod_1^N E(e^{t|X_j|^2})$$

$$= \exp\left(\sum_1^N \ln \frac{1}{1 - \sigma_j^2 t} - tx\right) \le \exp\left(C_0 \sum \sigma_j^2 t - tx\right).$$

It then suffices to take $t = (2s_1)^{-1}$. □

Recall that

$$Q_\omega u(x) = \sum_{|k|,|j| \le C_1/h} \alpha_{j,k}(\omega)(u|e^k) e^j(x), \quad e^k(x) = \frac{1}{\sqrt{2\pi}} e^{ikx}. \tag{3.4.3}$$

Since the Hilbert-Schmidt norm of Q_ω is given by $\|Q_\omega\|_{\mathrm{HS}}^2 = \sum |\alpha_{j,k}(\omega)|^2$, we get from the preceding proposition:

Proposition 3.4.2 *If $C > 0$ is large enough, then*

$$\|Q_\omega\|_{\mathrm{HS}} \leq \frac{C}{h} \text{ with probability } \geq 1 - e^{-\frac{1}{Ch^2}}. \tag{3.4.4}$$

Actually, we get a sharper statement: If $\lambda \geq \sqrt{2C_0}(1 + 2C_1/h)$, then $\|Q_\omega\|_{\mathrm{HS}} \leq \lambda$ with probability $\geq 1 - e^{-\lambda^2/4}$.

Now, we work under the assumption that $\|Q_\omega\|_{\mathrm{HS}} \leq C/h$ and recall that $\|Q_\omega\|_{\mathcal{L}(L^2,L^2)} \leq \|Q_\omega\|_{\mathrm{HS}}$. Assume that

$$0 \leq \delta \ll h^{3/2}, \tag{3.4.5}$$

so that $\|\delta Q_\omega\| \ll h^{1/2}$. Then, by simple perturbation theory, Proposition 3.2.1 shows that

$$\mathcal{P}_\delta(z) = \begin{pmatrix} P_\delta - z & R_- \\ R_+ & 0 \end{pmatrix} : H_h^1 \times \mathbf{C} \to L^2 \times \mathbf{C}$$

is bijective with the bounded inverse

$$\mathcal{E}_\delta = \begin{pmatrix} E^\delta & E_+^\delta \\ E_-^\delta & E_{-+}^\delta \end{pmatrix},$$

$$\begin{aligned} E^\delta &= E + \mathcal{O}(\frac{\delta}{h^2}) = \mathcal{O}(h^{-1/2}) \text{ in } \mathcal{L}(L^2, L^2), \\ E_+^\delta &= E_+ + \mathcal{O}(\frac{\delta}{h^{3/2}}) = \mathcal{O}(1) \text{ in } \mathcal{L}(\mathbf{C}, L^2), \\ E_-^\delta &= E_- + \mathcal{O}(\frac{\delta}{h^{3/2}}) = \mathcal{O}(1) \text{ in } \mathcal{L}(L^2, \mathbf{C}), \\ E_{-+}^\delta &= E_{-+} - \delta E_- Q E_+ + \mathcal{O}(\frac{\delta^2}{h^{5/2}}). \end{aligned} \tag{3.4.6}$$

In fact,

$$\mathcal{P}_\delta \circ \mathcal{E} = 1 + \mathcal{K}, \quad \mathcal{K} = \begin{pmatrix} \delta Q_\omega E & \delta Q_\omega E_+ \\ 0 & 0 \end{pmatrix},$$

$$\mathcal{K}^n = \begin{pmatrix} (\delta Q_\omega E)^n & (\delta Q_\omega E)^{n-1} \delta Q_\omega E_+ \\ 0 & 0 \end{pmatrix}$$

and Proposition 3.2.1 implies that $1+\mathcal{K}: H^0\times \mathbf{C}\to H^0\times \mathbf{C}$ is bijective with inverse $1-\mathcal{K}+\mathcal{K}^2\cdots$. Hence $\mathcal{P}_\delta$ has the right inverse

$$\mathcal{E}_\delta=\mathcal{E}(1+\mathcal{K})^{-1}=\begin{pmatrix}E^\delta & E_+^\delta\\ E_-^\delta & E_{-+}^\delta\end{pmatrix},$$

where

$$E^\delta=E\sum_0^\infty(-\delta Q_\omega E)^n,$$

$$E_+^\delta=\sum_0^\infty(-E\delta Q_\omega)^n E_+,$$

$$E_-^\delta=E_-\sum_0^\infty(-\delta Q_\omega E)^n,$$

$$E_{-+}^\delta=E_{-+}-E_-\delta Q_\omega E_+ + E_-\delta Q_\omega E\delta Q_\omega E_+\dots$$

Similarly, $\mathcal{P}_\delta$ has a left inverse, necessarily equal to $\mathcal{E}_\delta$.

As before, the eigenvalues of P_δ are the zeros of E_{-+}^δ and we have the d-bar equation

$$\partial_{\overline{z}}E_{-+}^\delta+f^\delta(z)E_{-+}^\delta=0,$$

$$f^\delta(z)=\partial_{\overline{z}}R_+E_+^\delta+E_-^\delta\partial_{\overline{z}}R_-=f(z)+\mathcal{O}(\frac{1}{h}\frac{\delta}{h^{3/2}}).$$

We can solve $\partial_{\overline{z}}F^\delta=f^\delta$ in Ω (making $e^{F^\delta}E_{-+}^\delta$ holomorphic) with

$$F^\delta=F+\mathcal{O}(\frac{\delta}{h^{5/2}})=F+\mathcal{O}(\frac{\delta}{h^{3/2}})\frac{1}{h}. \tag{3.4.7}$$

In fact, recall that Ω is open and relatively compact in $\{z\in\mathbf{C};\ \min\Im g<\Im z<\max\Im g\}$ and that $\partial_{\overline{z}}$ has the fundamental solution $1/(\pi z)$, so we can take

$$F^\delta(z)=\int_\Omega\frac{1}{\pi(z-w)}f^\delta(w)L(dw),$$

Remark 3.4.3 We define the multiplicity of a zero z_0 of $E_{-+}^\delta(z)$ as the multiplicity of z_0 as a zero of the holomorphic function $e^{F^\delta(z)}E_{-+}^\delta(z)$. As we observed after Proposition 3.2.1, the set of eigenvalues of P_δ in Ω and the set of zeros of $E_{-+}^\delta(z)$ agree; this follows from the formulae

$$(P_\delta - z)^{-1} = E^\delta(z) - E^\delta_+(z)\left(E^\delta_{-+}(z)\right)^{-1} E^\delta_-(z), \tag{3.4.8}$$

$$E^\delta_{-+}(z) = -R_+(P_\delta - z)^{-1}R_-, \tag{3.4.9}$$

valid respectively when $E^\delta_{-+}(z)$ and $P_\delta - z$ are bijective (and implying the equivalence of these two properties). If z_0 is an eigenvalue, its (algebraic) multiplicity is given by

$$m(z_0) = \operatorname{tr}\Pi(z_0),$$

where

$$\Pi(z_0) = \frac{1}{2\pi i}\int_\gamma (z - P_\delta)^{-1}dz$$

and γ is the oriented boundary of a small disc centered at z_0. Choose $\gamma = \gamma_r = \partial D(z_0, r)$ and let $0 < r \to 0$. (Here we follow an idea of Vogel [155], to exploit the fact that $\partial_{\overline{z}} E^\delta_{-+}(z_0) = 0$.) By (3.4.8),

$$\begin{aligned} m(z_0) &= \lim_{r\to 0} \operatorname{tr}\left(\frac{1}{2\pi i}\int_{\gamma_r} E^\delta_+\left(E^\delta_{-+}\right)^{-1} E^\delta_- dz\right) \\ &= \lim_{r\to 0}\frac{1}{2\pi i}\int_{\gamma_r} \operatorname{tr}\left(E^\delta_{-+}\right)^{-1} E^\delta_- E^\delta_+ dz. \end{aligned} \tag{3.4.10}$$

For the last identity we used the fact that the integrand is a composition of trace class operators, to move the trace inside the integral and then apply the cyclicity of the trace.

Differentiating the identity $\mathcal{E}^\delta \mathcal{P}^\delta = 1$, we get

$$\partial_z E^\delta_{-+} = E^\delta_- E^\delta_+ - ((\partial_z R_+)E^\delta_+ + E^\delta_- \partial_z R_-)E^\delta_{-+}.$$

If we insert this in (3.4.10), we see that the last term gives the contribution 0, so

$$m(z_0) = \lim_{r\to 0}\frac{1}{2\pi i}\int_{\gamma_r}\left(E^\delta_{-+}\right)^{-1}\partial_z E^\delta_{-+} dz. \tag{3.4.11}$$

On the other hand, the multiplicity $\widetilde{m}(z_0)$ of z_0 as a zero of E^δ_{-+} is given by

$$\begin{aligned} \widetilde{m}(z_0) &= \lim_{r\to 0}\frac{1}{2\pi i}\int_{\gamma_r}\left(e^{F^\delta}E^\delta_{-+}\right)^{-1}\partial_z\left(e^{F^\delta}E^\delta_{-+}\right)dz \\ &= \lim_{r\to 0}\frac{1}{2\pi i}\int_{\gamma_r}\left(E^\delta_{-+}\right)^{-1}\partial_z E^\delta_{-+}dz + \lim_{r\to 0}\frac{1}{2\pi i}\int_{\gamma_r}\partial_z F^\delta dz, \end{aligned} \tag{3.4.12}$$

and the last term in the last member vanishes, and hence

$$\widetilde{m}(z_0) = m(z_0).$$

Proposition 3.4.4 *Assume that* $0 < t \ll 1$, $\delta \ll h^{3/2}$,

$$\delta t \gg e^{-\frac{1}{C_0 h}}, \quad t \gg \frac{\delta}{h^{5/2}}, \tag{3.4.13}$$

where $C_0 \gg 1$ *is fixed. Then for* $h > 0$ *small enough we have with probability* $\geq 1 - e^{-\frac{1}{Ch^2}}$ *that*

$$|E_{-+}^{\delta}(z)| \leq e^{-\frac{1}{Ch}} + \frac{C\delta}{h}, \ \forall z \in \Omega. \tag{3.4.14}$$

For every $z \in \Omega$, *we have with probability* $\geq 1 - \mathcal{O}(t^2) - e^{-\frac{1}{Ch^2}}$, *that*

$$|E_{-+}^{\delta}(z)| \geq \frac{t\delta}{C}. \tag{3.4.15}$$

Proof By Proposition 3.2.1,

$$\begin{aligned} E_- Q_\omega E_+ &= (Q_\omega e_0 | f_0) \\ &= \sum_{|k|,\, |\ell| \leq \frac{C_1}{h}} \alpha_{\ell,k}(\omega)(e_0|e^k)(e^\ell|f_0) \\ &= \frac{1}{2\pi} \sum_{|k|,\, |\ell| \leq \frac{C_1}{h}} \alpha_{\ell,k}(\omega)\widehat{e_0}(k)\overline{\widehat{f_0}(\ell)}, \end{aligned}$$

where $\widehat{e_0}(k)$, $\widehat{f_0}(\ell)$ are the Fourier coefficients of e_0, f_0. This is a sum of independent Gaussian random variables $\sim \mathcal{N}_{\mathbf{C}}\left(0, \left(\frac{1}{2\pi}|\widehat{e_0}(k)| \cdot |\widehat{f_0}(\ell)|\right)^2\right)$. Now, if $X = \sum_1^M X_j$, where $X_j \sim \mathcal{N}_{\mathbf{C}}(0, \sigma_j^2)$ are independent Gaussian random variables, then $X \sim \mathcal{N}_{\mathbf{C}}(0, \sum_1^M \sigma_j^2)$.

Thus, $E_- Q_\omega E_+ \sim \mathcal{N}_{\mathbf{C}}(0, \sigma^2)$, where

$$\begin{aligned} \sigma^2 &= \frac{1}{(2\pi)^2} \sum_{|k|,\, |\ell| \leq \frac{C_1}{h}} |\widehat{e_0}(k)|^2 |\widehat{f_0}(\ell)|^2 \\ &= \left(\frac{1}{2\pi} \sum_{|k| \leq \frac{C_1}{h}} |\widehat{e_0}(k)|^2\right)\left(\frac{1}{2\pi} \sum_{|\ell| \leq \frac{C_1}{h}} |\widehat{f_0}(\ell)|^2\right). \end{aligned} \tag{3.4.16}$$

If C_1 is large enough, we get by repeated integration by parts that for every $N \in \mathbf{N}$,

$$|\widehat{e}_{\mathrm{wkb}}(k)| \le C_N \left(\frac{h}{\langle k\rangle}\right)^N, \quad |k| > \frac{C_1}{h},$$

and similarly for $\widehat{f}_{\mathrm{wkb}}$. Since $\|e_{\mathrm{wkb}}\| = \|f_{\mathrm{wkb}}\| = 1$, we get by Parseval's formula that

$$\frac{1}{2\pi}\sum_{|k|\le\frac{C_1}{h}} |\widehat{e}_{\mathrm{wkb}}(k)|^2 = 1 - \frac{1}{2\pi}\sum_{|k|>\frac{C_1}{h}} |\widehat{e}_{\mathrm{wkb}}(k)|^2 = 1 - \mathcal{O}(h^\infty),$$

and similarly for f_{wkb}. On the other hand, we know that

$$\|e_0 - e_{\mathrm{wkb}}\|,\ \|f_0 - f_{\mathrm{wkb}}\| = \mathcal{O}(h^\infty),$$

so

$$\frac{1}{2\pi}\sum_{|k|\le\frac{C_1}{h}} |\widehat{e}_0(k)|^2 = 1 - \mathcal{O}(h^\infty),$$

and similarly for f_0. Equation (3.4.16) now shows that

$$\sigma = 1 - \mathcal{O}(h^\infty). \tag{3.4.17}$$

To finish the proof, we combine this with the last relation in (3.4.6) and the fact that $|E_{-+}| \le e^{-1/Ch}$, to see that with probability $\ge 1 - e^{-1/(Ch^2)}$,

$$|E^\delta_{-+}(z)| \le e^{-\frac{1}{Ch}} + \frac{C\delta}{h} + \mathcal{O}\left(\frac{\delta^2}{h^{5/2}}\right), \quad z \in \Omega.$$

Since $\delta \ll h^{3/2}$, we get (3.4.14). Similarly by using again the last relation in (3.4.6), for every $z \in \Omega$, we have with probability $\ge 1 - \mathcal{O}(t^2) - e^{-1/(Ch^2)}$ that

$$|E^\delta_{-+}(z)| \ge t\delta - e^{-1/(Ch)} - \mathcal{O}\left(\frac{\delta^2}{h^{5/2}}\right),$$

which together with (3.4.13) implies (3.4.15). □

Proposition 3.4.5 *Let $\kappa > 5/2$ and fix $\epsilon_0 \in]0, 1[$ sufficiently small. Let $\delta = \delta(h)$ satisfy $e^{-\epsilon_0/h} \ll \delta \ll h^\kappa$, and put $\epsilon = \epsilon(h) = h \ln \frac{1}{\delta}$. Then with probability $\ge 1 - e^{-1/(Ch^2)}$ we have $|E^\delta_{-+}| \le 1$ for all $z \in \Omega$.*

Moreover, if $C \gg 1$, *then for any* $z \in \Omega$, *we have* $|E^{\delta}_{-+}| \geq e^{-C\epsilon/h}$ *with probability* $\geq 1 - \mathcal{O}(\delta^2/h^5)$.

This follows from Proposition 3.4.4 by choosing t such that

$$\max\left\{\frac{1}{\delta}e^{-\frac{1}{C_0 h}}, \frac{\delta}{h^{5/2}}, C\delta^{C-1}\right\} \ll t \leq \mathcal{O}(\frac{\delta}{h^{5/2}}),$$

which is possible because

$$\frac{1}{\delta}e^{-\frac{1}{C_0 h}},\ C\delta^{C-1} \ll \frac{\delta}{h^{5/2}}.$$

Under the same assumptions, we also have with probability $\geq 1 - e^{-1/(Ch^2)}$,

$$|F_\delta - F| \leq \mathcal{O}(\frac{\delta}{h^{3/2}})\frac{1}{h} \leq \mathcal{O}(\epsilon)\frac{1}{h}.$$

Thus for the holomorphic function $u(z) = e^{F_\delta(z)}E^{\delta}_{-+}(z)$:

- With probability $\geq 1 - e^{-1/(Ch^2)}$ we have $|u(z)| \leq \exp(\Re F(z) + C\epsilon/h)$ for all $z \in \Omega$.
- For every $z \in \Omega$, we have $|u(z)| \geq \exp(\Re F(z) - C\epsilon/h)$ with probability $\geq 1 - \mathcal{O}(\delta^2/h^5)$.

To conclude the proof of Theorem 3.0.6, we will use the following result of Hager [55, Proposition 6.1] with $\phi = h\Re F$.

Proposition 3.4.6 *Let* $\Gamma \Subset \mathbf{C}$ *have smooth boundary and let* ϕ *be a real valued* C^2*-function defined in a fixed neighborhood of* $\overline{\Gamma}$. *Let* $z \mapsto u(z; h)$ *be a family of holomorphic functions defined in a fixed neighborhood of* $\overline{\Gamma}$, *and let* $0 < \epsilon = \epsilon(h) \ll 1$. *Assume*

- $|u(z; h)| \leq \exp(\frac{1}{h}(\phi(z) + \epsilon))$ *for all* z *in a fixed neighborhood of* $\partial\Gamma$.
- *There exist* $z_1, \dots, z_N$ *depending on* h, *with* $N = N(h) \asymp \epsilon^{-1/2}$ *such that* $\partial\Gamma \subset \bigcup_1^N D(z_k, \sqrt{\epsilon})$ *and such that* $|u(z_k; h)| \geq \exp(\frac{1}{h}(\phi(z_k) - \epsilon))$, $1 \leq k \leq N(h)$.

Then, the number of zeros of u *in* Γ *satisfies*

$$\left|\#(u^{-1}(0) \cap \Gamma) - \frac{1}{2\pi h}\int_\Gamma \Delta\phi(z)dxdy\right| \leq C\frac{\sqrt{\epsilon}}{h}.$$

End of the Proof of Theorem 3.0.6 Consider the holomorphic function

$$u(z) = e^{F_\delta(z)}E^{\delta}_{-+}(z)$$

and put

$$\phi(z) = h\Re F.$$

Then with probability $\geq 1 - e^{-1/(Ch^2)}$, we have

$$|u(z)| \leq e^{\frac{1}{h}(\phi(z)+C\epsilon)} \text{ for all } z \in \Omega,$$

and for each $z \in \Omega$, we have

$$|u(z)| \geq e^{\frac{1}{h}(\phi(z)-C\epsilon)} \text{ with probability } \geq 1 - \mathcal{O}(\delta^2/h^5).$$

Choose $z_1, z_2, \dots, z_N \in \partial\Gamma$ such that $\partial\Gamma \subset \bigcup_1^N D(z_k, \sqrt{C\epsilon})$, $N = \mathcal{O}(\epsilon^{-1/2})$. Then with probability $\geq 1 - N\mathcal{O}(\delta^2/h^5) \geq 1 - \mathcal{O}(\delta^2/(\sqrt{\epsilon}h^5))$, we have

$$|u(z_j)| \geq e^{\frac{1}{h}(\phi(z_j)-C\epsilon)},\ j = 1, 2, \dots, N. \tag{3.4.18}$$

Having already identified the eigenvalues of P_δ with the zeros of u and their multiplicities in Remark 3.4.3, we conclude from Proposition 3.4.6 that with probability as in the theorem,

$$\left|\#(\sigma(P_\delta) \cap \Gamma) - \frac{1}{2\pi}\int_\Gamma \Delta(\Re F)(z)L(dz)\right| \leq \frac{C\sqrt{\epsilon}}{h}. \tag{3.4.19}$$

Returning to (3.3.13), we notice that the maps $z \mapsto (x_+(z), \xi_+(z))$ and $z \mapsto (x_-(z), \xi_-(z))$ have Jacobians of different sign, so when passing to densities, this relation shows that the direct image under p of $1/h$ times the symplectic volume density is equal to $\Re\Delta F L(dz) + \mathcal{O}(1)$. Consequently,

$$\mathcal{O}(1) + \int_\Gamma \Delta(\Re F)L(dz) = \frac{1}{h}\mathrm{vol}\, p^{-1}(\Gamma),$$

and the theorem follows. □

Proof of Proposition 3.4.6 Define $\phi_j(z)$ by $i\phi_j(z) = \phi(z_j) + 2\partial_z\phi(z_j)(z - z_j)$. Then

$$\phi(z) = \Re(i\phi_j(z)) + R_j(z), \quad R_j(z) = \mathcal{O}((z - z_j)^2)$$

$$\phi_j'(z) = \frac{2}{i}\partial_z\phi(z) + \mathcal{O}((z - z_j)).$$

Consider the holomorphic function

$$v_j(z;h) = u(z;h)e^{-\frac{i}{h}\phi_j(z)}.$$

Then $|v_j(z;h)| \le e^{\frac{1}{h}(\phi(z)-\Re i\phi_j(z))} = e^{\frac{1}{h}R_j} \le e^{\frac{C\epsilon}{h}}$, when $z - z_j = \mathcal{O}(\sqrt{\epsilon})$, while

$$|v_j(z_j;h)| \ge e^{-\frac{C\epsilon}{h}}.$$

In a $\sqrt{\epsilon}$-neighborhood of z_j we put $v = v_j$ and make the change of variables $w = (z - z_j)/\sqrt{\epsilon}$, $\widetilde{v}(w) = v(z)$, so that

$$|\widetilde{v}(w)| \le e^{C\epsilon/h} \text{ on } D(0,2), \quad |\widetilde{v}(0)| \ge e^{-C\epsilon/h}. \tag{3.4.20}$$

Using Jensen's formula (which we review at the end of Sect. 8.4, see also Sect. 13.5), we see that the number of zeros $w_1, \dots, w_N$ of $\widetilde{v}$ in $D(0, 3/2)$ (repeated according to their multiplicity) is $\mathcal{O}(\epsilon/h)$. Factorize

$$\widetilde{v}(w) = e^{g(w)}\prod_1^N (w - w_k). \tag{3.4.21}$$

In order to estimate g and g', we need to find circles on which we have good lower bounds on the product above. This is a classical argument in complex analysis and we follow [135, Section 5]. □

Lemma 3.4.7 *Let $x_1, x_2, \dots, x_N \in \mathbf{R}$ and let $I \subset \mathbf{R}$ be an interval of length $|I| \in]0, +\infty[$. Then there exists $x \in I$ such that*

$$\prod_{j=1}^N |x - x_j| \ge e^{-N\left(1+\ln\frac{2}{|I|}\right)}.$$

Proof Consider the function

$$F(x) = \sum_{j=1}^N \ln\frac{1}{|x - x_j|}.$$

We have

$$\int_I \ln\frac{1}{|x - x_j|}dx \le 2\int_0^{|I|/2} \ln\frac{1}{t} = |I|\left(1 + \ln\frac{2}{|I|}\right),$$

since the first integral takes its largest possible value when x_j is the midpoint of I. It follows that

$$\frac{1}{|I|}\int_I F(x)dx \le N\left(1+\ln\frac{2}{|I|}\right).$$

We can therefore find $x \in I$ such that $F(x) \le N(1+\ln(2/|I|))$, so

$$\prod_1^N |x-x_j| = e^{-F(x)} \ge e^{-N\left(1+\ln\frac{2}{|I|}\right)}.$$

□

The lemma shows that there exists $r \in [4/3, 3/2]$ such that, for every $w \in \partial D(0,r)$,

$$\left|\prod_1^N (w-w_k)\right| \ge \prod_1^N \big|r-|w_k|\big| \ge e^{-N\left(1+\ln\frac{4}{3}\right)} \ge e^{-\mathcal{O}(\epsilon)/h}.$$

Consequently, for $|w| = r$,

$$\left|e^{g(w)}\right| \le e^{\mathcal{O}(\epsilon)/h}|\widetilde{v}(w)| \le e^{\mathcal{O}(\epsilon)/h},$$

and by the maximum principle, this estimate extends to $\overline{D(0,r)}$:

$$\Re g(w) \le \frac{\mathcal{O}(\epsilon)}{h}, \quad |w| \le r.$$

If $C > 0$ is large enough, $\frac{C\epsilon}{h} - \Re g(w)$ is a non-negative harmonic function on $D(0,r)$ · that in view of (3.4.20), (3.4.21) satisfies

$$\frac{C\epsilon}{h} - \Re g(0) \le \frac{\mathcal{O}(\epsilon)}{h}.$$

We can then apply Harnack's inequality (see e.g., Ahlfors [4, Ch. 6, Section 6.3]) to conclude that

$$\frac{C\epsilon}{h} - \Re g(w) \le \frac{\mathcal{O}(\epsilon)}{h}, \quad |w| \le \frac{5}{4},$$

i.e.,

$$\Re g(w) \ge -\frac{\mathcal{O}(\epsilon)}{h}, \quad |w| \le \frac{5}{4},$$

and hence,

$$|\Re g(w)| \le \frac{\mathcal{O}(\epsilon)}{h}, \quad |w| \le \frac{5}{4}. \tag{3.4.22}$$

Representing $\Re g(w)$ by means of the Poisson kernel for $D(0, 5/4)$:

$$\Re g(w) = \int_{\partial D(0,5/4)} K(w, \omega)\Re g(\omega)|d\omega|,$$

and using the smoothness of the kernel $K(w, \omega)$ for w in the interior of the disc, we see that

$$\nabla \Re g = \frac{\mathcal{O}(\epsilon)}{h}, \quad |w| \le 6/5,$$

and hence by the Cauchy-Riemann equations,

$$g' = \frac{\mathcal{O}(\epsilon)}{h}, \quad |w| \le 6/5. \tag{3.4.23}$$

We now return to (3.4.21). If $\gamma : [0, 1] \ni t \mapsto \gamma(t) \in D(0, 6/5)$ with $\dot\gamma$ and $1/\dot\gamma$ uniformly bounded and which avoids the zeros w_k, then on one hand,

$$\frac{1}{2\pi}\mathrm{var}\ \arg_\gamma \widetilde{v} = \frac{1}{2\pi}\int_\gamma \Im d(\ln \widetilde{v}) = \Re\frac{1}{2\pi i}\int_\gamma d\ln\widetilde{v} = \Re\frac{1}{2\pi i}\int_\gamma \frac{\widetilde{v}'}{\widetilde{v}}dw,$$

and on the other hand,

$$\frac{1}{2\pi}\mathrm{var}\ \arg_\gamma \widetilde{v} = \Re\frac{1}{2\pi i}\int_\gamma g dw + \sum_1^N \mathrm{var}\ \arg_\gamma(w - w_k) = \frac{\mathcal{O}(\epsilon)}{h}.$$

We notice that these relations are invariant under the substitution $z = z_j + \sqrt{\epsilon}w$. Now cover $\partial\Gamma$ by $\asymp 1/\sqrt{\epsilon}$ discs $D(z_j, \sqrt{\epsilon})$ with $z_j \in \partial\Gamma$. Then there are at most $\mathcal{O}(\epsilon/h)$ zeros of u in each such disc and we assume for simplicity that none of them is situated on $\partial\Gamma$. We equip $\partial\Gamma$ with the positive orientation and partition it into segments $\widetilde{\gamma}_j$ so that $\widetilde{\gamma}_j \subset D(z_j, \sqrt{\epsilon})$. Then

$$\Re\frac{1}{2\pi i}\int_{\gamma_j}\frac{v_j'}{v_j}dz = \mathcal{O}\left(\frac{\epsilon}{h}\right).$$

Writing $u = v_j e^{i\phi_j/h}$ along $\widetilde{\gamma}_j$, we get

$$\begin{aligned}\Re \frac{1}{2\pi i}\int_{\widetilde{\gamma}_j} \frac{u'}{u} dz &= \Re \frac{1}{2\pi h}\int_{\widetilde{\gamma}_j} \phi_j' dz + \underbrace{\Re \frac{1}{2\pi i}\int_{\widetilde{\gamma}_j} \frac{v_j'}{v_j}}_{=\mathcal{O}(\frac{\epsilon}{h})} \\ &= \Re \frac{1}{2\pi h}\int_{\widetilde{\gamma}_j} \left(\frac{2}{i}\partial_z \phi(z) + \mathcal{O}(z - z_j)\right) dz + \mathcal{O}\left(\frac{\epsilon}{h}\right) \\ &= \Re \frac{1}{2\pi h}\int_{\widetilde{\gamma}_j} \frac{2}{i}\partial_z \phi(z) dz + \mathcal{O}\left(\frac{\epsilon}{h}\right).\end{aligned}$$

Summing over j, we get the number of zeros of u in Γ:

$$\begin{aligned}\Re \frac{1}{2\pi i}\int_{\partial\Gamma} \frac{u'(z)}{u(z)} dz &= \Re \frac{1}{2\pi h}\int_{\partial\Gamma} \frac{2}{i}\partial_z \phi(z) dz + \mathcal{O}(\frac{\sqrt{\epsilon}}{h}) \\ &= \frac{1}{2\pi h}\int \Delta\phi(x) dx dy + \mathcal{O}(\frac{\sqrt{\epsilon}}{h}).\end{aligned}$$

Here, we used Stokes's formula for the last equality:

$$d\frac{2}{i}\partial_z \phi(z) dz = \frac{2}{i}\partial_{\overline{z}}\partial_z \phi(z) d\overline{z} \wedge dz = 4\partial_{\overline{z}}\partial_z \phi(z) dx \wedge dy = \Delta\phi(z) dx \wedge dy.$$

□

Notes Hager [54] has studied generic perturbations of $hD + g(x)$ and random perturbations of more general operators in dimension 1 in [55]. In this chapter we have restricted our attention to random perturbations of $hD + g$, following rather closely the approach of [54, 55]. We have added some features (like the use of singular values) from later works, in particular [16] and [56]. Philosophically, one can say that the random perturbation generates Weyl asymptotics by creating an interaction between the points $\rho_+(z)$ and $\rho_-(z)$. For the present class of operators this cannot be achieved with multiplicative random perturbations.

Quasi-Modes and Spectral Instability in One Dimension

4

4.1 Asymptotic WKB Solutions

In this section we describe the general WKB construction of approximate "asymptotic" solutions to the ordinary differential equation

$$P(x, hD_x)u = \sum_{k=0}^{m} b_k(x)(hD_x)^k u = 0, \tag{4.1.1}$$

on an interval $\alpha < x < \beta$, where we assume that the coefficients $b_k \in C^\infty(]\alpha, \beta[)$. Here $h \in]0, h_0]$ is a small parameter and we wish to solve (4.1.1) up to any power of h. We look for u in the form

$$u(x; h) = a(x; h)e^{i\phi(x)/h}, \tag{4.1.2}$$

where $\phi \in C^\infty(]\alpha, \beta[)$ is independent of h. The exponential factor describes the oscillations of u, and when ϕ is complex valued it also describes the exponential growth or decay; $a(x; h)$ is the amplitude and should be of the form

$$a(x; h) \sim \sum_{\nu=0}^{\infty} a_\nu(x)h^\nu \text{ in } C^\infty(]\alpha, \beta[). \tag{4.1.3}$$

Here the sum is in general not convergent and (4.1.3) says that a is the asymptotic sum in the following sense:

J. Sjöstrand, *Non-Self-Adjoint Differential Operators, Spectral Asymptotics and Random Perturbations*, Pseudo-Differential Operators 14,
https://doi.org/10.1007/978-3-030-10819-9_4

$$\begin{aligned}&\text{For all } N, M \in \mathbf{N} \text{ and every interval } K \Subset]\alpha, \beta[,\\&\text{there is a constant } C = C_{K,N,M} \text{ such that}\\&\left|\left(\frac{d}{dx}\right)^M (a(x;h) - \sum_{\nu=0}^{N} a_\nu(x)h^\nu)\right| \le Ch^{N+1}, \forall x \in K.\end{aligned} \tag{4.1.4}$$

For any sequence $a_\nu \in C^\infty(]\alpha, \beta[)$ there exists an asymptotic sum by virtue of the following Borel lemma (see e.g., [41, Ch. 2]):

Lemma 4.1.1 *Let* $a_0, a_1, a_2, \ldots \in C^\infty(]\alpha, \beta[)$*. Then there exists* $a(x; h) \in C^\infty(]\alpha, \beta[)$, $0 < h \le h_0$, *such that* (4.1.3) *holds as defined in* (4.1.4).

Thus, $a(x; h)$ is smooth in x for every $h \in]0, h_0]$ and that is in general enough in practice. A closer look at the proof shows that we can choose a smooth also in h and even in both variables simultaneously; $a \in C^\infty(]\alpha, \beta[\times[0, h_0])$.

Let us make a remark about the uniqueness of a: If $\widetilde{a}(x; h)$ is a second function with $\widetilde{a} \sim \sum_0^\infty a_\nu h^\nu$, then

$$\begin{aligned}&\forall N, M \in \mathbf{N},\ K \Subset]\alpha, \beta[,\ \exists C_{K,N,M} \text{ such that}\\&\left|\left(\frac{d}{dx}\right)^N (a(x;h) - \widetilde{a}(x;h))\right| \le C_{K,N,M} h^{N+1},\ x \in K.\end{aligned} \tag{4.1.5}$$

We will write this more briefly as

$$\widetilde{a} = a + \mathcal{O}(h^\infty) \text{ locally uniformly on }]\alpha, \beta[$$
$$\text{and similarly for all the derivatives,}$$

or simply

$$\widetilde{a} = a + \mathcal{O}(h^\infty) \text{ in } C^\infty(]\alpha, \beta[).$$

To P we associate its semi-classical principal symbol:

$$p(x, \xi) = \sum_{k=0}^{m} b_k(x)\xi^k \in C^\infty(]\alpha, \beta[\times\mathbf{R}) \tag{4.1.6}$$

which is a polynomial of degree m in ξ with coefficients that depend smoothly on x. More generally, we can let b_k depend on h and consider the semi-classical differential operator

$$\begin{aligned} P = P(x, hD; h) &= \sum_{k=0}^{m} b_k(x; h)(hD)^k, \\ b_k(x; h) &\sim \sum_{\nu=0}^{\infty} b_{k,\nu}(x)h^{\nu}, \quad b_{k,\nu} \in C^{\infty}(]\alpha, \beta[) \end{aligned} \tag{4.1.7}$$

and its semi-classical principal symbol

$$p(x, \xi) = \sum_{k=0}^{m} b_{k,0}(x)\xi^k, \tag{4.1.8}$$

as well as its full symbol

$$P(x, \xi; h) = \sum_{k=0}^{m} b_k(x; h)\xi^k \sim p(x, \xi) + hp_1(x, \xi) + \cdots, \tag{4.1.9}$$

where $p_{\nu}(x, \xi) = \sum_{k=0}^{m} b_{k,\nu}(x)\xi^k$, and we will sometimes write $p = p_0$. The last asymptotic sum takes place in the Fréchet space of smooth functions on $]\alpha, \beta[\times \mathbf{R}$ that are polynomials of order m in ξ. If we consider a second semi-classical differential operator

$$Q = Q(x, hD; h) = \sum_{k=0}^{\widetilde{m}} c_k(x; h)(hD)^k, \text{ where } c_k \sim \sum_{\nu=0}^{\infty} c_{k,\nu}(x)h^{\nu} \text{ in } C^{\infty}, \tag{4.1.10}$$

then by using Leibnitz' formula we shall prove that the composition $R := P \circ Q$ is a semi-classical differential operator of order $m + \widetilde{m}$ with full symbol

$$R(x, \xi; h) = \sum_{\mu=0}^{\infty} \frac{h^{\mu}}{\mu!}(\partial_{\xi}^{\mu} P(x, \xi; h))D_x^{\mu} Q(x, \xi; h), \tag{4.1.11}$$

where the sum is finite. (When considering pseudodifferential operators, one encounters infinite sums of the same type and they can be defined as asymptotic sums.) Here $D_x = i^{-1}\partial_x$ denotes the partial derivative with respect to x. Writing $Q \sim \sum_0^k h^k q_k(x, \xi)$, $r \sim \sum_0^k h^k r_k(x, \xi)$ similarly to (4.1.9), we get from the composition formula above, that

$$r_0(x, \xi) = p_0(x, \xi)r_0(x, \xi), \tag{4.1.12}$$

$$r_1(x, \xi) = \frac{1}{i}\partial_{\xi} p_0(x, \xi)\partial_x q_0(x, \xi) + p_0(x, \xi)q_1(x, \xi) + p_1(x, \xi)q_0(x, \xi). \tag{4.1.13}$$

Proof of (4.1.11) It suffices to treat the case when $P = a(x)(hD_x)^k$, $Q = b(x)(hD_x)^\ell$. Leibnitz' formula gives

$$\begin{aligned} P \circ Qu(x) &= a(x)(hD_x)^k(b(x)(hD_x)^\ell u(x)) \\ &= \sum_{j=0}^{k} a(x)\frac{k!}{j!(k-j)!}((hD_x)^j b(x))(hD_x)^{k-j+\ell}u(x). \end{aligned}$$

Thus the symbol of $P \circ Q$ is

$$\begin{aligned} \sum_{j=0}^{k} \frac{h^j}{j!}a(x)\frac{k!}{(k-j)!}\xi^{k-j}D_x^j b(x)\xi^\ell &= \sum_{j=0}^{k}\frac{h^j}{j!}\partial_\xi^j(a(x)\xi^k)D_x^j(b(x)\xi^\ell) \\ &= \sum_{j=0}^{\infty}\frac{h^j}{j!}(\partial_\xi^j P)(x,\xi)D_x^j Q(x,\xi). \end{aligned}$$

□

Here is a more sophisticated, but perhaps more intuitive proof of (4.1.11): We observe that

$$e^{-ix\xi/h} \circ P(x, hD_x; h) \circ e^{ix\xi/h} = P(x, \xi + hD_x; h) = \sum_{j=0}^{\infty}\frac{h^j}{j!}\partial_\xi^j P(x,\xi;h)(hD_x)^j,$$

where the sum is finite. In particular, $P(x, \xi; h) = P(x, \xi + hD_x; h)(1)$. Moreover,

$$\begin{aligned} R(x, \xi + hD_x) &= e^{-ix\xi/h} \circ (P(x, hD)Q(x, hD)) \circ e^{ix\xi/h} \\ &= P(x, \xi + hD_x; h)Q(x, \xi + hD_x; h), \end{aligned}$$

and in particular,

$$\begin{aligned} R(x, \xi; h) &= P(x, \xi + hD_x)Q(x, \xi + hD_x; h)(1) \\ &= P(x, \xi + hD_x; h)(Q(x, \xi; h)) = \sum_{j=0}^{\infty}\frac{1}{j!}(\partial_\xi^j P)(x,\xi;h)h^j D_x^j Q(x,\xi;h), \end{aligned}$$

which proves (4.1.11).

If $\phi \in C^\infty(]\alpha, \beta[)$, we have

$$e^{-i\phi(x)/h} \circ hD_x \circ e^{i\phi(x)/h} = hD_x + \phi'(x),$$

which is a semi-classical differential operator of order 1 with full symbol $\xi + \phi'(x)$. From the composition result above we see that, more generally,

$$e^{-i\phi(x)/h} \circ (hD_x)^k \circ e^{i\phi(x)/h} = (hD_x + \phi'(x))^k, \quad k \in \mathbf{N},$$

is a semi-classical differential operator of order k. We can apply this to the conjugated operator

$$P^\phi := e^{-i\phi(x)/h} P(x, hD_x; h) e^{i\phi(x)/h} = \sum_{k=0}^{m} b_k(x; h)(hD_x + \phi'(x))^k \tag{4.1.14}$$

to see that P^ϕ is also a semi-classical differential operator of order m, with symbol

$$P^\phi(x, \xi) \sim \sum_{\nu=0}^{\infty} h^\nu p_\nu^\phi(x, \xi), \tag{4.1.15}$$

where $p_\nu(x, \xi)$ is a polynomial of order m in ξ with smooth coefficients. Moreover,

$$p^\phi(x, \xi) := p_0^\phi(x, \xi) = p(x, \phi'(x) + \xi). \tag{4.1.16}$$

Now assume that we have found a function ϕ as above such that

$$p(x, \phi'(x)) = 0, \ x \in]\alpha, \beta[\tag{4.1.17}$$

and, in addition,

$$p'_\xi(x, \phi'(x)) \neq 0, \ \text{on }]\alpha, \beta[\,, \tag{4.1.18}$$

where we use the notation $p'_\xi = \partial_\xi p$. Then for the conjugated operator above, we get

$$p^\phi(x, 0) = 0, \quad \partial_\xi p^\phi(x, 0) \neq 0. \tag{4.1.19}$$

We now look for $a(x; h) \sim \sum_{\nu=0}^{\infty} a_\nu(x) h^\nu$ with $a_0(x)$ non-vanishing on $]\alpha, \beta[$, such that

$$P^\phi(x, hD_x; h)(a(x; h)) \sim 0, \tag{4.1.20}$$

in the sense of asymptotic sums and for that we write P^ϕ as an asymptotic sum of h-*independent* differential operators of order $\leq m$,

$$P^\phi(x, hD_x; h) \sim Q_0(x, D_x) + hQ_1(x, D_x) + h^2 Q_2(x, D_x) + \cdots \tag{4.1.21}$$

Here $Q_0 = p^\phi(x, 0) = p(x, \phi'(x)) = 0$ and

$$Q_1 = \partial_\xi p^\phi(x, 0) D_x + p_1^\phi(x, 0) = p'_\xi(x, \phi'(x)) D_x + p_1^\phi(x, 0), \tag{4.1.22}$$

where we used (4.1.16). Applying (4.1.21) to $a(x; h) \sim \sum_{\nu=0}^{\infty} a_\nu(x) h^\nu$ and regrouping the terms in powers of h, we get the sequence of transport equations

$$\begin{aligned} &Q_1 a_0 = 0, \\ &Q_1 a_1 + Q_2 a_0 = 0, \\ &Q_1 a_2 + Q_2 a_1 + Q_3 a_0 = 0. \\ &\dots \end{aligned} \tag{4.1.23}$$

Each equation is of first order and can be solved explicitly. From the first one we get

$$a_0(x) = C_0 \exp\left(-i \int_{x_0}^{x} \frac{p_1^\phi(t, 0)}{\partial_\xi p^\phi(t, 0)} dt \right), \tag{4.1.24}$$

where $C_0 \in \mathbf{C}$ is an arbitrary integration constant and x_0 an arbitrary point in $]\alpha, \beta[$. We fix $C_0 \neq 0$.

Remark 4.1.2 Let $a(x; h) \sim a_0(x) + h a_1(x) + \cdots$ where $a_0, a_1, \dots$ are obtained by solving the transport equations (4.1.23) and where a_0 is non-vanishing. Then the general solution is of the form $\widetilde{a}(x; h) \sim c(h) a(x; h)$, where $c(h) \sim c_0 + h c_1 + \cdots$ for any complex numbers $c_0, c_1, \dots$

Example 4.1.3 Let

$$\begin{aligned} P &= -h^2 \left(\frac{d}{dx} \right)^2 + V(x) - E, \\ P(x, \xi) &= p(x, \xi) = \xi^2 + V(x) - E, \end{aligned} \tag{4.1.25}$$

where $V \in C^\infty(]\alpha, \beta[; \mathbf{R})$ and $E \in \mathbf{R}$.

(a) Assume that $V(x) < E$ on $]\alpha, \beta[$. We are in the classically allowed region and the equation $p(x, \xi) = 0$ has two real solutions $\xi = \xi_\pm = \pm\sqrt{E - V(x)}$, and we have two solutions $\pm\phi(x)$ of the eikonal equation (4.1.17), given by $\phi(x) = \int_{x_0}^{x} \sqrt{E - V(y)} dy$. The WKB constructions above can be applied and we find two *oscillatory* solutions

$$u_\pm(x; h) = a^\pm(x; h) e^{\pm i\phi(x)/h}, \tag{4.1.26}$$

with $u_- = \overline{u}_+$, that satisfy

$$Pu_\pm = \mathcal{O}(h^N) \tag{4.1.27}$$

locally uniformly on the interval for every $N \in \mathbf{N}$ together with all their derivatives. (We will express this more briefly by saying that $Pu_\pm = \mathcal{O}(h^\infty)$ locally uniformly together with all its derivatives.)

In this case, we have $P^\phi = (hD_x + \phi')^2 + V - E = h(2\phi'(x)D_x + \frac{1}{i}\phi''(x)) + (hD_x)^2$ and we get $a_0 = C(\phi')^{-\frac{1}{2}}$.

(b) Let $V(x) > E$ on $]\alpha, \beta[$. We are in the classically forbidden region (in the sense that there are no points in the real phase space where p vanishes). Now the equation $p(x, \xi) = 0$ has the two non-real solutions $\xi = \pm i\sqrt{V(x) - E}$. Let $\psi(x) = \int_{x_0}^x \sqrt{V(y) - E}dy$. The WKB method produces two functions,

$$v_\pm(x; h) = b^\pm(x; h)e^{\pm\psi(x)/h},$$

such that

$$P(x, hD_x; h)(b^\pm(x; h)e^{\psi(x)/h}) = r^\pm(x; h)e^{\pm\psi(x)/h},$$

where $r^\pm(x; h) = \mathcal{O}(h^\infty)$ locally uniformly with all their derivatives.

Again, $b_0 = C(\psi')^{-\frac{1}{2}}$.

Notice that if for some $(x_0, \xi_0) \in]\alpha, \beta[\times \mathbf{C}$

$$p(x_0, \xi_0) = 0, \;\; p'_\xi(x_0, \xi_0) \neq 0, \tag{4.1.28}$$

then by the implicit function theorem there is a small open interval $I \subset]\alpha, \beta[$ containing x_0 such that the equation $p(x, \xi) = 0$ has a solution $\xi = \xi(x) \in \mathbf{C}$ depending smoothly on $x \in I$ with $\xi(x_0) = \xi_0$, $p'_\xi(x, \xi(x)) \neq 0$. If we let $\phi \in C^\infty(I)$ be a primitive of $\xi(x)$, so that $\phi'(x) = \xi(x)$, then we get

$$p(x, \phi'(x)) = 0, \quad p'_\xi(x, \phi'(x)) \neq 0.$$

Summarizing the discussion, we have

Theorem 4.1.4 *Consider the semi-classical differential operator*

$$P = P(x, hD; h) = \sum_{k=0}^{m} b_k(x; h)(hD)^k, \quad x \in]\alpha, \beta[, \tag{4.1.29}$$

where $b_k(x;h) \sim \sum_{\nu=0}^{\infty} b_{k,\nu}(x)h^\nu$, $b_{k,\nu} \in C^\infty(]\alpha,\beta[)$ *and its semi-classical principal symbol*

$$p(x,\xi) = \sum_{k=0}^{m} b_{k,0}(x)\xi^k. \tag{4.1.30}$$

Let $(x_0,\xi_0) \in]\alpha,\beta[\times \mathbf{C}$ *be a point where* (4.1.28) *holds. Then there exist an open interval* I *with* $x_0 \in I \subset]\alpha,\beta[$ *and a function* $\phi \in C^\infty(I)$ *such that*

$$p(x,\phi'(x)) = 0, \quad p'_\xi(x,\phi'(x)) \neq 0, \quad \phi'(x_0) = \xi_0.$$

This function is uniquely determined up to a constant.

For any such (I,ϕ) *there exists* $a(x;h) \sim \sum_0^\infty a_\nu(x)h^\nu$ *with* $a_0 \neq 0$ *on* I *such that*

$$P(x,hD;h)(e^{i\phi(x)/h}a(x;h)) = r(x;h)e^{i\phi(x)/h}, \quad r = \mathcal{O}(h^\infty),$$

locally uniformly on I *with all its derivatives.*

4.2 Quasimodes in One Dimension

Davies [33] observed that for the one-dimensional Schrödinger operator we can construct quasimodes for values of the spectral parameter that may be quite far from the spectrum of the operator. Zworski [161] observed that this result can be viewed as a special case of more general and older results of Hörmander [78, 79] and generalized the result of Davies to more general operators in the semi-classical limit in arbitrary dimension, by adaptation of Hörmander's results via a known reduction. In [40] a direct proof was given. (See Chap. 9.) Here we explain the result in the simpler one-dimensional case.

Consider an operator as in (4.1.7). The formal complex adjoint is given by

$$P(x,hD;h)^* = \sum_0^m (hD_x)^k \overline{b}_k(x;h) = \sum_0^m c_k(x;h)(hD_x)^k, \tag{4.2.1}$$

where $c_k(x;h) \sim c_{k,0}(x) + hc_{k,1}(x) + \cdots$ in $C^\infty(]\alpha,\beta[)$ and $c_{k,0}(x) = \overline{b}_{k,0}(x)$. The (semi-classical) principal symbol of P^* is equal to $\overline{p}(x,\xi)$ (when ξ is real), where $p(x,\xi)$ is the one of P. Motivated by the notion of normal operators, we are interested in the commutator $[P,P^*]$.

Proposition 4.2.1 *Let*

$$Q(x, hD; h) = \sum_{0}^{\widetilde{m}} c_k(x; h)(hD_x)^k, \tag{4.2.2}$$

$$c_k(x; h) \sim c_{k,0}(x) + hc_{k,1}(x) + \cdots \text{ in } C^\infty(]\alpha, \beta[),$$

be a second semi-classical differential operator of the same type as P. *Then*

$$[P, Q] = h \sum_{0}^{m+\widetilde{m}-1} d_k(x; h)(hD_x)^k =: hR(x, hD_x; h), \tag{4.2.3}$$

where R *is a semi-classical differential operator of order* $m + \widetilde{m} - 1$ *of the same type and whose principal symbol is given by*

$$r(x, \xi) = \frac{1}{i}\{p, q\} = \frac{1}{i}(p'_\xi q'_x - p'_x q'_\xi). \tag{4.2.4}$$

This result follows easily from the composition formula (4.1.11) and especially from (4.1.12), (4.1.13). Notice that the Poisson bracket $\{a, b\} := a'_\xi(x, \xi)b'_x(x, \xi) - a'_x(x, \xi)b'_\xi(x, \xi)$ of two smooth functions a and b is anti-symmetric

$$\{a, b\} = -\{b, a\}, \tag{4.2.5}$$

and in particular

$$\{a, a\} = 0. \tag{4.2.6}$$

From the proposition and the fact that the principal symbol of P^* is equal to $\overline{p}(x, \xi)$, we get

$$[P, P^*] = hR, \quad r = \frac{1}{i}\{p, \overline{p}\}, \tag{4.2.7}$$

where r denotes the principal symbol of R. Notice that r is real-valued (reflecting the fact that $[P, P^*]$ is formally self-adjoint). We also have the formula

$$\frac{1}{i}\{p, \overline{p}\} = -2\{\Re p, \Im p\}. \tag{4.2.8}$$

Example 4.2.2 Let

$$P = -h^2 \left(\frac{d}{dx}\right)^2 + V(x),\ V \in C^\infty(]\alpha, \beta[). \tag{4.2.9}$$

Then $p(x, \xi) = \xi^2 + V(x)$ and an easy computation gives

$$\frac{1}{i}\{p, \overline{p}\} = -4\xi\, \Im V'(x). \tag{4.2.10}$$

Back to the general case, let $(x_0, \xi_0) \in]\alpha, \beta[\times \mathbf{R} = T^*\,]\alpha, \beta[$ be a point where

$$p(x_0, \xi_0) = 0, \quad \frac{1}{i}\{p, \overline{p}\} > 0. \tag{4.2.11}$$

In particular,

$$p'_\xi(x_0, \xi_0) \neq 0, \tag{4.2.12}$$

so we are in the situation of Theorem 4.1.4. Let $\xi(x) \in C^\infty(\mathrm{neigh}(x_0,]\alpha, \beta[))$ (using the notation "neigh(x, A)" for "some neighborhood of x in A") be the solution of $p(x, \xi(x)) = 0$ with $\xi(x_0) = \xi_0$ and $p'_\xi(x, \xi(x)) \neq 0$. Let

$$\phi(x) = \int_{x_0}^{x} \xi(y)dy \tag{4.2.13}$$

be the corresponding solution of

$$p(x, \phi'(x)) = 0, \quad \phi(x_0) = 0,\ \phi'(x_0) = \xi_0. \tag{4.2.14}$$

Proposition 4.2.3 *In the situation above,*

$$\Im\phi''(x_0) > 0. \tag{4.2.15}$$

Proof We differentiate (4.2.14) and get for $x = x_0$:

$$p'_x(x_0, \xi_0) + p'_\xi(x_0, \xi_0)\phi''(x_0) = 0,$$

$$\phi''(x_0) = -\frac{p'_x(x_0, \xi_0)}{p'_\xi(x_0, \xi_0)} = -\frac{p'_x \overline{p'_\xi}}{|p'_\xi|^2}.$$

Hence,

$$\Im\phi''(x_0) = |p'_\xi|^{-2}\frac{1}{2i}(-p'_x\overline{p}'_\xi + p'_\xi\overline{p}'_x)$$
$$= |p'_\xi|^{-2}\frac{1}{2i}\{p, \overline{p}\}(x_0, \xi_0) > 0.$$

□

Let

$$f(x; h) = h^{-\frac{1}{4}}a(x; h)e^{i\phi(x)/h}, \quad x \in \mathrm{neigh}\,(x_0,]\alpha, \beta[) =: J \tag{4.2.16}$$

be an asymptotic solution of

$$P(x, hD_x; h)f(x; h) = \mathcal{O}(h^\infty)e^{i\phi(x)/h} \tag{4.2.17}$$

with $a \sim a_0(x) + ha_1(x) + \cdots$ in $C^\infty(J)$ and a_0 non-vanishing. It follows from (4.2.14)–(4.2.16) that after an arbitrarily small shrinking of J, there exists a constant $C > 0$ such that

$$\frac{1}{C} \le \|f(x; h)\|^2_{L^2(J)} \le C, \tag{4.2.18}$$

and for every $\delta > 0$,

$$\int_{|x-x_0|\ge\delta} |f(x; h)|^2 dx = \mathcal{O}(h^{-\frac{1}{2}}e^{-\frac{\delta^2}{Ch}}), \quad 0 < h \ll 1. \tag{4.2.19}$$

From (4.2.17) we conclude that

$$\|P(x, hD_x; h)f\|_{L^2(J)} = \mathcal{O}(h^\infty). \tag{4.2.20}$$

Thanks to (4.2.19), the estimates (4.2.20), (4.2.18) remain valid if we replace f by χf, where $\chi \in C_0^\infty(J)$ is equal to 1 in a neighborhood of x_0.

Assume now that we have a closed operator $P = P_h : L^2(I) \to L^2(I)$ with $\mathcal{D}(P) \supset C_0^\infty(I)$, $I =]\alpha, \beta[$, such that $Pu = P(x, hD_x; h)u$ when $u \in C_0^\infty(I)$. Using the quasimode χf, we get

Theorem 4.2.4 *For every $N > 0$, we have $0 \in \sigma_{h^N}(P)$ when h is sufficiently small depending on N.*

We can have other values than 0 for the spectral parameter. It suffices to find $\rho(z) = (x(z), \xi(z)) \in T^*I$ such that

$$p(x(z), \xi(z)) = z, \quad \frac{1}{i}\{p, \overline{p}\}(x(z), \xi(z)) > 0. \tag{4.2.21}$$

Here we observe that if we identify p with the smooth map $F : T^*I \ni (x, \xi) \mapsto (\Re p(x, \xi), \Im p(x, \xi)) \in \mathbf{R}^2$, then $dF(x, \xi)$ is bijective precisely at the points where $i^{-1}\{p, \overline{p}\}(x, \xi) \neq 0$. By the implicit function theorem, the points for which we can solve (4.2.21) form an open set $\Sigma_+ \subset \mathbf{C}$ and locally we can choose the solution $(x(z), \xi(z))$ to depend smoothly on $z \in \Sigma_+$. The preceding theorem can be generalized as follows:

Theorem 4.2.5 *For every $N > 0$ and every compact set $K \subset \Sigma_+$, there exists $h = h(N, K) > 0$ such that $K \subset \sigma_{h^N}(P)$ for $0 < h \leq h(N, K)$.*

Exercise Show that if $z = p(x, \xi)$, then there exists $u \in C_0^\infty(\mathbf{R})$ which is normalized in L^2, such that $\|(P - z)u\| = \mathcal{O}(h^{1/2})$.

Identifying p again with the map $T^*I \ni (x, \xi) \mapsto (\Re p, \Im p) \in \mathbf{R}^2$, we see by Sard's theorem that the image $\mathcal{N}$ under p of $\{\rho \in T^*I;\ i^{-1}\{p, \overline{p}\} = 0\}$ is of (Lebesgue) measure 0. We get

Proposition 4.2.6 *Let $\Sigma_\pm = \{p(\rho);\ \rho \in T^*I,\ \pm i^{-1}\{p, \overline{p}\}(\rho) > 0\}$. Then*

$$\mathcal{R}(p) = \Sigma_+ \cup \Sigma_- \cup \mathcal{N}, \tag{4.2.22}$$

where $\mathcal{N}$ is of measure 0.

If $p(x, \xi)$ is an even function of ξ, then $i^{-1}\{p, \overline{p}\}$ is odd and we have $\Sigma_+ = \Sigma_-$, so (4.2.22) becomes $\mathcal{R}(p) = \Sigma_+ \cup \mathcal{N}$. In particular, this is the case for the semi-classical Schrödinger operator in Example 4.2.3.

Noticing that var $\arg_\gamma(p - z) = \pm 2\pi$ if $\gamma = \partial D(\rho, \epsilon)$ is the positively oriented boundary of a small disc centered at a point $\rho \in p^{-1}(z)$ where $\pm i^{-1}\{p, \overline{p}\}$ is > 0, it is possible to show that $\Sigma_+ = \Sigma_-$ under more general assumptions.

Now return to the non-self-adjoint harmonic oscillator $P = P_c$ in (2.5.1) with $c \in \mathbf{C}\setminus]-\infty, 0]$. We have $p(x, \xi) = \xi^2 + cx^2$ and $\Sigma := \mathcal{R}(p) = \{t + sc;\ t, s \geq 0\}$ and it is easy to see that $\Sigma_+ = \Sigma_- = \overset{\circ}{\Sigma}$ is the interior of Σ.

Assume that $\Im c > 0$ in order to fix the ideas. Davies [34] showed that if $0 < \theta < \arg(c)$, then in the case $h = 1$, we have

$$\|(P_{h=1} - E)^{-1}\| \to +\infty, \ E = e^{i\theta} F, \quad F \to +\infty.$$

It is now easy to see that we have the following improvement when E tends to infinity along the half-ray $e^{i\theta}[0,+\infty[$:

$$\begin{aligned}&\text{For every } N \in \mathbf{N}, \text{ there exists } c = c(N,\theta) > 0, \text{ such}\\ &\text{that } \|(P_{h=1}-E)^{-1}\| \geq c|E|^N,\ E = e^{i\theta}F,\ F \geq 1.\end{aligned} \tag{4.2.23}$$

Here, we adopt the convention that $\|(P_{h=1}-E)^{-1}\| = +\infty$ when $E \in \sigma(P_{h=1})$.
In fact, the equation $(P_{h=1}-E)u = v$ reads

$$(D_x^2 + cx^2 - E)u = v,$$

and the change of variable $x = |E|^{1/2}\widetilde{x}$ (which changes the L^2 norms of u and v by the same factor) gives

$$(|E|^{-1}D_{\widetilde{x}}^2 + c|E|\widetilde{x}^2 - E)u = v,$$

which we can write

$$((hD_{\widetilde{x}})^2 + c\widetilde{x}^2 - e^{i\theta})u = hv, \quad h = 1/|E|.$$

Since $e^{i\theta} \in \Sigma_+$, we can find $u \in B^2$ with $\|u\| = 1$ such that $\|v\| \leq C_N h^N$, and (4.2.23) follows.

Notes Section 4.1 is largely classical. As already mentioned, Davies [33] constructed quasimodes for the non-self-adjoint one-dimensional Schrödinger operator and Zworski [161] observed that this result can be viewed as a special case of more general and older results of Hörmander [78, 79] and deduced from that a generalization to more general semi-classical operators. A direct proof was given in [40]. (See Chap. 9.) In the analytic framework we obtain exponentially accurate quasimodes (cf. [40, 132]), implying that we can replace $c|E|^N$ by $c\exp|E|$ in (4.2.23). In the one-dimensional case more precise results are possible, using the complex WKB-method. Cf. [39]. See also [63,64] for related results.

Spectral Asymptotics for More General Operators in One Dimension

5

In this chapter, we generalize the results of Chap. 3. The results and the main ideas are close, but not identical, to the ones of Hager [55]. We will use some h-pseudodifferential machinery, see for instance [41, Ch. 7].

5.1 Preliminaries for the Unperturbed Operator

We will work in $L^2(\mathbf{R})$. Only minor modifications are required if we wish to replace $\mathbf{R}$ by the compact manifold $\mathbf{T} = S^1 = \mathbf{R}/(2\pi\mathbf{Z})$. Actually, the discussion in this section extends to the case of $\mathbf{R}^n$ and general smooth compact manifolds, respectively.

Let $m \in C^\infty(\mathbf{R}^2_{x,\xi};\,]0, +\infty[)$ be an order function in the sense that for some fixed $C_0 \geq 1$, $N_0 \geq 0$,

$$m(X) \leq C_0 \langle X - Y \rangle^{N_0} m(Y), \quad X, Y \in \mathbf{R}^2. \tag{5.1.1}$$

Here, for $X = (x, \xi)$, we write $\langle X \rangle = (1 + X^2)^{1/2}$, $X^2 = x^2 + \xi^2$. Basic examples of such functions are

- $m(X) = \langle \xi \rangle^{m_0}$, where $\langle \xi \rangle = (1 + \xi^2)^{1/2}$,
- $m(X) = \langle X \rangle^{m_0}$,

where $m_0 \in \mathbf{R}$.

J. Sjöstrand, *Non-Self-Adjoint Differential Operators, Spectral Asymptotics and Random Perturbations*, Pseudo-Differential Operators 14,
https://doi.org/10.1007/978-3-030-10819-9_5

We say that $P \in C^\infty(\mathbf{R}^2)$ belongs to the symbol space $S(m)$, where m is an order function, if for all $\alpha, \beta \in \mathbf{N}$, there exists $C_{\alpha,\beta} = C_{\alpha,\beta}(P)$ such that

$$|\partial_x^\alpha \partial_\xi^\beta P(x,\xi)| \le C_{\alpha,\beta} m(x,\xi),\ \forall (x,\xi) \in \mathbf{R}^2. \tag{5.1.2}$$

When P depends on additional parameters (like the semi-classical parameter $h \in]0, h_0]$, $h_0 > 0$), we say that $P \in S(m)$ if (5.1.2) holds uniformly. $S(m)$ is a Fréchet space with the smallest possible constants $C_{\alpha,\beta}(P)$ as seminorms. We say that $P \in h^N S(m) = S(h^N m)$, when P depends on h and (5.1.2) holds uniformly with m there replaced by $h^N m$.

If $P \in S(m)$, we define a corresponding h-pseudodifferential operator by using the Weyl quantization:

$$\begin{aligned} P(x,hD)u(x) &= \frac{1}{2\pi h} \iint e^{\frac{i}{h}(x-y)\cdot\theta} P(\frac{x+y}{2}, \theta) u(y) dy d\theta \\ &= \frac{1}{2\pi} \iint e^{i(x-y)\cdot\theta} P(\frac{x+y}{2}, h\theta) u(y) dy d\theta. \end{aligned} \tag{5.1.3}$$

As explained for instance in [41, Lemma 7.8], this gives a continuous operator $P : \mathcal{S}(\mathbf{R}) \to \mathcal{S}(\mathbf{R})$, which extends to $\mathcal{S}'(\mathbf{R}) \to \mathcal{S}'(\mathbf{R})$, and we use the same letter to denote the extension. The semi-classical version of the Calderón-Zygmund theorem tells us that when $m = 1$, then $P : L^2(\mathbf{R}) \to L^2(\mathbf{R})$ is bounded and the norm is bounded by a constant times a finite sum of the $C_{\alpha,\beta}(P)$.

In the h-dependent case, we say that $P = P_h \in S(m)$ is a classical symbol, and write $P \in S_{\mathrm{cl}}(m)$, if there exist $p_0, p_1, \ldots \in S(m)$ independent of h, such that

$$P \sim \sum_0^\infty h^k p_k \quad \text{in } S(m), \tag{5.1.4}$$

in the sense that for every $N \in \mathbf{N}$,

$$P - \sum_0^N h^k p_k \in h^{N+1} S(m). \tag{5.1.5}$$

The leading term $p := p_0$ is called the semi-classical principal symbol.

We assume that

$$m \ge 1 \tag{5.1.6}$$

and

$$\exists z_0 \in \mathbf{C} \text{ such that } p - z_0 \text{ is elliptic} \tag{5.1.7}$$

in the sense that $(p(X) - z_0)^{-1} = \mathcal{O}(1/m(X))$ uniformly for $X \in \mathbf{R}^2$. Then it is standard to check that $(p - z_0)^{-1} \in S(1/m)$ (noticing that $1/m$ is an order function). We also know (cf. [41, Ch. 8]) that when $h > 0$ is small enough, $P - z_0$ is bijective $\mathcal{S} \to \mathcal{S}$, $\mathcal{S}' \to \mathcal{S}'$ and the inverse $Q = (P - z_0)^{-1}$ is a classical h-pseudodifferential operator with symbol

$$Q \sim q_0 + hq_1 + \cdots \text{ in } S(1/m), \quad q := q_0 = \frac{1}{p - z}.$$

For h small enough, we can define the semi-classical Hilbert space $H(m) = (P - z_0)^{-1}L^2(\mathbf{R})$, which does not depend on the choice of $P - z_0$ as above: After regularization and without changing the order of magnitude of m, we can assume that $m \in S(m)$, then define $H(m) = m^{-1}L^2(\mathbf{R})$, where $m = m(x, hD)$ and $m^{-1} = (m(x, hD))^{-1}$. We also have

Proposition 5.1.1 *For h small enough, the closure of P as an unbounded operator in L^2 with domain $\mathcal{S}$ is given by $P : H(m) \to H(1) = L^2$ with domain $\mathcal{D}(P) = H(m)$. (We use the same letter to denote the closure.)*

Again, this follows from standard elliptic theory of h-pseudodifferential operators. Having now a closed operator in L^2, we can consider its spectrum $\sigma(P)$.

Let

$$\Sigma = \Sigma(p) = \overline{p(\mathbf{R}^2)}. \tag{5.1.8}$$

Proposition 5.1.2

- *$p - z$ is elliptic for every $z \in \mathbf{C} \setminus \Sigma(p)$,*
- *If $K \subset \mathbf{C}$ is compact, h-independent and disjoint from $\Sigma(p)$, then for h small enough, $P - z : H(m) \to L^2$ is bijective for every $z \in K$.*

Here, the first property is elementary and quite easy to verify. The second property is a standard elliptic result (as in Proposition 5.1.1).

Define

$$\Sigma_\infty(p) = \text{ the set of accumulation points of } p(\rho),\ \rho \to \infty, \tag{5.1.9}$$

so that

$$\Sigma_\infty(p) \subset \Sigma(p). \tag{5.1.10}$$

Proposition 5.1.3 *Let $W \Subset \mathbf{C}$ be independent of h with smooth boundary and assume that*

- *W is open and simply connected,*
- *$W \not\subset \Sigma$,*
- *$\overline{W} \cap \Sigma_\infty = \emptyset$.*

Then, for h small enough, the spectrum of P in W is discrete, i.e., it consists of a discrete set (in W) of eigenvalues of finite algebraic multiplicity.

Proof Let $z_0 \in W \setminus \Sigma$ and let $\kappa : \mathbf{C} \setminus \{z_0\} \to \mathbf{C} \setminus \{z_0\}$ be a diffeomorphism onto its image such that

- $\kappa(\overline{W} \setminus \{z_0\}) \cap \overline{W} = \emptyset$,
- $\kappa(z) = z$, for $z \in \mathbf{C} \setminus \widetilde{W}$, where $\widetilde{W}$ is a neighborhood of $\overline{W}$ that we can choose arbitrarily small.

We choose $\widetilde{W}$ with its closure disjoint from Σ_∞. Then $\widetilde{p} := p \circ \kappa$ is equal to p outside a compact set and $\widetilde{p} - z$ is elliptic for $z \in \overline{\widetilde{W}}$. Let

$$\widetilde{P}(x, hD) = P(x, hD) + (\widetilde{p} - p)(x, hD).$$

The symbol $\widetilde{p} - p$ has compact support and we know (cf. e.g., [41, Ch. 9]) that the corresponding operator is compact and even of trace class. For $h > 0$ small enough, $\widetilde{P} - z : H(m) \to H(1)$ is bijective for every $z \in \overline{W}$. Write on the operator level

$$P - z = (1 + (p - \widetilde{p})(\widetilde{P} - z)^{-1})(\widetilde{P} - z), \quad z \in W.$$

Here $W \ni z \mapsto 1 + (p - \widetilde{p})(\widetilde{P} - z)^{-1}$ is a holomorphic family of Fredholm operators, invertible for at least one value of z (in $W \setminus \Sigma$). By analytic Fredholm theory, reviewed in Chap. 8, we know that (for each h small enough) $1 + (p - \widetilde{p})(\widetilde{P} - z)^{-1}$ is bijective for $z \in W$ away from a discrete set $\Gamma(h)$, and since the spectrum of P in W has to be contained in $\Gamma(h)$, it is a discrete set. Moreover, the singularities of $1 + (p - \widetilde{p})(\widetilde{P} - z)^{-1}$ at $\Gamma(h)$ are poles of finite order with finite-rank Laurent coefficients, so we have the same property for

$$(P - z)^{-1} = (\widetilde{P} - z)^{-1}(1 + (p - \widetilde{p})(\widetilde{P} - z)^{-1})^{-1},$$

and in particular the spectral projections

$$\Pi(z) = \frac{1}{2\pi i} \int_{\partial D(z,\epsilon)} (z - P)^{-1} dz, \ 0 < \epsilon \ll 1,$$

have finite rank (equal to the algebraic multiplicity of z as an eigenvalue of P) for every $z \in \Gamma(h)$. $\square$

5.2 The Result

We treat the case of operators on $\mathbf{R}$ and indicate later the modifications to be made in the case of $\mathbf{T}$. Let W be as in Proposition 5.1.3. Then $p^{-1}(\overline{W})$ is compact. Let $\Omega \subset W$ be open, h-independent, simply connected with smooth boundary. We assume

$$\Omega \subset \Sigma, \text{ and for every } \rho \in p^{-1}(\overline{\Omega}), \text{ we have } \frac{1}{i}\{p, \overline{p}\}(\rho) \neq 0. \tag{5.2.1}$$

This means that $d\Re p$, and $d\Im p$ are linearly independent at every point of $p^{-1}(\overline{\Omega})$, and by the implicit function theorem, we conclude that $p^{-1}(z)$ is a finite set for every $z \in \overline{\Omega}$, whose cardinality is independent of z. Since Ω is simply connected, we can write $p^{-1}(z) = \{\rho_1(z), \dots, \rho_M(z)\}$, where $\rho_j(z)$ are mutually distinct and depend smoothly on $z \in \Omega$.

Proposition 5.2.1 *The cardinality of $p^{-1}(z)$ is even $= 2N$ for every $z \in \Omega$, and we can write*

$$p^{-1}(z) = \{\rho_1^+(z), \dots, \rho_N^+(z), \rho_1^-(z), \dots, \rho_N^-(z)\},$$

where $\rho_j^\pm(z)$ depend smoothly on z and $\pm i^{-1}\{p, \overline{p}\}(\rho_j^\pm(z)) > 0$.

Proof We only have to prove that the number points in $p^{-1}(z)$ with $i^{-1}\{p, \overline{p}\}(\rho) > 0$ is equal to the number of points with $i^{-1}\{p, \overline{p}\}(\rho) < 0$. For that, we notice that if $\rho \in p^{-1}(z)$ and γ is a simple closed positively oriented loop around ρ in $\mathbf{R}^2 \setminus \{\rho\}$, confined to a small neighborhood of ρ, then

$$\frac{1}{2\pi} \text{var arg}_\gamma (p - z) = \text{sign}\left(\frac{1}{i}\{p, \overline{p}\}(\rho)\right). \tag{5.2.2}$$

Here, to define positive orientation of a closed contour, we take for granted that notion in the case of contours in $\mathbf{C}$ and make the corresponding convention that $(2i)^{-1} dz \wedge d\overline{z} = dy \wedge dx > 0$. As for $\mathbf{R}^2_{x,\xi}$, the corresponding positive form is the symplectic one, $\sigma = d\xi \wedge dx$. (Or more down-to-earth, we identify $\mathbf{R}^2_{x,\xi} \simeq \mathbf{R}^2_{\Re z, \Im z} \simeq \mathbf{C}$.) Then (5.2.2) follows by direct calculation, or by the identity,

$$\frac{1}{2i} dp \wedge d\overline{p} = \frac{1}{2i}\{p, \overline{p}\} d\xi \wedge dx, \tag{5.2.3}$$

cf. Proposition 3.3.3.

Let γ be the positively oriented boundary of $\partial B_{\mathbf{R}^2}(0, R)$, for $R \gg 1$. Then,

$$\frac{1}{2\pi}\text{var arg}_\gamma (p - z) = \sum_1^M \frac{1}{2\pi}\text{var arg}_{\gamma_j} (p - z), \tag{5.2.4}$$

where γ_j are the oriented boundaries of $B_{\mathbf{R}^2}(\rho_j(z), \epsilon)$, $0 < \epsilon \ll 1$. On the other hand, $p = \widetilde{p}$ along γ when R is large enough, so

$$\frac{1}{2\pi}\text{var arg}_\gamma (p - z) = \frac{1}{2\pi}\text{var arg}_\gamma (\widetilde{p} - z) = 0. \tag{5.2.5}$$

Indeed, $B_{\mathbf{R}^2}(0, R)$ is contractible and z is not in its image under $\widetilde{p}$. The result follows from (5.2.2), (5.2.4), (5.2.5). □

From the proposition and its proof, in particular (5.2.3), we get

$$p_*(|\sigma|) = \sum_1^N \left(\frac{2i}{\{p, \overline{p}\}(\rho_j^+(z))} - \frac{2i}{\{p, \overline{p}\}(\rho_j^-(z))} \right) L(dz), \tag{5.2.6}$$

where $L(dz) = |d\Im z \wedge d\Re z|$ is the Lebesgue measure and $|\sigma| = |d\xi \wedge dx|$ is the symplectic volume element.

Let $L > 0$ be large enough, so that $\pi_x \left(p^{-1}(\overline{\Omega})\right) \subset]-\pi L, \pi L[$, where $\pi_x : \mathbf{R}^2_{x,\xi} \to \mathbf{R}_x$ is the natural projection. Let $\chi \in C_0^\infty(\mathbf{R}; [0, 1])$ be equal to 1 on $[-\pi L, \pi L]$. Our perturbed operator is given by

$$P_\delta = P(x, hD; h) + \delta Q_\omega, \tag{5.2.7}$$

$$Q_\omega u(x) = \sum_{|k|,\, |\ell| \le \frac{C_1}{h}} \alpha_{\ell,k}(\omega)(\chi u|e^k)(\chi e^\ell)(x), \tag{5.2.8}$$

where $C_1 > 0$ is sufficiently large, $e^k(x) = (2\pi L)^{-1/2} e^{ikx/L}$, $k \in \mathbf{Z}$, and $\alpha_{j,k} \sim \mathcal{N}_{\mathbf{C}}(0, 1)$ are independent complex Gaussian random variables.

Notice that the functions e^k form an orthonormal family in $L^2([-\pi L, \pi L[)$. Alternatively, we could have taken

$$Q_\omega u = \sum_{k,\ell \le \frac{C_1}{h}} \alpha_{\ell,k}(\omega)(u|\widetilde{e}_k)\widetilde{e}_\ell,$$

where $\widetilde{e}_j(x; h) = h^{-1/4} e_j(x/\sqrt{h})$ and e_j is the sequence of L^2-normalized Hermite functions (reviewed in Sect. 2.5), so that $\widetilde{e}_j$ is an orthonormal family of eigenfunctions of the semi-classical harmonic oscillator.

Theorem 5.2.2 *Let* $\Gamma \Subset \Omega$ *have smooth boundary, and left* $e^{-\epsilon_0/h} \leq \delta \ll h^{7/2}$ *where* $\epsilon_0 > 0$ *is a sufficiently small constant. For* $2Nh \ln(1/\delta) \leq \epsilon \leq 1$*, we have with probability* $\geq 1 - \mathcal{O}(\epsilon^{-1/2})e^{-\epsilon/2h}$ *that*

$$\left| \#(\sigma(P_\delta) \cap \Gamma) - \frac{\mathrm{vol}\, p^{-1}(\Gamma)}{2\pi h} \right| \leq \mathcal{O}(1) \frac{\sqrt{\epsilon}}{h}. \tag{5.2.9}$$

If instead we let Γ vary in a set of subsets that satisfy the assumptions uniformly, then with probability $\geq 1 - \mathcal{O}(\epsilon^{-1})e^{-\epsilon/2h}$ we have (5.2.9) uniformly for all Γ in that subset. The remainder of the chapter is devoted to the proof of this result that will follow the general strategy in Chap. 3, with an improvement of the probabilistic discussion.

Example 5.2.3 Let P_c be the semi-classical variant of the non-self-adjoint harmonic oscillator in (2.5.1),

$$P_c = (hD)^2 + cx^2, \text{ with } \Im c > 0.$$

This operator fulfills the general assumptions with order function $m(x, \xi) = 1 + x^2 + \xi^2$ and

$$\Sigma_\infty = \emptyset, \quad \Sigma = [0, +\infty[+ [0, +\infty[\times c.$$

Let Ω be open, bounded, simply connected, h-independent with smooth boundary, such that

$$\Omega \subset \overset{\circ}{\Sigma}.$$

Then we can find W as in Proposition 5.1.3, containing Ω.

If $z \in \Omega$, then $p^{-1}(z)$ consists of the four points $(\pm x, \pm\xi)$, where $x = x(z) > 0$, and $\xi = \xi(z) > 0$ are determined by

$$(\Im c)x^2 = \Im z, \quad \xi^2 + (\Re c)x^2 = \Re z.$$

The Poisson bracket $\{p, \overline{p}\}/(2i)$ is positive at two of these points, namely at $(x, -\xi)$ and $(-x, \xi)$, and negative at the other two. Theorem 5.2.2 shows that after adding a small random perturbation, the eigenvalues will obey a Weyl law in Ω, while the eigenvalues of the unperturbed operator are confined to the bisector of Σ, as we saw in Proposition 2.5.1.

5.3 Preparations for the Unperturbed Operator

To each point $\rho_j^+(z) = (x_j^+(z), \xi_j^+(z))$, $z \in \Omega$, we associate a quasimode $e^j(x, z; h) = a^j(x, z; h)e^{\frac{i}{h}\phi^j(x,z)}$ as in Theorem 4.1.4, Proposition 4.2.3, so that

$$(P - z)(e^j) = \mathcal{O}(h^\infty) \text{ in } L^2(\text{neigh}\,(x_j^+)). \tag{5.3.1}$$

Here ϕ^j satisfies the eikonal equation

$$p(x, \partial_x \phi^j(x)) - z = 0, \ x \in \text{neigh}\,(x_j^+), \tag{5.3.2}$$

and

$$\phi^j(x_j^+(z), z) = 0, \ \partial_x \phi^j(x_j^+(z), z) = \xi_j^+(z), \ \Im \phi^j(x, z) \asymp (x - x_j^+(z))^2. \tag{5.3.3}$$

Actually, this follows from the quoted results only when P is a semi-classical differential operator and we need to appeal to some more microlocal analysis in the pseudodifferential case. Equation (5.3.2) makes sense when p is analytic in ξ, since $\partial_x \phi^j$ is complex-valued, and for a more general pseudodifferential operator considered here, we have to weaken that equation to

$$p(x, \partial_x \phi^j(x)) - z = \mathcal{O}\left((x - x_j^+(z))^\infty\right), \tag{5.3.4}$$

in the sense of formal Taylor expansions. We refer to [41, 102] for more details.

To get a globally well-defined quasimode we let χ be a standard smooth cutoff, equal to one near 0, and notice that by the last property in (5.3.3), the above properties remain unchanged if we insert the cutoff $\chi(x - x_j^+(z))$ and obtain

$$e^j(x, z; h) = \chi(x - x_j^+(z))h^{-1/4}a^j(x, z; h)e^{\frac{i}{h}\phi^j(x,z)}. \tag{5.3.5}$$

We can then arrange so that a^j is a classical symbol of order 0, with

$$\|e^j\|_{L^2(\mathbf{R})} = 1, \tag{5.3.6}$$

and so that (5.3.1) holds globally:

$$(P - z)(e^j) = \mathcal{O}(h^\infty) \text{ in } L^2(\mathbf{R}). \tag{5.3.7}$$

In the construction, we can arrange so that

$$\partial_z^k \partial_{\overline{z}}^\ell e^j = \mathcal{O}(h^{-k-\ell}) \text{ in } L^2. \tag{5.3.8}$$

Naturally, we have the analogous constructions of quasimodes for $(P-z)^*$, where we replace the ρ_j^+ by the ρ_j^-. We have quasimodes of the form

$$f^j(x,z;h)=\chi(x-x_j^-)h^{-\frac{1}{4}}b^j(x,z;h)e^{\frac{i}{h}\psi^j(x,z)}, \tag{5.3.9}$$

where b_j is a classical symbol of order 0,

$$p^*(x,\partial_x\psi^j)-\overline{z}=0 \text{ near } x_j^-, \tag{5.3.10}$$

$$\psi^j(x_j^-(z),z)=0,\ \partial_x\psi^j(x_j^-(z),z)=\xi_j^-(z),\ \Im\psi^j(x_j^-(z),z)\asymp(x-x_j^-(z))^2, \tag{5.3.11}$$

$$\|f_j\|_{L^2}=1,\ \partial_z^k\partial_{\overline{z}}^\ell f^j=\mathcal{O}(h^{-k-\ell}) \text{ in } L^2, \tag{5.3.12}$$

$$(P-z)^*(f^j)=\mathcal{O}(h^\infty) \text{ in } L^2. \tag{5.3.13}$$

Here we write $p^*(x,\xi)=\overline{p(\overline{x},\overline{\xi})}$, to be understood in the sense of Taylor series, when p is not analytic. Also when p is not analytic in ξ, (5.3.10) should be weakened to

$$p^*(x,\partial_x\psi^j(x,z),z)-\overline{z}=\mathcal{O}((x-x_j^-)^\infty), \tag{5.3.14}$$

in the sense of Taylor expansions at $x_j^-(z)$.

By construction,

$$(e_j|e_k),\ (f_j|f_k),\ (e_\nu|f_\mu)\ =\mathcal{O}(h^\infty),\ \ j\neq k, \tag{5.3.15}$$

so approximately we can say that $\{e_j\}_1^N$, $\{f_j\}_1^N$ are orthonormal families, which are mutually orthogonal.

Define $R_+:L^2\to\mathbf{C}^N$, $R_-:\mathbf{C}^N\to L^2$, by

$$R_+u(j)=(u|e_j),\ u\in L^2, \tag{5.3.16}$$

$$R_-u_-=\sum_1^N u_-(j)f_j,\ u_-\in\mathbf{C}^N. \tag{5.3.17}$$

Note that $R_+R_+^*-1$, $R_-^*R_--1$ are $\mathcal{O}(h^\infty)$ in $\mathcal{L}(\mathbf{C}^N,\mathbf{C}^N)$ and that, in particular, $\|R_+\|_{\mathcal{L}(L^2,\mathbf{C}^N)}=1+\mathcal{O}(h^\infty)$, $\|R_-\|_{\mathcal{L}(\mathbf{C}^N,L^2)}=1+\mathcal{O}(h^\infty)$.

Define

$$\mathcal{P}(z)=\begin{pmatrix}P-z & R_-\\ R_+ & 0\end{pmatrix}:\ H(m)\times\mathbf{C}^N\to L^2\times\mathbf{C}^N, \tag{5.3.18}$$

and notice that $\mathcal{P}(z)$ is uniformly bounded in $\mathcal{L}(H(m) \times \mathbf{C}^N, L^2 \times \mathbf{C}^N)$ when h is small and z varies in any fixed compact subset of Ω.

Proposition 5.3.1 *$\mathcal{P}(z)$ is bijective for $z \in \Omega$, with an inverse*

$$\mathcal{E}(z) = \begin{pmatrix} E(z) & E_+(z) \\ E_-(z) & E_{-+}(z) \end{pmatrix}, \tag{5.3.19}$$

which satisfies,

$$\begin{aligned} &E(z) = \mathcal{O}(h^{-\frac{1}{2}}) : L^2 \to H(m), \\ &E_+(z) - R_+^*(z) = \mathcal{O}(h^\infty) : \mathbf{C}^N \to H(m), \\ &E_-(z) - R_-^*(z) = \mathcal{O}(h^\infty) : L^2 \to \mathbf{C}^N, \\ &E_{-+}(z) = \mathcal{O}(h^\infty) : \mathbf{C}^N \to \mathbf{C}^N. \end{aligned} \tag{5.3.20}$$

The proof involves some microlocal analysis that we shall not develop in detail here. Using the h-pseudodifferential calculus [41] we first see that

$$Q_+ := (P-z)^*(P-z), \quad Q_- := (P-z)(P-z)^*$$

are essentially self-adjoint with domain $H(m^2)$. Here we make use of the ellipticity assumption (5.1.7). (Here and in what follows, it is tacitly assumed that h is small enough.) Moreover, because of the ellipticity assumption and the fact that $m \geq 1$, we know that these operators have purely discrete spectrum in some h-independent neighborhood of 0. We also notice that $Q_\pm \geq 0$, so the spectra are contained in $[0, +\infty[$.

For a given fixed constant $C > 0$, the eigenvalues in $[0, Ch[$ have complete asymptotic expansions in integer powers of h and can be described in the following way: The operators $Q_\pm$ have the same principal symbol $q(x, \xi, z) = |p(x, \xi) - z|^2$, which vanishes precisely at the points $\rho_j^\pm$, $j = 1, 2, \ldots, N$, that we will also label ρ_ν, $1 \leq \nu \leq 2N$ whenever convenient. These points are also nondegenerate minima for $q(\cdot, z)$. Therefore, as for semi-classical Schrödinger operators [59, 129] and in later works for more general pseudodifferential operators [65, 67], each such point generates a sequence of eigenvalues of $Q_\pm$,

$$\lambda_{\nu,k}^\pm(z; h) \sim h\lambda_{\nu,k}^{1,\pm}(z) + h^2\lambda_{\nu,k}^{2,\pm}(z) + \cdots, \quad k = 0, 1, 2, \ldots$$

where $\lambda_{\nu,k}^{1,\pm}(z)$ are the eigenvalues (arranged in increasing arithmetic progression) of a harmonic oscillator, appearing as a quadratic approximation of $Q_\pm$ at ρ_ν. These quadratic approximations are $P_\nu^* P_\nu$ and $P_\nu P_\nu^*$ for Q_+ and Q_-, respectively, where

$P_\nu = \partial_x p(\rho_\nu)x + \partial_\xi p(\rho_\nu)D_x$. When $\rho_\nu = \rho_j^+$, then P_j is of annihilation type (surjective $H(\langle(x,\xi)\rangle) \to L^2$ with a one-dimensional kernel, generated by a Gaussian function), while P_j^* is of creation type (injective), and we then know that $\lambda_{\nu,0}^{1,+} = 0$, $\lambda_{\nu,k}^{1,+} > 0$ for $k \geq 1$, while $\lambda_{\nu,k}^{1,-} > 0$ for $k \geq 0$. When $\rho_\nu = \rho_j^-$, then P_j^*, P_j are annihilation and creation operators, respectively, and we get $\lambda_{\nu,k}^{1,+} > 0$ for $k \geq 0$, $\lambda_{\nu,0}^{1,-} = 0$, $\lambda_{\nu,k}^{1,-} > 0$ for $k \geq 1$.

Thus, if $C > 0$ is large enough, each of Q_+, Q_- has precisely N eigenvalues in $[0, h/C]$ and these eigenvalues are $\mathcal{O}(h^2)$. However, the existence of approximately orthonormal systems $\{e_j\}_1^N$, $\{f_j\}_1^N$ with $Q_+ e_j = \mathcal{O}(h^\infty)$, $Q_+ f_j = \mathcal{O}(h^\infty)$, shows that these eigenvalues are actually $\mathcal{O}(h^\infty)$.

Let $S_\pm$ be the N-dimensional subspace of L^2 corresponding to the eigenvalues of $Q_\pm$ that are $\mathcal{O}(h^\infty)$. We observe that $S_\pm \subset H(m^k)$, $\forall k \in \mathbf{N}$, by the ellipticity of $Q_\pm$ near ∞. Let $t_1^2 \leq t_2^2 \leq \cdots \leq t_N^2$, with $0 \leq t_j \leq \mathcal{O}(h^\infty)$ be the corresponding eigenvalues of Q_+ and let $0 \leq N_0 \leq N$ be the number of zero eigenvalues, so that (when $N_0 \geq 1$) $0 = t_1 = t_2 = \cdots = t_{N_0} < t_{N_0+1} \leq \cdots \leq t_N$. N_0 is then also equal to the dimension of the kernel of $P - z$. By Fredholm theory and ellipticity, we know that $P - z : H(m) \to H(1)$ is Fredholm of index 0 for $z \in \Omega$ and this implies that dim $\mathrm{Ker}(P^* - \overline{z}) =$ dim $\mathrm{Ker}(P - z)$. It follows that Q_- has precisely N_0 vanishing eigenvalues.

We have the intertwining property

$$Q_-(P - z) = (P - z)Q_+,$$

and a similar one with $(P^* - \overline{z})$. It follows that if $Q_+\epsilon_j = t_j^2\epsilon_j$ for some $j \geq N_0 + 1$, then $Q_-(P - z)\epsilon_j = t_j^2(P - z)\epsilon_j$, and hence $(P - z)\epsilon_j$ is an eigenvector for Q_- with the same eigenvalue t_j^2. Since $P - z$ is injective on the eigenspace corresponding to (Q_+, t_j^2), we see that the multiplicity of t_j^2 as an eigenvalue Q_- is $\geq$ that of t_j^2 as an eigenvalue of Q_+. The sum of all the multiplicities is equal to N for each of Q_+ and Q_-, so the multiplicities have to agree.

Pursuing this argument, we see that we can find orthonormal bases $\{\epsilon_j\}_1^N$ in S_+ and $\{\gamma_j\}_1^N$ in S_- such that

$$(P - z)\epsilon_j = t_j\gamma_j, \quad (P - z)^*\gamma_j = t_j\epsilon_j.$$

Let $\{\epsilon_j\}_1^N$ and $\{\gamma_j\}_1^N$ be orthonormal bases in S_+ and S_-, respectively. For instance, we can take

$$\begin{aligned} &\{\epsilon_j\}_1^N \text{ equal to the Gram orthonormalization of } \{\Pi_{S_+} e_j\}_1^N, \\ &\{\gamma_j\}_1^N \text{ equal to the Gram orthonormalization of } \{\Pi_{S_-} f_j\}_1^N, \end{aligned} \tag{5.3.21}$$

where $\Pi_{S_\pm}$ denotes the orthogonal projection onto $S_\pm$. With that choice, we have

$$\epsilon_j - e_j,\ \gamma_j - f_j = \mathcal{O}(h^\infty) \text{ in } L^2. \tag{5.3.22}$$

With ϵ_j, γ_j as in (5.3.21), (5.3.22) we define $\widetilde{R}_+ : L^2 \to \mathbf{C}^N$, $\widetilde{R}_- : \mathbf{C}^N \to L^2$ by

$$\widetilde{R}_+ u(j) = (u|\epsilon_j),\ \ \widetilde{R}_- u_- = \sum_1^N u_-(j)\gamma_j, \tag{5.3.23}$$

so that

$$\|R_+ - \widetilde{R}_+\|_{\mathcal{L}(L^2,\mathbf{C}^N)},\ \|R_- - \widetilde{R}_-\|_{\mathcal{L}(\mathbf{C}^N,L^2)} = \mathcal{O}(h^\infty). \tag{5.3.24}$$

Using the spectral and orthogonal decompositions $L^2 = S_+ \oplus S_+^\perp$, $L^2 = S_- \oplus S_-^\perp$, we see that

$$\widetilde{\mathcal{P}}(z) = \begin{pmatrix} P - z & \widetilde{R}_- \\ \widetilde{R}_+ & 0 \end{pmatrix} : H(m) \times \mathbf{C}^N \to L^2 \times \mathbf{C}^N \tag{5.3.25}$$

is bijective, with an "explicit" inverse

$$\widetilde{\mathcal{E}}(z) = \begin{pmatrix} \widetilde{E}(z) & \widetilde{E}_+(z) \\ \widetilde{E}_-(z) & \widetilde{E}_{-+}(z) \end{pmatrix}, \tag{5.3.26}$$

where

$$\begin{aligned} \widetilde{E}(z) &\simeq (P - z)^{-1} : S_-^\perp \to S_+^\perp, \\ \widetilde{E}_\pm(z) &= \widetilde{R}_\pm^*, \\ \widetilde{E}_{-+}(z) &= \big(((P - z)\epsilon_k|\gamma_j)\big)_{j,k=1}^N \simeq P - z : S_+ \to S_-. \end{aligned}$$

It follows that

$$\begin{aligned} \widetilde{E} &= \mathcal{O}(h^{-\frac{1}{2}}) : L^2 \to H(m), \\ \widetilde{E}_+ &= \mathcal{O}(1) : \mathbf{C}^N \to H(m), \\ \widetilde{E}_- &= \mathcal{O}(1) : H(1) \to \mathbf{C}^N, \\ \widetilde{E}_{-+} &= \mathcal{O}(h^\infty). \end{aligned}$$

Using this together with (5.3.24), we see that $\mathcal{P}(z)$ has an inverse $\mathcal{E}(z)$ such that $\mathcal{E} - \widetilde{\mathcal{E}} = \mathcal{O}(h^\infty)$, and we get Proposition 5.3.1.

5.4 d-Bar Equation for E_{-+}

Here we generalize the discussion in Sect. 3.3. The proof of Proposition 3.3.1 leads to the d-bar equation,

$$\partial_{\overline{z}} E_{-+} + E_{-+}(\partial_{\overline{z}} R_+)E_+ + E_-(\partial_{\overline{z}} R_-)E_{-+} = 0. \tag{5.4.1}$$

We are interested in the eigenvalues of P as the zeros of $\det E_{-+}$. Using the identity $\partial_{\overline{z}} \ln \det E_{-+} = \mathrm{tr}\left(E_{-+}^{-1}\partial_{\overline{z}} E_{-+}\right)$, the cyclicity of the trace and (5.4.1), we get

$$\begin{aligned} &\partial_{\overline{z}}(\ln \det E_{-+}) + f(z) = 0, \text{ where} \\ &f(z) = \mathrm{tr}\, ((\partial_{\overline{z}} R_+)E_+ + E_-\partial_{\overline{z}} R_-)\,, \end{aligned} \tag{5.4.2}$$

away from the zeros of $\det E_{-+}(z)$, which can also be written,

$$\partial_{\overline{z}}(\det E_{-+}(z)) + f(z)\det E_{-+} = 0, \tag{5.4.3}$$

valid also near the zeros of $\det E_{-+}$, either by continuity, or by direct computation without explicit use of logarithms. Again, $e^F \det E_{-+}$ is holomorphic, if F solves

$$\partial_{\overline{z}} F = f. \tag{5.4.4}$$

The multiplicity $m(z_0)$ of a zero z_0 of E_{-+} is by definition equal to the multiplicity of z_0 as a zero of $e^F \det E_{-+}$. We have (using an observation of Vogel [155]),

$$m(z_0) = \lim_{\epsilon\to 0} \int_{\partial D(z_0,\epsilon)} \frac{\partial_z(e^F \det E_{-+})}{e^F \det E_{-+}} \frac{dz}{2\pi i} = \lim_{\epsilon\to 0} \int_{\partial D(z_0,\epsilon)} \frac{\partial_z(\det E_{-+})}{\det E_{-+}} \frac{dz}{2\pi i}, \tag{5.4.5}$$

since $\int_{\partial D(z_0,\epsilon)} \partial_z F dz \to 0$, $\epsilon \to 0$. On the other hand, in view of the standard identity

$$(z-P)^{-1} = -E(z) + E_+(z)E_{-+}(z)^{-1}E_-(z),$$

the algebraic multiplicity $\widetilde{m}(z_0)$ of z_0 as an eigenvalue of P is equal to

$$\begin{aligned} &\mathrm{tr} \lim_{\epsilon\to 0} \int_{\partial D(z_0,\epsilon)} (z-P)^{-1}\frac{dz}{2\pi i} = \mathrm{tr} \lim_{\epsilon\to 0} \int_{\partial D(z_0,\epsilon)} E_+(z)E_{-+}(z)^{-1}E_-(z)\frac{dz}{2\pi i} \\ &= \lim_{\epsilon\to 0} \int_{\partial D(z_0,\epsilon)} \mathrm{tr}\left(E_+E_{-+}^{-1}E_-\right)\frac{dz}{2\pi i} = \lim_{\epsilon\to 0} \int_{\partial D(z_0,\epsilon)} \mathrm{tr}\left(E_{-+}^{-1}E_-E_+\right)\frac{dz}{2\pi i}. \end{aligned} \tag{5.4.6}$$

Now by the same proof as for (5.4.1),

$$E_-E_+ = \partial_z(E_{-+}) + E_{-+}(\partial_z R_+)E_+ + E_-(\partial_z R_-)E_{-+},$$

and using this in the last integral together with the cyclicity of the trace in one of the resulting terms, we get

$$\widetilde{m}(z_0) = \lim_{\epsilon\to 0}\int_{\partial D(z_0,\epsilon)} \operatorname{tr}\left(E_{-+}^{-1}\partial_z E_{-+}\right)\frac{dz}{2\pi i} = m(z_0). \tag{5.4.7}$$

Here we also used the classical identity,

$$(\det E_{-+})^{-1}\,\partial_z\,(\det E_{-+}) = \operatorname{tr}\left(E_{-+}^{-1}\partial_z E_{-+}\right).$$

The formula (3.3.4) and its derivation have straight forward generalizations, and we get

Proposition 5.4.1 *We have* (5.4.2), *so that* e^F *a bit closer* $\det E_{-+}$ *is holomorphic when* (5.4.4) *holds. The zeros of* $\det E_{-+}$ *coincide with the eigenvalues of* P *in* Ω *and the multiplicities agree, cf.* (5.4.5)–(5.4.7). *Moreover,*

$$\Re\Delta F(z) = \frac{1}{h}\sum_{j=1}^{N}\left(\frac{2i}{\{p,\overline{p}\}(\rho_j^+(z))} - \frac{2i}{\{p,\overline{p}\}(\rho_j^-(z))}\right) + \mathcal{O}(1). \tag{5.4.8}$$

5.5 Adding the Random Perturbation

We still have Proposition 3.4.2 and we make the assumption (3.4.5),

$$0 \le \delta \ll h^{\frac{3}{2}}, \tag{5.5.1}$$

to be strengthened later on. Proposition 3.4.2 is applicable and we shall work under the assumption

$$\|Q_\omega\|_{HS} \le C/h.$$

With R_+ as above, we introduce the perturbed Grushin matrix,

$$\mathcal{P}_\delta(z) = \begin{pmatrix} P_\delta - z & R_- \\ R_+ & 0 \end{pmatrix} : \; H(m)\times \mathbf{C}^N \to L^2 \times \mathbf{C}^N, \tag{5.5.2}$$

which is bijective with a bounded inverse (cf. (3.4.6)),

$$\mathcal{E}_\delta = \begin{pmatrix} E^\delta & E_+^\delta \\ E_-^\delta & E_{-+}^\delta \end{pmatrix},$$

such that

$$\begin{aligned} E^\delta &= E + \mathcal{O}(\frac{\delta}{h^2}) = \mathcal{O}(h^{-1/2}) \text{ in } \mathcal{L}(L^2, H(m)), \\ E_+^\delta &= E_+ + \mathcal{O}(\frac{\delta}{h^{3/2}}) = \mathcal{O}(1) \text{ in } \mathcal{L}(\mathbf{C}^N, H(m)), \\ E_-^\delta &= E_- + \mathcal{O}(\frac{\delta}{h^{3/2}}) = \mathcal{O}(1) \text{ in } \mathcal{L}(L^2, \mathbf{C}^N), \\ E_{-+}^\delta &= E_{-+} - \delta E_- Q E_+ + \mathcal{O}(\frac{\delta^2}{h^{5/2}}) \text{ in } \mathcal{L}(\mathbf{C}^N, \mathbf{C}^N). \end{aligned} \tag{5.5.3}$$

The eigenvalues of P_δ in Ω coincide with the zeros there of $\det E_{-+}^\delta$ and the multiplicities agree. We have the d-bar equation,

$$\begin{aligned} &\partial_{\bar{z}} \left(\det E_{-+}^\delta(z)\right) + f^\delta \det E_{-+}^\delta(z) = 0, \\ &f^\delta(z) = \operatorname{tr}\left((\partial_{\bar{z}} R_+) E_+^\delta + E_-^\delta \partial_{\bar{z}} R_-\right), \end{aligned} \tag{5.5.4}$$

and from (5.5.3) and the boundedness of $E_\pm$, we have

$$f^\delta(z) = f(z) + \mathcal{O}\left(\frac{\delta}{h^{\frac{5}{2}}}\right). \tag{5.5.5}$$

We can solve $\partial_{\bar{z}} F^\delta = f^\delta$ (making $e^{F^\delta} \det E_{-+}^\delta$ holomorphic) with (cf. (3.4.7))

$$F^\delta = F + \frac{1}{h}\mathcal{O}\left(\frac{\delta}{h^{\frac{3}{2}}}\right). \tag{5.5.6}$$

We next look at the second term in the expression for E_{-+}^δ in (5.5.3). From (5.3.20), (5.3.16), (5.3.17), we get for the matrix $E_- Q E_+$,

$$E_- Q E_+(j,k) = (Q e_k | f_j) + \mathcal{O}(h^\infty). \tag{5.5.7}$$

Here, by (5.2.8),

$$
\begin{aligned}
(Qe_k|f_j) &= \sum_{|\nu|,\,|\mu|\le \frac{C_1}{h}} \alpha_{\nu,\mu}(\omega)(\chi e_k|e^\mu)(e^\nu|\chi f_j)\\
&= \sum_{|\nu|,\,|\mu|\le \frac{C_1}{h}} \alpha_{\nu,\mu}(\omega)\widehat{\chi e_k}(\mu)\overline{\widehat{\chi f_j}}(\nu),
\end{aligned}
$$

where the last equality is an implicit definition of the Fourier coefficients of χe_k and χf_j with respect to the orthonormal family $\{e^\nu\}$, where we can assume that e_j and f_j are supported in $]-\pi L, \pi L[$. We have seen that $\chi e_j, \dots, \chi e_N, \chi f_1, \dots, \chi f_N$ is an orthonormal system up to $\mathcal{O}(h^\infty)$ and if C_1 is large enough, we get the same properties for $\widehat{\chi e_j}, \dots, \widehat{\chi e_N}$ and $\widehat{\chi f_1}, \dots, \widehat{\chi f_N}$ in $\mathbf{C}^{1+2[C_1/h]}$. Then $\overline{\widehat{\chi e_k}} \otimes \widehat{\chi f_j}$ is an orthonormal family in $\mathbf{C}^M$ up to $\mathcal{O}(h^\infty)$, where $M := (1+2[C_1/h])^2$. Let $E_1, \dots, E_{N^2}$ be the Gram orthonormalization of this family, and complete it to an orthonormal basis $E_1, \dots, E_M$ in $\mathbf{C}^M$. Write,

$$
\alpha = \sum_1^{N^2} \alpha_j E_j + \sum_{N^2+1}^{M} \alpha_j E_j, \quad \alpha' = (\alpha_1, \dots, \alpha_{N^2}),\ \alpha'' = (\alpha_{N^2+1}, \dots, \alpha_M).
$$

By (5.5.7) and the following discussion, we have with $Q = Q(\alpha)$ in (5.2.8),

$$
E_- Q E_+ = \alpha' + \mathcal{O}(h^\infty) \text{ in } \mathbf{C}^{N^2}. \tag{5.5.8}
$$

Our change of coordinates depends on z, but the estimates are uniform.

For notational reasons, we change the sign of Q, so we get rid of the minus sign in the last equation in (5.5.3), which we then write as

$$
\begin{aligned}
&E_{-+}^\delta = E_{-+} + \delta A,\\
&A = E_- Q E_+ + \mathcal{O}\left(\frac{\delta}{h^{5/2}}\right) = \alpha' + \mathcal{O}\left(h^\infty + \frac{\delta}{h^{5/2}}\right),
\end{aligned} \tag{5.5.9}
$$

for the new α-coordinates, obtained from the old ones by a z-dependent unitary transformation. Applying the Cauchy inequality to complex lines in $\mathbf{C}^M$, we get

$$
d_\alpha A = d\alpha' + \mathcal{O}\left(h^\infty + \frac{\delta}{h^{3/2}}\right) \text{ in } \mathcal{L}\left(\mathbf{C}^M; \mathbf{C}^{N^2}\right). \tag{5.5.10}
$$

By the implicit function theorem, the map

$$
B_{\mathbf{C}^M}(0, C/h) \ni \alpha \mapsto (A, \alpha'') \in \mathbf{C}^M \tag{5.5.11}
$$

is a holomorphic diffeomorphism onto its image $\widetilde{B}$, which is sandwiched between

$$B_{\mathbf{C}^M}\left(0, \frac{C}{h} - \mathcal{O}\left(h^\infty + \frac{\delta}{h^{\frac{5}{2}}}\right)\right) \text{ and } B_{\mathbf{C}^M}\left(0, \frac{C}{h} + \mathcal{O}\left(h^\infty + \frac{\delta}{h^{\frac{5}{2}}}\right)\right).$$

It also follows from (5.5.10) that

$$L(d\alpha) = L(d\alpha')L(d\alpha'') = \left(1 + \mathcal{O}\left(h^\infty + \frac{\delta}{h^{\frac{3}{2}}}\right)\right) L(dA)L(d\alpha''), \tag{5.5.12}$$

where $L(d\alpha)$, $L(d\alpha')$, and $L(d\alpha'')$ denote the Lebesgue measure on $\mathbf{C}^M$, $\mathbf{C}^{N^2}$ and $\mathbf{C}^{M-N^2}$, respectively.

By (5.5.9), we have

$$|\alpha|^2 = |A|^2 + |\alpha''|^2 + \mathcal{O}\left(h^\infty + \frac{\delta}{h^{\frac{7}{2}}}\right). \tag{5.5.13}$$

Now, strengthen the assumption (5.5.1) to

$$\delta \ll h^{\frac{7}{2}}. \tag{5.5.14}$$

Then from (5.5.12), (5.5.13), we get

$$\pi^{-M} e^{-|\alpha|^2} L(d\alpha) = \left(1 + \mathcal{O}\left(h^\infty + \frac{\delta}{h^{\frac{7}{2}}}\right)\right) \pi^{-M} e^{-|A|^2 - |\alpha''|^2} L(dA)L(d\alpha''). \tag{5.5.15}$$

The probability distribution μ of A is the direct image of $\pi^{-M}e^{-|\alpha|^2}L(d\alpha)$ under the map

$$B_{\mathbf{C}^M}(0, C/h) \ni \alpha \mapsto A(\alpha) \in \mathbf{C}^{N^2}.$$

Letting $\widetilde{B}$ denote the image of $B(0, 1/h)$ under $\alpha \mapsto (A(\alpha), \alpha'')$ and $\pi_A(\widetilde{B})$ its projection in $\mathbf{C}_A^{N^2}$, we conclude that if $f(A)$ is a continuous function, then

$$\begin{aligned}
&\int f(A)\mu(dA) = \pi^{-M} \int_{|\alpha| \le C/h} f(A(\alpha))e^{-|\alpha|^2} L(d\alpha) \\
&= \pi^{-M} \iint_{(A,\alpha'') \in \widetilde{B}} f(A) \left(1 + \mathcal{O}\left(h^\infty + \frac{\delta}{h^{7/2}}\right)\right) e^{-|A|^2 - |\alpha''|^2} L(dA)L(d\alpha'') \\
&\le \pi^{-N^2} \int_{\pi_A(\widetilde{B})} f(A)e^{-|A|^2} \left(1 + \mathcal{O}\left(h^\infty + \frac{\delta}{h^{7/2}}\right)\right) L(dA),
\end{aligned}$$

and we conclude that

$$\mu \leq \left(1 + \mathcal{O}\left(h^\infty + \frac{\delta}{h^{7/2}}\right)\right) 1_{\pi_A(\widetilde{B})}(A)\pi^{-N^2} e^{-|A|^2} L(dA). \tag{5.5.16}$$

Combining this with (5.A.21) below, we see that

$$P(\ln|\det(D + A)|) \leq a) \leq \mathcal{O}(1)e^{a/2} + e^{-1/(Ch^2)}, \text{ for } a \leq 0, \tag{5.5.17}$$

uniformly with respect to $D \in \mathbf{C}^{N^2}$. Equivalently,

$$\ln|\det(D + A)| \geq a, \text{ with probability } \geq 1 - \mathcal{O}(1)e^{a/2} - e^{-1/(Ch^2)}.$$

Applying this to $E_{-+}^\delta = \delta(A + \delta^{-1}E_{-+})$, we see that for every $z \in \Omega$,

$$\ln|\det E_{-+}^\delta(z)| - N \ln \delta \geq a \text{ with probability } \geq 1 - \mathcal{O}(1)e^{a/2} - e^{-1/(Ch^2)},$$

i.e.,

$$\ln|\det E_{-+}^\delta(z)| \geq a - N \ln \frac{1}{\delta} \text{ with probability } \geq 1 - \mathcal{O}(1)e^{a/2} - e^{-1/(Ch^2)}, \tag{5.5.18}$$

uniformly for $z \in \Omega$, $a \leq 0$.

On the other hand, by the last equation in (5.5.3), we know that with probability $\geq 1 - e^{-1/(Ch^2)}$, we have

$$\ln|\det E_{-+}^\delta(z)| \leq 0, \text{ for all } z \in \Omega. \tag{5.5.19}$$

Now, consider the holomorphic function

$$u^\delta(z; h) = e^{F^\delta(z)} \det E_{-+}^\delta(z), \tag{5.5.20}$$

and recall (5.5.6). Then with probability $\geq 1 - e^{-1/(Ch^2)}$,

$$\ln|u^\delta(z)| \leq \Re F(z) + \frac{\mathcal{O}(\delta)}{h^{5/2}}, \ z \in \Omega. \tag{5.5.21}$$

Moreover, for every $z \in \Omega$, $b \geq 0$ we have

$$\ln|u^\delta| \geq \Re F(z) - \left(b + N \ln \frac{1}{\delta} + \frac{\mathcal{O}(\delta)}{h^{5/2}}\right),$$

with probability $\geq 1 - \mathcal{O}(1)e^{-b/2} - e^{-1/(Ch^2)}$. By the assumption (5.5.14), we have $N \ln(1/\delta) \gg \delta/h^{5/2}$, and restricting b to the interval $2N \ln(1/\delta) \leq b \leq 1/h$, we get for every $z \in \Omega$:

$$\ln |u^\delta| \geq \Re F(z) - 2b, \text{ with probability } \geq 1 - \mathcal{O}(1)e^{-b/2} - e^{-1/(Ch^2)}. \tag{5.5.22}$$

Let $\Gamma \Subset \Omega$ be independent of h and have smooth boundary. In view of (5.5.21), (5.5.22), we can apply Proposition 3.4.6 to $u = u^\delta$, with $\phi(z)/h = \Re F(z) + b$, $\epsilon = hb$ and with $\asymp \epsilon^{-1/2}$ boundary points, where the lower bound (5.5.22) is required. We conclude that with probability $\geq 1 - \mathcal{O}(1/\sqrt{hb})e^{-b/2} - e^{-1/(Ch^2)}$ we have

$$\left| \# \left((u^\delta)^{-1}(0) \cap \Gamma \right) - \frac{1}{2\pi} \int_\Gamma \Delta \Re F(z) L(dz) \right| \leq \mathcal{O}(1) \frac{\sqrt{b}}{\sqrt{h}}. \tag{5.5.23}$$

By (5.2.6) and (5.4.8),

$$\Delta \Re F(z) L(dz) = \frac{1}{h} p_*(|\sigma|) + \mathcal{O}(1),$$

so (5.5.23) gives,

$$\left| \# \left((u^\delta)^{-1}(0) \cap \Gamma \right) - \frac{1}{2\pi h} \mathrm{vol}_{T^*\mathbf{R}}\, p^{-1}(\Gamma) \right| \leq \mathcal{O}(1) \frac{\sqrt{b}}{\sqrt{h}}, \tag{5.5.24}$$

where we also used that $\sqrt{b}/\sqrt{h} \gg 1$, since $2N \ln(1/\delta) \leq b \leq 1/h$. Now, $(u^\delta)^{-1}(0) \cap \Gamma = \sigma(P_\delta) \cap \Gamma$ and we get Theorem 5.2.2 with $\epsilon = bh$. □

5.A Appendix: Estimates on Determinants of Gaussian Random Matrices

In this appendix, we follow Section 7 in [56]. Consider first a random vector

$$u(\omega)^{\mathrm{t}} = (\alpha_1(\omega), \dots, \alpha_N(\omega)) \in \mathbf{C}^N, \tag{5.A.1}$$

where $\alpha_1, \dots, \alpha_N$ are independent complex Gaussian random variables with a $\mathcal{N}_{\mathbf{C}}(0, 1)$ law:

$$(\alpha_j)_*(P) = \frac{1}{\pi} e^{-|z|^2} L(dz) =: f(z) L(dz). \tag{5.A.2}$$

The distribution of u is

$$u_*(P) = \frac{1}{\pi^N} e^{-|u|^2} L_{\mathbf{C}^N}(du). \tag{5.A.3}$$

If $U : \mathbf{C}^N \to \mathbf{C}^N$ is unitary, then Uu has the same distribution as u.

We next compute the distribution of $|u(\omega)|^2$. The distribution of $|\alpha_j(\omega)|^2$ is $\mu(r)dr$, where

$$\mu(r) = -H(r)\frac{d}{dr}e^{-r} = e^{-r}H(r),$$

and $H(r) = 1_{[0,\infty[}(r)$. The Fourier transform of μ is given by

$$\widehat{\mu}(\rho) = \int_0^\infty e^{-r}e^{-ir\rho}dr = \frac{1}{1+i\rho}.$$

We have $|u(\omega)|^2 = \sum_1^N |\alpha_j(\omega)|^2$, and since $|\alpha_j(\omega)|^2$ are independent and identically distributed, the distribution of $|u(\omega)|^2$ is $\mu * \cdots * \mu\, dr = \mu^{*N}dr$, where $*$ indicates convolution. For $r > 0$, we get

$$\begin{aligned}
\mu^{*N}(r) &= \frac{1}{2\pi}\int e^{ir\rho}\frac{1}{(1+i\rho)^N}d\rho \\
&= \frac{1}{(N-1)!2\pi}\int_\gamma e^{ir\rho}\left(-\frac{1}{i}\frac{d}{d\rho}\right)^{N-1}\left(\frac{1}{1+i\rho}\right)d\rho \\
&= \frac{1}{(N-1)!2\pi}\int_\gamma \left(\frac{1}{i}\frac{d}{d\rho}\right)^{N-1}(e^{ir\rho})\left(\frac{1}{1+i\rho}\right)d\rho \\
&= \frac{r^{N-1}}{(N-1)!2\pi}\int_\gamma e^{ir\rho}\frac{1}{1+i\rho}d\rho \\
&= \frac{r^{N-1}e^{-r}}{(N-1)!},
\end{aligned}$$

where γ is a small simple positively oriented loop around the pole $\rho = i$. Hence

$$\mu^{*N}dr = \frac{r^{N-1}e^{-r}}{(N-1)!}H(r)dr. \tag{5.A.4}$$

Recall here that

$$\int_0^\infty r^{N-1}e^{-r}dr = \Gamma(N) = (N-1)!,$$

so μ^{*N} is indeed normalized.

The expectation value $E(|\alpha_j|^2) = \langle|\alpha_j|^2\rangle$ of each $|\alpha_j(\omega)|^2$ is equal to 1, so

$$\langle|u(\omega)|^2\rangle = N. \tag{5.A.5}$$

We next estimate the probability that $|u(\omega)|^2$ is very large. It will be convenient to pass to the variable $\ln(|u(\omega)|^2)$ which has the distribution obtained from (5.A.4) by replacing r by $t = \ln r$, so that $r = e^t$, $dr/r = dt$. Thus $\ln(|u(\omega)|^2)$ has the distribution

$$\frac{r^N e^{-r}}{(N-1)!}H(r)\frac{dr}{r} = \frac{e^{Nt-e^t}}{(N-1)!}dt =: \nu_N(t)dt. \tag{5.A.6}$$

Now consider a random matrix

$$\left(u_1\ u_2\ \dots\ u_N\right), \tag{5.A.7}$$

where $u_k(\omega)$ are random vectors in $\mathbf{C}^N$ (here viewed as column vectors) of the form

$$u_k(\omega)^{\mathrm{t}} = (\alpha_{1,k}(\omega), \dots, \alpha_{N,k}(\omega)),$$

and all the $\alpha_{j,k}$ are independent with the same law (5.A.2).

Then

$$\det(u_1\, u_2 \dots u_N) = \det(u_1\, \widetilde{u}_2 \dots \widetilde{u}_N), \tag{5.A.8}$$

where $\widetilde{u}_j$ are obtained in the following way (assuming the u_j to be linearly independent, as they are almost surely): $\widetilde{u}_2$ is the orthogonal projection of u_2 in the orthogonal complement $(u_1)^\perp$, $\widetilde{u}_3$ is the orthogonal projection of u_3 in $(u_1, u_2)^\perp = (u_1, \widetilde{u}_2)^\perp$, etc.

If u_1 is fixed, then $\widetilde{u}_2$ can be viewed as a random vector in $\mathbf{C}^{N-1}$ of the type (5.A.1), (5.A.2), and with u_1, u_2 fixed, we can view $\widetilde{u}_3$ as a random vector of the same type in $\mathbf{C}^{N-2}$, etc. On the other hand

$$|\det(u_1\, u_2 \dots u_N)|^2 = |u_1|^2|\widetilde{u}_2|^2 \cdots |\widetilde{u}_N|^2. \tag{5.A.9}$$

The squared lengths $|u_1|^2, |\widetilde{u}_2|^2, \dots, |\widetilde{u}_N|^2$ are independent random variables with distributions $\mu^{*N}dr, \mu^{*(N-1)}dr, \dots, \mu dr$. This reduction plays an important role in [47].

Taking the logarithm of (5.A.9), we get in the right-hand side a sum of independent random variables with distributions $\nu_N dt, \dots, \nu_1 dt$, so the distribution of the random variable $\ln|\det(u_1\, u_2 \cdots u_N)|^2$ is equal to

$$(\nu_1 * \nu_2 * \cdots * \nu_N)dt, \tag{5.A.10}$$

with ν_j defined in (5.A.6).

We have

$$\nu_N(t) \le \widetilde{\nu}_N(t) := \frac{1}{(N-1)!} e^{Nt}.$$

Choose $x(N) \in \mathbf{R}$ such that

$$\int_{-\infty}^{x(N)} \widetilde{\nu}_N(t)dt = 1. \tag{5.A.11}$$

More explicitly, we have $x(N) \ge 0$ and

$$\frac{1}{N!} e^{Nx(N)} = 1, \quad x(N) = \frac{1}{N} \ln(N!) = \frac{1}{N} \ln \Gamma(N+1). \tag{5.A.12}$$

In [56] we used Stirling's formula, to get

$$x(N) = \ln N + \frac{1}{2N} \ln N - 1 + \frac{C_0}{N} + \mathcal{O}(\frac{1}{N^2}), \tag{5.A.13}$$

where $C_0 = (\ln 2\pi)/2 > 0$. Here we shall not need the large-N limit.

With this choice of $x(N)$, we put

$$\rho_N(t) = 1_{]-\infty, x(N)]}(t) \widetilde{\nu}_N(t),$$

so that $\rho_N(t)dt$ is a probability measure "obtained from $\nu_N(t)dt$ by transferring mass to the left", in the sense that

$$\int f \nu_N dt \le \int f \rho_N dt, \tag{5.A.14}$$

whenever f is a bounded decreasing function. Equivalently,

$$g * \nu_N \le g * \rho_N,$$

when g is a bounded increasing function. Now, for such a g, both $g * \nu_N$ and $g * \rho_N$ are bounded increasing functions, so by iteration,

$$g * \nu_1 * \cdots * \nu_N \le g * \rho_1 * \cdots * \rho_N.$$

In particular, taking $g = H$ we get

$$\int_{-\infty}^{t} \nu_1 * \cdots * \nu_N(s)ds \le \int_{-\infty}^{t} \rho_1 * \cdots * \rho_N(s)ds, \ t \in \mathbf{R}. \tag{5.A.15}$$

We have by (5.A.12)

$$\begin{aligned}\widehat{\rho}_N(\tau) &= \int_{-\infty}^{x(N)} \frac{1}{(N-1)!} e^{t(N-i\tau)} dt = \frac{1}{(N-1)!(N-i\tau)} e^{Nx(N)-ix(N)\tau} \\ &= \frac{e^{-ix(N)\tau}}{1-i\frac{\tau}{N}}.\end{aligned} \tag{5.A.16}$$

This function has a pole at $\tau = -iN$.

Similarly,

$$\widehat{1_{]-\infty,a]}}(\tau) = \frac{i}{\tau + i0} e^{-ia\tau}. \tag{5.A.17}$$

By Parseval's formula, we get

$$\begin{aligned}\int_{-\infty}^{a} \rho_1 * \cdots * \rho_N dt &= \frac{1}{2\pi} \int_{-\infty}^{\infty} \mathcal{F}(\rho_1 * \cdots * \rho_N)(\tau) \overline{\mathcal{F} 1_{]-\infty,a]}}(\tau) d\tau \\ &= \frac{1}{2\pi} \int_{-\infty}^{+\infty} e^{-i\tau(\sum_1^N x(j) - a)} \frac{-i}{\tau - i0} \prod_1^N \frac{1}{1-\frac{i\tau}{j}} d\tau.\end{aligned}$$

We deform the contour to $\Im\tau = -1/2$ (half-way between $\mathbf{R}$ and the first pole in the lower half-plane). It follows that for $a \le \sum_1^N x(j)$:

$$\int_{-\infty}^{a} \rho_1 * \cdots * \rho_N dt \le C(N) e^{a/2}. \tag{5.A.18}$$

In view of (5.A.15), the right-hand side is an upper bound for the probability that $\ln|\det(u_1 \cdots u_N)|^2 \le a$. Hence, for $a \le 0$,

$$\mathbf{P}(\ln|\det(u_1 \dots u_N)|^2 \le a) \le C(N) e^{a/2}. \tag{5.A.19}$$

We shall next extend our probabilistic bounds to determinants of the form

$$\det(D+Q)$$

where $Q = (u_1 \dots u_N)$ is as before, and $D = (d_1 \dots d_N)$ is a fixed complex $N \times N$ matrix. As before, we can write

$$|\det((d_1+u_1)\dots(d_N+u_N))|^2 = |d_1+u_1|^2 |\widetilde{d}_2 + \widetilde{u}_2|^2 \cdots |\widetilde{d}_N + \widetilde{u}_N|^2,$$

where $\widetilde{d}_2 = \widetilde{d}_2(u_1)$, $\widetilde{u}_2 = \widetilde{u}_2(u_1, u_2)$ are the orthogonal projections of d_2, u_2 on $(d_1 + u_1)^\perp$, $\widetilde{d}_3 = \widetilde{d}_3(u_1, u_2)$, $\widetilde{u}_3 = \widetilde{u}_3(u_1, u_2, u_3)$ are the orthogonal projections of d_3, u_3 on $(d_1 + u_1, d_2 + u_2)^\perp$, and so on.

Let $\nu_d^{(N)}(t)dt$ be the probability distribution of $\ln |d + u|^2$, when $d \in \mathbf{C}^N$ is fixed and $u \in \mathbf{C}^N$ is random as in (5.A.1), (5.A.2). Notice that $\nu_0^{(N)}(t) = \nu^{(N)}(t)$ is the density we have already studied.

Lemma 5.A.1 *For every $a \in \mathbf{R}$, we have*

$$\int_{-\infty}^{a} \nu_d^{(N)}(t)dt \leq \int_{-\infty}^{a} \nu^{(N)}(t)dt.$$

Proof Equivalently, we have to show that $\mathbf{P}(|d+u|^2 \leq \widetilde{a}) \leq \mathbf{P}(|u|^2 \leq \widetilde{a})$ for every $\widetilde{a} > 0$. For this, we may assume that $d = (c, 0, \dots, 0)$, $c > 0$. We then only have to prove that

$$\mathbf{P}(|c + \Re u_1|^2 \leq b^2) \leq \mathbf{P}(|\Re u_1|^2 \leq b^2),\ b > 0,$$

and here we may replace P by the corresponding probability density

$$\mu(t)dt = \frac{1}{\sqrt{\pi}} e^{-t^2} dt$$

for $\Re \mu_1$. Thus, we have to show that

$$\frac{1}{\sqrt{\pi}} \int_{|c+t|\leq b} e^{-t^2} dt \leq \frac{1}{\sqrt{\pi}} \int_{|t|\leq b} e^{-t^2} dt. \tag{5.A.20}$$

Fix b and rewrite the left-hand side as

$$I(c) = \frac{1}{\sqrt{\pi}} \int_{-b-c}^{b-c} e^{-t^2} dt.$$

The derivative satisfies

$$I'(c) = \frac{1}{\sqrt{\pi}} (e^{-(b+c)^2} - e^{-(b-c)^2}) \leq 0.$$

hence $c \mapsto I(c)$ is decreasing and (5.A.20) follows, since it holds when $c = 0$. □

Now consider the probability that $\ln|\det(D+Q)|^2 \leq a$. If $\chi_a(t) = H(a-t)$, this probability becomes

$$\int \cdots \int \mathbf{P}(du_1)\cdots \mathbf{P}(du_N)\chi_a\big(\ln|d_1+u_1|^2 + \ln|\widetilde{d}_2(u_1)+\widetilde{u}_2(u_1,u_2)|^2+\cdots$$
$$+\ln|\widetilde{d}_N(u_1,\cdots,u_{N-1})+\widetilde{u}_N(u_1,\cdots,u_N)|^2\big).$$

Here we first carry out the integration with respect to u_N, noticing that with the other $u_1,\dots,u_{N-1}$ fixed, we may consider $\widetilde{d}_N(u_1,\dots,u_{N-1})$ as a fixed vector in $\mathbf{C}\simeq(d_1+u_1,\dots,d_{N-1}+u_{N-1})^\perp$ and $\widetilde{u}_N$ as a random vector in $\mathbf{C}$. Using also the lemma, we get

$$\mathbf{P}(\ln|\det(D+Q)|^2\leq a)$$
$$=\int\cdots\int \nu^{(1)}_{\widetilde{d}_N}(t_N)dt_N\mathbf{P}(du_{N-1})\cdots\mathbf{P}(du_1)\times$$
$$\chi_a\left(\ln|d_1+u_1|^2+\cdots+\ln|\widetilde{d}_{N-1}(u_1,\cdots,u_{N-2})+\widetilde{u}_{N-1}(u_1,\cdots,u_{N-1})|^2+t_N\right)$$
$$\leq\int\cdots\int \nu^{(1)}(t_N)dt_N\mathbf{P}(du_{N-1})\cdots\mathbf{P}(du_1)\times$$
$$\chi_a\left(\ln|d_1+u_1|^2+\cdots+\ln|\widetilde{d}_{N-1}(u_1,\cdots,u_{N-2})+\widetilde{u}_{N-1}(u_1,\cdots,u_{N-1})|^2+t_N\right).$$

We next estimate the u_{N-1}- integral in the same way, and so on. Eventually, we get

Proposition 5.A.2 *Under the assumptions above,*

$$\mathbf{P}(\ln|\det(D+Q)|^2\leq a)$$
$$\leq\int\cdots\int \chi_a(t_1+\cdots+t_N)\nu^{(1)}(t_N)\nu^{(2)}(t_{N-1})\cdots\nu^{(N)}(t_1)$$
$$=\mathbf{P}(\ln|\det Q|^2\leq a).$$

In particular, the estimate (5.A.19) *extends to random perturbations of constant matrices:*

$$\mathbf{P}(\ln|\det(D+Q)|^2\leq a)\leq C(N)e^{a/2},\ \textit{for}\ a\leq 0. \tag{5.A.21}$$

Notes In this chapter we have generalized the results of Chap. 3 and tried to explain the role of the points ρ in the cotangent space where $i^{-1}\{p,\overline{p}\}(\rho)$ is positive or negative. In higher dimension this simple reduction via a Grushin seems to break down. The results are close to the ones in [55], but we have not included the case of multiplicative random perturbations since we will deal with that case in later chapters, directly in higher dimensions. It was convenient to use some elements from later works, like singular values. In particular, in Sect. 5.A, we have followed [56].

Resolvent Estimates Near the Boundary of the Range of the Symbol

6

6.1 Introduction and Statement of the Main Result

The purpose of this chapter is to give quite explicit bounds on the resolvent near the boundary of $\Sigma(p)$ (or more generally, near certain "generic boundary-like" points). The result is due (up to a small generalization) to Montrieux [18] and improves earlier results by Martinet [98] about upper and lower bounds for the norm of the resolvent of the complex Airy operator, which has empty spectrum [7]. There are more results about upper bounds, and some of them will be recalled in Chap. 10 when dealing with such bounds in arbitrary dimension. To fix the ideas, we will consider operators on $\mathbf{R}$ and indicate later the minor modifications needed for operators on S^1.

Let $P \in S(m)$ be as in Chap. 5 and assume Eqs. (5.1.1), (5.1.2), (5.1.4)–(5.1.7). Depending on the context, P may also denote the corresponding pseudodifferential operator in (5.1.3). Define $\Sigma = \Sigma(p)$, $\Sigma_\infty = \Sigma_\infty(p)$ as in (5.1.8), (5.1.9) with $p = p_0$ in (5.1.4), and recall (5.1.10). Let $z_0 \in \Sigma(p) \setminus \Sigma_\infty(p)$ and assume that

$$\text{for every } \rho \in p^{-1}(z_0), \text{ we have } \frac{1}{2i}\{p, \overline{p}\}(\rho) = 0, \ \{p, \{p, \overline{p}\}\}(\rho) \neq 0. \tag{6.1.1}$$

This is in some sense the generic situation for $z_0 \in \partial\Sigma(p) \setminus \Sigma_\infty(p)$.

Example 6.1.1

a) Let $P = hD + g(x)$ be Hager's operator, $g \in C^\infty(S^1)$, and assume that $\Im g$ has a unique minimum $x_{\min} \in S^1$ and that this minimum is nondegenerate. Notice that the extension of the definition of Σ and Σ_∞ to the case of semi-classical differential operators on compact manifolds is straightforward. In our case $p(x, \xi) = \xi + g(x)$, either as a

J. Sjöstrand, *Non-Self-Adjoint Differential Operators, Spectral Asymptotics and Random Perturbations*, Pseudo-Differential Operators 14,
https://doi.org/10.1007/978-3-030-10819-9_6

symbol on $T^*\mathbf{R}$, or as one on T^*S^1. In both cases, $\Sigma = \mathbf{R} + i[\min \Im g, \max \Im g]$, and we can take $m = \langle \xi \rangle$. We have $\Sigma_\infty = \Sigma$ in the case of $\mathbf{R}$ and $\Sigma_\infty = \emptyset$ in the S^1-case. Let $z_0 \in \partial\Sigma$ belong to the lower part $\mathbf{R} + i \min \Im g$ of $\partial\Sigma$.

When the underlying manifold is S^1, then $p^{-1}(z_0) = \{\rho_0\}$, where $\rho_0 = (x_{\min}, \xi_0)$, $\Im g(x_{\min}) = \min \Im g$, and $\xi_0 = \Re z_0 - \Re g(x_{\min})$. We have

$$\frac{1}{2i}\{p, \overline{p}\}(\rho_0) = -\Im g'(x_{\min}) = 0, \ \{p, \{p, \overline{p}\}\}(\rho_0) = -2i \Im g''(x_{\min}) \neq 0.$$

Our main result (Theorem 6.1.4) applies in this case.

When P acts on $L^2(\mathbf{R})$, then $x_{\min}$ is unique up to a multiple of 2π, and using the periodicity, we can check that the conclusion of the main result still holds.

b) The non-self-adjoint Airy operator, $P = (hD)^2 + ix$, with symbol $p = \xi^2 + ix$. We have $\Sigma = [0, +\infty[+ i\mathbf{R}$, $\Sigma_\infty = \emptyset$, and we can take $m = \xi^2 + \langle x \rangle$. Let $z_0 = iy_0 \in \partial\Sigma$, so that $p^{-1}(z_0) = \{\rho_0\}$, where $\rho_0 = (y_0, 0)$. We have $(2i)^{-1}\{p, \overline{p}\}(x, \xi) = -2\xi$ which vanishes at ρ_0. Further, $\{p, (2i)^{-1}\{p, \overline{p}\}\} = 2i \neq 0$.

c) The non-self-adjoint harmonic oscillator $P = \frac{1}{2}((hD)^2 + ix^2)$. We have $\Sigma = [0, +\infty[+ i[0, +\infty[$, $\Sigma_\infty = \emptyset$ and we can take $m = 1 + x^2 + \xi^2$. The boundary of Σ is the union of $[0, +\infty[$ and $i[0, +\infty[$ and by Fourier transform, the study near one of these half-rays is equivalent to that near the other. Let $0 \neq z_0 \in \partial\Sigma$ and take $z_0 > 0$ to fix the ideas. Then $p^{-1}(z_0) = \{\rho_+, \rho_-\}$, where $\rho_\pm = (0, \pm\xi_0)$ and $\xi_0 > 0$ is given by $\xi_0^2/2 = z_0$. We have $(2i)^{-1}\{p, \overline{p}\}(x, \xi) = -x\xi$ which vanishes at $\rho_\pm$. Further, $\{p, \{p, \overline{p}\}\} = -2i(\xi^2 - ix^2)$ is $\neq 0$ at $\rho_\pm$.

As we shall see below, the assumptions imply that $p^{-1}(z_0)$ is a finite set. Let $\rho_0 \in p^{-1}(z_0)$. Then

$$\gamma := \{\rho \in \text{neigh}\,(\rho_0);\ \frac{1}{2i}\{p, \overline{p}\}(\rho) = 0\} \tag{6.1.2}$$

is a smooth curve, since $\frac{1}{2i}\{p, \overline{p}\}$ is real-valued with a non-vanishing differential near ρ_0. Along γ the vectors $H_{\Re p}$ are $H_{\Im p}$ are collinear and never both equal to zero. Hence there is a smooth function $\gamma \ni \rho \mapsto \theta(\rho) \in \mathbf{R}$ such that $e^{-i\theta(\rho)} H_p$ is real and non-zero; $\theta(\rho)$ is unique up to a multiple of π.

We can parametrize γ by

$$\gamma(t) = \exp\left(tH_{\frac{1}{2i}\{p,\overline{p}\}}\right)(\rho_0). \tag{6.1.3}$$

Then,

$$\frac{d}{dt}p(\gamma(t)) = H_{\frac{1}{2i}\{p,\overline{p}\}}(p)(\gamma(t)) = -\{p, \frac{1}{2i}\{p, \overline{p}\}\}(\gamma(t)) \neq 0,$$

so

$$\delta := p \circ \gamma \tag{6.1.4}$$

is a smooth curve in neigh $(z_0, \mathbf{C})$ and

$$e^{-i\theta(\gamma(t))}\dot{\delta}(t) = -e^{-i\theta(\gamma(t))} H_p\left(\frac{1}{2i}\{p, \overline{p}\}\right) \in \mathbf{R} \setminus \{0\}. \tag{6.1.5}$$

In order to simplify the local geometric discussion, it will be helpful to replace p by $f \circ p$, where

$$f : \text{neigh}\,(z_0, \mathbf{C}) \to \text{neigh}\,(0, \mathbf{C}) \tag{6.1.6}$$

is almost holomorphic (cf. Appendix 6.A) along the curve δ:

$$\partial_{\overline{z}} f = \mathcal{O}(\text{dist}\,(z, \delta)^{\infty}). \tag{6.1.7}$$

When p is real-analytic, δ is a real-analytic curve and we will be able to choose f holomorphic. Notice that

$$H_{f\circ p}(\rho) = (\partial_z f)(p(\rho))H_p(\rho) + \mathcal{O}(\text{dist}\,(\rho, \gamma)^{\infty}), \tag{6.1.8}$$

$$\frac{1}{2i}\{f \circ p, \overline{f \circ p}\}(\rho) = |\partial_z f|^2 \frac{1}{2i}\{p, \overline{p}\}(\rho) + \mathcal{O}(\text{dist}\,(\rho, \gamma)^{\infty}), \tag{6.1.9}$$

so $\frac{1}{2i}\{f \circ p, \overline{f \circ p}\}$ vanishes on γ and

$$\{f \circ p, \frac{1}{2i}\{f \circ p, \overline{f \circ p}\}\} = \partial_z f\, |\partial_z f|^2 \{p, \frac{1}{2i}\{p, \overline{p}\}\} \neq 0 \text{ on } \gamma. \tag{6.1.10}$$

Thus, $\widetilde{p} = f \circ p$ satisfies the same general assumptions as p, with z_0 replaced by $0 = f(z_0)$.

Now, choose f mapping (the image of) δ to a real interval. Then $\widetilde{p}$ is real-valued on γ, and since $d\widetilde{p}$ is a complex multiple of a real differential when $\rho \in \gamma$, we conclude that $d\widetilde{p}$ is real, and hence that $H_{\widetilde{p}}$ is real on γ. By the choice of f, we know that $\Im \widetilde{p} = 0$ on γ and it follows that

$$\Im \widetilde{p} = \mathcal{O}(\text{dist}\,(\rho, \gamma)^2). \tag{6.1.11}$$

The natural parametrization in (6.1.3) induces an orientation of γ such that the region where $(2i)^{-1}\{p, \overline{p}\} > 0$ is to the left when walking along γ in the positive direction. Replacing p by $\widetilde{p}$ will change the parametrization in (6.1.3), but not the orientation of γ.

We now choose smooth local symplectic coordinates (x, ξ) (i.e., for which the symplectic form is $d\xi \wedge dx$) centered at ρ_0 such that γ becomes the ξ-axis $\{(0, \xi)\}$ with it natural orientation, namely that of increasing values of ξ. Then, since $\Im \widetilde{p} = \mathcal{O}(x^2)$, we can write

$$\widetilde{p}(x, \xi) = d(x, \xi) + ir(x, \xi) \tag{6.1.12}$$

where d, r are smooth real-valued functions such that

$$\partial_\xi d(0, \xi) \neq 0, \quad \Im r = \mathcal{O}(x^2).$$

We have

$$\frac{1}{2i}\{\widetilde{p}, \overline{\widetilde{p}}\} = -\{d, r\} = -d'_\xi r'_x + d'_x r'_\xi = -d'_\xi r'_x + \mathcal{O}(x^2),$$

and at $x = 0$

$$0 \neq \{\widetilde{p}, \frac{1}{2i}\{\widetilde{p}, \overline{\widetilde{p}}\}\} = -(d'_\xi)^2 r''_{xx}. \tag{6.1.13}$$

Having chosen the symplectic coordinates so that the orientation of the ξ-axis is the same as the one of γ (when identifying the two sets), we know that $(2i)^{-1}\{\widetilde{p}, \overline{p}\}$ has the same sign as $-x$. In order to fix the ideas, we can impose an additional condition on the map f, namely that the orientation of $f \circ \delta = f \circ p \circ \gamma$ on $\mathbf{R}$ is the one of increasing real values. This means that $\widetilde{p}(0, \xi)$ should be an increasing function of ξ, or in other terms, that $0 < \{(2i)^{-1}\{\widetilde{p}, \widetilde{p}\}, \widetilde{p}\}$, i.e. that the quantity in (6.1.13) is < 0. Hence

$$\partial_\xi d(0, \xi) > 0, \quad r''_{xx}(0, \xi) > 0. \tag{6.1.14}$$

We can refine the choice of symplectic coordinates above so that $d(x, \xi) = \xi$. Then with a new r as in (6.1.14), we get

$$\widetilde{p}(x, \xi) = \xi + ir(x, \xi), \quad 0 \leq r(x, \xi) \asymp x^2 \tag{6.1.15}$$

and

$$\frac{1}{2i}\{\widetilde{p}, \overline{\widetilde{p}}\} = -\partial_x r, \tag{6.1.16}$$

which is positive for $x < 0$ (i.e., to the left of γ) and negative to the right. The map $\widetilde{p}$ is orientation preserving to the left and orientation reversing to the right, when we adopt the standard orientations on $\mathbf{R}^2_{x,\xi}$ and on $\mathbf{C}_w \equiv \mathbf{R}^2_{\Re w, \Im w}$. Indeed,

$$d\Re\widetilde{p} \wedge d\Im\widetilde{p} = -\{\Re\widetilde{p}, \Im\widetilde{p}\}dx \wedge d\xi = \frac{1}{2i}\{\widetilde{p}, \overline{\widetilde{p}}\}dx \wedge d\xi.$$

We can say that $\widetilde{p}$ has a simple fold along γ: The equation $\widetilde{p}(\rho) = w$, for $\rho \in \mathrm{neigh}((0,0), \mathbf{R}^2)$, $w \in \mathrm{neigh}(0, \mathbf{C})$

- has two simple solutions when $\Im w > 0$,
- has one degenerate solution when $\Im w = 0$,
- has no solution when $\Im w < 0$.

The equation $p(\rho) = z$, $z \in \mathrm{neigh}(z_0, \mathbf{C})$, is equivalent to $\widetilde{p}(\rho) = w$ when $w = f(z)$ and locally, the inverse image under f of the upper half-plane is the region situated to the left of the oriented curve δ. p has a simple fold along γ and is orientation preserving to the left and orientation reversing to the right of γ as $\frac{1}{2i}\{p, \overline{p}\}$ is positive to the left and negative to the right. The range of p is the region situated to the left of δ. Thus the equation $p(\rho) = z$

- has two simple solutions when $z \notin \delta$ is to the left of δ,
- has one degenerate solution when $z \in \delta$,
- has no solution when $z \notin \delta$ is to the right.

For both equations we denote the solutions (when they exist) by ρ_+, ρ_- with ρ_+ and ρ_- to the left and to the right of δ, respectively.

Let $\widetilde{\rho}_\pm$ be the corresponding solutions of $\widetilde{p}(\rho) = w$, so that $\widetilde{\rho}_\pm(f(z)) = \rho_\pm(z)$. We get $\widetilde{\rho}_\pm(w) = (x_\pm, \xi_\pm)$, where

$$\xi_\pm = \Re w \tag{6.1.17}$$

and $x_\pm$ are given by

$$r(x, \Re w) = \Im w, \quad \mp x_\pm \le 0. \tag{6.1.18}$$

We have

$$x_\pm = k(\mp\sqrt{\Im w}, \Re w), \tag{6.1.19}$$

where $k(s,t)$ is the smooth function given by

$$\sqrt{r}(k(s,t), t) = s$$

and $\sqrt{r(x,\xi)}$ is the smooth branch of the square root of r which has the same sign as x. To leading order,

$$k(s,t) = \frac{s}{(r''_{xx}(0,t)/2)^{1/2}} + \mathcal{O}(s^2),$$

so

$$x_\pm = \mp (r''_{xx}(0,\Re w))^{-1/2}\sqrt{2\Im w} + \mathcal{O}(\Im w). \tag{6.1.20}$$

Naturally, (6.1.17), (6.1.19), (6.1.20) also describe $\rho_\pm$. All we have to do, is to work in the same symplectic coordinates and represent z by the local coordinates $(s,t) = (\Im f, \Re f) = (\Im w, \Re w)$.

Combining (6.1.10), where $\widetilde{p} = f \circ p$, with (6.1.13), where now $d'_\xi = 1$, we get

$$0 < r''_{xx}(0,\xi) = -|\partial_z f|^2 \partial_z f\{p, \frac{1}{2i}\{p,\overline{p}\}\}(\rho),\ \ \partial_z f = \partial_z f(\rho), \tag{6.1.21}$$

where $\rho \in \gamma$ is determined by $f(p(\rho)) = \xi$. We may choose f so that $|f'| = 1$ on δ. Then (6.1.21) simplifies to

$$r''_{xx}(0,\xi) = \epsilon(\xi),\ \ \epsilon(\xi) := \left|\{p, \frac{1}{2i}\{p,\overline{p}\}\}(\rho)\right|. \tag{6.1.22}$$

Combining (6.1.20), (6.1.22) with (6.1.15), we get for $\Im w \geq 0$,

$$\frac{1}{2i}\{\widetilde{p},\overline{\widetilde{p}}\}(\rho_\pm) = \pm(2\epsilon(\Re w))^{1/2}(\Im w)^{1/2} + \mathcal{O}(\Im w).$$

As in (6.1.21), we have

$$\frac{1}{2i}\{\widetilde{p},\overline{\widetilde{p}}\}(\rho) = |\partial_z f(p(\rho))|^2 \frac{1}{2i}\{p,\overline{p}\}(\rho) + \mathcal{O}(\mathrm{dist}\,(\rho,\gamma)^\infty),$$

$$\frac{1}{2i}\{p,\overline{p}\}(\rho_\pm) = \pm(2\epsilon(\Re w))^{1/2}(\Im w)^{1/2} + \mathcal{O}(\Im w),\ \Im w \geq 0. \tag{6.1.23}$$

Assume for the moment that p is analytic near ρ_0, so that f, $\widetilde{p}$ are analytic also. If $\Im w > 0$, then $\{\rho \in \mathrm{neigh}\,(\rho_0, \mathbf{C}^2);\ p(\rho) = z\}$ is a complex curve $\Gamma(z)$ which intersects the real phase space at the two points $\rho_\pm(z)$ and we can therefore define the action

$$J(z) = \int_{\Gamma(z,\rho_+,\rho_-)} \xi dx, \tag{6.1.24}$$

where $\Gamma(z, \rho_+, \rho_-)$ is a real curve in $\Gamma(z)$ which starts at $\rho_+(z)$ and ends at $\rho_-(z)$, and which stays in a small neighborhood of ρ_0, so that all such curves are homotopic to one another. All curves in $\mathbf{R}^2$ are Lagrangian manifolds and this remains true when we pass to the complex-holomorphic category. Hence, $\xi dx|_{\Gamma(z)}$ is closed (and even exact since we work locally) and the value of $J(z)$ does not depend on the choice of $\Gamma(z, \rho_+, \rho_-)$. If (y, η) are some other real and analytic symplectic coordinates, then $\xi dx - \eta dy$ is closed and hence exact, so $\xi dx - \eta dy = dG$, where G is real and analytic. Consequently,

$$\int_{\Gamma(z,\rho_+,\rho_-)} \xi dx - \int_{\Gamma(z,\rho_+,\rho_-)} \eta dy = G(\rho_-) - G(\rho_+)$$

is real and we conclude that

$$\begin{aligned} I(z) := \Im J(z) \text{ is invariant under changes of real} \\ \text{and analytic symplectic coordinates.} \end{aligned} \tag{6.1.25}$$

This means that we can work with $\widetilde{p}$ in the form (6.1.15) and write

$$I(z) = \widetilde{I}(w), \quad w = f(z),$$

with the analogous definition of $\widetilde{I}(w)$. Parametrize z by

$$w = f(z) = is + t, \quad (s, t) \in \operatorname{neigh}((0, 0), \mathbf{R}^2), \tag{6.1.26}$$

and write $I(z) = I(s, t)$.

Working in the coordinates of (6.1.15), $\Gamma(z)$ is given by $\widetilde{p}(x, \xi) = w$, i.e., $\xi + ir(x, \xi) = w$ and

$$I(z) = \Im \int_{x_+}^{x_-} \lambda(x, w) dx, \tag{6.1.27}$$

where λ is the solution of

$$\widetilde{p}(x, \lambda(x, w)) = w. \tag{6.1.28}$$

By successive differentiations of this equation, we get

$$\lambda'_x = -\frac{\widetilde{p}'_x(x, \lambda)}{\widetilde{p}'_\xi(x, \lambda)}, \quad \lambda'_w = \frac{1}{\widetilde{p}'_\xi(x, \lambda)},$$

$$\lambda''_{xx} = -\frac{\widetilde{p}''_{xx}(x, \lambda)}{\widetilde{p}'_\xi(x, \lambda)} + \frac{2\widetilde{p}''_{x\xi}(x, \lambda)\widetilde{p}'_x(x, \lambda)}{\widetilde{p}'_\xi(x, \lambda)^2} - \frac{\widetilde{p}'_x(x, \lambda)^2 \widetilde{p}''_{\xi\xi}(x, \lambda)}{\widetilde{p}'_\xi(x, \lambda)^3}.$$

In particular, $\lambda(0, w) = w$, $\lambda'_x(0, w) = 0$,

$$\lambda''_{xx}(0, w) = -\widetilde{p}''_{xx}(0, w) = -ir''_{xx}(0, w). \tag{6.1.29}$$

We get the leading Taylor expansion in x,

$$\lambda(x, w) = w - \frac{i}{2} r''_{xx}(0, w)x^2 + \mathcal{O}(x^3), \tag{6.1.30}$$

where we recall that $r''_{xx}(0, w) = \epsilon(w)$ for real w and we can extend this relation and (6.1.30) holomorphically in w. Thus, with $w = t + is$ as in (6.1.26),

$$\Im\lambda(x, w) = s - \frac{\epsilon(t)}{2}x^2 + \mathcal{O}(s)x^2 + \mathcal{O}(x^3). \tag{6.1.31}$$

By (6.1.20),

$$x_\pm = \mp\sqrt{\frac{2s}{\epsilon(t)}} + \mathcal{O}(s) \tag{6.1.32}$$

are smooth functions of $\mp\sqrt{s}$, for $s \geq 0$ and we notice that (6.1.32) also follows from (6.1.31) and the fact that $\Im\lambda(x_\pm, w) = 0$. $I(z)$ is a smooth function of $\sqrt{s}$ and

$$I(z) = \mathcal{O}(s^2) + \int_{-\sqrt{2s/\epsilon(t)}}^{\sqrt{2s/\epsilon(t)}} \left(s - \frac{\epsilon(t)}{2}x^2\right) dx = \frac{2}{3}\frac{(2s)^{\frac{3}{2}}}{\epsilon(t)^{\frac{1}{2}}} + \mathcal{O}(s^2). \tag{6.1.33}$$

With z represented as in (6.1.26), we have

$$I(z) = I(s, t) = \iota(\sqrt{s}, t), \tag{6.1.34}$$

where $\iota(\sigma, t)$ is smooth and $= \mathcal{O}(\sigma^3)$.

We shall now drop the analyticity assumption and pause for a general discussion: Let p be a smooth function defined in a neighborhood of $0 \in \mathbf{R}^2$. We assume that p depends smoothly on a finite number of real parameters $w \in \operatorname{neigh}(0, \mathbf{R}^k)$ that we do not write out. Assume that $dp \neq 0$ and that (for w in a closed subset of the parameter space, containing 0), there are two real points $x_\pm \in \mathbf{R}^2$ where p vanishes and that $x_\pm$ are Hölder continuous as functions of the parameter with $x_+ = x_- = 0$ for $w = 0$.

Let p also denote an almost holomorphic extension of p to a complex neighborhood of $0 \in \mathbf{C}^2$. This means that $\overline{\partial} p(x) = \mathcal{O}((\Im x)^\infty)$, and we know (see e.g. [103] and Appendix 6.A) that such an extension is unique up to a term which is $\mathcal{O}((\Im x)^\infty)$. After a permutation of the variables we may assume that $\partial_{x_2} p \neq 0$. Then by the real implicit function theorem we have in a neighborhood of $x = 0$ in $\mathbf{C}^2$

$$p(x) = 0 \Longleftrightarrow x_2 = \lambda(x_1),$$

where $\lambda(x_1)$ is a smooth function of x_1. Next, differentiating the equation $p(x_1, \lambda(x_1)) = 0$ we see that

$$\partial_{\overline{x}_1} \lambda = \mathcal{O}((\Im x_1, \Im \lambda(x_1))^\infty).$$

Let $\omega(x)dx$ be a smooth 1-form and denote by the same symbol an almost holomorphic extension (as a (1,0) form). Consider

$$J_\gamma = \int_\gamma \omega(x)dx, \tag{6.1.35}$$

where γ is a smooth curve from x_+ to x_- inside the complex set $p^{-1}(0)$, with the property that

$$\text{length}\,(\gamma) = \mathcal{O}(|x_+ - x_-|).$$

We have $\sup |\Im \gamma| = \mathcal{O}(|x_+ - x_-|)$, we can write $\gamma : [0, 1] \ni t \mapsto (\gamma_1(t), \lambda(\gamma_1(t)))$ where γ_1 is a curve from $x'_{+,1}$ to $x_{-,1}$ with length $(\gamma_1) = \mathcal{O}(|x_{+,1} - x_{-,1}|)$. In the given coordinates, the simplest choice of γ would be to take $\gamma = \gamma_0(t) = (\gamma_{0,1}(t), \lambda(\gamma_{0,1}(t)),$ where $\gamma_{0,1}(t) = (1 - t)x_{+,1} + t x_{-,1}$.

Using Stokes' formula and the observation that

$$d(\omega dx|_{p^{-1}(0)}) = \mathcal{O}(|\Im x|^\infty),$$

we get

Proposition 6.1.2 *In the general situation described above, and up to a term $\mathcal{O}(|x_+ - x_-|^\infty)$, $J = J_\gamma$ does not depend on the choice of γ, nor on the choices of almost holomorphic extensions of p and ω. We have the same invariance under smooth changes of local coordinates in $\mathbf{R}^2$.*

Now return to p in (6.1.1) and $\widetilde{p} = f \circ p$ in (6.1.15) and let "ωdx" above be the fundamental 1-form ξdx. By taking almost holomorphic extensions we can still define the almost complex curve[1] $\Gamma(z)$ by $p(\rho) = z$ as prior to (6.1.24), and when $\Im w \geq 0$ we define the action $J(z)$ as in (6.1.24), choosing the curve from ρ_+ to ρ_- of length $\mathcal{O}(|\rho_+ - \rho_-|)$. Then $J(z)$ is well defined up to $\mathcal{O}(|\rho_+ - \rho_-|^\infty)$, or equivalently up to $\mathcal{O}(|\Im w|^\infty)$. The imaginary part $I(z)$ is invariant under real canonical transformations (see e.g. [51, p. 100]) up to $\mathcal{O}(|\Im w|^\infty)$ and we have (6.1.27), where λ solves (6.1.28), in which $\widetilde{p}$ also denotes an almost holomorphic extension.

Proposition 6.1.3 *Under the assumptions above, let $w = f(z)$ be defined in* (6.1.6) *and after, so that $|f'| = 1$ on the curve $\delta = p \circ \gamma$. Recall the definition of $\epsilon(t) \simeq \epsilon(f \circ p \circ (t))$ in* (6.1.22). *Then $I(z) = \Im J(w)$ is well defined (cf.* (6.1.24), (6.1.27)) *up to $\mathcal{O}(|\Im w|^\infty)$ in the region $\Im w \geq 0$. Writing $w = t + is$, we have* (6.1.34): *$I(s, t) = \iota(\sqrt{s}, t)$, where $\iota(\sigma, t)$ is a smooth function, well-defined* mod (σ^∞), *such that*

$$\iota(\sigma, t) = \frac{4\sqrt{2}}{3} \frac{\sigma^3}{\epsilon(t)^{\frac{1}{2}}} + \mathcal{O}(\sigma^4) \tag{6.1.36}$$

We now start to formulate the main result of this chapter. Recall that $z_0 \in \Sigma(p) \backslash \Sigma_\infty(p)$ and that we work under the assumptions (5.1.1), (5.1.2), (5.1.4)–(5.1.7), and (6.1.1). It follows from the discussion above that the points of $p^{-1}(z_0)$ are isolated and hence that this set is finite,

$$p^{-1}(z_0) = \{\rho_0^1, \ldots, \rho_0^N\}. \tag{6.1.37}$$

The arguments above then apply with ρ_0 equal to ρ_0^j, $1 \leq j \leq N$. Through ρ_0^j we have a curve γ_j, defined as in (6.1.2), (6.1.1) with ρ_0 replaced by ρ_0^j. Let $\delta_j = p \circ \gamma_j$ be the image curves, all passing through z_0. We have smooth functions f_j : neigh $(z_0, \mathbf{C}) \rightarrow$ neigh $(0, \mathbf{C})$, almost holomorphic at δ_j, such that $|f_j'| = 1$ on δ_j and $\widetilde{p}_j = f_j \circ p$ is real-valued on γ_j. Put

$$f_j(z) = w_j = t_j + is_j. \tag{6.1.38}$$

When $s_j \geq 0$ we have two real points $\rho_j^\pm \in$ neigh $(\rho_0^j, T^*\mathbf{R})$ determined by $p(\rho_j^\pm) = z$, $\pm(2i)^{-1}\{p, \overline{p}\}(\rho_j^\pm) \geq 0$, with strict inequality (and $\rho_j^+ \neq \rho_j^-$) when $s_j > 0$.

As in (6.1.23),

$$\frac{1}{2i}\{p, \overline{p}\}(\rho_\pm^j) = \pm(2\epsilon_j(t_j))^{\frac{1}{2}}\sqrt{s_j} + \mathcal{O}(s_j), \quad s_j \geq 0, \tag{6.1.39}$$

[1] We say that the smooth submanifold $\Gamma \subset \mathbf{C}^n$ is an almost complex curve, if $\dim_{\mathbf{R}}\Gamma = 2$ and dist $(T_x\Gamma, i(T_x\Gamma)) = \mathcal{O}(|\Im x|^\infty)$ locally uniformly for $x \in \Gamma$. See [103] for a more systematic discussion of almost holomorphic geometry.

where $\epsilon_j(t) = \epsilon_j(\gamma_j(t))$ is defined as in (6.1.22),

$$\epsilon_j(t) = \left| \{p, \frac{1}{2i}\{p, \overline{p}\}\}(\gamma_j(t)) \right|. \tag{6.1.40}$$

As in Proposition 6.1.3, we can define for $s_j \geq 0$,

$$I_j(z) = \iota_j(\sqrt{s_j}, t_j), \quad \iota_j(\sigma, t) = \frac{4\sqrt{2}}{3} \frac{\sigma^3}{\epsilon_j(t)^{1/2}} + \mathcal{O}(\sigma^4), \tag{6.1.41}$$

where $\iota_j(\sigma, t)$ is smooth and real-valued.

Theorem 6.1.4 *We make the assumptions* (5.1.1), (5.1.2), (5.1.4)–(5.1.7), (6.1.1), *and define the quantities* ρ_j, $\rho_\pm^j$, I_j, ι_j *as above. Let* $z \in \text{neigh}(z_0, \mathbf{C})$, *so that we have the representation* (6.1.38) *for each* $j = 1, \ldots, N$. *For* $C_0 > 0$ *large enough, put*

$$M_j(z; h) = \begin{cases} (h^{2/3} + |s_j|)^{-1}, & \text{when } s_j \leq C_0 h^{2/3}, \\ \dfrac{\left(\frac{\pi}{h}\right)^{1/2} \exp(I_j(z)/h)}{\left(\frac{1}{2i}\{p, \overline{p}\}(\rho_+^j)\right)^{1/4} \left(\frac{1}{2i}\{\overline{p}, p\}(\rho_-^j)\right)^{1/4}}, & \text{when } s_j \geq C_0 h^{2/3}. \end{cases} \tag{6.1.42}$$

Assume that for some arbitrarily small and fixed $\delta_0 > 0$,

$$-h^{\delta_0} \leq s_j \leq \mathcal{O}(1) \left(h \ln \frac{1}{h}\right)^{\frac{2}{3}}, \; j = 1, \ldots, N. \tag{6.1.43}$$

Put

$$\mathcal{N} = \{j \in \{1, 2, \ldots, N\}; \; s_j \geq C_0 h^{\frac{2}{3}}\}.$$

Then for $h > 0$ *small enough, we have uniformly with respect to* z *that* $P - z : H(m) \to H(1)$ *is bijective and*

$$\|(P - z)^{-1}\|_{H(1) \to H(1)} \begin{cases} = \max_{j \in \mathcal{N}} (1 + \mathcal{O}(h/s_j^{3/2})) M_j(z; h), & \text{when } \mathcal{N} \neq \emptyset, \\ \leq \mathcal{O}(1) \max_j M_j, & \text{in general.} \end{cases} \tag{6.1.44}$$

The lower bound in (6.1.43) has been introduced for convenience only and could probably be removed without too much extra work.

6.2 Preparations and Reductions

In this section, we discuss the situation locally. Let $\rho_0 \in T^*\mathbf{R} = \mathbf{R}^2$, $p \in C^\infty(\mathrm{neigh}\,(\rho_0, \mathbf{R}^2))$ satisfy (6.1.1):

$$\frac{1}{2i}\{p, \overline{p}\}(\rho_0) = 0, \tag{6.2.1}$$

$$\{p, \frac{1}{2i}\{p, \overline{p}\}\}(\rho_0) \neq 0. \tag{6.2.2}$$

Let $z_0 = p(\rho_0)$. We are interested in $p - z$ (and in a corresponding pseudodifferential operator $P(x, hD; h) - z$) when $z - z_0$ is small. Let $f : \mathrm{neigh}\,(z_0, \mathbf{C}) \to \mathrm{neigh}\,(0, \mathbf{C})$ be the map for which $\widetilde{p} = f \circ p$, where $\widetilde{p}$ is as in (6.1.15). One could probably develop a functional calculus, allowing to work with "$f(P)$", but we opt for a more direct approach, where we linearize f at z_0: $f(z) = e^{-i\theta(\rho_0)}(z - z_0) + \mathcal{O}((z - z_0)^2)$ and, redefine $\widetilde{p}$ by

$$p - z = e^{i\theta(\rho_0)}(\widetilde{p}(\rho) - \omega),$$

where

$$\widetilde{p}(\rho) = e^{-i\theta(\rho_0)}(p(\rho) - z_0), \quad \omega = e^{-i\theta(\rho_0)}(z - z_0).$$

We are then interested in $\widetilde{p}(\rho) - \omega$ when ω is small. $\widetilde{p}$ satisfies the same assumptions, now with z_0 replaced by 0.

Remark 6.2.1 $z_0 = p(z_0)$ corresponds to $w = w_0 = 0$ and hence to $\omega = 0$. By Taylor expansion we get $\omega = w + \mathcal{O}(w^2)$ for $z \in \mathrm{neigh}\,(z_0)$. In the application later on, we will choose the point $\rho_0 \in \gamma$ and the corresponding point $z_0 \in \delta$ as functions of the spectral parameter z. Thus, for a given z, we can choose the new base point $\rho_0 = \rho(z) \in \gamma$ and the new $z_0 = p(\rho(z))$, so that $w = w(z)$ has a vanishing real part. Then

$$\omega = i\Im w + \mathcal{O}((\Im w)^2).$$

The image curve $\widetilde{p} \circ \gamma$ is tangent to $\mathbf{R}$ at 0 and oriented in the positive real direction at that point. It is of the form

$$\widetilde{\delta} : \Im\omega = k(\Re\omega), \ \Re\omega \in \mathrm{neigh}\,(0), \ k'(0) = 0.$$

- The range under $\widetilde{p}$ of neigh $(\rho_0, \mathbf{R}^2)$ is equal to $\{\omega \in \text{neigh}\,(0, \mathbf{C});\ \Im\omega \geq k(\Re\omega)\}$. (It is situated to the left of $\widetilde{p} \circ \gamma$.)
- The map $\widetilde{p}$: neigh $(\rho_0, \mathbf{R}^2) \rightarrow$ neigh $(0, \mathbf{C})$ has a simple fold along γ. It is a local diffeomorphism on each side of γ, orientation preserving to the left and orientation reversing to the right.

Until further notice, we discuss $\widetilde{p}$, but drop the tilde and simply write p. After composing p with an affine canonical transformation, we may assume that

$$\rho_0 = 0,\ dp(0) = d\xi(0), \tag{6.2.3}$$

in addition to the fact that $p(0) = 0$. Then,

$$p(x, \xi) = \xi + ir(x, \xi),\ \ r(x, \xi) = \mathcal{O}(x^2 + \xi^2). \tag{6.2.4}$$

Moreover,

$$\frac{1}{2i}\{p, \overline{p}\} = -\{\Re p, \Im p\} = -\partial_x \Re r + \mathcal{O}(x^2 + \xi^2),$$

and (cf. (6.1.22))

$$\partial_x^2 \Re r(0, 0) = \epsilon(0) > 0. \tag{6.2.5}$$

Using Malgrange's preparation theorem (see [55] for the use in the present context and for further references), we know that

$$p(x, \xi) - \omega = a(x, \xi, \omega)(\xi + g(x, \omega)),\ (x, \xi, \omega) \in \text{neigh}\,((0, 0, 0), \mathbf{R}^2 \times \mathbf{C}), \tag{6.2.6}$$

where a, g are smooth functions (and g is basically equal to $-\lambda$, as discussed prior to (6.1.35) with $n = 2$, $(x_1, x_2) = (x, \xi)$). Actually, for our purposes it suffices to have (6.2.6) at the level of formal Taylor series at $(x, \xi, \omega) = (0, 0, 0)$, and hence to settle for the more elementary fact that

$$p(x, \xi) - \omega = a(x, \xi, \omega)(\xi + g(x, \omega)) + \mathcal{O}\left((x, \xi, \omega)^{\infty}\right). \tag{6.2.7}$$

Looking at the Taylor expansions for r, a, g up to second order, we first write

$$r(x, \xi) = r_{2,0}x^2 + r_{1,1}x\xi + r_{0,2}\xi^2 + \mathcal{O}((x, \xi)^3), \tag{6.2.8}$$

and get after some calculations

$$a(x,\xi,\omega) = 1 + ir_{1,1}x + ir_{0,2}(\xi+\omega) + \mathcal{O}((x,\xi,\omega)^2), \tag{6.2.9}$$

$$g(x,\omega) = -\omega + ir_{2,0}x^2 + ir_{1,1}x\omega + ir_{0,2}\omega^2 + \mathcal{O}((x,\omega)^3). \tag{6.2.10}$$

Here we know that $\Re r_{2,0} = \epsilon(0)/2$.

Let

$$P(x, hD_x; h) = hD_x + ir(x, hD_x) + hQ(x, hD_x; h)$$

be a classical h-pseudodifferential operator with symbol defined in a neighborhood of $(0,0)$. Using Malgrange's preparation theorem repeatedly, we find classical symbols

$$\begin{aligned} A(x,\xi,\omega;h) &\sim a(x,\xi,\omega) + ha_1(x,\xi,\omega) + \cdots, \\ G(x,\omega;h) &\sim g(x,\omega) + hg_1(x,\omega) + \cdots, \end{aligned} \tag{6.2.11}$$

defined in a neighborhood of $(x,\xi,\omega) = (0,0,0)$, such that in the sense of formal composition of h-pseudodifferential operators,

$$P(x, hD_x; h) - \omega = A(x, hD_x, \omega; h)(hD_x + G(x,\omega;h)). \tag{6.2.12}$$

When using the softer version of Taylor series division, we have to add an error term to the right-hand side of the form $S(x, hD_x, \omega; h)$, where

$$S(x,\xi,\omega;h) \sim s_0(x,\xi,\omega) + hs_1(x,\xi,\omega) + \ldots,\ \ s_j(x,\xi,\omega) = \mathcal{O}((x,\xi,\omega)^\infty),$$

and this will still suffice for our purposes.

We now concentrate on the region

$$|\Re\omega| \le \Im\omega \gg h^{\frac{2}{3}} \tag{6.2.13}$$

and study the factor

$$B(x, hD_x, \omega; h) = hD_x + G(x,\omega;h). \tag{6.2.14}$$

We introduce the scaling

$$\omega = \alpha\widetilde{\omega},\ x = \sqrt{\alpha}\widetilde{x}, \tag{6.2.15}$$

where we let

$$\alpha \asymp \Im\omega, \text{ so that } \Im\widetilde{\omega} \asymp 1,\ |\Re\widetilde{\omega}| \le \Im\widetilde{\omega}. \tag{6.2.16}$$

Then, with $\widetilde{h} = h/\alpha^{3/2} \ll 1$, we get

$$B = \alpha\widetilde{B},\ \widetilde{B} = \widetilde{h}D_{\widetilde{x}} + \frac{1}{\alpha}G(\sqrt{\alpha}\widetilde{x}, \alpha\widetilde{\omega}; h) =: \widetilde{h}D_{\widetilde{x}} + \widetilde{G}(\widetilde{x}, \widetilde{\omega}, \alpha; \widetilde{h}). \tag{6.2.17}$$

From (6.2.11) we get

$$\widetilde{G}(\widetilde{x}, \widetilde{\omega}, \alpha; \widetilde{h}) \sim \widetilde{g}(\widetilde{x}, \widetilde{\omega}, \alpha) + \sum_{k=1}^{\infty} \alpha^{\frac{3k}{2}-1}\widetilde{h}^k g_k(\sqrt{\alpha}\widetilde{x}, \alpha\widetilde{\omega}), \tag{6.2.18}$$

where

$$\widetilde{g}(\widetilde{x}, \widetilde{\omega}, \alpha) := \frac{1}{\alpha}g(\sqrt{\alpha}\widetilde{x}, \alpha\widetilde{\omega}) \sim -\widetilde{\omega} + \sum_{j+k\ge 2} g_{j,k}\alpha^{\frac{j}{2}+k-1}\widetilde{x}^j\widetilde{\omega}^k,$$

and we write the Taylor expansion

$$g(x, \omega) = -\omega + \sum_{j+k\ge 2} g_{j,k}x^j\omega^k.$$

Recall here that, by (6.2.10), $g_{j,k} = ir_{j,k}$ for $j + k = 2$.

Hence,

$$\widetilde{g}(\widetilde{x}, \widetilde{\omega}, \alpha) \sim \sum_{0}^{\infty} \alpha^{\frac{\ell}{2}}\widetilde{g}_\ell(\widetilde{x}, \widetilde{\omega}), \tag{6.2.19}$$

where

$$\widetilde{g}_0(\widetilde{x}, \widetilde{\omega}) = -\widetilde{\omega} + ir_{2,0}\widetilde{x}^2, \quad \widetilde{g}_1(\widetilde{x}, \widetilde{\omega}) = g_{3,0}\widetilde{x}^3 + ir_{1,1}\widetilde{x}\widetilde{\omega}, \tag{6.2.20}$$

and in general, for $\ell \ge 1$,

$$\widetilde{g}_\ell(\widetilde{x}, \widetilde{\omega}) = \sum_{\frac{j}{2}+k=1+\frac{\ell}{2}} g_{j,k}\widetilde{x}^j\widetilde{\omega}^k. \tag{6.2.21}$$

Recall that $\Re r_{2,0} = \epsilon(0)/2$.

6.3 The Factor $hD_x + G(x, \omega; h)$

In this section we study $hD_x + G(x, \omega; h)$ and its inverse. We mainly concentrate on the region (6.2.13), where we add the restriction that $\Im\omega \ll h^\delta$, for $\delta > 0$ arbitrarily small and fixed; this assumption will be strengthened later. From (6.2.10), (6.2.11) we see that

$$\Im G''_{xx} \asymp 1, \tag{6.3.1}$$

$$\Im G'_x = 2\Re r_{2,0}x + (\Re r_{1,1})\omega + \mathcal{O}(|(x, \omega)|^2 + h), \tag{6.3.2}$$

for x in a small fixed neighborhood of 0. Hence $\Im G(\cdot, \omega)$ has a nondegenerate minimum at

$$x = x_{\min}(\omega) = \mathcal{O}(|\omega| + h), \tag{6.3.3}$$

with

$$\Im G(x_{\min}, \omega; h) = -\Im\omega + \mathcal{O}(|\omega|^2 + h). \tag{6.3.4}$$

We can extend the definition of G to $x \in \mathbf{R}$ in such a way that (6.2.10), (6.2.11), (6.3.1), (6.3.2) become valid for $x \in \mathbf{R}$ and so that $G = -\omega + ir_{2,0}x^2$ outside a small neighborhood of $x = 0$. We have

$$\Im G(x, \omega) - \Im G(x_{\min}(\omega), \omega) \asymp (x - x_{\min}(\omega))^2, \tag{6.3.5}$$

$$\Im \partial_x G(x, \omega) \asymp x - x_{\min}(\omega), \tag{6.3.6}$$

and we see that the equation

$$\Im G(x, \omega) = 0 \tag{6.3.7}$$

has exactly two solutions, $x = x_\pm(\omega)$, with

$$\mp (x_\pm(\omega) - x_{\min}(\omega)) \asymp (\Im\omega)^{\frac{1}{2}}. \tag{6.3.8}$$

Let $\xi = \xi_\pm(\omega) = \mathcal{O}(\Im\omega)$ be the solutions of $\xi + \Re G(x_\pm(\omega), \omega) = 0$, so that

$$\xi_\pm + G(x_\pm(\omega), \omega) = 0. \tag{6.3.9}$$

Apart from the fact that $\Im\omega \ll 1$, we are very much in the situation of Chap. 3. To remedy for the smallness of ω, we make the scalings (6.2.15)–(6.2.19) and work with

$$\widetilde{B} = hD_{\widetilde{x}} + \widetilde{G}(\widetilde{x}, \widetilde{\omega}, \alpha; h),$$

where we recall that,

$$\widetilde{G}(\widetilde{x}, \widetilde{\omega}, \alpha; \widetilde{h}) = \frac{1}{\alpha} G(\sqrt{\alpha}\widetilde{x}, \alpha\widetilde{\omega}; h) = \widetilde{g}(\widetilde{x}, \widetilde{\omega}) + \alpha^{\frac{1}{2}} \widetilde{h} \mathcal{O}(1).$$

Here $\mathcal{O}(1)$ stands for a function which is bounded with all its $\widetilde{x}$-derivatives. (The asymptotic expansion (6.2.18) will be used in a region $\widetilde{x} = \mathcal{O}(1)$.) Then

$$\Im \partial_{\widetilde{x}}^2 \widetilde{G} \asymp 1,$$

and $\Im\widetilde{G}(\cdot, \widetilde{\omega})$ has a nondegenerate minimum at $\widetilde{x}_{\min}(\widetilde{\omega}) = x_{\min}(\omega)/\sqrt{\alpha}$, so

$$\Im \partial_{\widetilde{x}} \widetilde{G} \asymp \widetilde{x} - \widetilde{x}_{\min}(\widetilde{\omega}).$$

Further, the equation

$$\Im\widetilde{G}(\widetilde{x}, \widetilde{\omega}) = 0, \tag{6.3.10}$$

has exactly two solutions $\widetilde{x} = \widetilde{x}_{\pm}(\widetilde{\omega}) = x_{\pm}(\omega)/\sqrt{\alpha}$, with

$$\mp(\widetilde{x}_{\pm}(\widetilde{\omega}) - \widetilde{x}_{\min}(\widetilde{\omega})) \asymp (\Im\widetilde{\omega})^{\frac{1}{2}} \asymp 1.$$

Let $\widetilde{\xi} = \widetilde{\xi}_{\pm}(\widetilde{\omega}, \alpha)$ be the solution of $\widetilde{\xi} + \Re\widetilde{G}(\widetilde{x}_{\pm}, \widetilde{\omega}, \alpha) = 0$, so that

$$\widetilde{\xi}_{\pm} + \widetilde{G}(\widetilde{x}_{\pm}, \widetilde{\omega}) = 0. \tag{6.3.11}$$

Notice that $\widetilde{\xi}_{\pm} = \xi_{\pm}/\alpha = \mathcal{O}(1)$.

Under the assumption

$$\widetilde{h} \ll 1, \text{ i.e., } \alpha \gg h^{\frac{2}{3}}, \tag{6.3.12}$$

we can now study $\widetilde{B} = \widetilde{h} D_{\widetilde{x}} + \widetilde{G}$ as in Sects. 3.1 and 3.2. Since we will work in the rescaled variables for a while, we will drop the tildes until further notice and simply recall their existence by adding $^{\sim}$ after each formula. As in Sect. 3.1, we have a function

$$e_{\mathrm{wkb}}(x) = h^{-\frac{1}{4}} a(h) \chi(x - x_{+}(\omega, \alpha)) e^{\frac{i}{h}\phi_{+}(x)}, \quad \sim \tag{6.3.13}$$

such that

$$\chi \in C_0^\infty(]-C, x_-(\omega,\alpha) - \frac{1}{C}[,\ \chi = 1 \text{ on }]-\frac{C}{2}, x_-(\omega,\alpha) - \frac{2}{C}[, \quad \sim \tag{6.3.14}$$

where $C \gg 1$ ~,

$$a \sim a_0(\omega,\alpha) + h a_1(\omega,\alpha) + \cdots, \quad a_0(\omega,\alpha) = \left(\frac{\Im \partial_x^2 \phi_+(x_+)}{\pi}\right)^{\frac{1}{4}}, \quad \sim \tag{6.3.15}$$

$$\|e_{\mathrm{wkb}}\|_{L^2} = 1, \quad \sim \tag{6.3.16}$$

$$\phi_+(x) = \phi_+(x,\omega,\alpha) = -\int_{x_+(\omega,\alpha)}^{x} G(y,\omega,\alpha)dy, \quad \sim \tag{6.3.17}$$

$$\Im\phi_+(x) \asymp (x - x_+)^2, \quad \sim \tag{6.3.18}$$

uniformly on any fixed compact subset of $]-\infty, x_-[$ ~. By construction,

$$(hD_x + G)e_{\mathrm{wkb}}(x) = \frac{h}{i}\chi'(x - x_+(\omega,\alpha))h^{-\frac{1}{4}}a(h)e^{\frac{i}{h}\phi_+(x)}, \quad \sim \tag{6.3.19}$$

which decays exponentially in L^2.

Define the (ω,α) ~-dependent self-adjoint operators

$$\Box = (hD + G)^*(hD + G), \quad \Box^\dagger = (hD + G)(hD + G)^* : L^2(\mathbf{R}) \to L^2(\mathbf{R}) \quad \sim$$

with domain

$$\mathcal{D}(\Box) = \mathcal{D}(\Box^\dagger) = \{u \in L^2(\mathbf{R});\ x^{2j}(hD)^k \in L^2(\mathbf{R}),\ \text{for } j + k \le 2\}. \quad \sim$$

These operators can be viewed as h ~-pseudodifferential operators with symbol in $S(m)$, $m(x,\xi) = \left(\langle\xi\rangle + \langle x\rangle^2\right)^2$.

Again,

$$\sigma(\Box) = \sigma(\Box^\dagger) = \{t_0^2,\ t_1^2, \ldots\}, \quad 0 \le t_j \nearrow +\infty \quad \sim$$

and Proposition 3.1.1 applics so that t_0^2 ~ is a simple eigenvalue which is exponentially small and $t_1^2 - t_0^2 \ge h/C$ ~. The second fact can be proved by modifying slightly the proof of Lemma 3.1.2, or else we can apply more general microlocal analysis as indicated after Proposition 5.3.1.

Let Π_0 ~ be the spectral projection corresponding to $(\Box, t_0^2)$ ~ and let $e_0 = \|\Pi_0 e_{\mathrm{wkb}}\|^{-1}\Pi_0 e_{\mathrm{wkb}}$ ~ be the normalized eigenstate with $(\Box - t_0^2)e_0 = 0$ ~ (unique up to a factor $e^{i\theta}$, for some $\theta \in \mathbf{R}$). By (6.3.19),

$$\Box e_{\rm wkb} = r, \quad r = (hD + G)^* \frac{h}{i} \chi'(x - x_+) h^{-\frac{1}{4}} a(h) e^{\frac{i}{h}\phi_+}, \tag{6.3.20}$$ ~

and we see that for any given $\epsilon > 0$, we can ensure that

$$r = \mathcal{O}\left(e^{-\frac{1}{h}(S_0 - \epsilon)}\right) \text{ in } L^2, \tag{6.3.21}$$ ~

by choosing the cutoff "wide enough", i.e., with C ~in (6.3.14) large enough. Here $S_0 > 0$ ~is the tunneling action between (x_+, ξ_+) ~and (x_-, ξ_-) ~, given by

$$S_0 = -\Im \int_{x_+}^{x_-} G(x, \omega, \alpha) dx, \tag{6.3.22}$$ ~

or equivalently by

$$S_0 = \Im \int_\gamma \xi dx, \tag{6.3.23}$$ ~

where $\gamma = \{(x, \xi);\ x_+ \leq x \leq x_-,\ \xi + G(x, \omega, \alpha) = 0\}$ ~is oriented from (x_+, ξ_+) ~to (x_-, ξ_-) ~.

For $|z| = h\delta$ ~, $0 < \delta \ll 1$ ~, we write

$$(z - \Box) e_{\rm wkb} = z e_{\rm wkb} - r,$$ ~

leading to

$$(z - \Box)^{-1} e_{\rm wkb} = z^{-1} e_{\rm wkb} + z^{-1} (z - \Box)^{-1} r.$$ ~

Using that

$$\Pi_0 = \frac{1}{2\pi i} \int_{|z| = \delta h} (z - \Box)^{-1} dz,$$ ~

we get

$$\Pi_0 e_{\rm wkb} = e_{\rm wkb} + \frac{1}{2\pi i} \int_{|z| = \delta h} z^{-1} (z - \Box)^{-1} r dz = e_{\rm wkb} + \mathcal{O}\left(\frac{1}{h} e^{-\frac{1}{h}(S_0 - \epsilon)}\right) \text{ in } L^2.$$ ~

Here we use (6.3.21) in the last step. It follows that

$$\|\Pi_0 e_{\rm wkb}\| = 1 + \mathcal{O}(h^{-1} e^{-\frac{1}{h}(S_0 - \epsilon)}),$$ ~

so

$$e_0 = e_{\mathrm{wkb}} + \mathcal{O}(h^{-1} e^{-\frac{1}{h}(S_0 - \epsilon)}) \text{ in } L^2. \tag{6.3.24}$$

Similarly, we have a WKB state for $(hD + G)^* = hD + \overline{G}$ given by

$$f_{\mathrm{wkb}}(x) = h^{-\frac{1}{4}} b(h) \widehat{\chi}(x - x_-(\omega, \alpha)) e^{\frac{i}{h}\phi_-(x)}, \tag{6.3.25}$$

$$\phi_-(x) = \phi_-(x, \omega, \alpha) = -\int_{x_-(\omega,\alpha)}^{x} \overline{G(y, \omega, \alpha)} dy, \tag{6.3.26}$$

$$\Im \phi_-(x) \asymp (x - x_-)^2, \tag{6.3.27}$$

uniformly on any fixed compact subset of $]x_+, \infty[$,

$$b \sim b_0(\omega, \alpha) + h b_1(\omega, \alpha) + \dots, \; b_0(\omega, \alpha) = \left(\frac{\Im \partial_x^2 \phi_-(x_-)}{\pi} \right)^{\frac{1}{4}}, \tag{6.3.28}$$

$$\| f_{\mathrm{wkb}} \|_{L^2} = 1, \tag{6.3.29}$$

$$(hD_x + \overline{G}) f_{\mathrm{wkb}}(x) = \frac{h}{i} \widehat{\chi}'(x - x_-(\omega, \alpha)) h^{-\frac{1}{4}} b(h) e^{\frac{i}{h}\phi_-(x)}, \tag{6.3.30}$$

which is exponentially decaying in L^2. The properties of $\widehat{\chi}(x - x_-)$ are analogous to those of $\chi(x - x_+)$.

As above, we get a corresponding normalized eigenstate f_0 of $\Box^\dagger$:

$$(\Box^\dagger - t_0^2) f_0 = 0, \tag{6.3.31}$$

with

$$f_0 = f_{\mathrm{wkb}} + \mathcal{O}(h^{-1} e^{-\frac{1}{h}(S_0 - \epsilon)}) \text{ in } L^2. \tag{6.3.32}$$

We have

$$(hD + G) e_0 = \alpha_0 f_0, \;\; (hD + \overline{G}) f_0 = \overline{\alpha}_0 e_0, \;\; |\alpha_0| = t_0. \tag{6.3.33}$$

We now go beyond Chap. 3 and study the precise asymptotics of t_0. Use (6.3.33) to write

$$\alpha_0 = ((hD + G) e_0 | f_0) = (\chi_+ (hD + G) e_0 | f_0) + (\chi_- (hD + G) e_0 | f_0),$$

where $\chi_\pm \in C^\infty(\mathbf{R})$, $\operatorname{supp} \chi_+ \subset]-\infty, x_-[$, $\operatorname{supp} \chi_- \subset]x_+, +\infty[$, $1 = \chi_+ + \chi_-$. We get

$$\begin{aligned}\alpha_0 &= (\chi_+(hD+G)e_0|f_0) + (\chi_- e_0|(hD+G)^* f_0) + ([\chi_-, hD]e_0|f_0)\\ &= \alpha_0\left((\chi_+ f_0|f_0) + (\chi_- e_0|e_0)\right) + ih(\chi_-' e_0|f_0).\end{aligned}$$

Here we use (6.3.24), (6.3.32) and the exponential decay of f_{wkb} and e_{wkb} away from x_- and x_+, respectively, to see that

$$(\chi_+ f_0|f_0),\ (\chi_- e_0|e_0) = \mathcal{O}\left(e^{-\frac{1}{Ch}}\right).$$

Hence,

$$\left(1 + \mathcal{O}\left(e^{-\frac{1}{Ch}}\right)\right)\alpha_0 = ih(\chi_-' e_0|f_0). \tag{6.3.34}$$

Using again (6.3.24), (6.3.32) and the fact that $\operatorname{supp} \chi_-'$ is contained in a compact subset of $]x_+, x_-[$, where e_{wkb} and f_{wkb} are exponentially decaying, we get

$$(\chi_-' e_0|f_0) = (\chi_-' e_{\mathrm{wkb}}|f_{\mathrm{wkb}}) + \mathcal{O}(1)e^{-\frac{1}{h}(S_0 + \frac{1}{C})}.$$

Then (6.3.34) shows that

$$\alpha_0 = ih(\chi_-' e_{\mathrm{wkb}}|f_{\mathrm{wkb}}) + \mathcal{O}(1)e^{-\frac{1}{h}(S_0 + \frac{1}{C})}. \tag{6.3.35}$$

Recall that $t_0 \geq 0$ by construction and that we have inserted an x-independent factor of modulus one in front of f_0 and f_{wkb}.

Combining (6.3.35), (6.3.25), (6.3.28), (6.3.13), (6.3.15), we get

$$\alpha_0 = (1 + \mathcal{O}(h))h^{\frac{1}{2}}\left(\frac{\Im \partial_x^2 \phi_+(x_+)}{\pi}\right)^{\frac{1}{4}}\left(\frac{\Im \partial_x^2 \phi_-(x_-)}{\pi}\right)^{\frac{1}{4}} e^{i\widetilde{\theta} - \frac{S_0}{h}},\ \widetilde{\theta} \in \mathbf{R}.$$

Here we also used that

$$e^{\frac{i}{h}(\phi_+(x) - \overline{\phi_-(x)})} = e^{\frac{i}{h}\int_{x_+}^{x_-}(-G(y))dy} = e^{i\widetilde{\theta} - \frac{1}{h}S_0},$$

where $\widetilde{\theta} \in \mathbf{R}$ is independent of x and the fact that $\int_{-\infty}^{+\infty} \chi_-'(y)dy = 1$. It follows that

$$t_0 = |\alpha_0| = (1 + \mathcal{O}(h))h^{\frac{1}{2}}\left(\frac{\Im \partial_x^2 \phi_+(x_+)}{\pi}\right)^{\frac{1}{4}}\left(\frac{\Im \partial_x^2 \phi_-(x_-)}{\pi}\right)^{\frac{1}{4}} e^{-\frac{S_0}{h}}. \tag{6.3.36}$$

From (6.3.17) and (6.3.26), we get

$$\Im\partial_x^2\phi_+(x_+) = -\Im\partial_x G(x_+), \ \sim$$
$$\Im\partial_x^2\phi_-(x_-) = -\Im\partial_x \overline{G}(x_-). \ \sim$$

Introducing $q = \xi + g(x, \omega, \alpha)$ ~, we get

$$\begin{aligned}\Im\partial_x^2\phi_+(x_+) &= \frac{1}{2i}\{q, \overline{q}\}(\rho_+) + \mathcal{O}(h) \asymp 1, \ \sim\\ \Im\partial_x^2\phi_-(x_-) &= \frac{1}{2i}\{\overline{q}, q\}(\rho_-) + \mathcal{O}(h) \asymp 1, \ \sim\end{aligned} \tag{6.3.37}$$

where $\rho_\pm = (x_\pm, \xi_\pm)$ ~. Then (6.3.36) becomes

$$t_0 = (1 + \mathcal{O}(h))\left(\frac{h}{\pi}\right)^{\frac{1}{2}}\left(\frac{1}{2i}\{q, \overline{q}\}(\rho_+)\right)^{\frac{1}{4}}\left(\frac{1}{2i}\{\overline{q}, q\}(\rho_-)\right)^{\frac{1}{4}} e^{-\frac{1}{h}S_0}. \ \sim \tag{6.3.38}$$

After replacing f_0, f_{wkb} by $e^{i\theta} f_0$, $e^{i\theta} f_{\mathrm{wkb}}$ for a suitable real θ, we may assume that α_0 in (6.3.33) is equal to t_0.

Proposition 3.2.1 carries over to the present situation, where we write out explicitly the various tildes.

Proposition 6.3.1 *Let $m = \langle\xi\rangle + \langle x\rangle^2$ and define the semi-classical Sobolev space $H_{\widetilde{h}}(m)$ as in Sect. 5.1, with $\widetilde{h}$ as the semi-classical parameter. Define $\widetilde{R}_+ : H^1_{\widetilde{h}}(\mathbf{R}) \to \mathbf{C}$, $\widetilde{R}_- : \mathbf{C} \to L^2(\mathbf{R})$ by*

$$\widetilde{R}_+ u = (u|\widetilde{e}_0), \quad \widetilde{R}_- u_- = u_-\widetilde{f}_0.$$

Then

$$\widetilde{\mathcal{P}}(z) := \begin{pmatrix}\widetilde{h}D + \widetilde{G} & \widetilde{R}_-\\ \widetilde{R}_+ & 0\end{pmatrix} : H_{\widetilde{h}}(m) \times \mathbf{C} \to L^2 \times \mathbf{C}$$

is bijective with the bounded inverse

$$\widetilde{\mathcal{E}}(z) = \begin{pmatrix}\widetilde{E} & \widetilde{E}_+\\ \widetilde{E}_- & \widetilde{E}_{-+}\end{pmatrix} = \begin{pmatrix}\mathcal{O}(\widetilde{h}^{-\frac{1}{2}}) & \mathcal{O}(1)\\ \mathcal{O}(1) & \mathcal{O}(e^{-1/C\widetilde{h}})\end{pmatrix} : L^2 \times \mathbf{C} \to H_{\widetilde{h}}(m) \times \mathbf{C}.$$

Here,

$$\widetilde{E}_+ v_+ = v_+\widetilde{e}_0, \ \widetilde{E}_- v = (v|\widetilde{f}_0), \ \widetilde{E}_{-+} = -\widetilde{t}_0.$$

We turn to the corresponding result for the unscaled operator $hD+G$ in (6.2.12). Recall that

$$hD_x + G = \alpha(\widetilde{h}D_{\widetilde{x}} + \widetilde{G}), \quad \widetilde{h} = h/\alpha^{3/2}, \; x = \sqrt{\alpha}\widetilde{x}. \tag{6.3.39}$$

Thus the common lowest eigenvalue of $\Box = (hD + G)^*(hD + G)$ and $\Box^\dagger = (hD + G)(hD+G)^*$ is of the form $t_0^2 = \alpha^2\widetilde{t}_0^2$, where $t_0 = \alpha\widetilde{t}_0$, and the corresponding normalized eigenfunctions are

$$e_0(x) = \alpha^{-\frac{1}{4}}\widetilde{e}_0(\widetilde{x}), \quad f_0(x) = \alpha^{-\frac{1}{4}}\widetilde{f}_0(\widetilde{x}). \tag{6.3.40}$$

They are approximated by

$$e_{\mathrm{wkb}}(x) = \alpha^{-\frac{1}{4}}\widetilde{e}_{\mathrm{wkb}}(\widetilde{x}), \quad f_{\mathrm{wkb}}(x) = \alpha^{-\frac{1}{4}}\widetilde{f}_{\mathrm{wkb}}(\widetilde{x}). \tag{6.3.41}$$

The common gap $t_1^2 - t_0^2$ between the first two eigenvalues of $\Box$, $\Box^\dagger$ is $\asymp \alpha^2\widetilde{h} = \sqrt{\alpha}h$. Recalling the proof of Proposition 3.2.1 by spectral decomposition, we get

Proposition 6.3.2 *With m as in the preceding proposition, we define the semi-classical Sobolev space $H(m)$ as there, with h as the semi-classical parameter. Define $R_+ : H^1(\mathbf{R}) \to \mathbf{C}$, $R_- : \mathbf{C} \to L^2(\mathbf{R})$ by*

$$R_+u = (u|e_0), \quad R_-u_- = u_-f_0.$$

Then

$$\mathcal{P}(z) := \begin{pmatrix} hD + G & R_- \\ R_+ & 0 \end{pmatrix} : H_h(m) \times \mathbf{C} \to L^2 \times \mathbf{C}$$

is bijective with the bounded inverse

$$\begin{aligned}\mathcal{E}(z) &= \begin{pmatrix} E & E_+ \\ E_- & E_{-+} \end{pmatrix} \\ &= \begin{pmatrix} \mathcal{O}(\alpha^{-\frac{1}{4}}h^{-\frac{1}{2}}) & \mathcal{O}(1) \\ \mathcal{O}(1) & \mathcal{O}(\alpha^{\frac{1}{4}}h^{\frac{1}{2}})e^{-\frac{\alpha^{3/2}}{Ch}} \end{pmatrix} : L^2 \times \mathbf{C} \to H_h(m) \times \mathbf{C}.\end{aligned}$$

Here,

$$E_+v_+ = v_+e_0, \; E_-v = (v|f_0), \; E_{-+} = -t_0.$$

Let us finally derive an asymptotic formula for t_0 with the help of (6.3.38), where all the quantities carry invisible tildes. Now with the tildes written out, we recall that $\widetilde{x}_\pm$ are given by

$$\Im \widetilde{G}(\widetilde{x}_\pm) = 0, \quad \widetilde{x}_+ < \widetilde{x}_-.$$

This corresponds to $x_\pm = \sqrt{\alpha}\widetilde{x}_\pm$ satisfying $\Im G(x_\pm) = 0$. We get,

$$\frac{\widetilde{S}_0}{\widetilde{h}} = -\frac{\alpha^{\frac{3}{2}}}{h}\int_{\widetilde{x}_+}^{\widetilde{x}_-} \Im \widetilde{G}(\widetilde{y})d\widetilde{y} = -\frac{\alpha^{\frac{3}{2}}}{h}\int_{x_+}^{x_-} \frac{\Im G(y)}{\alpha}\frac{dy}{\alpha^{\frac{1}{2}}} = \frac{S_0}{h},$$

where,

$$S_0 = -\Im \int_{x_+}^{x_-} G(y)dy. \tag{6.3.42}$$

Recalling that $\widetilde{\phi}_+(\widetilde{x}) = -\int_{\widetilde{x}_+}^{\widetilde{x}} \widetilde{G}(\widetilde{y})d\widetilde{y}$, we put $\phi_+(x) = -\int_{x_+}^{x} G(y)dy$. Then by the change of variables above, we get $\widetilde{\phi}_+ = \alpha^{-3/2}\phi_+(x)$, or equivalently,

$$\frac{\widetilde{\phi}_+}{\widetilde{h}} = \frac{\phi_+}{h}.$$

Similarly,

$$\frac{\widetilde{\phi}_-}{\widetilde{h}} = \frac{\phi_-}{h},$$

where (cf. (6.3.26)), $\phi_-(x) = -\int_{x_-}^{x} \overline{G(y)}dy$. Clearly,

$$\partial_{\widetilde{x}}^2\widetilde{\phi}_\pm = \alpha\partial_x^2\widetilde{\phi}_\pm = \alpha^{-\frac{1}{2}}\partial_x^2\phi_\pm.$$

Let $q = \xi + g$, $\widetilde{q} = \widetilde{\xi} + \widetilde{g}$, so

$$\{\widetilde{q}, \overline{\widetilde{q}}\}(\widetilde{\rho}_\pm) = \alpha^{-\frac{1}{2}}\{q, \overline{q}\}(\rho_\pm).$$

We use this in (6.3.38) and get

$$t_0 = \alpha\widetilde{t}_0 =$$

$$\alpha\left(1 + \mathcal{O}\left(\frac{h}{\alpha^{\frac{3}{2}}}\right)\right)\left(\frac{h}{\pi\alpha^{\frac{3}{2}}}\right)^{\frac{1}{2}}\left(\alpha^{-\frac{1}{2}}\frac{1}{2i}\{q, \overline{q}\}(\rho_+)\right)^{\frac{1}{4}}\left(\alpha^{-\frac{1}{2}}\frac{1}{2i}\{\overline{q}, q\}(\rho_-)\right)^{\frac{1}{4}} e^{-\frac{1}{h}S_0}.$$

We already know that the precise choice of $\alpha \asymp \Im\omega$ should not appear in the final result, and indeed the powers of α cancel out and we end up with

$$t_0 = \left(1+\mathcal{O}\left(h/(\Im\omega)^{\frac{3}{2}}\right)\right)\left(\frac{h}{\pi}\right)^{\frac{1}{2}}\left(\frac{1}{2i}\{q,\overline{q}\}(\rho_+)\right)^{\frac{1}{4}}\left(\frac{1}{2i}\{\overline{q},q\}(\rho_-)\right)^{\frac{1}{4}}e^{-\frac{1}{h}S_0}.$$

Here $\rho_\pm$ and S_0 are determined from the full symbol $Q = \xi + G(x,\omega;h) = q + \mathcal{O}(h)$. There are unique (real) points $\rho_\pm^0 = \rho_\pm + \mathcal{O}(h/\sqrt{\alpha})$ such that $q(\rho_\pm^0) = 0$, and moreover

$$\{q,\overline{q}\}(\rho_\pm) = \{q,\overline{q}\}(\rho_\pm^0) + \mathcal{O}(h/\sqrt{\alpha}) = (1+\mathcal{O}(h/\alpha))\{q,\overline{q}\}(\rho_\pm^0). \tag{6.3.43}$$

Similarly, we can compare S_0 with

$$I = -\int_{x_+^0}^{x_-^0} \Im g(x,\omega)dx, \quad \rho_\pm^0 = (x_\pm^0, \xi_\pm^0). \tag{6.3.44}$$

Since

$$x_\pm^0 - x_\pm = \mathcal{O}(h/\sqrt{\alpha}), \quad |x_\pm^0| \asymp \sqrt{\alpha}, \quad G - g = \mathcal{O}(h),$$

we get

$$I - S_0 = \mathcal{O}(1)\left(\frac{h\alpha}{\sqrt{\alpha}} + h\sqrt{\alpha}\right) = \mathcal{O}(h\sqrt{\alpha}).$$

Here we also used that $G = \mathcal{O}(\alpha)$ when $x = \mathcal{O}(\sqrt{\alpha})$. Hence,

$$e^{-S_0/h} = (1+\mathcal{O}(\sqrt{\alpha}))e^{-I/h} \tag{6.3.45}$$

and here $\sqrt{\alpha} \le h/\alpha^{3/2}$ if we work with α in the range

$$C_0 h^{2/3} \le \alpha \le h^{1/2}, \tag{6.3.46}$$

which will be the case when in addition to (6.2.13), we have $\Im w \simeq \alpha \le \mathcal{O}(1)(h\ln(1/h))^{2/3}$. Using (6.3.43) and (6.3.45) in the formula for t_0 above, we get

$$t_0 = \left(1+\mathcal{O}\left(h/(\Im\omega)^{\frac{3}{2}}\right)\right)\left(\frac{h}{\pi}\right)^{\frac{1}{2}}\left(\frac{1}{2i}\{q,\overline{q}\}(\rho_+)\right)^{\frac{1}{4}}\left(\frac{1}{2i}\{\overline{q},q\}(\rho_-)\right)^{\frac{1}{4}}e^{-\frac{1}{h}I}, \tag{6.3.47}$$

where from now on $\rho_\pm$ denote the real zeros of $q = \xi + g(x,\omega)$.

As we have already seen,

$$t_0 \asymp h^{\frac{1}{2}}\alpha^{\frac{1}{4}}e^{-I/h} \ll h^{\frac{1}{2}}, \tag{6.3.48}$$

$$I \asymp \alpha^{\frac{3}{2}}. \tag{6.3.49}$$

So far in this section, we have worked under the assumption (6.2.13) in conjunction with $\Im w \leq \mathcal{O}(1)(h\ln(1/h))^{2/3}$. Let us consider two more regions:

1. Let

$$|\omega| \leq \mathcal{O}(h^{2/3}). \tag{6.3.50}$$

We check that $hD_x + G : H_h(m) \to L^2$ has the two-sided inverse

$$Ev(x) = \frac{i}{h}\int_{+\infty}^{x} e^{\frac{i}{h}(\phi(x)-\phi(y))}v(y)dy, \tag{6.3.51}$$

where ϕ is determined up to a constant by the eikonal equation

$$\phi'_x + G(x,\omega;h) = 0. \tag{6.3.52}$$

We have

$$|-\Im\phi'_x| \begin{cases} = \mathcal{O}(h^{2/3}), & \text{if } |x| \leq \mathcal{O}(h^{1/3}), \\ \asymp h^{2/3}+x^2, & \text{if } |x| \gg h^{1/3}. \end{cases}$$

It follows that for $y \geq x$,

$$\begin{aligned} \frac{1}{h}(\Im\phi(x) - \Im\phi(y)) &\geq -\mathcal{O}(1) + \frac{1}{Ch}\int_x^y (h^{\frac{2}{3}} + t^2)dt \\ &= -\mathcal{O}(1) + \frac{1}{Ch}\left(h^{\frac{2}{3}}(y-x) + \frac{y^3}{3} - \frac{x^3}{3}\right) \\ &\geq -\mathcal{O}(1) + \frac{1}{\widetilde{C}h}(y-x)\left(h^{\frac{2}{3}} + x^2 + y^2\right), \end{aligned}$$

for some $C, \widetilde{C} > 0$. Thus in (6.3.51)

$$\left|e^{\frac{i}{h}(\phi(x)-\phi(y))}\right| \leq \mathcal{O}(1)e^{-\frac{1}{\widetilde{C}h}(h^{\frac{2}{3}}+|x|^2+|y|^2)(y-x)}, \tag{6.3.53}$$

and applying Schur's lemma to E we can conclude that

$$(h^{2/3} + x^2)E = \mathcal{O}(1) : L^2 \to L^2. \tag{6.3.54}$$

Proof of (6.3.54) The distribution kernel of $(h^{2/3} + x^2)E$ is bounded in modulus by a constant times

$$K(x, y; h) = \frac{h^{2/3} + x^2}{h} e^{-\frac{1}{Ch}(h^{2/3}+x^2+y^2)(y-x)} 1_{y \geq x},$$

and if $K(h)$ denotes the corresponding integral operator, it suffices to show that the $\mathcal{L}(L^2, L^2)$-norm $\|K(h)\|$ is $\mathcal{O}(1)$, uniformly in h. We observe that

$$K(h^{1/3}\widetilde{x}, h^{1/3}\widetilde{y}; h)h^{1/3}d\widetilde{y} = K(\widetilde{x}, \widetilde{y}; 1)d\widetilde{y},$$

so $\|K(h)\| = \|K(1)\|$. It then suffices to consider the case when $h = 1$. By Schur's lemma, $\|K(1)\| \leq m_x^{1/2} m_y^{1/2}$, where

$$m_x = \sup_x \int K(x, y; 1)dy, \quad m_y = \sup_y \int K(x, y; 1)dx,$$

and it suffices to show that these quantities are bounded.

Here

$$m_x \leq \sup_x \int_x^{+\infty} (1 + x^2)e^{-\frac{1}{C}(1+x^2)(y-x)}dy \leq \mathcal{O}(1).$$

The estimate of m_y is slightly more difficult. Write,

$$m_y = \sup_y \int_{-\infty}^{y} (1 + x^2)e^{-\frac{1}{C}(1+x^2+y^2)(y-x)}dx \leq \mathrm{I} + \mathrm{II},$$

where

$$\mathrm{I} = \sup_y \int_{-\infty}^{y} (1 + y^2)e^{-\frac{1}{C}(1+x^2+y^2)(y-x)}dx \leq \mathcal{O}(1),$$

as for m_x, and

$$\begin{aligned}\mathrm{II} &= \sup_y \int_{-\infty}^{x} |x^2 - y^2|e^{-\frac{1}{C}(1+x^2+y^2)(y-x)}dx \\ &= \sup_y \int_{-\infty}^{x} |x + y|(x - y)e^{-\frac{1}{C}(1+x^2+y^2)(y-x)}dx.\end{aligned}$$

Since $|x+y| \le \mathcal{O}(1)(1+x^2+y^2)$, the integrand in II is

$$\le |x+y|(x-y)e^{-\frac{1}{2C}(1+x^2+y^2)(y-x)}e^{-\frac{1}{2C}(y-x)} \le \mathcal{O}(1)e^{-\frac{1}{2C}(y-x)},$$

and it follows that II $= \mathcal{O}(1)$. Hence $m_y = \mathcal{O}(1)$. □

From $(hD+G)E = 1$, we see that $hDE = \mathcal{O}(1) : L^2 \to L^2$. It follows that

$$E = \mathcal{O}(h^{-2/3}) : L^2 \to H_h(m). \tag{6.3.55}$$

2. When $\Im\omega \ll -h^{2/3}$, $|\Re\omega| \le \mathcal{O}(1)|\Im\omega|$, we do essentially the same scaling as above, $x = \alpha^{1/2}\widetilde{x}$, $\alpha \asymp -\Im\omega$, to see that

$$hD + G = \alpha(\widetilde{h}D_{\widetilde{x}} + \widetilde{G}(\widetilde{x},\omega;h)),\quad \widetilde{G}(\widetilde{x},\omega;h) = \frac{1}{\alpha}G(\sqrt{\alpha}x,\omega;h),$$

which is an elliptic $\widetilde{h}$-pseudodifferential operator. We conclude by elliptic theory that $\widetilde{h}D_{\widetilde{x}} + \widetilde{G} : H_{\widetilde{h}}(m) \to L^2$ has a bounded inverse and, after returning to the x-coordinates, that

$$hD + G : H_h(m) \to L^2$$

is bijective with inverse E satisfying,

$$E = \mathcal{O}(\alpha^{-1}) : L^2 \to H_h(m). \tag{6.3.56}$$

6.4 Global Grushin Problem, End of the Proof

As a preparation for the cutting and pasting in the global situation, we establish some microlocal properties for E, $E_\pm$ in Proposition 6.3.2. This will mainly concern the eigenfunctions e_0, f_0. Assume (cf. (6.3.46)) that

$$|\Re\omega| \le \Im\omega,\quad h^{\frac{2}{3}} \ll \Im\omega \le h^{\frac{1}{2}}.$$

The symbol $\xi + G(x,\omega;h)$ belongs to $S(m)$, $m = \langle\xi\rangle + \langle x\rangle^2$. For $0 < \delta \ll 1$, $|x| \gg h^{\delta/2}$, we have $\Im G \ge h^\delta + |x|^2/\mathcal{O}(1) \ge h^\delta\langle x\rangle^2/\mathcal{O}(1)$ and hence $|\xi + G| \ge h^\delta m(x,\xi)/\mathcal{O}(1)$ for all ξ. The same conclusion is valid when $|\xi| \gg h^\delta$. Thus,

$$|\xi + G| \ge h^\delta m(x,\xi),\ \text{when } |\xi| + |x|^2 \gg h^\delta. \tag{6.4.1}$$

Hence $\xi + G(x)$ is a slightly degenerate elliptic symbol in the region $|\xi| + |x|^2 \gg h^\delta$, and following the standard construction of parametrices for elliptic operators (cf. [41], Ch. 8) we get a symbol $J(x, \xi; h)$ (depending also on ω) such that

$$\partial_{x,\xi}^\alpha J = \mathcal{O}_\alpha(1) h^{-\delta(1+|\alpha|)} m(x, \xi)^{-1}, \ (x, \xi) \in \mathbf{R}^2, \tag{6.4.2}$$

$$(\xi + G)\sharp J, \ J\sharp(\xi + G) \ \sim 1 \text{ in } S(1), \ \text{when } |\xi| + |x|^2 \gg h^\delta. \tag{6.4.3}$$

Here $\sharp$ denotes the composition of symbols in the semi-classical Weyl calculus [41], corresponding to composition of h-pseudodifferential operators.

Letting J also denote the corresponding h-pseudodifferential operators, this means that

$$(hD + G) \circ J = 1 + K, \quad J \circ (hD + G) = 1 + L, \tag{6.4.4}$$

where K and L are h-pseudodifferential operators with symbols $K(x, \xi; h)$ and $L(x, \xi; h)$ that satisfy

$$\partial_{(x,\xi)}^\alpha K, \ \partial_{(x,\xi)}^\alpha L = \begin{cases} \mathcal{O}(h^{-\delta|\alpha|}), \text{ on } \mathbf{R}^2, \\ \mathcal{O}(h^\infty), \text{ for } |\xi| + |x|^2 \gg h^\delta. \end{cases} \tag{6.4.5}$$

Recall that $\Box = (hD + G)^*(hD + G)$, $\Box^\dagger = (hD + G)(hD + G)^*$, and use the same letters for the symbols of these operators. Both symbols are $\equiv |\xi + G|^2 \bmod S(hm^2)$. By (6.3.48), we have $t_0^2 \ll h$. We conclude that

$$|\Box(x, \xi, \omega; h) - t_0^2|, \ |\Box^\dagger(x, \xi, \omega; h) - t_0^2| \geq h^{2\delta} m(x, \xi)^2, \ \text{when } |\xi| + |x|^2 \gg h^\delta, \tag{6.4.6}$$

provided that $\delta > 0$ is small enough ($< 1/2$ suffices here). If $\delta < 1/4$, we can construct symbols $R, \widetilde{R}$ such that

$$\partial_{(x,\xi)}^\alpha R, \ \partial_{(x,\xi)}^\alpha \widetilde{R} = \mathcal{O}_\alpha(h^{-2\delta(1+|\alpha|)} m(x, \xi)^{-2}), \ (x, \xi) \in \mathbf{R}^2, \tag{6.4.7}$$

$$\begin{aligned} &(\Box - t_0^2)\sharp R, \ R\sharp(\Box - t_0^2), \ (\Box^\dagger - t_0^2)\sharp\widetilde{R}, \ \widetilde{R}\sharp(\Box^\dagger - t_0^2) \ \sim 1 \\ &\text{in } S(1), \ \text{in the region } |\xi| + |x|^2 \gg h^\delta. \end{aligned} \tag{6.4.8}$$

Passing to the corresponding h-pseudodifferential operators, this means that

$$\begin{aligned} (\Box - t_0^2) \circ R = 1 + K, \quad R \circ (\Box - t_0^2) = 1 + L, \\ (\Box^\dagger - t_0^2) \circ \widetilde{R} = 1 + \widetilde{K}, \quad \widetilde{R} \circ (\Box^\dagger - t_0^2) = 1 + \widetilde{L}, \end{aligned} \tag{6.4.9}$$

where $K,\ L,\ \widetilde{K},\ \widetilde{L}$ are h-pseudodifferential operators with symbols satisfying

$$\partial^{\alpha}_{(x,\xi)}K,\ \partial^{\alpha}_{(x,\xi)}\widetilde{K},\ \partial^{\alpha}_{(x,\xi)}L,\ \partial^{\alpha}_{(x,\xi)}\widetilde{L} = \begin{cases} \mathcal{O}(h^{-2\delta(1+|\alpha|)}), & \text{on } \mathbf{R}^2, \\ \mathcal{O}(h^{\infty}), & \text{for } |\xi|+|x|^2 \gg h^{\delta}. \end{cases} \tag{6.4.10}$$

Applying the 2nd and the 4th equations in (6.4.9) to e_0 and f_0, respectively, we get

$$e_0 = -Le_0,\quad f_0 = -\widetilde{L}f_0, \tag{6.4.11}$$

showing that $e_0,\ f_0$ are microlocally concentrated in a region $|\xi|+|x|^2 \le \mathcal{O}(h^{\delta})$. This gives a corresponding localization for $R_{\pm} = E_{\mp}$ in Proposition 6.3.2:

$$E_+ = -LE_+,\quad E_- = -E_-\widetilde{L}^*. \tag{6.4.12}$$

Let $\chi \in C_0^{\infty}(\mathbf{R})$ be a standard cutoff function, equal to 1 on $[-1/2, 1/2]$ and with support in $]-1, 1[$. We claim that

$$E = \chi_2 E\chi_1 + J\left((1-\chi_2\chi_1) - [hD+G, \chi_2]E\chi_1\right) + \mathcal{O}(h^{\infty}) \text{ in } \mathcal{L}(L^2, H_h(m)), \tag{6.4.13}$$

where J is the h-pseudodifferential operator in (6.4.4) and

$$\chi_j(x,\xi) = \chi\left((\xi^2+x^4)/(C_j h^{2\delta})\right),\quad C_1, C_2 \gg 1.$$

(In (6.4.13), χ_j denote the corresponding h-pseudodifferential operators.) In fact, let

$$F = \chi_2 E\chi_1 + J((1-\chi_2\chi_1) - [hD+G, \chi_2]E\chi_1).$$

Then, since $(hD+G)E + R_-E_- = 1$ and $K(1-\chi_2\chi_1),\ K[hD+G, \chi_2],\ R_-E_- - \chi_2 R_-E_-\chi_1$ are $= \mathcal{O}(h^{\infty})$, we have

$$(hD+G)F = 1 - R_-E_- + \mathcal{O}(h^{\infty}) \text{ in } \mathcal{L}(L^2, L^2),$$

so

$$(hD+G)F + R_-E_- = 1 + \mathcal{O}(h^{\infty}) \text{ in } \mathcal{L}(L^2, L^2). \tag{6.4.14}$$

Moreover,

$$R_+F = \mathcal{O}(h^{\infty}) \text{ in } \mathcal{L}(L^2, \mathbf{C}). \tag{6.4.15}$$

Apply E to the left in (6.4.14) and use that $E(hD+G) = 1 - E_+R_+$, $ER_- = 0$. Then

$$E = (1 - E_+R_+)F + \mathcal{O}(h^\infty) \text{ in } \mathcal{L}(L^2, H_h(m)).$$

From (6.4.15) we then infer that $E = F + \mathcal{O}(h^\infty)$ and (6.4.13) follows.

Let $\widetilde{\chi}_j$ be narrower cutoffs with the same properties as χ_j so that $\widetilde{\chi}_j \sharp \chi_j \equiv \chi_j \sharp \widetilde{\chi}_j \equiv \widetilde{\chi}_j$ modulo $\mathcal{O}(h^\infty)$ in $S(1)$ and so that (6.4.13) remains valid with χ_j replaced by $\widetilde{\chi}_j$:

$$E = \widetilde{\chi}_2 E \widetilde{\chi}_1 + J\left((1 - \widetilde{\chi}_2\widetilde{\chi}_1) - [hD+G, \widetilde{\chi}_2]E\widetilde{\chi}_1\right) + \mathcal{O}(h^\infty) \text{ in } \mathcal{L}(L^2, H_h(m)).$$

Then modulo $\mathcal{O}(h^\infty)$ in $\mathcal{L}(L^2, H_h(m))$,

$$\begin{aligned}
&[hD+G, \chi_2]E\chi_1 \\
&\equiv [hD+G, \chi_2]J(1 - \widetilde{\chi}_2\widetilde{\chi}_1)\chi_1 - [hD+G, \chi_2]J[hD+G, \widetilde{\chi}_2]E\widetilde{\chi}_1\chi_1 \\
&\equiv [hD+G, \chi_2]J(1 - \widetilde{\chi}_2\widetilde{\chi}_1)\chi_1 \\
&\equiv [hD+G, \chi_2]J\chi_1,
\end{aligned}$$

and (6.4.13) simplifies to

$$E = \chi_2 E \chi_1 + J\left((1 - \chi_2\chi_1) - [hD+G, \chi_2]J\chi_1\right) + \mathcal{O}(h^\infty) \text{ in } \mathcal{L}(L^2, H_h(m)). \tag{6.4.16}$$

We can now set up the global Grushin problem for $P - z$ in Theorem 6.1.4. Recall that $p^{-1}(z_0)$ consists of N points $\rho_0^1, \dots, \rho_0^N$ by (6.1.37). For $z \in \text{neigh}(z_0, \mathbf{C})$, we introduce the new base points $\rho_j(z)$ as in Remark 6.2.1, so that (6.1.38) reduces to $w_j = is_j$. Let κ_j (depending on z) be an affine canonical transformation, mapping $(0, 0)$ to $\rho_j(z)$, such that as in the discussion around Remark 6.2.1,

$$p \circ \kappa_j - p(\rho_j(z)) = e^{i\theta_j(\rho_j(z))}(\xi + ir^j(x, \xi)),$$

or equivalently,

$$p \circ \kappa_j - z = e^{i\theta_j(\rho_j(z))}(\xi + ir^j(x, \xi) - \omega_j),$$

where $r^j(x, \xi) = \mathcal{O}(x^2 + \xi^2)$, $\omega_j = is_j + \mathcal{O}(s_j^2)$

On the operator level, this means that

$$\begin{aligned}
U_j^{-1}(P - z)U_j &= e^{i\theta_j(\rho_j(z))}(P_j - \omega_j) \\
&= \widetilde{A}_j\left(hD_x + G_j(x, \omega_j; h)\right) + S_j, \qquad (6.4.17)\\
\widetilde{A}_j &= e^{i\theta_j(\rho_j(z))}A_j,
\end{aligned}$$

as in (6.2.12) and where S_j has the properties of "S", explained after that relation. Here U_j is a unitary metaplectic Fourier integral operator associated to κ_j (see the notes at the end of this chapter).

Most of the work in this section concerns the case when $j \in \mathcal{N}$, where we recall that

$$\mathcal{N} = \{j \in \{1, \dots, N\};\ C_0 h^{2/3} \le s_j \le \mathcal{O}(1)\,(h \ln(1/h))^{2/3}\}.$$

For each j, we consider the Grushin problem in Proposition 6.3.2:

$$\mathcal{P}_j = \begin{pmatrix} hD + G_j & R_-^j \\ R_+^j & 0 \end{pmatrix}$$

with inverse

$$\mathcal{E}_j = \begin{pmatrix} E^j & E_+^j \\ E_-^j & E_{-+}^j \end{pmatrix},$$

where $R_+^j u = (u|e_0^j)$, $R_-^j u_- = u_- f_0^j$. Recall that $E_\pm^j = (R_\pm^j)^*$ and that $E_{-+}^j = -t_j$, where t_j fulfills the j-dependent version of (6.3.47).

Let $\psi \in C_0^\infty(\mathbf{R}^2)$ be a standard cutoff function, equal to 1 near $(0,0)$. For $0 < \delta_0 \ll 1$ fixed, we put $\psi_j(x, \xi; h) = \psi(h^{-\delta_0}((x,\xi) - \rho_j))$ and also write ψ_j for the corresponding h-pseudodifferential operator. Our global Grushin problem is then

$$\mathcal{P}(z) = \begin{pmatrix} P - z & R_- \\ R_+ & 0 \end{pmatrix} : H_h(m) \times \mathbf{C}^{\mathcal{N}} \to L^2 \times \mathbf{C}^{\mathcal{N}}, \tag{6.4.18}$$

where m now denotes the original order function, associated to P and

$$(R_+ u)(j) = R_+^j U_j^{-1} u,\ j \in \mathcal{N}, \tag{6.4.19}$$

$$R_- u_- = \sum_{j \in \mathcal{N}} U_j \widetilde{A}_j R_-^j u_-(j). \tag{6.4.20}$$

The localization properties of e_0^j, f_0^j imply that R_- is well defined mod $\mathcal{O}(h^\infty)$ despite the presence of $\widetilde{A}_j$, which is only defined microlocally near $(0,0)$. It is independent of the choice of ψ_j up to operators that are $\mathcal{O}(h^\infty)$.

Let $J(x, \xi; h)$ be a parametrix of $P - z$ in the region,

$$\operatorname{dist}((x,\xi), \{\gamma_1, \dots, \gamma_N\}) \ge h^{\widetilde{\delta}}, \quad \gamma_j = \gamma_j(\tau_j(z)), \tag{6.4.21}$$

where $\delta_0 < \widetilde{\delta} \ll 1$, such that

$$\partial^{\alpha}_{(x,\xi)} J = \mathcal{O}(h^{-\widetilde{\delta}(1+|\alpha|)} m(x,\xi)^{-1}), \tag{6.4.22}$$

$$(P-z)\sharp J,\ J\sharp(P-z) \sim 1 \text{ in the region (6.4.21)}. \tag{6.4.23}$$

As usual, J will also denote the corresponding h-pseudodifferential operator, depending on the context.

To construct a right inverse of $\mathcal{P}(z)$ amounts to finding a solution $(u, u_-) \in H(m)\times\mathbf{C}^N$ of the system

$$\begin{cases} (P-z)u + R_- u_- = v, \\ R_+ u = v_+, \end{cases} \tag{6.4.24}$$

for any given $(v, v_+) \in L^2 \times \mathbf{C}^N$. Take first $u_0 = J(1 - \sum_1^N \psi_j)v$, so that

$$\begin{cases} (P-z)u_0 = (1 - \sum_1^N \psi_j)v + \mathcal{O}(h^\infty \|v\|) \text{ in } L^2, \\ R_+ u_0 = \mathcal{O}(h^\infty \|v\|) \text{ in } \mathbf{C}^N. \end{cases}$$

When $j \in \mathcal{N}$, we look for a solution u_j microlocally concentrated to a small neighborhood of ρ_j, and $u_-(j) \in \mathbf{C}$, so that

$$\begin{cases} (P-z)u_j + R_-^j u_-(j) = \psi_j v, \\ (R_+ u_j) = v_+(j)\delta_j, \end{cases}$$

up to small errors, where we let δ_j denote the jth canonical basis vector in $\mathbf{C}^N$, so that $\delta_j(k) = \delta_{j,k}$, the latter being the Kronecker delta. The concentration of u_j to a small neighborhood of ρ_j will then imply that $(R_+ u_j)(k) = \mathcal{O}(h^\infty)$ for $k \neq j$ and we try to solve

$$\begin{cases} U_j \widetilde{A}_j (hD + G_j) U_j^{-1} u_j + U_j \widetilde{A}_j R_-^j u_-(j) = \psi_j v, \\ R_+^j U_j^{-1} u_j = v_+(j), \end{cases}$$

which formally would follow from

$$\begin{cases} (hD + G_j) U_j^{-1} u_j + R_-^j u_-(j) = \widetilde{A}_j^{-1} U_j^{-1} \psi_j v, \\ R_+^j U_j^{-1} u_j = v_+(j), \end{cases}$$

so we are led to the choice

$$\begin{cases} u_j = U_j E^j \widetilde{A}_j^{-1} U_j^{-1} \psi_j v + U_j E_+^j v_+(j), \\ u_-(j) = E_-^j \widetilde{A}_j^{-1} U_j^{-1} \psi_j v + E_{-+}^j v_+(j). \end{cases}$$

Notice here that $\widetilde{A}_j$ is well-defined and elliptic only in a fixed neighborhood of (0, 0), but the operator $\widetilde{A}_j^{-1} U_j^{-1} \psi_j$ is well-defined modulo $\mathcal{O}(h^\infty)$, thanks to the cutoff ψ_j.

When

$$j \in \{1, 2, \ldots, N\} \setminus \mathcal{N} =: \mathcal{N}^c,$$

we know from the discussion at the end of Sect. 6.3 that $hD_x + G_j : H_h(m) \to L^2(\mathbf{R})$ is bijective, with inverse $E^j : L^2 \to H_h(m)$ of norm $\mathcal{O}((h^{2/3} + |s_j|)^{-1})$. The obvious j-dependent version of (6.4.16) also holds.

As an approximate right inverse of $\mathcal{P}(z)$, we try

$$\widetilde{\mathcal{E}} = \begin{pmatrix} \widetilde{E} & \widetilde{E}_+ \\ \widetilde{E}_- & \widetilde{E}_{-+} \end{pmatrix}, \tag{6.4.25}$$

where

$$\begin{cases} \widetilde{E} v = J(1 - \sum_1^N \psi_j) v + \sum_1^N U_j E^j \widetilde{A}_j^{-1} U_j^{-1} \psi_j v, \\ \widetilde{E}_+ v_+ = \sum_{j \in \mathcal{N}} U_j E_+^j v_+(j), \\ (\widetilde{E}_- v)(j) = E_-^j \widetilde{A}_j^{-1} U_j^{-1} \psi_j v, \ j \in \mathcal{N}, \\ \widetilde{E}_{-+} = \mathrm{diag}\,(E_{-+}^j). \end{cases} \tag{6.4.26}$$

Recall that we work under the assumptions of Theorem 6.1.4, so

$$-h^{\delta_0} \le s_j \le \mathcal{O}(1)(h \ln(1/h))^{2/3}, \ \forall j,$$

where $\delta_0 > 0$ is arbitrarily small and fixed.

Proposition 6.4.1 *We have,*

$$\begin{cases} \widetilde{E} = \mathcal{O}(1) \max \left\{ \max_{j \in \mathcal{N}} h^{-\frac{1}{2}} s_j^{-\frac{1}{4}}, \max_{j \in \mathcal{N}^c} (h^{2/3} + |s_j|)^{-1} \right\}, \\ \widetilde{E}_+ = \mathcal{O}(1) : \mathbf{C}^{\mathcal{N}} \to H_h(m), \\ \widetilde{E}_- = \mathcal{O}(1) : L^2 \to \mathbf{C}^{\mathcal{N}}, \\ \widetilde{E}_{-+} = \mathcal{O}(1) \max_{j \in \mathcal{N}} s_j^{\frac{1}{4}} h^{\frac{1}{2}} e^{-s_j^{3/2}/\mathcal{O}(h)} : \mathbf{C}^{\mathcal{N}} \to \mathbf{C}^{\mathcal{N}}. \end{cases} \tag{6.4.27}$$

Moreover,

$$\begin{cases}(P-z)\widetilde{E}+R_-\widetilde{E}_-=1+\mathcal{O}(h^\infty):\ L^2\to L^2,\\(P-z)\widetilde{E}_++R_-\widetilde{E}_{-+}=\mathcal{O}(h^\infty):\ \mathbf{C}^{\mathcal{N}}\to L^2,\\R_+\widetilde{E}=\mathcal{O}(h^\infty):\ L^2\to\mathbf{C}^{\mathcal{N}},\\R_+\widetilde{E}_+=1+\mathcal{O}(h^\infty):\ \mathbf{C}^{\mathcal{N}}\to\mathbf{C}^{\mathcal{N}}.\end{cases}\tag{6.4.28}$$

Proof (6.4.27) follows from Proposition 6.3.2, (6.3.54), (6.3.56) (where we recall that $\alpha\asymp s_j$ when $|s_j|\gg h^{2/3}$) and the bounds $J=\mathcal{O}(h^{-\delta})$, U_j, $\widetilde{A}_j^{-1}=\mathcal{O}(1)$.

The proof of (6.4.28) is just a long calculation with some attention to terms that disappear because of the localization properties. Using (6.4.26) we get modulo $\mathcal{O}(h^\infty)$: $L^2\to L^2$,

$$(P-z)\widetilde{E}\equiv 1-\sum_1^N\psi_j+\sum_1^N(P-z)U_jE^j\widetilde{A}_j^{-1}U_j^{-1}\psi_j.\tag{6.4.29}$$

By (6.4.17),

$$(P-z)U_jE^j\widetilde{A}_j^{-1}U_j^{-1}\psi_j\equiv U_j\widetilde{A}_j(hD+G_j)E^j\widetilde{A}_j^{-1}U_j^{-1}\psi_j+U_jS_jE^j\widetilde{A}_j^{-1}U_j^{-1}\psi_j.\tag{6.4.30}$$

By Egorov's theorem, $\widetilde{A}_j^{-1}U_h^{-1}\psi_j\ \equiv\ \widetilde{\psi}_j\widetilde{A}_j^{-1}U_j^{-1}$, where $\widetilde{\psi}_j$ is a pseudodifferential operator of the same class as ψ_j and whose symbol is supported in an $\mathcal{O}(h^{\widetilde{\delta}})$-neighborhood of (0, 0). Now by the localization properties of E_j, we see that $E^j\widetilde{\psi}_j\equiv\widehat{\psi}_jE^j\widetilde{\psi}_j$, where $\widehat{\psi}_j$ has the same properties as $\widetilde{\psi}_j$. Recalling that every term in the h-asymptotic expansion of S_j vanishes to infinite order at (0, 0), we conclude that $S_jE^j\widetilde{\psi}_j\equiv 0$, and hence the last term in (6.4.30) is $\equiv 0$. Using also that

$$\begin{cases}(hD+G_j)E^j+R_-^jE_-^j=1,\ j\in\mathcal{N},\\(hD+G_j)E^j=1,\ j\in\mathcal{N}^{\mathrm{c}}\end{cases}$$

we get

$$(P-z)U_jE^j\widetilde{A}_j^{-1}U_j^{-1}\psi_j\equiv\begin{cases}\psi_j-U_j\widetilde{A}_jR_-^jE_-^j\widetilde{A}_j^{-1}U_j^{-1}\psi_j, & \text{if } j\in\mathcal{N},\\\psi_j, \quad \text{if } j\in\mathcal{N}^{\mathrm{c}}.\end{cases}\tag{6.4.31}$$

On the other hand, using (6.4.26) and (6.4.20),

$$R_-\widetilde{E}_- = \sum_{\mathcal{N}} U_j \widetilde{A}_j R_-^j E_-^j \widetilde{A}_j^{-1} U_j^{-1} \psi_j. \tag{6.4.32}$$

Summing (6.4.31) over j and adding (6.4.32), we get

$$\sum (P-z) U_j E^j \widetilde{A}_j^{-1} U_j^{-1} \psi_j + R_-\widetilde{E}_- \equiv \sum_1^N \psi_j,$$

which together with (6.4.29) gives the first equation in (6.4.28).

Next look at

$$(P-z)\widetilde{E}_+ v_+ = \sum_{\mathcal{N}} (P-z) U_j E_+^j v_+(j).$$

Modulo $\mathcal{O}(h^\infty \|v_+\|)$ in L^2, we get

$$(P-z)\widetilde{E}_+ v_+ \equiv \sum_{\mathcal{N}} U_j \widetilde{A}_j (hD + G_j) E_+^j v_+(j) \equiv -\sum_{\mathcal{N}} U_j \widetilde{A}_j R_-^j E_{-+}^j v_+(j).$$

On the other hand,

$$R_-\widetilde{E}_{-+} v_+ = \sum_{\mathcal{N}} U_j \widetilde{A}_j R_-^j E_{-+}^j v_+(j).$$

Adding the two equations then gives the second equation in (6.4.28).

Next, look at

$$R_+\widetilde{E} v(j) \equiv R_+^j U_j^{-1} U_j E^j \widetilde{A}_j^{-1} U_j^{-1} \psi_j v,$$

which simplifies to

$$R_+\widetilde{E} v(j) \equiv \underbrace{R_+^j E^j}_{\equiv 0} \widetilde{A}_j^{-1} U_j^{-1} \psi_j v \equiv 0.$$

This gives the third equation in (6.4.28).

Finally, we turn to

$$(R_+\widetilde{E}_+ v_+)(j) = R_+^j U_j^{-1} \sum_{k \in \mathcal{N}} U_k E_+^k v_+(k).$$

Because of the localization in E_+^k, we get modulo $\mathcal{O}(h^\infty \|v_+\|)$:

$$(R_+ \widetilde{E}_+ v_+)(j) \equiv R_+^j U_j^{-1} U_j E_+^j v_+(j) = R_+^j E_+^j v_+(j) = v_+(j)$$

and the fourth equation in (6.4.28) follows. □

A similar discussion of the uniqueness of the solutions of (6.4.24) leads to the approximate left inverse of $\mathcal{P}(z)$:

$$\widehat{\mathcal{E}} = \begin{pmatrix} \widehat{E} & \widehat{E}_+ \\ \widehat{E}_- & \widehat{E}_{-+} \end{pmatrix}, \tag{6.4.33}$$

where

$$\begin{cases} \widehat{E} v = (1 - \sum_1^N \psi_j) J v + \sum_1^N \psi_j U_j E_j \widetilde{A}_j^{-1} U_j^{-1} v, \\ \widehat{E}_+ = \sum_{\mathcal{N}} U_j E_+^j v_+(j), \\ (\widehat{E}_- v)(j) = E_-^j \widetilde{A}_j^{-1} U_j^{-1} v, \ j \in \mathcal{N}, \\ \widehat{E}_{-+} = \operatorname{diag}(E_{-+}^j). \end{cases} \tag{6.4.34}$$

Using the localization properties of E^j, E_{-+}^j, we can check directly that $\widehat{\mathcal{E}} = \widetilde{\mathcal{E}} + \mathcal{O}(h^\infty)$. This fact also follows from

Proposition 6.4.2 *$\widehat{\mathcal{E}}$ satisfies* (6.4.27). *Moreover,*

$$\begin{cases} \widehat{E}(P - z) + \widehat{E}_+ R_+ = 1 + \mathcal{O}(h^\infty) : \ H_h(m) \to H_h(m), \\ \widehat{E}_-(P - z) + \widehat{E}_{-+} R_+ = \mathcal{O}(h^\infty) : \ H_h(m) \to \mathbf{C}^{\mathcal{N}}, \\ \widehat{E} R_- = \mathcal{O}(h^\infty) : \ \mathbf{C}^{\mathcal{N}} \to H_h(m), \\ \widehat{E}_- R_- = 1 : \ \mathbf{C}^{\mathcal{N}} \to \mathbf{C}^{\mathcal{N}}. \end{cases} \tag{6.4.35}$$

We omit the proof, which is merely a variation of the proof of Proposition 6.4.1. An immediate consequence of the two results is

Proposition 6.4.3 *For $h > 0$ small enough, $\mathcal{P}(z) : \ H_h(m) \times \mathbf{C}^{\mathcal{N}} \to L^2 \times \mathbf{C}^{\mathcal{N}}$ is bijective with a bounded inverse $\mathcal{E}(z) = \mathcal{P}(z)^{-1}$ which satisfies,*

$$\mathcal{E}(z) = \widetilde{\mathcal{E}}(z) + \mathcal{O}(h^\infty) = \widehat{\mathcal{E}}(z) + \mathcal{O}(h^\infty) : \ L^2 \times \mathbf{C}^{\mathcal{N}} \to H_h(m) \times \mathbf{C}^{\mathcal{N}}. \tag{6.4.36}$$

We can finally look at the norm of the resolvent,

$$(P-z)^{-1} = E(z) - E_+(z)E_{-+}(z)^{-1}E_-(z). \tag{6.4.37}$$

Here $\|E(z)\|$ can be estimated as in (6.4.27), in view of (6.4.36), and we concentrate on the second term. This term is of rank $\#\mathcal{N}$ and we pause to consider in general, the norm of a finite rank operator $A : L^2 \to L^2$, given by

$$Au = \sum_{j,k=1}^{n} a_{j,k}(u|f_k)e_j,$$

where $f_1, \ldots, f_n$ and $e_1, \ldots, e_n$ are two linearly independent families in L^2. We orthonormalize the two families,

$$\begin{aligned}\left(\widetilde{f}_1 \ldots \widetilde{f}_n\right) &= \left(f_1 \ldots f_n\right) G_f^{-\frac{1}{2}},\\ \left(\widetilde{e}_1 \ldots \widetilde{e}_n\right) &= \left(e_1 \ldots e_n\right) G_e^{-\frac{1}{2}}.\end{aligned}$$

Here the Gramian G_e is given by

$$G_e = \left((e_k|e_j)\right) = \begin{pmatrix} e_1^* \\ e_2^* \\ \vdots \\ e_n^* \end{pmatrix} \left(e_1\ e_2 \ldots e_n\right),$$

identifying e_j with the operator $\mathbf{C} \ni z \mapsto ze_j$ and letting e_j^* be the adjoint. G_f is defined the same way. A straightforward calculation yields

$$Au = \sum_{\widetilde{j},\widetilde{k}=1}^{n} b_{\widetilde{j},\widetilde{k}}(u|\widetilde{f}_{\widetilde{k}})\widetilde{e}_{\widetilde{j}}, \tag{6.4.38}$$

where

$$(b_{\widetilde{j},\widetilde{k}}) = G_e^{\frac{1}{2}} \circ \left(a_{j,k}\right) \circ (G_f^{\frac{1}{2}})^* = G_e^{\frac{1}{2}} \circ \left(a_{j,k}\right) \circ G_f^{\frac{1}{2}}. \tag{6.4.39}$$

Since the families $\widetilde{e}_1, \ldots, \widetilde{e}_n$ and $\widetilde{f}_1, \ldots, \widetilde{f}_n$ are orthonormal, it is clear from (6.4.38) that

$$\|A\|_{L^2 \to L^2} = \|(b_{\widetilde{j},\widetilde{k}})\|_{\mathbf{C}^n \to \mathbf{C}^n},$$

and (6.4.39) gives

$$\|A\|_{L^2\to L^2} = \|G_e^{\frac{1}{2}} \circ \left(a_{j,k}\right) \circ G_f^{\frac{1}{2}}\|_{\mathbf{C}^n\to\mathbf{C}^n}. \tag{6.4.40}$$

We apply this to $E_+E_{-+}^{-1}E_-$. From (6.4.26) (where we can drop ψ_j because of the localization property of E_-^j) and Proposition 6.4.3 we have in L^2,

$$E_+E_{-+}^{-1}E_-u = \underbrace{\sum_{\mathcal{N}} U_j E_+^j (E_{-+}^j)^{-1} E_-^j \widetilde{A}_j^{-1} U_j^{-1} u}_{=:Au} + \mathcal{O}(h^\infty \|u\|_{L^2}), \tag{6.4.41}$$

From the discussion after (6.4.17), we recall that

$$E_+^j v_+(j) = v_+(j) e_0^j, \quad (E_-^j v)(j) = (v|f_0^j),$$

and so

$$Au = \sum_{\mathcal{N}} (E_{-+}^j)^{-1} (u|U_j(\widetilde{A}_j^*)^{-1} f_0^j) U_j e_0^j. \tag{6.4.42}$$

To apply (6.4.40), note that the Gramian G_{Ue_0} of $U_1e_0^1, \ldots, U_Ne_0^N$ is equal to $1+\mathcal{O}(h^\infty)$ and that the Gramian $G_{U(\widetilde{A}^*)^{-1}f_0}$ of $U_1(\widetilde{A}_1^*)^{-1}f_0^1, \ldots, U_N(\widetilde{A}_N^*)^{-1}f_0^N$ is of the form $D+\mathcal{O}(h^\infty)$, where $D = \mathrm{diag}\,(d_j)$ with

$$d_j = (U_j(\widetilde{A}_j^*)^{-1} f_0^j | U_j(\widetilde{A}_j^*)^{-1} f_0^j) = ((\widetilde{A}_j^*)^{-1} f_0^j | (\widetilde{A}_j^*)^{-1} f_0^j). \tag{6.4.43}$$

By the complex stationary phase method in the rescaled variables,

$$((\widetilde{A}_j^*)^{-1} f_0^j | (\widetilde{A}_j^*)^{-1} f_0^j) = (1 + \mathcal{O}(h/s_j^{3/2}))|a_j(\rho_j^-(z))|^{-2},$$

and since $\rho_j^- = \rho_j(z) + \mathcal{O}(s_j^{1/2})$ and $s_j^{1/2} \leq \mathcal{O}(h/s_j^{3/2})$ by the upper bound (6.1.43), we can replace $\rho_j^-(z)$ by $\rho_j(z)$ in the above formula. Here a_j denotes the leading symbol in A_j, given in (6.4.17),

$$p \circ \kappa_j - z = e^{i\theta_j(\gamma_j(z))} a_j(\xi + g_j) + \widehat{s}_j, \tag{6.4.44}$$

where $\widehat{s}_j = \mathcal{O}((\rho - \rho_j(z)^\infty)$ and $\rho_j(z) \in \gamma_j$ still denotes the z-dependent base point, introduced in Remark 6.2.1.

Combining (6.4.40)–(6.4.43), we get when $\mathcal{N} \neq \emptyset$,

$$\begin{aligned}\|E_+E_{-+}^{-1}E_-\| &= \mathcal{O}(h^\infty) + \|D^{1/2} \circ \operatorname{diag}((E_{-+}^j)^{-1})\| \\ &= \mathcal{O}(h^\infty) + \max_j \frac{\left(1+\mathcal{O}(h/s_j^{3/2})\right)}{|a_j(\rho_j(z))E_{-+}^j(z)|}.\end{aligned} \tag{6.4.45}$$

Here we recall that $t_j := |E_{-+}^j(z)|$ satisfies the j-dependent version of (6.3.47),

$$t_j = \left(1+\mathcal{O}\left(\frac{h}{s_j^{3/2}}\right)\right)\left(\frac{h}{\pi}\right)^{\frac{1}{2}}\left(\frac{1}{2i}\{q_j,\overline{q}_j\}(\rho_+^j)\frac{1}{2i}\{\overline{q}_j,q_j\}(\rho_-^j)\right)^{\frac{1}{4}} e^{-\frac{I_j}{h}}. \tag{6.4.46}$$

From (6.4.44) we see that

$$\{q_j,\overline{q}_j\}(\rho_\pm^j) = \frac{1}{|a_j(\rho_\pm^j)|^2}\{p,\overline{p}\}(\rho_\pm^j) + \mathcal{O}(s_j^\infty),$$

and again $a_j(\rho_\pm^j) = (1+\mathcal{O}(h/s_j^{3/2})a_j(\rho_j(z))$. Then (6.4.46) gives

$$|a_j(\rho_j(z))|t_j = \left(1+\mathcal{O}\left(\frac{h}{s_j^{3/2}}\right)\right)\left(\frac{h}{\pi}\right)^{\frac{1}{2}}\left(\frac{1}{2i}\{p,\overline{p}\}(\rho_+^j)\frac{1}{2i}\{\overline{p},p\}(\rho_-^j)\right)^{\frac{1}{4}} e^{-\frac{I_j}{h}}. \tag{6.4.47}$$

Using this in (6.4.45) gives, when $\mathcal{N} \neq \emptyset$,

$$\begin{aligned}&\|E_+E_{-+}^{-1}E_-\| \\ &= \max_{j\in\mathcal{N}}\left(1+\mathcal{O}\left(\frac{h}{s_j^{3/2}}\right)\right)\left(\frac{\pi}{h}\right)^{\frac{1}{2}}\left(\frac{1}{2i}\{p,\overline{p}\}(\rho_+^j)\frac{1}{2i}\{\overline{p},p\}(\rho_-^j)\right)^{-\frac{1}{4}} e^{\frac{I_j}{h}}.\end{aligned} \tag{6.4.48}$$

From (6.4.27) and the fact that $\widetilde{E} = E + \mathcal{O}(h^\infty)$ in (6.4.26), we have

$$\|E\| = \mathcal{O}(1)\max\left\{\max_{j\in\mathcal{N}} s_j^{-\frac{1}{4}}h^{-\frac{1}{2}},\ \max_{j\in\mathcal{N}^c}(h^{\frac{2}{3}}+|s_j|)^{-1})\right\}. \tag{6.4.49}$$

Combining (6.4.37), (6.4.48), (6.4.49), we get the conclusion in Theorem 6.1.4.

6.A Appendix: Quick Review of Almost Holomorphic Functions

Almost analytic functions have been introduced by Hörmander [81], Nirenberg [108] and Dynkin [42]. We prefer the terminology "almost holomorphic functions" to emphasize that these functions are defined in the complex domain. The work [81] has been very important for the author's works and we refer to the more recent commentary by the author (JS) at the end of the online version of [81], with pointers to more recent works.

If $\omega \subset \mathbf{R}^n$ and $\Omega \subset \mathbf{C}^n$ are open with $\Omega \cap \mathbf{R}^n = \omega$, and if $u \in C^\infty(\omega)$, there exists a function $\widetilde{u} \in C^\infty(\Omega)$ such that

$$\widetilde{u}|_\omega = u, \tag{6.A.1}$$

$$\overline{\partial}_z \widetilde{u}(z) = \mathcal{O}(|\Im z|^\infty), \text{ locally uniformly on } \Omega. \tag{6.A.2}$$

Here in the last equation, we write

$$\overline{\partial} u = \overline{\partial}_z \widetilde{u} = \partial_{\overline{z}_1} \widetilde{u} d\overline{z}_1 + \cdots + \partial_{\overline{z}_n} \widetilde{u} d\overline{z}_n \simeq (\partial_{\overline{z}_1} \widetilde{u}, \ldots, \partial_{\overline{z}_n} \widetilde{u})$$

with $\partial_{\overline{z}_j} = (1/2)(\partial_{\Re z_j} + i\partial_{\Im z_j})$ and the meaning of the estimate is that $\overline{\partial}\widetilde{u}(z) = \mathcal{O}(|\Im z|^N)$ locally uniformly on Ω for every $N \in \mathbf{N}$. We call $\widetilde{u}$ an almost holomorphic (rather than almost analytic) extension of u. The Taylor expansion of $\widetilde{u}$ is uniquely determined by that of u at every real point, so if $\widehat{u}$ is another almost holomorphic extension of the same function u, then $\widetilde{u} - \widehat{u} = \mathcal{O}(|\Im z|^\infty)$ locally uniformly on Ω (or on the intersection of the domains of the two extensions, when these domains are different). The extension $\widetilde{u}$ can be constructed by means of truncated Taylor expansions and if $0 < r \in C(\omega)$, we can choose $\widetilde{u}$ with support in $\{z \in \Omega;\ |\Im z| < r(\Re z)\}$. See [81, 102] for further details.

In the case $n = 1$, let $\gamma : [a, b] \to \mathbf{C}$ be a smooth injective curve with injective differential, $-\infty < a < b < +\infty$, and let u be a smooth function on $\gamma(\,]a, b[\,)$. Then we can find a smooth function $\widetilde{u}$ on $\mathrm{neigh}\,(\gamma(\,]a, b[\,), \mathbf{C})$ such that

$$\begin{aligned} \widetilde{u}|_{\gamma(\,]a,b[\,)} &= u, \\ \overline{\partial}\widetilde{u} &= \mathcal{O}(\mathrm{dist}\,(\cdot, \gamma(\,]a, b[\,))^\infty), \end{aligned} \tag{6.A.3}$$

locally uniformly, on the neighborhood of definition. Again, $\widetilde{u}$ is unique mod $\mathcal{O}(\mathrm{dist}\,(\cdot, \gamma(\,]a, b[\,))^\infty)$ locally uniformly, and we call $\widetilde{u}$ an almost holomorphic extension.

Indeed, if $\widetilde{\gamma}$ is an almost holomorphic extension of γ, then by the implicit function theorem, $\widetilde{\gamma}$ is a diffeomorphism: $\mathrm{neigh}\,(\,]a, b[\,, \mathbf{C}) \to \mathrm{neigh}\,(\gamma(\,]a, b[\,), \mathbf{C})$ and letting $\widetilde{\gamma}^{-1}$ denote the inverse, we can take

$$\widetilde{u} = \widetilde{u \circ \gamma} \circ \widetilde{\gamma}^{-1}, \tag{6.A.4}$$

where $\widetilde{u \circ \gamma}$ denotes an almost holomorphic extension of $u \circ \gamma$. This can be generalized to higher dimensions, $n \geq 2$, by replacing the range of a curve by a totally real manifold of maximal dimension. (See [102].)

Notes The main result of this chapter, Theorem 6.1.4, is a small generalization of a result of Montrieux [18]. In Chap. 10 we will review more general results in all dimensions, which are less precise however.

Metaplectic (Fourier integral) operators (in the semi-classical limit) are Fourier integral operators $L^2(\mathbf{R}^n) \to L^2(\mathbf{R}^n)$ with quadratic phase functions and constant amplitudes. The phase function is assumed to satisfy standard nondegeneracy assumptions which imply that the operator can be associated to an affine canonical transformation $\kappa : T^*\mathbf{R}^n \to T^*\mathbf{R}^n$. Such an operator U with non-zero amplitude, associated to κ has a bounded inverse U^{-1} associated to κ^{-1}. If $A = a^w(x, hD; h)$ is the h-Weyl quantization of the symbol a and U is a metaplectic operator as above, then $B := U^{-1}AU$ is the h-Weyl quantization of the symbol $b = a \circ \kappa$. This is an exact form of Egorov's theorem. It suffices that $a \in \mathcal{S}'(T^*\mathbf{R}^n)$. See e.g., [41, Chapter 7, Appendix].

7 The Complex WKB Method

In this chapter we shall study the exponential growth and asymptotic expansions of exact solutions of second-order differential equations in the semi-classical limit. As an application, we establish a Bohr-Sommerfeld quantization condition for Schrödinger operators with real-analytic complex-valued potentials.

7.1 Estimates on an Interval

In this section we derive some basic estimates for differential equations on an interval. Let $I = [a, b]$ be a bounded interval and consider the problem

$$(h\partial_x - A(x))u(x) = 0, \ x \in I, \tag{7.1.1}$$

where $A \in C^\infty(I; \mathrm{Mat}\,(n, n))$ and we let $\mathrm{Mat}\,(n, n)$ denote the space of complex $n \times n$-matrices. Using the basic result in the theory of linear ODEs about the well-posedness of the Cauchy problem, we can introduce the *fundamental matrix* $E(x, y) \in C^\infty(I \times I; \mathrm{Mat}\,(n, n))$, determined by

$$(h\partial_x - A(x))E(x, y) = 0, \ E(y, y) = 1. \tag{7.1.2}$$

Let $W(A(x))$ denote the numerical range of $A(x)$ as in Definition 2.3.1.

Proposition 7.1.1 *Let*

$$\mu_+(A(x)) = \sup_{\lambda \in W(A(x))} \Re\lambda, \ \mu_-(A(x)) = \inf_{\lambda \in W(A(x))} \Re\lambda$$

J. Sjöstrand, *Non-Self-Adjoint Differential Operators, Spectral Asymptotics and Random Perturbations*, Pseudo-Differential Operators 14,
https://doi.org/10.1007/978-3-030-10819-9_7

Then

$$\|E(x,y)\| \leq \begin{cases} \exp(\int_y^x (\mu_+(A(t))dt/h), & \text{if } x \geq y, \\ \exp(\int_y^x (\mu_-(A(t))dt/h), & \text{if } x \leq y. \end{cases} \tag{7.1.3}$$

Proof If $u = u(x)$ is a solution of $(h\partial_x - A(x))u = 0$, we have $u(x) = E(x,y)u(y)$. Moreover,

$$h\partial_x(u(x)|u(x)) = (A(x)u(x)|u(x)) + (u(x)|A(x)u(x)) = 2\Re(A(x)u(x)|u(x)),$$

so

$$h\partial_x \|u(x)\|^2 \begin{cases} \leq 2\mu_+(A(x))\|u(x)\|^2 \\ \geq 2\mu_-(A(x))\|u(x)\|^2. \end{cases}$$

Integrating these differential inequalities, we get

$$\|u(x)\|^2 \leq \begin{cases} \exp(2\int_y^x \mu_+(A(t))dt/h)\|u(y)\|^2, & \text{if } x \geq y, \\ \exp(2\int_y^x \mu_-(A(t))dt/h)\|u(y)\|^2, & \text{if } x \leq y, \end{cases}$$

and (7.1.3) follows, since $u(y)$ can be chosen arbitrarily in $\mathbf{C}^n$. □

Remark 7.1.2 For $x = y$ we have $h\partial_y E = -h\partial_x E$ and hence $h\partial_y E(x,y) + E(x,y)A(y) = 0$ when $x = y$. On the other hand,

$$(h\partial_x - A(x))(h\partial_y E(x,y)) = 0,$$
$$(h\partial_x - A(x))(E(x,y)A(y)) = 0,$$

so

$$(h\partial_x - A(x))(h\partial_y E(x,y) + E(x,y)A(y)) = 0 \text{ on } I \times I.$$

From the uniqueness in the Cauchy problem, we deduce the second differential equation for the fundamental matrix,

$$h\partial_y E(x,y) + E(x,y)A(y) = 0, \; x, y \in I. \tag{7.1.4}$$

Differentiating (7.1.2), (7.1.4) several times, we see that $(h\partial_x)^j(h\partial_y)^k E(x,y)$ is a linear combination of terms

$$(h\partial_x)^{j_1}A(x)\circ\dots\circ(h\partial_x)^{j_\nu}A(x)\circ E(x,y)\circ(h\partial_y)^{k_1}A(y)\circ\dots\circ(h\partial_y)^{k_\mu}A(y),$$

where

$$j_\ell,\ k_m\geq 0,\ \ \nu+j_1+\ldots+j_\nu=j,\ \ \mu+k_1+\ldots+k_\mu=k.$$

It follows that

$$\|(h\partial_x)^j(h\partial_y)^k E(x,y)\|\leq C_{j,k}\times \text{ the RHS of (7.1.3).}$$

We now assume, in order to fix the ideas, that $n=2$. Assume that

$$\sigma(A(x))=\{\lambda_1(x),\lambda_2(x)\},\ \lambda_1(x)\neq\lambda_2(x),\ x\in I. \tag{7.1.5}$$

We then know that $A(x)$ is diagonizable and more precisely that there exists

$$U_0(x)\in C^\infty(I;\mathrm{GL}\,(n)), \tag{7.1.6}$$

where $\mathrm{GL}\,(n)\subset\mathrm{Mat}\,(n,n)$ is the group of invertible complex $n\times n$ matrices, such that

$$U_0^{-1}(x)A(x)U_0(x)=\begin{pmatrix}\lambda_1(x) & 0\\ 0 & \lambda_2(x)\end{pmatrix}=:\Lambda_0(x). \tag{7.1.7}$$

Then

$$\begin{aligned}U_0(x)^{-1}(h\partial_x-A(x))U_0(x)&=h\partial_x-\Lambda_0(x)+h\underbrace{U_0(x)^{-1}\partial_x(U_0(x))}_{=:-B_1(x)}\\ &=h\partial_x-\begin{pmatrix}\lambda_1(x)+hb_{11}(x) & hb_{12}(x)\\ hb_{21}(x) & \lambda_2(x)+hb_{22}(x)\end{pmatrix}.\end{aligned} \tag{7.1.8}$$

Naturally, we have the equivalence

$$(h\partial_x-A(x))(U_0(x)u)=0\Leftrightarrow(h\partial_x-(\Lambda_0(x)+hB_1(x)))u=0.$$

If $F(x, y; h)$ is the fundamental matrix for $h\partial_x - (\Lambda_0(x) + hB_1(x))$, then we have the (equivalent) relations,

$$\begin{aligned} E(x, y; h) &= U_0(x)F(x, y; h)U_0(y)^{-1}, \\ F(x, y; h) &= U_0(x)^{-1}E(x, y; h)U_0(y). \end{aligned} \tag{7.1.9}$$

In addition to (7.1.5), we now assume that

$$\Re\lambda_1(x) \geq \Re\lambda_2(x), \ x \in I. \tag{7.1.10}$$

Then,

$$\begin{aligned} \mu_+(\Lambda_0(x) + hB_1(x)) &= \Re\lambda_1(x) + \mathcal{O}(h), \\ \mu_-(\Lambda_0(x) + hB_1(x)) &= \Re\lambda_2(x) + \mathcal{O}(h), \end{aligned}$$

and Proposition 7.1.1 gives

$$\|F(x, y; h)\| \leq \begin{cases} \exp\left(\frac{1}{h}\int_y^x \Re\lambda_1(t)dt + \mathcal{O}(|x-y|)\right), & \text{if } x \geq y \\ \exp\left(\frac{1}{h}\int_y^x \Re\lambda_2(t)dt + \mathcal{O}(|x-y|)\right), & \text{if } x \leq y \end{cases} \tag{7.1.11}$$

As before, we get

$$\|(h\partial_x)^j(h\partial_y)^k F(x, y)\| \leq C_{j,k} \times \text{ the RHS of (7.1.11).} \tag{7.1.12}$$

Back to $h\partial_x - A(x)$, we get

Theorem 7.1.3 *Under the assumptions (7.1.5), (7.1.18), we have*

$$\|(h\partial_x)^j(h\partial_y)^k E(x, y; h)\| \leq C_{j,k} \begin{cases} \exp h^{-1}\int_y^x \Re\lambda_1(t)dt, & \text{if } x \geq y, \\ \exp h^{-1}\int_y^x \Re\lambda_2(t)dt, & \text{if } x \leq y. \end{cases} \tag{7.1.13}$$

This follows from (7.1.9), (7.1.11), (7.1.12).

We can eliminate the off-diagonal elements in (7.1.8) to any order in h by means of additional conjugations. Let $C = \begin{pmatrix} c_{11} & c_{12} \\ c_{21} & c_{22} \end{pmatrix} \in C^\infty(I; \mathrm{Mat}(n,n))$ and consider $1+hC(x)$, which is invertible for h small enough, with inverse

$$(1 + hC(x))^{-1} = 1 - hC(x) + h^2 C(x)^2 - \cdots$$

where the series is convergent but will be viewed as an asymptotic one. Then,

$$\begin{aligned}
&(1 + hC(x))^{-1}(h\partial_x - (\Lambda_0(x) + hB_1(x)))(1 + hC(x)) \\
&\quad = h\partial_x - (\Lambda_0(x) + h(-C(x)\Lambda_0(x) + \Lambda_0(x)C(x) + B_1(x))) + \mathcal{O}(h^2) \\
&\quad\quad = h\partial_x - (\Lambda_0(x) + h[\Lambda_0(x), C(x)] + B_1(x)) + \mathcal{O}(h^2),
\end{aligned}$$

and we see that

$$[\Lambda_0, C] = \begin{pmatrix} 0 & (\lambda_1 - \lambda_2)c_{12} \\ (\lambda_2 - \lambda_1)c_{21} & 0 \end{pmatrix}.$$

Choose $c_{11} = c_{22} = 0$, $c_{j,k} = -b_{j,k}/(\lambda_j - \lambda_k)$, $k \neq j$. Then with $\widetilde{U}_1(x) = U_0(x)(1 + hC(x))$, we find that

$$\begin{aligned}
&\widetilde{U}_1(x; h)^{-1}(h\partial_x - A(x))\widetilde{U}_1(x; h) \\
&= h\partial_x - \begin{pmatrix} \lambda_1(x) + hb_{11}(x) & 0 \\ 0 & \lambda_2(x) + hb_{22}(x) \end{pmatrix} + \mathcal{O}(h^2),
\end{aligned} \tag{7.1.14}$$

where the last term has an asymptotic expansion in powers of h. Again we can kill the leading off diagonal entries (of the form $h^2 b_{jk}^{(2)}(x)$) by conjugating with a matrix $1 + h^2 D(x)$, and so on. We get

Proposition 7.1.4 *Under the assumption* (7.1.5), *we can find*

$$U(x; h) \sim U_0(x) + hU_1(x) + h^2 U_2(x) + \cdots \in C^\infty(I; Mat\,(n, n)) \tag{7.1.15}$$

(with the asymptotic sum in the sense of (4.1.3), (4.1.4)) *with* $U_0(x)^{-1} \in C^\infty(I; Mat\,(n, n))$, *such that*

$$U(x; h)^{-1}(h\partial_x - A(x))U(x; h) = h\partial_x - \Lambda(x; h), \tag{7.1.16}$$

where

$$\Lambda(x;h) \sim \Lambda_0(x) + h\Lambda_1(x) + h^2\Lambda_2(x) + \cdots \text{ in } C^\infty(I; Mat(n,n)),$$

and each matrix Λ_j *is diagonal, so*

$$\Lambda(x;h) = \begin{pmatrix} \widetilde{\lambda}_1(x;h) & r_{1,2}(x;h) \\ r_{2,1}(x;h) & \widetilde{\lambda}_2(x;h) \end{pmatrix}, \quad r_{j,k}(x;h) \sim 0$$
$$\widetilde{\lambda}_j(x;h) \sim \lambda_j(x) + h\lambda_{j,1}(x) + h^2\lambda_{j,2}(x) + \cdots . \tag{7.1.17}$$

Using this result it is easy to find formal asymptotic solutions.

The discussion in this section can be applied to the scalar Schrödinger equation

$$(-(h\partial_x)^2 + V(x))v = 0 \tag{7.1.18}$$

if the potential V is smooth on I. Indeed, introducing

$$u = \begin{pmatrix} u_0(x) \\ u_1(x) \end{pmatrix} = \begin{pmatrix} v(x) \\ h\partial_x v(x) \end{pmatrix},$$

we see that (7.1.18) is equivalent to

$$(h\partial_x - A(x))u = 0, \text{ where } A(x) = \begin{pmatrix} 0 & 1 \\ V & 0 \end{pmatrix}. \tag{7.1.19}$$

The eigenvalues of A are $\pm V(x)^{1/2}$ and the condition (7.1.5) is equivalent to the fact that $V(x) \neq 0$ for all $x \in I$ or, in other words that there is no turning point in I at energy 0. (In general, we say that $x_0 \in I$ is a turning point of V at energy E, if $V(x_0) - E = 0$. This turning point is simple if in addition $V'(x_0) \neq 0$.)

7.2 The Schrödinger Equation in the Complex Domain

Let $\Omega \Subset \mathbf{C}$ be open and simply connected. Let $A(z) \in \mathrm{Hol}(\Omega; \mathrm{Mat}(2,2))$. As in the case of an interval, the very basic result is that the Cauchy problem is well-posed: Let $w \in \mathcal{O}$ and let $u_0 \in \mathbf{C}^2$. Then there is a unique holomorphic solution, $u = u(z) \in \mathrm{Hol}(\Omega; \mathbf{C}^2)$, of the problem

$$(h\partial_z - A(z))u(z) = 0 \text{ in } \Omega, \quad u(w) = u_0, \tag{7.2.1}$$

which can be written

$$u(z) = E(z, w)u_0, \tag{7.2.2}$$

where $E(z, w)$ is the fundamental matrix.

As in (7.1.5), we now assume that $A(z)$ has distinct eigenvalues:

$$\sigma(A(z)) = \{\lambda_1(z), \lambda_2(z)\}, \text{ where } \lambda_1(z) \neq \lambda_2(z), \ \forall z \in \Omega. \tag{7.2.3}$$

Let $E_1(z),\ E_2(z) \in \mathbf{C}^2$ be the corresponding one-dimensional eigenspaces that depend holomorphically on z. Locally, we can find non-vanishing holomorphic sections $e_j(z) \in E_j(z)$. The choice can be made global if we impose that $\partial_z e_1(z) \in E_2(z)$, $\partial_z e_2(z) \in E_1(z)$ everywhere. In fact, this leads to simple differential equations that have global holomorphic solutions: Choose local holomorphic sections $e_j^0(z) \in E_j(z)$. Then $\partial_z e_1^0(z) = a_1(z)e_1^0(z) + a_2(z)e_2^0(z)$ for some holomorphic coefficients. If we put $e_1(z) = u_1(z)e_1^0(z)$, then the condition that $\partial_z e_1(z) \in E_2(z)$ is equivalent to the differential equation $\partial_z u_1 + a_1(z)u_1 = 0$, which locally has a unique non-vanishing solution, if we prescribe $u_1(z_0)$ in $\mathbf{C} \setminus \{0\}$ at some point z_0. Since Ω is simply connected, this leads to a unique non-vanishing holomorphic section e_1 in E_1 over Ω. The same works for e_2 of course.

With such a global choice, we let $U_0(z)$ be the invertible matrix with $e_1(z)$ and $e_2(z)$ as the two columns. Then as in (7.1.7) we have

$$U_0^{-1}(z)A(z)U_0(z) = \begin{pmatrix} \lambda_1(z) & 0 \\ 0 & \lambda_2(z) \end{pmatrix} =: \Lambda_0(z),\ z \in \Omega. \tag{7.2.4}$$

From this we obtain the following analogue of Proposition 7.1.4

Proposition 7.2.1 *Under the assumption* (7.2.3), *we can find*

$$U(x; h) \sim U_0(x) + hU_1(x) + h^2U_2(x) + \cdots \in Hol\,(\Omega; Mat\,(n, n)) \tag{7.2.5}$$

with $U_0(x)^{-1} \in Hol\,(\Omega; Mat\,(n, n))$, *such that*

$$U(z; h)^{-1}(h\partial_z - A(z))U(z; h) = h\partial_z - \Lambda(z; h), \tag{7.2.6}$$

where

$$\Lambda(z; h) \sim \Lambda_0(z) + h\Lambda_1(z) + h^2\Lambda_2(z) + \cdots \text{ in } C^\infty(I; Mat\,(n, n)),$$

and each matrix Λ_j *is diagonal, so*

$$\Lambda(z;h) = \begin{pmatrix} \widetilde{\lambda}_1(z;h) & r_{1,2}(z;h) \\ r_{2,1}(z;h) & \widetilde{\lambda}_2(z;h) \end{pmatrix}, \quad r_{j,k}(z;h) \sim 0$$
$$\widetilde{\lambda}_j(z;h) \sim \lambda_j(z) + h\lambda_{j,1}(z) + h^2\lambda_{j,2}(z) + \cdots . \tag{7.2.7}$$

Strictly speaking, for every $K \Subset \Omega$, *the inverse of* $U(x;h)$ *exists for* $x \in K$, $0 < h \leq h(K)$ *for some* $h(K) > 0$ *small enough.*

Corollary 7.2.2 *Let* $\phi_j(z)$ *be holomorphic in* Ω *with* $\phi_j'(z) = \lambda_j(z)$. *Then there exits* $a(z;h) \sim a_0(z) + ha_1(z) + \cdots$ *in Hol* (Ω) *such that* $0 \neq a_0(z) \in \mathcal{N}(A(z) - \lambda_j(z))$ *for all* $z \in \Omega$, *and*

$$(h\partial_z - A(z))(a(z;h)e^{\phi_j(z)/h}) = r(z;h)e^{\phi_j(z)/h}, \quad r \sim 0. \tag{7.2.8}$$

Proof Let $\widetilde{\phi}_j(z;h) \sim \phi_j(z) + h\phi_{j,1}(z) + \cdots$ be a holomorphic primitive of $\widetilde{\lambda}_j(z;h)$. Then $e^{\widetilde{\phi}_j(z;h)/h} = \widetilde{a}_j(z;h)e^{\phi_j(z)/h}$, $\widetilde{a}_j(z;h) = 1 + h\widetilde{a}_{j,1}(z) + \cdots$, and if ν_1, ν_2 is the canonical basis in $\mathbf{C}^2$, we have locally uniformly,

$$(h\partial_z - \Lambda)(e^{\widetilde{\phi}_j(z)/h}\nu_j) = \mathcal{O}(h^\infty)e^{\phi_j(z)/h}.$$

It then suffices to define a (depending also on j) by

$$a(z;h)e^{\phi_j(z)/h} = U(z;h)(e^{\widetilde{\phi}_j(z;h)/h}\nu_j).$$

□

By examining directly the equations for a_0 and a_1 that follow from (7.2.8), we see that $a_0(z) = \text{Const}\, e_j(z)$ respectively when $j = 1, 2$, where $e_j(z)$ are the non-vanishing sections of $\mathcal{N}(A(z) - \lambda_j)$ that we constructed prior to (7.2.4).

Let $\gamma : [a,b] \ni t \mapsto \gamma(t) \in \Omega$ be a smooth curve with $\dot{\gamma}(t) \neq 0$. If we restrict the equation $(h\partial_z - A(z))u = 0$ to γ, we get

$$(h\partial_t - \dot{\gamma}(t)A(\gamma(t)))u(\gamma(t)) = 0, \tag{7.2.9}$$

from which we deduce the general estimate for the fundamental matrix:

$$\|E(\gamma(t),\gamma(s);h)\| \leq \mathcal{O}(1)\exp\left\{\frac{1}{h}\int_s^t \max_{j=1,2}(\Re(\dot{\gamma}(\tau)\lambda_j(\gamma(\tau)))d\tau\right\}, \tag{7.2.10}$$

for $a \leq s \leq t \leq b$.

Now assume that

$$\Re(\dot{\gamma}(t)\lambda_1(\gamma(t))) \geq \Re(\dot{\gamma}(t)\lambda_2(\gamma(t))),\ a \leq t \leq b. \tag{7.2.11}$$

Then the integral in the exponent in (7.2.10) simplifies to

$$\int_s^t (\Re\dot{\gamma}(\tau)\lambda_1(\gamma(\tau)))d\tau = \Re\int_s^t \frac{d}{d\tau}(\phi_1(\gamma(\tau)))d\tau = \Re(\phi_1(\gamma(t)) - \phi_1(\gamma(s))),$$

and (7.2.10) becomes,

$$\|E(\gamma(t), \gamma(s); h)\| \leq \mathcal{O}(1)\exp\left\{\frac{1}{h}(\Re\phi_1(\gamma(t)) - \Re\phi_1(\gamma(s)))\right\}. \tag{7.2.12}$$

Similarly (still with $s \leq t$)

$$\|E(\gamma(s), \gamma(t); h)\| \leq \mathcal{O}(1)\exp\left\{\frac{1}{h}(\Re\phi_2(\gamma(s)) - \Re\phi_2(\gamma(t)))\right\}. \tag{7.2.13}$$

Theorem 7.2.3 *Under the assumption* (7.2.11), *let*

$$u_{\mathrm{WKB}}(z; h) = a_{\mathrm{WKB}}(z; h)e^{\phi_j(z)/h}$$

be an asymptotic solution of $(h\partial_z - A(z))u_{\mathrm{WKB}}(z; h) \approx 0$ *as in Corollary* 7.2.2. *Let* $u(z; h)$ *be the exact solution of* $(h\partial_z - A(z))u = 0$ *in* neigh $([a, b], \mathbf{C})$ *such that* $u(\gamma(a)) = u_{\mathrm{WKB}}(\gamma(a))$ *when* $j = 1$, *and* $u(\gamma(b)) = u_{\mathrm{WKB}}(\gamma(b))$ *when* $j = 2$. *Then* $u(z; h) - u_{\mathrm{WKB}}(z; h) = \mathcal{O}(h^\infty)e^{\phi_j(z)/h}$ *with all its derivatives on* $\gamma([a, b])$.

If we strengthen the assumption (7.2.11) *to*

$$\Re\dot{\gamma}(t)\lambda_1(\gamma(t)) > \Re\dot{\gamma}(t)\lambda_2(\gamma(t)),\ a \leq t \leq b, \tag{7.2.14}$$

then

$$u(z; h) - u_{\mathrm{WKB}}(z; h) = \mathcal{O}(h^\infty)e^{\phi_j(z)/h}$$

in neigh $(\gamma(]a, b]), \Omega)$ *and* neigh $(\gamma([a, b[), \Omega)$ *for* j *equal to* 1 *and* 2, *respectively.*

Proof The cases $j = 1$ and $j = 2$ are basically equivalent and we choose $j = 1$ in order to fix the ideas. Let $u(z; h)$ be the unique exact solution such that $u(\gamma(a)) = u_{\mathrm{WKB}}(\gamma(a))$. Recall that

$$(h\partial_z - A)u_{\mathrm{WKB}} = r(z; h)e^{\phi_1(z)/h},\ r \sim 0,$$

so that

$$u(z) - u_{\mathrm{WKB}}(z) = -\int_{\gamma(a)}^{z} E(z, w; h) r(w; h) e^{\phi_1(w)/h} dw$$

(line integral). If $z = \gamma(t)$ and we integrate along γ and use (7.2.12), we get the desired conclusion under the assumption (7.2.11).

Under the stronger assumption (7.2.14), it suffices to take a smooth family of curves $\gamma_s : [a, b + \epsilon] \to \Omega$, $s \in \mathrm{neigh}\,(0, \mathbf{R})$, starting at $\gamma(a)$, with $\gamma_0|_{[a,b]} = \gamma$ so that the images of the γ_s fill up a neighborhood of $\gamma(\,]a, b]\,)$. □

Remark 7.2.4 Assume (7.2.14) and normalize the choice of ϕ_1, ϕ_2 so that $\phi_1(\gamma(a)) = \phi_2(\gamma(a))$. Let

$$u_{\mathrm{WKB}}(z; h) = a_{\mathrm{WKB}}(z; h) e^{\phi_1(z)/h} + b_{\mathrm{WKB}}(z; h) e^{\phi_2(z)/h}$$

be the sum of two asymptotic null solutions as in Corollary 7.2.2. Then we have the same conclusion as in Theorem 7.2.3, namely that the exact solution u with the "initial condition" $u(\gamma(a)) = u_{\mathrm{WKB}}(\gamma(a))$ satisfies

$$u(z; h) - u_{\mathrm{WKB}}(z; h) = \mathcal{O}(h^{\infty}) e^{\phi_1(z)/h} \text{ in neigh}\,(\gamma(\,]a, b]), \Omega).$$

Further, notice that

$$u_{\mathrm{WKB}}(z; h) = a_{\mathrm{WKB}}(z; h) e^{\phi_1(z)/h} + \mathcal{O}(h^{\infty}) e^{\phi_1(z)/h} \text{ in neigh}\,(\gamma(]a, b]), \Omega).$$

Now consider the scalar Schrödinger equation

$$(-(h\partial)^2 + V(z))u = 0, \tag{7.2.15}$$

where V is holomorphic in the open simply connected domain Ω. Writing

$$\widetilde{u} = \begin{pmatrix} u \\ h\partial u \end{pmatrix}$$

we get the equivalent first-order system

$$(h\partial - A)\widetilde{u} = 0, \tag{7.2.16}$$

where

$$A(z) = \begin{pmatrix} 0 & 1 \\ V & 0 \end{pmatrix}. \tag{7.2.17}$$

The eigenvalues of $A(z)$ are $\pm V(z)^{1/2}$, so (7.2.3) is equivalent to the assumption that $V(z) \neq 0$ everywhere in Ω, i.e., that there are no turning points in Ω. The earlier discussion can be applied with $\phi_1' = V(z)^{1/2}$, $\phi_2' = -V(z)^{1/2}$ after fixing a holomorphic branch of the square root of $V(z)$. Notice that this gives an alternative approach to the construction of asymptotic solutions to (7.2.15).

Definition 7.2.5 A Stokes line is a curve along which $\Re\phi$ is constant. An anti-Stokes line is a curve along which $\Im\phi$ is constant. Here $\phi = \phi_1$ or ϕ_2.

Study Near a Simple Turning Point Let $z_0 \in \Omega$ be a simple turning point (at energy 0) in the sense that

$$V(z_0) = 0, \ V'(z_0) \neq 0. \tag{7.2.18}$$

In order to simplify the notation, assume that $z_0 = 0$. Consider the eikonal equation

$$\phi'(z) = V(z)^{\frac{1}{2}} \tag{7.2.19}$$

in a neighborhood of 0. Clearly, $\phi(z)$ will have to be multivalued and in order to understand this better, we pass to the double covering of a pointed neighborhood of 0, by putting $z = w^2$. Then

$$\frac{\partial}{\partial z} = \frac{1}{2w}\frac{\partial}{\partial w},$$

and if we put $\widetilde{V}(w) = V(z) = F(z)z = F(w^2)w^2$, $\phi(z) = \widetilde{\phi}(w)$, where $F(0) \neq 0$, the eikonal equation becomes

$$\partial_w \widetilde{\phi} = F(w^2)^{\frac{1}{2}} 2w^2,$$

and the right-hand side is an even holomorphic function of w. If we also require that $\phi(0) = \widetilde{\phi}(0) = 0$, we see that $\widetilde{\phi}(w)$ is an odd holomorphic function of the form

$$\widetilde{\phi}(w) = \frac{2}{3}\widetilde{F}(w^2)w^3, \ \text{where } \widetilde{F}(0) = F(0)^{\frac{1}{2}} = V'(0)^{\frac{1}{2}}.$$

In the original coordinates, we get the double-valued solution

$$\phi(z) = \frac{2}{3}\widetilde{F}(z)z^{\frac{3}{2}}. \tag{7.2.20}$$

Now look for Stokes and anti-Stokes lines that reach 0. On such curves, we have $\Re\phi = 0$ or $\Im\phi = 0$, i.e., $\Im\phi^2 = 0$: $\Im\widetilde{F}(z)^2 z^3 = 0$. In other words $\widetilde{F}(z)^2 z^3 = t^3$ for some $t \in \mathrm{neigh}\,(0, \mathbf{R})$, and taking the cubic root we get three curves γ_k,

$$\widetilde{F}(z)^{\frac{2}{3}} z = e^{2\pi i k/3} t, \; k \in \{0, 1, 2\} \simeq \mathbf{Z}/3\mathbf{Z}.$$

We get the following picture, where we have taken $V'(0) > 0$ for simplicity.

Each curve γ_k splits into a Stokes curve γ_k^- ending at the turning point and including that point by convention, and an anti-Stokes curve γ_k^+, which does not include the turning point. Restricting the attention to a small suitably shaped neighborhood W of 0, the three Stokes curves delimit three closed Stokes "sectors" Σ_k in that neighborhood. In Fig. 7.1 we also draw some Stokes curves inside each sector.

Let ϕ_k be the branch of ϕ in $\mathrm{neigh}\,(0) \setminus \gamma_k^-$ such that $\Re\phi_k < 0$ in $\overset{\circ}{\Sigma}_k$ (and such that $\phi_k(0) = 0$ since ϕ_k is odd). Notice that ϕ_{k+1} and ϕ_k are both well defined in $\Sigma_k \cup \Sigma_{k+1}$ and satisfy $\phi_{k+1} = -\phi_k$ there.

Let W be a sufficiently small open disc centered at 0. Then (as can be seen by working in the coordinate $z^{3/2}$) each level set $\{z \in W;\ \Re\phi_j(z) = \text{const.} \neq 0\}$ is connected and hence equal to a Stokes line. Let $A_j \in \overset{\circ}{\Sigma}_j$, and notice that every point in

$$\{z \in W \setminus \gamma_j^-;\ \Re\phi_j(z) > \Re\phi_j(A_j)\} \tag{7.2.21}$$

can be reached by a curve in $W \setminus \gamma_j^-$, starting at A_j and ending at z, along which $\Re\phi_j$ is strictly increasing. By Theorem 7.2.3, it is clear that we have a holomorphic solution to $(-(h\partial)^2 + V)u_j = 0$ in W, such that

$$\begin{cases} u_j(z; h) = a_j(z; h) e^{\phi_j(z)/h}, \\ a_j(z; h) \sim a_{j,0}(z) + h a_{j,1}(z) + \cdots \end{cases} \quad \text{in the set (7.2.21).}$$

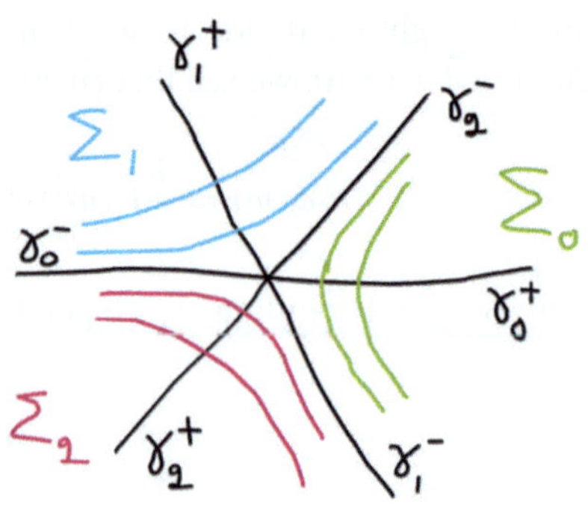

Fig. 7.1 Stokes sectors and Stokes lines near a simple turning point

Here, as in Example 4.1.3 (which extends to the complex case), $a_{j,0}(z)$ is unique up to a constant factor and we can choose

$$a_{j,0}(z) = (\phi_j'(z))^{-\frac{1}{2}},$$

where we have not chosen any preferred sign. Further, by Remark 7.2.4, we may arrange so that

$$u_j(A_j; h) = 0.$$

In the following we replace the disc W by

$$W \setminus \bigcup_{-1}^{1} \{z \in \Sigma_j;\ \Re\phi_j(z) \leq \Re\phi_j(A_j)\}$$

and decrease Σ_j accordingly.

Recall that if u, v are solutions to our homogeneous Schrödinger equation, then the Wronskian

$$\mathrm{Wr}\,(u, v) = (h\partial u)v - uh\partial v$$

is constant. Applying the asymptotics of u_0 and u_1 at some point in the interior of $\Sigma_0 \cup \Sigma_1$, we see that $\mathrm{Wr}\,(u_0, u_1)$ has an asymptotic expansion in powers of h:

$$\begin{aligned}\mathrm{Wr}\,(u_0, u_1) &= 2a_{0,0}a_{1,0}\partial\phi_0 + \mathcal{O}(h)\\ &= 2\frac{\phi_0'}{\sqrt{\phi_0'\phi_1'}} + \mathcal{O}(h)\\ &= 2\frac{\sqrt{\phi_0'}}{\sqrt{\phi_1'}} + \mathcal{O}(h).\end{aligned}$$

Similarly,

$$\mathrm{Wr}\,(u_1, u_{-1}) = 2\frac{\sqrt{\phi_1'}}{\sqrt{\phi_{-1}'}} + \mathcal{O}(h),$$

$$\mathrm{Wr}\,(u_{-1}, u_0) = 2\frac{\sqrt{\phi_{-1}'}}{\sqrt{\phi_0'}} + \mathcal{O}(h).$$

This can be further determined in the following way: Let us fix a branch of $(\phi'_j)^{1/2}$ as above for $j = 0, 1, -1 \bmod 4\mathbf{Z}$. Then for any two different Stokes sectors, $j \neq k$, we have in the interior of $\Sigma_j \cup \Sigma_k$ that

$$(\phi'_j)^{1/2} = i^{\nu_{j,k}}(\phi'_k)^{1/2}, \tag{7.2.22}$$

where $\nu_{j,k} \in \mathbf{Z}/4\mathbf{Z}$ is odd and $\nu_{j,k} = -\nu_{k,j}$.

Starting in Σ_0 we make a tour around 0 in the positive direction and write

$$\begin{aligned}(\phi'_1)^{1/2} &= i^{\nu_{1,0}}(\phi'_0)^{1/2},\\ (\phi'_{-1})^{1/2} &= i^{\nu_{-1,1}}(\phi'_1)^{1/2},\\ (\phi'_0)^{1/2} &= i^{\nu_{0,-1}}(\phi'_{-1})^{1/2}.\end{aligned}$$

This means that if we follow the continuous branch of $(\phi'_0)^{1/2}$ around 0 in the positive sense, then after one tour, we get the branch

$$i^{-(\nu_{0,-1}+\nu_{-1,1}+\nu_{1,0})}(\phi'_0)^{1/2}.$$

But $(\phi'_0)^{1/2} = V^{1/4}$ for a suitable branch of the fourth root and following this function around 0 once in the positive sense, we get $iV^{1/4}$. Hence we get the cocycle condition

$$\nu_{0,-1} + \nu_{-1,1} + \nu_{1,0} \equiv 1 \bmod 4\mathbf{Z}. \tag{7.2.23}$$

We can now specify the signs in the computations of the Wronskians above:

$$\mathrm{Wr}\,(u_j, u_k) = 2\frac{\sqrt{\phi'_j}}{\sqrt{\phi'_k}} + \mathcal{O}(h) = 2i^{\nu_{j,k}} + \mathcal{O}(h). \tag{7.2.24}$$

The space of null solutions is of dimension 2 and any two of u_{-1}, u_0, u_1 are linearly independent, so we have a relation

$$\alpha_{-1}u_{-1} + \alpha_0 u_0 + \alpha_1 u_1 = 0, \tag{7.2.25}$$

where the vector $(\alpha_{-1}, \alpha_0, \alpha_1)^{\mathrm{t}} \in \mathbf{C}^3 \setminus \{0\}$ is well defined up to a scalar factor. Applying $\mathrm{Wr}\,(u_j, \cdot)$ to this relation, we get

$$(\mathrm{Wr}\,(u_j, u_k))_{j,k}\begin{pmatrix}\alpha_{-1}\\ \alpha_0\\ \alpha_1\end{pmatrix} = 0, \tag{7.2.26}$$

or more explicitly,

$$\begin{pmatrix} 0 & a & b \\ -a & 0 & c \\ -b & -c & 0 \end{pmatrix} \begin{pmatrix} \alpha_{-1} \\ \alpha_0 \\ \alpha_1 \end{pmatrix} = 0. \tag{7.2.27}$$

We can take

$$\begin{pmatrix} \alpha_{-1} \\ \alpha_0 \\ \alpha_1 \end{pmatrix} = \begin{pmatrix} c \\ -b \\ a \end{pmatrix}. \tag{7.2.28}$$

Using (7.2.24) we can specify the values of a, b, c and of $\alpha_{-1}, \alpha_0, \alpha_1$:

$$\begin{aligned} a &= \mathrm{Wr}\,(u_{-1}, u_0) = 2i^{\nu_{-1,0}} + \mathcal{O}(h), \\ b &= \mathrm{Wr}\,(u_{-1}, u_1) = 2i^{\nu_{-1,1}} + \mathcal{O}(h), \\ c &= \mathrm{Wr}\,(u_0, u_1) = 2i^{\nu_{0,1}} + \mathcal{O}(h), \end{aligned} \tag{7.2.29}$$

which gives a choice

$$\begin{pmatrix} \alpha_{-1} \\ \alpha_0 \\ \alpha_1 \end{pmatrix} = \begin{pmatrix} i^{\nu_{0,1}} \\ -i^{\nu_{-1,1}} \\ i^{\nu_{-1,0}} \end{pmatrix} + \mathcal{O}(h) = \begin{pmatrix} i^{\nu_{0,1}} \\ i^{\nu_{1,-1}} \\ i^{\nu_{-1,0}} \end{pmatrix} + \mathcal{O}(h). \tag{7.2.30}$$

Bohr-Sommerfeld Quantization Condition for a Potential Well Let V_0 be a real-valued and analytic function on neigh $([A, B], \mathbf{R})$, where $-\infty < A < B < +\infty$. Let $E_0 \in \mathbf{R}$ and assume that there exist $A < \alpha_0 < \beta_0 < B$ such that

$$V_0 - E_0 \begin{cases} > 0, & \text{on } [A, \alpha_0[\cup]\beta_0, B], \\ < 0, & \text{on }]\alpha_0, \beta_0[\,. \end{cases} \tag{7.2.31}$$

Also assume that α_0, β_0 are simple turning points for $V_0(x) - E_0$:

$$V_0'(\alpha_0) < 0, \quad V_0'(\beta_0) > 0. \tag{7.2.32}$$

Then the situation is stable under small perturbations of the real energy E: For $E \in$ neigh $(E_0, \mathbf{R})$ we have simple turning points $\alpha(E) < \beta(E)$ in $]A, B[$ such that $V_0 - E > 0$ on $[A, \alpha(E)[\cup]\beta(E), B]$ and $V_0 - E < 0$ on $]\alpha(E), \beta(E)[$. In Fig. 7.2 we draw the Stokes lines of $V_0 - E$ in a complex neighborhood of $[A, B]$ that have α or β as an end point.

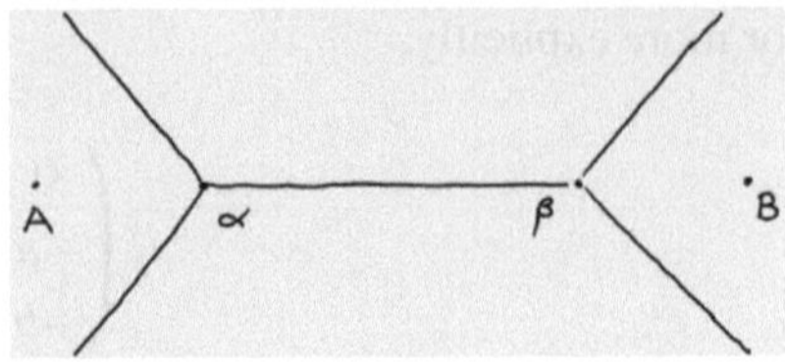

Fig. 7.2 A real potential well with nondegenerate turning points at the ends

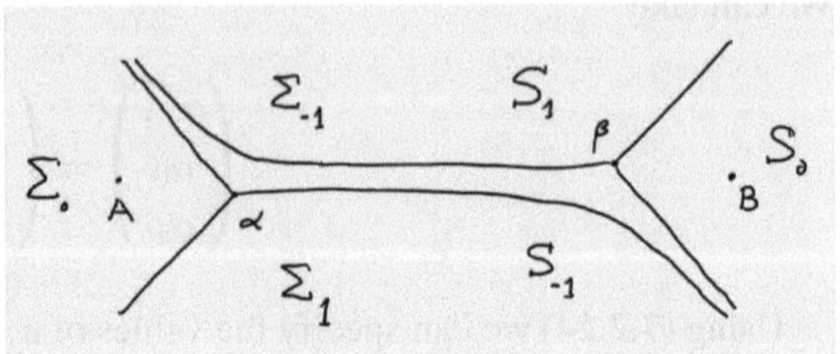

Fig. 7.3 The fate of a real potential well after a complex perturbation of the potential and/or the energy

In particular, the real interval between α and β is a Stokes line which connects the two turning points.

Let $\Omega \Subset \mathbf{C}$ be a complex neighborhood of $[A, B]$ to which V_0 extends holomorphically. Let $V(x) = V_0(x) + W(x)$, where W is holomorphic in Ω and

$$|W(x)| < \epsilon, \ x \in \Omega. \tag{7.2.33}$$

If $\epsilon > 0$ is small enough and E belongs to a small complex neighborhood of E_0, we still have two simple (in general complex) turning points $\alpha = \alpha(V, E)$ and $\beta = \beta(V, E)$ close to α_0 and β_0, respectively, and in general α and β will not be connected by a Stokes line.

The drawing in Fig. 7.3 indicates the three Stokes sectors Σ_k near α and the three Stokes sectors S_k near β for $k = -1, 0, 1$. Note that $A \in \Sigma_0$, $B \in S_0$. For each Σ_k we have an exact solution u_j which is of the form

$$u_j = a_j(z; h)e^{\phi_j(z)/h} \quad \text{in } \overset{\circ}{\Sigma}_j$$

with $\phi_j(\alpha) = 0$ and u_j subdominant (i.e., $|u_j| \ll |u_k|$ for all $k \neq j$) in the interior of Σ_j. Similarly, we have the exact solutions v_j associated to the sectors S_j of the form

$$v_j = b_j(z; h)e^{\psi_j(z)/h} \quad \text{in } \overset{\circ}{S}_j,$$

subdominant in the interior of S_j, $\psi_j(\beta) = 0$.

When dealing with v_j we think of $-z$ as the new independent variable, rather than z (giving a new Schrödinger operator). Consequently, the leading term in b_j becomes $b_{j,0} = (-\psi_j')^{-1/4}$. In analogy with (7.2.22), we have

$$(-\psi_j')^{1/2} = i^{\nu_{j,k}}(-\psi_k')^{1/2}, \tag{7.2.34}$$

where we choose *the same* $\nu_{j,k}$.

u_j satisfy (7.2.25) with α_j as in (7.2.30). v_j satisfy the analogous relations with coefficients, that we denote by β_j and we have $\beta_j = \alpha_j + \mathcal{O}(h)$.

We may arrange so that $u_0(A) = 0$, $v_0(B) = 0$. Also we may arrange so that u_j, v_j, a_j, b_j depend holomorphically on E.

Now consider the Dirichlet problem

$$(-(h\partial)^2 + V - E)u = 0, \ u(A) = u(B) = 0. \tag{7.2.35}$$

In other words, we are looking for the spectrum in neigh $(E_0, \mathbf{C})$ of the unbounded operator

$$P = -(h\partial_x)^2 + V : L^2(]A, B[) \to L^2(]A, B[),$$

with domain

$$\mathcal{D}(P) = \{u \in H_h^2(]A, B[); \ u(A) = u(B) = 0\}.$$

It is well known that the spectrum of P is purely discrete and we see that

$$E \in \sigma(P) \Longleftrightarrow \mathrm{Wr}\,(u_0, v_0) = 0. \tag{7.2.36}$$

In the construction of the subdominant solutions, we may arrange so that

$$u_j = f_j(h)v_{-j} \text{ for } j = \pm 1. \tag{7.2.37}$$

In order to determine the asymptotics of f_j, we compare the asymptotic expansions for u_j and v_{-j} in $\Sigma_j \cap S_{-j}$.

We first look at the exponential factors. In Σ_0, we have $\phi_0(x) = \int_\alpha^x (V(y) - E)^{1/2} dy$ with the continuous branch of the square root which is positive for $x < \alpha$ when V, E are real. Thus,

$$\phi_j(x) = -\int_\alpha^x (V - E)^{1/2} dy, \ j = \pm 1,$$

where the branch of the square root is given by the continuous extension of the one in Σ_0 to the adjacent sectors (and with a cut along the curve that separates Σ_1 and Σ_{-1}). When V, E are real, we get for $j = \pm 1$,

$$\phi_j(x) = -ji \int_\alpha^x (E - V)^{1/2} dy, \ \alpha < x < \beta, \tag{7.2.38}$$

where $(E - V)^{1/2}$ denotes the natural branch of the square root (> 0 on $]\alpha, \beta[$ when V, E are real).

Similarly, $\psi_0(x) = -\int_\beta^x (V(y) - E)^{1/2} dy$ in S_0 with the natural branch so

$$\psi_j(x) = \int_\beta^x (V - E)^{1/2} dy \text{ in } S_j, \ j = \pm 1,$$

with the continuous branch, having a cut along the curve separating S_1 from S_{-1}. Hence with the same branch as in (7.2.38),

$$\psi_j(x) = ji \int_\beta^x (E - V)^{1/2} dy. \tag{7.2.39}$$

It follows that

$$\begin{aligned} \phi_1 &= \psi_{-1} - i \int_\alpha^\beta (E - V)^{1/2} dy, \\ \phi_{-1} &= \psi_1 + i \int_\alpha^\beta (E - V)^{1/2} dy. \end{aligned} \tag{7.2.40}$$

Next compare the leading amplitudes. From the eikonal equations, we know that

$$\phi_0' = (V - E)^{1/2}, \quad -\psi_0' = (V - E)^{1/2}$$

in Σ_0, S_0 respectively, with the natural branches of the square root. Hence,

$$(\phi_0')^{1/2} = (V - E)^{1/4}, \quad (-\psi_0')^{1/2} = (V - E)^{1/4}$$

with the natural "positive" branches of the square and quartic roots.

It follows that

$$(\phi_{\pm 1}')^{1/2} = i^{\nu_{\pm 1,0}} (V - E)^{1/4} = i^{\nu_{\pm 1,0}} e^{\pm i\pi/4} (E - V)^{1/4} \quad \text{in } \Sigma_{\pm 1},$$

where $(E - V)^{1/4}$ denotes the branch which is > 0 on $]\alpha, \beta[$ when V, E are real. Thus for $j = \pm 1$,

$$a_{\pm 1,0} = i^{-\nu_{\pm 1,0}} e^{\mp i\pi/4} (E - V)^{-1/4} \quad \text{in } \Sigma_{\pm 1}. \tag{7.2.41}$$

Similarly,

$$b_{\pm 1,0} = i^{-\nu_{\pm 1,0}} e^{\mp i\pi/4} (E - V)^{-1/4} \text{ in } S_{\pm 1}. \tag{7.2.42}$$

It follows that

$$\begin{aligned} f_1(h) &= g_1 \exp\left(-\frac{i}{h}\int_\alpha^\beta (E-V)^{1/2}dy\right), \\ f_{-1}(h) &= g_{-1} \exp\left(\frac{i}{h}\int_\alpha^\beta (E-V)^{1/2}dy\right), \end{aligned} \tag{7.2.43}$$

where, $g_1 = a_1/b_{-1}$, $g_{-1} = a_{-1}/b_1$ have complete asymptotic expansions in powers of h, with leading terms

$$\begin{aligned} g_{1,0} &= \frac{a_{1,0}}{b_{-1,0}} = i^{-\nu_{1,0}+\nu_{-1,0}} e^{-i\pi/4-i\pi/4} = i^{-\nu_{1,0}+\nu_{-1,0}-1}, \\ g_{-1,0} &= \frac{a_{-1,0}}{b_{1,0}} = i^{-\nu_{-1,0}+\nu_{1,0}} e^{i\pi/4+i\pi/4} = i^{\nu_{1,0}-\nu_{-1,0}+1}. \end{aligned} \tag{7.2.44}$$

From (7.2.25), (7.2.30) and the analogous relations for v_j with coefficients $\beta_j = \alpha_j + \mathcal{O}(h)$, we get

$$u_0 = \widetilde{\alpha}_1 u_1 + \widetilde{\alpha}_{-1} u_{-1}, \ v_0 = \widetilde{\beta}_1 v_1 + \widetilde{\beta}_{-1} v_{-1}, \tag{7.2.45}$$

$$\widetilde{\alpha}_j = -\alpha_j/\alpha_0, \ \widetilde{\beta}_j = -\beta_j/\beta_0, \ \widetilde{\alpha}_j = \widetilde{\beta}_j + \mathcal{O}(h) = \pm 1 + \mathcal{O}(h). \tag{7.2.46}$$

From this and (7.2.37), we get

$$\begin{aligned} \mathrm{Wr}\,(u_0, v_0) &= \widetilde{\alpha}_1\widetilde{\beta}_1 \mathrm{Wr}\,(u_1, v_1) + \widetilde{\alpha}_{-1}\widetilde{\beta}_{-1}\mathrm{Wr}\,(u_{-1}, v_{-1}) \\ &= \widetilde{\alpha}_1\widetilde{\beta}_1 f_1 \mathrm{Wr}\,(v_{-1}, v_1) + \widetilde{\alpha}_{-1}\widetilde{\beta}_{-1} f_{-1}\mathrm{Wr}\,(v_1, v_{-1}) \\ &= (\widetilde{\alpha}_1\widetilde{\beta}_1 f_1 - \widetilde{\alpha}_{-1}\widetilde{\beta}_{-1} f_{-1})\mathrm{Wr}\,(v_{-1}, v_1). \end{aligned}$$

Here $\widetilde{\alpha}_j\widetilde{\beta}_j = 1 + \mathcal{O}(h)$ has a complete asymptotic expansion in powers of h (as well as similar quantities below) and $\mathrm{Wr}\,(v_{-1}, v_1) = 2i^{\nu_{-1,1}} + \mathcal{O}(h) \neq 0$ by the analogue of (7.2.29). Using also (7.2.43), (7.2.44), we get

$$\begin{aligned} \mathrm{Wr}\,(u_0, v_0) = 2i^{\nu_{-1,1}}\Big((1+\mathcal{O}(h)) i^{-\nu_{1,0}+\nu_{-1,0}-1} e^{-\frac{i}{h}\int_\alpha^\beta (E-V)^{1/2}dy} \\ - (1+\mathcal{O}(h)) i^{\nu_{1,0}-\nu_{-1,0}+1} e^{\frac{i}{h}\int_\alpha^\beta (E-V)^{1/2}dy}\Big). \end{aligned}$$

Here $\nu_{1,0} - \nu_{-1,0} + 1$ is odd,

$$\begin{aligned} \mathrm{Wr}\,(u_0, v_0) = 2(1+\mathcal{O}(h)) i^{\nu_{-1,1}+\nu_{1,0}-\nu_{-1,0}+1} e^{-\frac{i}{h}\int_\alpha^\beta (E-V)^{1/2}dy} \\ \left(-e^{\frac{2i}{h}\int_\alpha^\beta (E-V)^{1/2}dy}(1+\mathcal{O}(h)) - 1\right), \end{aligned}$$

and using (7.2.23) we get

$$\mathrm{Wr}\,(u_0, v_0) = -2(1+\mathcal{O}(h))e^{-\frac{i}{h}\int_\alpha^\beta (E-V)^{1/2}dy}\left(e^{\frac{i}{h}(I(E;h)-\pi h)} - 1\right), \tag{7.2.47}$$

where

$$I(E;h) = I_0(E) + \mathcal{O}(h^2),\ \ I_0(E) := 2\int_{\alpha(E)}^{\beta(E)} (E - V(y))^{1/2}dy. \tag{7.2.48}$$

We have

$$I_0(E) = \int_{\widetilde{\gamma}} \zeta dz,$$

where $\widetilde{\gamma} \subset p^{-1}(E)$ is the closed curve which is the concatenation of $\widetilde{\gamma}_+$ and $\widetilde{\gamma}_-$, where

$$\begin{cases} \widetilde{\gamma}_+(t) = ((1-t)\alpha + t\beta, (E - V((1-t)\alpha + t\beta))^{1/2}), \\ \widetilde{\gamma}_-(t) = ((1-t)\beta + t\alpha, -(E - V((1-t)\alpha + t\beta)))^{1/2}), \end{cases} \quad t \in [0, 1[\,.$$

When V and E are real, $\widetilde{\gamma}$ is the real energy curve $p = E$ with the orientation of the Hamiltonian field, and from the Stokes formula it follows that

$$I(E) = \mathrm{vol}_{\mathbf{R}\times\mathbf{R}}\, p^{-1}(]-\infty, E]).$$

It is classical (and rather easy to show) that

$$\partial_E I(E) = T(E) > 0$$

is the primitive period of the real energy curve $p^{-1}(E)$ as a closed H_p-trajectory. In the non-real case, one can still give a sense to and establish the same formula, where now $\Re T(E) > 0$ and $|\Im T(E)| \ll 1$.

From (7.2.47) we see that the eigenvalues of P near E_0 are given by

$$e^{\frac{1}{ih}(I(E;h)-\pi h)} = 1, \tag{7.2.49}$$

or equivalently by the Bohr-Sommerfeld quantization condition

$$I(E;h) = 2\pi(k + \frac{1}{2})h,\ \ k \in \mathbf{Z}. \tag{7.2.50}$$

From the construction it follows that $I(E;h)$ is a holomorphic function of E in a fixed neighborhood of E_0 in $\mathbf{C}$ and has a complete asymptotic expansion in powers of h in the space of such functions with the two leading terms given in (7.2.48). Moreover, $\partial_E I(E; h) = T(E) + \mathcal{O}(h^2) \neq 0$, so $I(\cdot; h)$ is biholomorphic from neigh$(E_0, \mathbf{C})$ onto neigh$(I(E_0), \mathbf{C})$ and (7.2.50) gives a sequence of eigenvalues

$$E_k = I^{-1}\left(2\pi\left(k + \frac{1}{2}\right)h; h\right) = I_0^{-1}\left(2\pi\left(k + \frac{1}{2}\right)h\right) + \mathcal{O}(h^2),$$

all situated on the curve given by the condition that $I(E; h) \in \mathbf{R}$, which is an $\mathcal{O}(h^2)$ deformation of the curve $I_0(E) \in \mathbf{R}$. Notice that the reality of $I_0(E)$ is equivalent to the condition that $\alpha(E)$ and $\beta(E)$ are connected by a Stokes line.

It is not hard to show that the eigenvalues E_k are simple, either by deformation of P to the harmonic oscillator, or by appealing to general facts about Grushin problems. The latter argument has been carried out in [101] (see also [23]). Under suitable assumptions on the behaviour of V near $\pm\infty$, the spectrum of $-(h\partial)^2 + V$ as a closed operator in $L^2(\mathbf{R})$ is still given by a Bohr-Sommerfeld quantization condition [23], and for the proof we just need to complement the WKB-arguments above with a completely analogous study near infinity. (See [101] for details.)

Notes It is from André Voros' lecture notes [157] that we have learnt the basic principle of complex WKB analysis, namely to follow the solutions in the directions of growth. See also the monograph of Fedoryuk [44].

There is a large literature on the complex WKB-method, often with lots of sophisticated and deep considerations (like resurgence), far beyond the scope of this book. Also, there are many spectral results for operators in one dimension based on the complex WKB method. See for instance Redparth [115], Shkalikov [127, 128], Nekhaev and Shafarevich [107], and in the special case of $\mathcal{PT}$-symmetry, [23, 101].

Review of Classical Non-self-adjoint Spectral Theory

8

The first section of this chapter deals with Fredholm theory in the spirit of Appendix A in [60], see also an appendix in [105] and [148]. The remaining sections give a brief account of the very beautiful classical theory of non-self-adjoint operators, taken from a section in [136] which is a brief account of parts of the classical book by Gohberg and Krein [48].

8.1 Fredholm Theory Via Grushin Problems

Most of this section follows Appendix A in [60] quite closely. For simplicity we only consider the case of (separable) Hilbert spaces. Let $\mathcal{H}_1, \mathcal{H}_2$ be two such spaces. Recall that a bounded operator $P : \mathcal{H}_1 \to \mathcal{H}_2$ ($P \in \mathcal{L}(\mathcal{H}_1, \mathcal{H}_2)$) is a Fredholm operator if the null space $\mathcal{N}(P)$ is of finite dimension and the range $\mathcal{R}(P)$ is closed and of finite codimension, where $\operatorname{codim} \mathcal{R}(P) := \dim \mathcal{R}(P)^{\perp}$. Equivalently, P is (a) Fredholm (operator) if there exists $Q \in \mathcal{L}(\mathcal{H}_2, \mathcal{H}_1)$, such that $PQ = 1 + R$, $QP = 1 + L$, where $L : \mathcal{H}_1 \to \mathcal{H}_1$, $R : \mathcal{H}_2 \to \mathcal{H}_2$ are compact. If P is Fredholm, we define its index by

$$\operatorname{ind} P = \dim \mathcal{N}(P) - \operatorname{codim} \mathcal{R}(P). \tag{8.1.1}$$

Let $\Omega \subset \mathbf{C}$ be an open connected set (or an open interval in $\mathbf{R}$) and let

$$\Omega \ni z \mapsto P_z \in \mathcal{L}(\mathcal{H}_1, \mathcal{H}_2) \tag{8.1.2}$$

be a continuous family (i.e., continous for the operator norm; uniformly continuous).

J. Sjöstrand, *Non-Self-Adjoint Differential Operators, Spectral Asymptotics and Random Perturbations*, Pseudo-Differential Operators 14,
https://doi.org/10.1007/978-3-030-10819-9_8

Proposition 8.1.1 *If P_{z_0} is Fredholm for some z_0, then there is a neighborhood $V \subset \Omega$ of z_0 such that*

$$P_z \text{ is Fredholm,} \tag{8.1.3}$$

$$\text{ind } P_z = \text{ind } P_{z_0}, \tag{8.1.4}$$

for every $z \in V$.

Proof Let $n_+^0 = \dim \mathcal{N}(P_{z_0})$, $n_-^0 = \operatorname{codim} \mathcal{R}(P_{z_0})$ and define

$$R_+^0 : \mathcal{H}_1 \to \mathbf{C}^{n_+^0}, \ R_-^0 : \mathbf{C}^{n_-^0} \to \mathcal{H}_2,$$

by

$$R_+^0 u(j) = (u|e_j), \ j = 1, 2, .., n_+^0, \ R_- u_- = \sum_1^{n_-^0} u_-(j) f_j,$$

where $e_1, \dots, e_{n_+^0}$ and $f_1, \dots, f_{n_-^0}$ are orthonormal bases for $\mathcal{N}(P_{z_0})$ and $\mathcal{R}(P_{z_0})^\perp$, respectively. Put

$$\mathcal{P}_z = \begin{pmatrix} P_z & R_-^0 \\ R_+^0 & 0 \end{pmatrix} : \mathcal{H}_1 \times \mathbf{C}^{n_-^0} \to \mathcal{H}_2 \times \mathbf{C}^{n_+^0}. \tag{8.1.5}$$

Using the orthogonal decompositions

$$\mathcal{H}_1 = \mathcal{N}(P_{z_0})^\perp \oplus \mathcal{N}(P), \ \ \mathcal{H}_2 = \mathcal{R}(P_{z_0}) \oplus \mathcal{R}(P)^\perp,$$

we see that $\mathcal{P}_{z_0}$ is bijective with bounded inverse. By continuity, $\mathcal{P}_z$ has the same property for all z in a neighborhood of z_0 and the proposition follows from Proposition 8.1.2 below. □

Proposition 8.1.2 *Let $P \in \mathcal{L}(\mathcal{H}_1, \mathcal{H}_2)$ and let $R_+ : \mathcal{H}_1 \to \mathbf{C}^{n_+}$, $R_- : \mathbf{C}^{n_-} \to \mathcal{H}_2$ be bounded linear operators of maximal ranks $n_+, n_- \in \mathbf{N}$. If*

$$\mathcal{P} = \begin{pmatrix} P & R_- \\ R_+ & 0 \end{pmatrix} : \mathcal{H}_1 \times \mathbf{C}^{n_-} \to \mathcal{H}_2 \times \mathbf{C}^{n_+}$$

is bijective with a bounded inverse, then P is Fredholm of index $n_+ - n_-$. *(When* $n_+ = 0$, *then* R_+, $\mathbf{C}^{n_+}$ *are absent as well as the last line in the matrix for* $\mathcal{P}$, *and similarly when* $n_- = 0$.*)*

Proof Let

$$\mathcal{E} = \begin{pmatrix} E & E_+ \\ E_- & E_{-+} \end{pmatrix} : \mathcal{H}_2 \times \mathbf{C}^{n_+} \to \mathcal{H}_1 \times \mathbf{C}^{n_-}$$

be the inverse of $\mathcal{P}$, so that the system

$$\begin{cases} Pu + R_- u_- = v, \\ R_+ u = v_+, \end{cases}$$

has the unique solution

$$\begin{cases} u = Ev + E_+ v_+ \in \mathcal{H}_1, \\ u_- = E_- v + E_{-+} v_+ \in \mathbf{C}^{n_-}, \end{cases}$$

for any given $v \in \mathcal{H}_2$, $v_+ \in \mathbf{C}^{n_+}$.

The equation, $Pu = v$ can be written $Pu + R_- 0 = v$ and if we introduce the unknown $v_+ = R_+ u$, we get the equivalent system

$$\begin{cases} u = Ev + E_+ v_+, \\ 0 = E_- v + E_{-+} v_+. \end{cases}$$

Thus,

$$\mathcal{R}(P) = \{v \in \mathcal{H}_2;\ \exists v_+ \in \mathbf{C}^{n_+},\ E_- v = -E_{-+} v_+\}, \tag{8.1.6}$$

$$\mathcal{N}(P) = \{E_+ v_+;\ E_{-+} v_+ = 0\}. \tag{8.1.7}$$

The fact that $\mathcal{E} = \mathcal{P}^{-1}$ is equivalent to the following two systems of equations:

$$\begin{aligned} &PE + R_- E_- = 1, \ PE_+ + R_- E_{-+} = 0, \\ &R_+ E = 0, \qquad\qquad R_+ E_+ = 1, \end{aligned} \tag{8.1.8}$$

$$\begin{aligned} &EP + E_+ R_+ = 1, \qquad ER_- = 0, \\ &E_- P + E_{-+} R_+ = 0, \ E_- R_- = 1. \end{aligned} \tag{8.1.9}$$

From the last equation, we see that E_- is surjective and (8.1.6) shows that $\mathcal{R}(P)$ is closed and that $\operatorname{codim}\mathcal{R}(P) = \operatorname{codim}\mathcal{R}(E_{-+})$. Further, (8.1.7) shows that $\mathcal{N}(P)$ is finite-dimensional and

$$\dim \mathcal{N}(P) = \dim \mathcal{N}(E_{-+}).$$

Here we also used the injectivity of E_+, provided by the last equation in (8.1.8).

Thus P is a Fredholm operator and

$$\operatorname{ind} P = \dim \mathcal{N}(E_{-+}) - \operatorname{codim} \mathcal{R}(E_{-+}) = n_+ - n_-,$$

where the last equality is a general fact for the index of any $n_- \times n_+$-matrix. □

The following result can be proved by straightforward computations (cf. (8.1.8), (8.1.9)):

Proposition 8.1.3 *Let P, R_+, R_-, $\mathcal{P}$ be as in Proposition 8.1.2 and assume that $\mathcal{P}$ is bijective with a bounded inverse $\mathcal{E}$ as in the beginning of the proof of that result.*

- *If P is bijective, then E_{-+} is bijective (necessarily P is of index 0, so $n_+ = n_-$) and*

$$E_{-+}^{-1} = -R_+ P^{-1} R_-. \tag{8.1.10}$$

- *If E_{-+} is bijective, then P is bijective and*

$$P^{-1} = E - E_+ E_{-+}^{-1} E_-. \tag{8.1.11}$$

See also Proposition 4.1 in [144] for a characterization of the invertibility of $\mathcal{P}$.

We next review analytic Fredholm theory. Assume that the family P_z in (8.1.2) is not only continuous, but also holomorphic (for the operator norm topology) and that P_z is Fredholm for every $z \in \Omega$. Then we know that $\operatorname{ind} P_z$ is constant and we assume that it is equal to 0.

Proposition 8.1.4 *Assume in addition that P_w is bijective for some $w \in \Omega$. Then*

$$\Sigma := \{z \in \Omega;\ P_z \text{ is not bijective}\}$$

is discrete.

If $z_0 \in \Sigma$, then $z \mapsto P_z^{-1}$ has a pole of order $N_0 < \infty$ at z_0:

$$P_z^{-1} = \sum_{-N_0}^{-1} (z - z_0)^j A_j + Q(z), \tag{8.1.12}$$

where $Q(z) \in \mathcal{L}(\mathcal{H}_2, \mathcal{H}_1)$ is holomorphic in a neighborhood of z_0 and the operators $A_{-N_0}, \dots, A_{-1} \in \mathcal{L}(\mathcal{H}_2, \mathcal{H}_1)$ are of finite rank.

Proof If $z_1 \in \Omega$, we can define

$$\mathcal{P}_z^{z_1} = \begin{pmatrix} P_z & R_-^{z_1} \\ R_+^{z_1} & 0 \end{pmatrix} : \mathcal{H}_1 \times \mathbf{C}^{n_0(z_1)} \to \mathcal{H}_2 \times \mathbf{C}^{n_0(z_1)}$$

with $R_\pm = R_\pm^{z_1}$ independent of z, such that $\mathcal{P}_z^{z_1}$ is bijective for z in a connected neighborhood V_{z_1} of z_1 in Ω. Let

$$\mathcal{E}^{z_1} = \begin{pmatrix} E^{z_1}(z) & E_+^{z_1}(z) \\ E_-^{z_1}(z) & E_{-+}^{z_1}(z) \end{pmatrix}$$

be the inverse, so that $E_{-+}^{z_1}(z)$ is a holomorphic function of $z \in V_{z_1}$ with values in the $n_0(z_1) \times n_0(z_1)$ matrices. Now, $\Sigma \cap V_{z_1}$ coincides with the set of zeros of the holomorphic function $V_{z_1} \ni z \mapsto \det E_{-+}^{z_1}(z)$, which is either a discrete set or equal to V_{z_1}. Covering Ω with such V_{z_1}, we conclude that Σ is either discrete or equal to all of Ω. But the latter possibility is excluded by the assumption that $w \notin \Sigma$ for some $w \in \Omega$.

Now, let $z_0 \in \Sigma$ and choose $\mathcal{P}_z, \mathcal{E}_z$ as above with $z_1 = z_0$. Then

$$P_z^{-1} = E(z) - E_+(z)E_{-+}(z)^{-1}E_-(z), \ z \in V_{z_0}.$$

Here $E_{-+}(z)^{-1}$ has a pole at z_0:

$$E_{-+}(z)^{-1} = \frac{R_{-N_0}}{(z - z_0)^{N_0}} + \cdots + \frac{R_{-1}}{(z - z_0)} + \mathrm{hol}\,(z),$$

$1 \le N_0 < \infty$, $\operatorname{rank} R_{-j} \le n_0 := n_0(z_0)$. Using that $E(z)$, $E_\pm(z)$ are holomorphic we get (8.1.12), where $A_{-N_0}, \dots, A_{-1}$ can be expressed in terms of $R_{-N_0}, \dots, R_{-1}$ and $E_+^{(j)}(z_0)$, $E_-^{(j)}(z_0)$, for $0 \le j \le N_0$. □

Let $P : \mathcal{H} \to \mathcal{H}$ be a closed densely defined operator with domain $\mathcal{D} = \mathcal{D}(P)$. With Ω as above, we assume that

$$\begin{aligned} &P - z : \mathcal{D} \to \mathcal{H} \text{ is Fredholm of index 0 for } z \in \Omega \\ &\text{and bijective for some } z_0 \in \Omega. \end{aligned} \tag{8.1.13}$$

If $\rho(P)$ denotes the resolvent set of P, we know from the above discussion that $\rho(P) \cap \Omega = \Omega \setminus \Sigma$, where $\Sigma = \sigma(P) \cap \Omega$ is discrete.

Proposition 8.1.5 *Write the Laurent series in* (8.1.12) *as*

$$(z-P)^{-1} = (z-z_0)^{-N_0} A_{-N_0} + \cdots + (z-z_0)^{-1} A_{-1} + Q(z) \text{ in } \mathcal{L}(\mathcal{H}, \mathcal{D}), \qquad (8.1.14)$$

where Q is holomorphic near $z = z_0$ and A_{-j} are of finite rank.

The operator $\pi_{-1} := A_{-1}$ is a projection which commutes with P. This implies that the finite-dimensional space $\mathcal{R}(\pi_{-1})$ is contained in the domains of all powers P^k, $k \in \mathbf{N}$, and is invariant under P.

The restriction of $z_0 - P$ to $\mathcal{R}(\pi_{-1})$ is nilpotent. Indeed,

$$A_{-j} = (P-z_0)^{j-1}\pi_{-1}, \ 1 \le j \le N_0, \qquad (8.1.15)$$

$$(P-z_0)^{N_0}\pi_{-1} = 0. \qquad (8.1.16)$$

Let $\gamma = \gamma_r = \partial D(z_0, r)$ be the oriented boundary of the disc $D(z_0, r)$ for $0 < r \ll 1$ small enough. Then

$$\pi_{-1} = \frac{1}{2\pi i}\int_\gamma (z-P)^{-1} dz, \qquad (8.1.17)$$

$$A_{-j} = \frac{1}{2\pi i}\int_\gamma (z-z_0)^{j-1}(z-P)^{-1} dz. \qquad (8.1.18)$$

Proof Equations (8.1.17), (8.1.18) are standard formulae for the Laurent coefficients, obtained by multiplying (8.1.14) by $(z-z_0)^{j-1}$ and then integrating along γ.

Knowing that $\pi_{-1} : \mathcal{H} \to \mathcal{D}$, we apply $P - z_0$ to the left in (8.1.17) and integrate:

$$(P-z_0)\pi_{-1} = \frac{1}{2\pi i}\int_\gamma (P-z_0)(z-P)^{-1} dz.$$

Here the first P in the integral can be replaced by z, since the corresponding difference of integrals is

$$\frac{1}{2\pi i}\int_\gamma (P-z)(z-P)^{-1} dz,$$

which is zero, since the integrand is holomorphic near z_0. Thus,

$$(P-z_0)\pi_{-1} = \frac{1}{2\pi i}\int_\gamma (z-z_0)(z-P)^{-1} dz = A_{-2}.$$

By the same argument, we see that the range of π_{-1} is contained in the domain of P^k for all $k \in \mathbf{N}$ and we get (8.1.16). Of course, if $u \in \mathcal{D}$, then $\pi_{-1}P = P\pi_{-1}$, so π_1 and P commute.

Let us finally recall why π_{-1} is a projection. Let $0 < r_1 < r_2 \ll 1$ and use the resolvent identity, to get with $\gamma_j = \gamma_{r_j}$:

$$\begin{aligned}\pi_{-1}^2 &= \int_{\gamma_2}\int_{\gamma_1}(w-P)^{-1}(z-P)^{-1}\frac{dz}{2\pi i}\frac{dw}{2\pi i}\\ &= \int_{\gamma_2}\int_{\gamma_1}\frac{1}{w-z}(z-P)^{-1}\frac{dz}{2\pi i}\frac{dw}{2\pi i} - \int_{\gamma_2}\int_{\gamma_1}\frac{1}{w-z}(w-P)^{-1}\frac{dz}{2\pi i}\frac{dw}{2\pi i}\\ &=: \mathrm{I}+\mathrm{II}.\end{aligned}$$

Here,

$$\mathrm{I} = \int_{\gamma_1}\left(\int_{\gamma_2}\frac{1}{w-z}\frac{dw}{2\pi i}\right)(z-P)^{-1}\frac{dz}{2\pi i} = \int_{\gamma_1}(z-P)^{-1}\frac{dz}{2\pi i} = \pi_{-1},$$

$$\mathrm{II} = -\int_{\gamma_2}(w-P)^{-1}\left(\int_{\gamma_1}\frac{1}{w-z}\frac{dz}{2\pi i}\right)\frac{dw}{2\pi i} = 0.$$

Hence $\pi_{-1}^2 = \pi_{-1}$. □

One can also follow multiplicities through Grushin reductions, and we refer to [105, 148] and many other papers for such discussions. Under the assumptions of the last proposition, the (algebraic) multiplicity of the eigenvalue z_0 of P is by definition,

$$m(P, z_0) = \dim \mathcal{R}(\pi_{-1}). \tag{8.1.19}$$

Similarly, if

$$\mathcal{P}(z) = \begin{pmatrix} P-z & R_- \\ R_+ & 0 \end{pmatrix} : \mathcal{D}\times\mathbf{C}^{n_0} \to \mathcal{H}\times\mathbf{C}^{n_0}$$

is bijective for $z \in \mathrm{neigh}\,(z_0)$, with inverse

$$\mathcal{E}(z) = \begin{pmatrix} E(z) & E_+(z) \\ E_-(z) & E_{-+}(z) \end{pmatrix},$$

we let $m(E_{-+}, z_0)$ be the multiplicity of z_0 as a zero of $z \mapsto \det(E_{-+}(z))$.

Proposition 8.1.6 *Under the above assumptions, $m(P, z_0) = m(E_{-+}, z_0)$.*

Proof Anticipating the treatment of traces later in this chapter, we have

$$\begin{aligned} m(P, z_0) &= \operatorname{tr} \pi_{-1} = \operatorname{tr} \int_\gamma E_+(z) E_{-+}(z)^{-1} E_-(z) \frac{dz}{2\pi i} \\ &= \int_\gamma \operatorname{tr}\Big(E_+(z) E_{-+}(z)^{-1} E_-(z) \Big) \frac{dz}{2\pi i} \\ &= \int_\gamma \operatorname{tr}\Big(E_{-+}(z)^{-1} E_-(z) E_+(z) \Big) \frac{dz}{2\pi i} \end{aligned}$$

since E_{-+} is a finite-rank operator.

From the relation

$$\partial_z \mathcal{E}(z) = -\mathcal{E}(z) \partial_z \mathcal{P}(z) \mathcal{E}(z),$$

we get

$$\partial_z \mathcal{E}(z) = \begin{pmatrix} E & E_+ \\ E_- & E_{-+} \end{pmatrix} \begin{pmatrix} 1 & 0 \\ 0 & 0 \end{pmatrix} \begin{pmatrix} E & E_+ \\ E_- & E_{-+} \end{pmatrix}$$

and hence,

$$\partial_z E_{-+}(z) = E_-(z) E_+(z).$$

Thus,

$$\begin{aligned} m(P, z_0) &= \int_\gamma \operatorname{tr}\Big(E_{-+}(z)^{-1} \partial_z E_{-+}(z) \Big) \frac{dz}{2\pi i} \\ &= \int_\gamma \frac{\partial_z \det E_{-+}(z)}{\det E_{-+}(z)} \frac{dz}{2\pi i} \\ &= m(E_{-+}, z_0). \end{aligned}$$

□

8.2 Singular Values

In the remainder of this chapter we give a brief account of abstract non-self-adjoint theory, following [48] closely.

Let $\mathcal{H}$ be a separable complex Hilbert space. If $A \in \mathcal{L}(\mathcal{H}, \mathcal{H})$ is compact, we let $s_1(A) \geq s_2(A) \geq \cdots \searrow 0$ be the eigenvalues of the compact self-adjoint operator

$(A^*A)^{1/2}$. They are called the singular values of A. We notice that $s_j(A^*) = s_j(A)$. In fact, this follows from the intertwining relations:

$$A(A^*A) = (AA^*)A, \ (A^*A)A^* = A^*(AA^*).$$

The singular values appear naturally in the *polar decomposition:* If $A \in \mathcal{L}(\mathcal{H}, \mathcal{H})$, then

$$\|Au\|^2 = (Au|Au) = (A^*Au|u) = ((A^*A)^{1/2}u|(A^*A)^{1/2}u) = \|(A^*A)^{1/2}u\|^2.$$

Let $e_1, e_2, \ldots$ be an orthonormal family of eigenvectors of $(A^*A)^{1/2}$ associated to the eigenvalues $s_1(A), s_2(A), \ldots$ that are > 0. It is an orthonormal basis for $\overline{\mathcal{R}((A^*A)^{1/2})}$, and $\mathcal{R}((A^*A)^{1/2})^{\perp} = \mathcal{N}((A^*A)^{1/2}) = \mathcal{N}(A)$. The operator

$$U : \mathcal{R}((A^*A)^{1/2}) \ni (A^*A)^{1/2}u \mapsto Au \in \mathcal{R}(A)$$

is isometric and bijective. It extends to a unitary operator, which we also denote by U, from $\overline{\mathcal{R}((A^*A)^{1/2})}$ to $\overline{\mathcal{R}(A)}$, and to a partial isometry if we put $U = 0$ on the orthogonal space $\mathcal{N}(A)$. We get the polar decomposition

$$A = U(A^*A)^{1/2}. \tag{8.2.1}$$

This leads to the Schmidt decomposition of A

$$Au = \sum s_j(A)(u|e_j)f_j, \tag{8.2.2}$$

where $f_j = Ue_j$ is also an orthonormal family.

Recall the mini-max characterization of the s_j ([48], p. 25):

$$s_j(A) = \inf_{\substack{E\subset\mathcal{H};\ E \text{ is a closed} \\ \text{subspace of } \mathcal{H} \\ \text{of codimension } \leq j-1}} \ \sup_{u\in E\setminus\{0\}} \frac{((A^*A)^{1/2}u|u)}{\|u\|^2}. \tag{8.2.3}$$

From that we get the following characterization of the singular values due to Allahverdiev [48, pp. 28, 29]:

Theorem 8.2.1 *Let $A \in \mathcal{L}(\mathcal{H}, \mathcal{H})$ be compact. Then*

$$s_{n+1}(A) = \min_{\substack{K\in\mathcal{L}(\mathcal{H},\mathcal{H}) \\ K \text{ of rank} \leq n}} \|A - K\|, \quad n = 0, 1, \ldots$$

The minimum is achieved by an operator K for which $s_1(K) = s_1(A)$,...,$s_n(K) = s_n(A)$, $s_{n+1}(K) = 0$, $s_1(A - K) = s_{n+1}(A)$, $s_2(A - K) = s_{n+2}(A)$, ...

Proof If K is of rank $\le n$, then $\mathcal{N}(K)$ is of codimension $\le n$ and

$$s_{n+1}(A) \le \sup_{0\neq u\in\mathcal{N}(K)} \frac{\|Au\|}{\|u\|} = \sup_{0\neq u\in\mathcal{K}} \frac{\|(A-K)u\|}{\|u\|} \le \|A-K\|.$$

To obtain the minimizing operator consider the polar decomposition $A = U(A^*A)^{1/2}$ and take $K = U(A^*A)^{1/2}P_n$, where P_n is the orthogonal projection onto the space spanned by $e_1, \dots, e_n$. Then

$$A - K = U(A^*A)^{1/2}(1-P_n),$$

$$(A-K)^*(A-K) = (1-P_n)(A^*A)^{1/2}U^*U(A^*A)^{1/2}(1-P_n) = A^*A(1-P_n),$$

and we get $s_{n+1}(A) = \|A-K\|$ as well as the statement about the singular values of K. □

The following corollary is due to Ky Fan.

Corollary 8.2.2 *Let $A, B \in \mathcal{L}(\mathcal{H})$ be compact. Then for $n, m \ge 1$,*

$$s_{m+n-1}(A+B) \le s_m(A) + s_n(B), \tag{8.2.4}$$

$$s_{m+n-1}(AB) \le s_m(A)s_n(B). \tag{8.2.5}$$

Proof Let K_A, K_B be operators of rank $\le m-1$ and $\le n-1$, respectively, such that

$$s_m(A) = \|A - K_A\|, \; s_n(B) = \|B - K_B\|.$$

Then

$$\begin{aligned} s_{m+n-1}(A+B) &\le \|A+B-(K_A+K_B)\| \le \|A-K_A\| + \|B-K_B\| \\ &= s_m(A) + s_n(B). \end{aligned}$$

Since

$$(A-K_A)(B-K_B) = AB - ((A-K_A)K_B + K_AB),$$

where the last term in the parentheses is of rank $\le m+n-2$, we have

$$s_{m+n-1}(AB) \le \|(A-K_A)(B-K_B)\| \le \|A-K_A\|\|B-K_B\|| \le s_m(A)s_n(B).$$

□

Corollary 8.2.3 $|s_n(A) - s_n(B)| \leq \|A - B\|$.

Proof Let K be an operator of rank $n-1$. Then

$$s_n(A) \leq \|A - K\| = \|B - K + A - B\| \leq \|B - K\| + \|A - B\|.$$

Varying K, we get $s_n(A) \leq s_n(B) + \|A - B\|$, and we have the same inequality with A and B exchanged. □

We now discuss Weyl inequalities, and start with the following result of Weyl (see [48, pp. 35, 36]):

Theorem 8.2.4 *Let $A \in \mathcal{L}(\mathcal{H}, \mathcal{H})$ be compact and let $\lambda_1(A), \lambda_2(A), \dots$ be the non-vanishing eigenvalues of A arranged in such a way that $|\lambda_1| \geq |\lambda_2| \geq \cdots$ and repeated according to their multiplicity (which by definition is the rank of the associated spectral projection). Then for every $n \geq 1$ for which $\lambda_n(A)$ is defined, we have*

$$|\lambda_1(A) \cdots \lambda_n(A)| \leq s_1(A) \cdots s_n(A). \tag{8.2.6}$$

Proof For $n = 1$, (8.2.6) just says that $|\lambda_1(A)| \leq \|A\|$. Approaching A by a sequence of finite rank operators, we can assume that A is of finite rank and replace $\mathcal{H}$ by the finite-dimensional space $\mathcal{R}(A) + (\mathcal{N}(A))^\perp$, which we denote by $\mathcal{H}$ from now on. (Indeed, λ_j and s_j depend continuously on A.) Introduce the space

$$\bigwedge^n \mathcal{H} = \mathcal{H} \wedge \cdots \wedge \mathcal{H} \tag{8.2.7}$$

generated by n-fold exterior products of vectors in $\mathcal{H}$. $\bigwedge^n \mathcal{H}$ is a Hilbert space with a scalar product that satisfies

$$(u_1 \wedge .. \wedge u_n | v_1 \wedge .. \wedge v_n) = \det((u_j | v_k)), \ u_j, v_j \in \mathcal{H}. \tag{8.2.8}$$

Further, there is a linear operator $\wedge^n A : \bigwedge^n \mathcal{H} \to \bigwedge^n \mathcal{H}$ which is uniquely determined by the condition

$$(\wedge^n A)(u_1 \wedge \cdots \wedge u_n) = Au_1 \wedge \cdots \wedge Au_n, \ u_j \in \mathcal{H}. \tag{8.2.9}$$

Using a basis of generalized eigenvectors, we see that the eigenvalues of $\wedge^n A$ are the values $\lambda_{j_1} \cdots \lambda_{j_n}$, with $j_\nu \neq j_\mu$, for $\nu \neq \mu$. The eigenvalue of greatest modulus is then $\lambda_1 \cdots \lambda_n$. On the other hand, the adjoint of $\wedge^n A$ is $\wedge^n A^*$. We also have

$(\wedge^n A)(\wedge^n B) = \wedge^n(AB)$. Then $(\wedge^n A)^*(\wedge^n A) = \wedge^n(A^*A)$ and this operator has the eigenvalues $s_{j1}^2(A)\cdots s_{jn}(A)^2$, out of which the largest one is

$$(s_1(A)\cdots s_n(A))^2 = \| \wedge^n A\|^2 \geq |\lambda_1|^2 \cdots |\lambda_n|^2.$$

The proof is complete. □

In the same spirit we have the inequality of Horn (see [48, p. 48]):

$$\prod_{j=1}^{n} s_j(AB) \leq \prod_{j=1}^{n} (s_j(A)s_j(B)). \tag{8.2.10}$$

Proof As before it suffices to treat the case when $\mathcal{H}$ is of finite dimension. The largest eigenvalue of

$$(\wedge^n AB)^*(\wedge^n AB) = \wedge^n((AB)^*AB)$$

is equal to $(s_1(AB)\cdots s_n(AB))^2$. On the other hand,

$$\begin{aligned}((\wedge^n AB)^*(\wedge^n AB)u|u) &= \|(\wedge^n AB)u\|^2 = \|(\wedge^n A)\circ(\wedge^n B)u\|^2 \\ &\leq \| \wedge^n A\|^2 \| \wedge^n B\|^2 \|u\|^2 \leq (s_1(A)\cdots s_n(A))^n (s_1(B)\cdots s_n(B))^2 \|u\|^2,\end{aligned}$$

and taking the supremum over all normalized u, we obtain the required inequality. □

We next need a convexity inequality, due to Weyl and Hardy, Littlewood, and Pólya [48, p. 37].

Lemma 8.2.5 *Let $\Phi(x)$ be a convex function on $\mathbf{R}$, which tends to 0, when $x \to -\infty$. Let $a_1 \geq \cdots \geq a_N$, $b_1 \geq \cdots \geq b_N$ be real numbers satisfying*

$$\sum_{j=1}^{k} a_j \leq \sum_{j=1}^{k} b_j,\ 1 \leq k \leq N.$$

Then,

$$\sum_{j=1}^{k} \Phi(a_j) \leq \sum_{j=1}^{k} \Phi(b_j),\ 1 \leq k \leq N.$$

Proof Approaching Φ by a sequence of smooth functions, we can reduce the proof to the case when $\Phi \in C^\infty$. Then $\Phi' \geq 0$,

$$\Phi'(x) \to 0, \; x \to -\infty. \tag{8.2.11}$$

Letting $y \to -\infty$ in the identity

$$\Phi(x) = \Phi(y) + \int_y^x \Phi'(t)dt, \tag{8.2.12}$$

we get

$$\Phi(x) = \int_{-\infty}^x \Phi'(t)dt.$$

From the convergence of the last integral, we conclude that $\int_y^0 \Phi'(t)dt \leq C$, $y \leq 0$, implying that $|y|\Phi'(y)$ is a bounded function for $y \leq 0$. Also, $|x-y|\Phi'(y) \leq \Phi(x)$ for $y \leq x \leq 0$, so $|y|\Phi'(y) \leq 2\Phi(x)$ when $|y| \geq 2|x|$. Hence $|y|\Phi'(y) \to 0$ when $y \to -\infty$.

Integration by parts in (8.2.12) gives

$$\begin{aligned}\Phi(x) &= \Phi(y) + [(t-x)\Phi'(t)]_{t=y}^x - \int_y^x (t-x)\Phi''(t)dt \\ &= \Phi(y) + (x-y)\Phi'(y) + \int_y^x (x-t)\Phi''(t)dt.\end{aligned}$$

Letting y tend to $-\infty$, we get

$$\Phi(x) = \int_{-\infty}^x (x-t)\Phi''(t)dt = \int (x-t)_+\Phi''(t)dt.$$

Hence

$$\sum_1^k \Phi(a_j) = \int \left(\sum_{j=1}^k (a_j - t)_+\right) \Phi''(t)dt.$$

Let $k(t) \leq k$ be the largest $\widetilde{k} \leq k$ with $a_{\widetilde{k}} \geq t$. Then

$$\sum_{j=1}^k (a_j - t)_+ = \sum_{j=1}^{k(t)} (a_j - t) \leq \sum_{j=1}^{k(t)} (b_j - t) \leq \sum_{j=1}^k (b_j - t)_+.$$

Hence $\sum_1^k \Phi(a_j) \leq \sum_1^k \Phi(b_k)$. □

As a consequence we get the following result of H. Weyl [48, pp. 39, 40]:

Theorem 8.2.6 *Let $A : \mathcal{H} \to \mathcal{H}$ be a compact operator, and let $f(x) \geq 0$ be a continuous function on $[0, \infty[$ with $f(0) = 0$ such that $f(e^t)$ is convex. Let λ_j and s_j be the eigenvalues and singular values of A, arranged as $|\lambda_1| \geq |\lambda_2| \geq \cdots$, $s_1 \geq s_2 \geq \cdots$. Then for every $k \geq 1$,*

$$\sum_{j=1}^{k} f(|\lambda_j|) \leq \sum_{j=1}^{k} f(s_j). \tag{8.2.13}$$

Proof We know from Theorem 8.2.4 that

$$\sum_{j=1}^{k} \log|\lambda_j| \leq \sum_{j=1}^{k} \log s_j,$$

and it suffices to apply the preceding convexity lemma. □

Corollary 8.2.7 *For every $p > 0$, we have*

$$\sum_{j=1}^{n} |\lambda_j(A)|^p \leq \sum_{j=1}^{n} s_j(A)^p.$$

For every $r > 0$, we have

$$\prod_{j=1}^{n} (1 + r|\lambda_j(A)|) \leq \prod_{j=1}^{n} (1 + r s_j(A)).$$

Let A, B be compact operators. With $\Phi(t) = e^t$, $a_j = \log s_j(AB)$, $b_j = \log(s_j(A)s_j(B))$, we get from Horn's inequality (8.2.10) and Lemma 8.2.5:

Corollary 8.2.8 $\sum_{j=1}^{n} s_j(AB) \leq \sum_1^n s_j(A)s_j(B)$.

Let $C_\infty \subset \mathcal{L}(\mathcal{H})$ be the subspace of compact operators. The following Lemma is due to Ky Fan [48, p. 47].

Lemma 8.2.9 *Let $A \in C_\infty$. Then for every $1 \leq n \in \mathbf{N}$, we have*

$$\sum_{j=1}^{n} s_j(A) = \max \sum_{j=1}^{n} (UA\phi_j|\phi_j),$$

where the maximum is taken over the set of all unitary operators U and all orthonormal systems $\phi_1, \ldots, \phi_n$.

For the next result, see [48, p. 48]:

Corollary 8.2.10 *If $A, B \in \mathcal{S}_\infty$, then*

$$\sum_{j=1}^{n} s_j(A+B) \leq \sum_{j=1}^{n} s_j(A) + \sum_{j=1}^{n} s_j(B).$$

8.3 Schatten–von Neumann Classes

We are now ready to discuss the Schatten–von Neumann classes (still following [48]).

Definition 8.3.1 For $1 \leq p \leq \infty$, we put

$$C_p = \{A \in C_\infty;\ \|(s_j(A))_1^\infty\|_{\ell^p} < \infty\},$$

where

$$\|(s_j(A))_1^\infty\|_{\ell^p} = \begin{cases} \left(\sum_1^\infty s_j(A)^p\right)^{1/p}, & \text{if } p < \infty, \\ \sup s_j(A), & \text{if } p = \infty. \end{cases}$$

Theorem 8.3.2 *C_p is a closed two-sided ideal in $\mathcal{L}(\mathcal{H}, \mathcal{H})$ equipped with the norm*

$$\|A\|_{C_p} = \|(s_j(A))_1^\infty\|_{\ell^p}.$$

If $p_1 \leq p_2$, then $C_{p_1} \subset C_{p_2}$. The space of finite rank operators is dense in C_p for every p.

We will only recall the proof of the fact that $\|\cdot\|_{C_p}$ satisfies the triangle inequality. Let $A, B \in C_p$, and put $\xi_j = s_j(A+B)$, $\eta_j = s_j(A) + s_j(B)$. According to Corollary 8.2.10, we have $\sum_1^n \xi_j \leq \sum_1^n \eta_j$, $\forall n$, and hence

$$\|A+B\|_{C_1} \leq \|A\|_{C_1} + \|B\|_{C_1},$$

by letting n tend to ∞. The case $p = \infty$ follows from the triangle inequality for the ordinary operator norm.

It remains to treat the case $1 < p < \infty$. Both sequences ξ_j and η_j are decreasing. It suffices to show that

$$\|\xi\|_{\ell^p} \leq \|\eta\|_{\ell^p}. \tag{8.3.1}$$

We have

$$\|\xi\|_{\ell^p} = \sup_{\substack{\zeta\in\ell^q,\\ \|\zeta\|_{\ell^q}=1}} \langle \xi, \zeta\rangle,$$

by Hölder's inequality, where $q \in [1, +\infty[$ is the conjugate index, given by $p^{-1}+q^{-1} = 1$ and $\langle\cdot|\cdot\rangle$ denotes the real scalar product on $\ell^2(\{1, 2, \dots\})$. We also know that the supremum is attained by a $\zeta = \zeta^0$ of the form $\zeta_j^0 = (\text{const.} > 0)\xi_j^{p/q}$, and in particular ζ_j^0 is a decreasing sequence. We use partial summation with $\Xi_j = \sum_1^j \xi_k$, $\Xi_0 = 0$:

$$\begin{aligned}
\langle \xi|\zeta^0\rangle^{(n)} &:= \sum_{j=1}^{n} \xi_j\zeta_j^0 = \sum_{j=1}^{n}(\Xi_j - \Xi_{j-1})\zeta_j^0\\
&= \sum_{j=1}^{n} \Xi_j\zeta_j^0 - \sum_{j=1}^{n-1} \Xi_j\zeta_{j+1}^0 = \Xi_n\zeta_n^0 + \sum_{j=1}^{n-1} \Xi_j(\zeta_j^0 - \zeta_{j+1}^0)\\
&= \left(\sum_{j=1}^{n} \xi_k\right)\zeta_n^0 + \sum_{j=1}^{n-1}\sum_{k=1}^{j} \xi_k \underbrace{(\zeta_j^0 - \zeta_{j+1}^0)}_{\geq 0}
\end{aligned}$$

The last expression is $\leq$ the same expression with ξ replaced by η, and running the same calculation backwards, the latter expression is equal to $\langle\eta|\zeta^0\rangle^{(n)}$. Hence $\langle\xi|\zeta^0\rangle^{(n)} \leq \langle\eta|\zeta^0\rangle \leq \|\eta\|_{\ell^p}$. Letting $n \to \infty$, we get (8.3.1), and this completes the proof of the triangle inequality for the C_p-norms. □

We notice that $\|A\| = s_1(A) \leq \|A\|_{C_p}$. The space C_1 is the space of nuclear or trace-class operators, and C_2 is the space of Hilbert–Schmidt operators. We have the following Hölder type result:

Theorem 8.3.3 *Let $p, q \in [1, \infty]$ be conjugate indices; $p^{-1} + q^{-1} = 1$. If $A \in C_p$, $B \in C_q$, then $AB \in C_1$ and $\|AB\|_{C_1} \leq \|A\|_{C_p}\|B\|_{C_q}$.*

Proof By Corollary 8.2.8, we know that

$$\sum_{j=1}^{n} s_j(AB) \leq \sum_{j=1}^{n} s_j(A)s_j(B),$$

and letting n tend to ∞, we get from the usual Hölder inequality:

$$\sum_{j=1}^{\infty} s_j(AB) \le \sum_{j=1}^{\infty} s_j(A)s_j(B) \le \|s.(A)\|_{\ell^p} \|s.(B)\|_{\ell^q} = \|A\|_{C_q} \|B\|_{C_q}.$$

□

Remark 8.3.4 When $p = \infty$, the assumption $A \in C_\infty$ can be weakened to $A \in \mathcal{L}(\mathcal{H}, \mathcal{H})$, with $\|A\|_{C_\infty}$ replaced by $\|A\|_{\mathcal{L}(\mathcal{H},\mathcal{H})}$. Idem for B when $q = \infty$.

8.4 Traces and Determinants

We next discuss *the trace* of a nuclear operator. If $A \in \mathcal{L}(\mathcal{H}, \mathcal{H})$ is of finite rank, we choose a finite-dimensional subspace $\mathcal{H}' \subset \mathcal{H}$ such that $\mathcal{N}(A)^\perp$, $\mathcal{R}(A) \subset \mathcal{H}'$. We then define the trace of A, $\operatorname{tr} A$ as the trace $\operatorname{tr} A|_{\mathcal{H}'}$ of the restriction of A to $\mathcal{H}'$. We can check that this does not depend on the choice of $\mathcal{H}'$ and that $\operatorname{tr} A$ is the sum of the finitely many non-vanishing eigenvalues of A (each counted with its algebraic multiplicity). We see that

$$A \mapsto \operatorname{tr} A \tag{8.4.1}$$

is a linear functional on the space of finite rank operators. Moreover, by Corollary 8.2.7,

$$|\operatorname{tr} A| \le \sum |\lambda_j(A)| \le \|A\|_{C_1}. \tag{8.4.2}$$

We can then extend (8.4.1) to a continuous linear functional on C_1, and we still have

$$|\operatorname{tr} A| \le \|A\|_{C_1}. \tag{8.4.3}$$

In the case of finite rank operators, we also have the cyclicity of the trace

$$\operatorname{tr} AB = \operatorname{tr} BA. \tag{8.4.4}$$

Let now $A \in C_p$, $B \in C_q$, where $p, q \in [1, \infty]$ are conjugate indices, and choose A_ν, B_ν, $\nu = 1, 2, \dots$ of finite rank, so that $\|A - A_\nu\|_{C_p} \to 0$, $\|B - B_\nu\|_{C_q} \to 0$. Then

$$\begin{aligned}\|AB - A_\nu B_\nu\|_{C_1} &= \|(A - A_\nu)B + A_\nu(B - B_\nu)\|_{C_1} \\ &\le \|A - A_\nu\|_{C_p} \|B\|_{C_q} + \|A_\nu\|_{C_p} \|B - B_\nu\|_{C_q} \to 0, \ \nu \to \infty.\end{aligned}$$

Using this also for BA and the cyclicity of the trace (8.4.4) for finite rank operators, we obtain (8.4.4) also in the case $A \in C_p$, $B \in C_q$, where $p, q \in [1, \infty]$ are conjugate indices.

When $A \in \mathcal{L}(\mathcal{H}, \mathcal{H})$ we still have (8.4.4). Indeed, if B is of finite rank, let $A_1, A_2, \dots$ be a sequence of finite-rank operators converging to A weakly. Then,

$$\mathrm{tr}\,(AB) \leftarrow \mathrm{tr}\,(A_\nu B) = \mathrm{tr}\,(BA_\nu) \to \mathrm{tr}\,(BA), \quad \nu \to \infty.$$

If A is bounded and $B \in C_1$, let $B_\nu \in C_1$ be a sequence of operators converging to B in C_1. Then

$$\mathrm{tr}\,(AB) \leftarrow \mathrm{tr}\,(AB_\nu) = \mathrm{tr}\,(B_\nu A) \to \mathrm{tr}\,(BA), \quad \nu \to \infty,$$

and the claim follows.

Remark 8.4.1 One simple way of extending most of the theory to the case of operators $A : \mathcal{H}_1 \to \mathcal{H}_2$, where $\mathcal{H}_1, \mathcal{H}_2$ are two different Hilbert spaces, is the following. Consider the corresponding operator

$$\begin{pmatrix} 0 & 0 \\ A & 0 \end{pmatrix} : \mathcal{H}_1 \oplus \mathcal{H}_2 \to \mathcal{H}_1 \oplus \mathcal{H}_2, \tag{8.4.5}$$

and say that A belongs to C_p if the operator in (8.4.5) does. The cyclicity of the trace still holds in this setting, namely if $A : \mathcal{H}_1 \to \mathcal{H}_2$ and $B : \mathcal{H}_2 \to \mathcal{H}_1$ belong to C_p and C_q, respectively, where p and q are conjugate indices.

We next discuss determinants of trace-class perturbations of the identity operator. Let first $A \in \mathcal{L}(\mathcal{H}, \mathcal{H})$ be of finite rank and chose a finite dimensional Hilbert space as above. Then we define

$$\det(1 - A) = \det((1 - A)|_{\mathcal{H}'}) = \prod_j (1 - \lambda_j(A)), \tag{8.4.6}$$

where $\lambda_j(A)$ denote the non-zero eigenvalues, repeated according to their multiplicity. We remark that

$$|\det(1 - A)| \le \prod_j (1 + |\lambda_j(A)|) \le \prod_j (1 + s_j(A)) \le e^{\sum_j s_j(A)},$$

where the second inequality follows from Corollary 8.2.7.

We want to extend the definition to the case when $A \in C_1$. Let first I be a compact interval and let $I \ni t \mapsto A_t$ be a C^1 family of finite-rank operators, with

$\mathcal{N}(A_t)^\perp$, $\mathcal{R}(A_t) \subset \mathcal{H}'$, for some finite-dimensional subspace $\mathcal{H}'$ which is independent of t. We first assume that $1 - A_t$ is invertible for all $t \in I$, or in other words, that $1 - \lambda_j(A_t) \neq 0$ for all t and j. Then $\det(1 - A_t) \neq 0$, and by a classical formula,

$$\frac{\partial}{\partial t} \log\det(1 - A_t) = -\mathrm{tr}\left((1 - A_t)^{-1} \frac{\partial}{\partial t}(A_t)\right) = -\mathrm{tr}\left(\left(\frac{\partial}{\partial t} A_t\right)(1 - A_t)^{-1}\right).$$

Hence,

$$\left|\frac{\frac{\partial}{\partial t}\det(1 - A_t)}{\det(1 - A_t)}\right| = \left|\frac{\partial}{\partial t}\log\det(1 - A_t)\right| \leq \|(1 - A_t)^{-1}\| \|\frac{\partial}{\partial t} A_t\|_{C_1}.$$

In particular, if $I = [0, 1]$ and $A_t = tA_1 + (1 - t)A_0$, we get

$$|\log\det(1 - A_1) - \log\det(1 - A_0)| \leq \sup_{0 \leq t \leq 1} \|(1 - (tA_1 + (1 - t)A_0))^{-1}\| \|A_1 - A_0\|_{C_1}. \tag{8.4.7}$$

Now let $A \in C_1$. If $1 - A$ is not invertible, we put $\det(1 - A) = 0$. Assume then that $1 - A$ is invertible. Let A_ν be a sequence of finite-rank operators which converges to A in C_1. For ν, μ large enough, we have $\|(1 - A_\nu)^{-1}\| \leq C_0$ for some fixed constant C_0, and more generally $\|(1 - (tA_\nu + (1 - t)A_0))^{-1}\| \leq C_0$. Then

$$|\log\det(1 - A_\nu) - \log\det(1 - A_\mu)| \leq C_0\|A_\nu - A_\mu\|_{C_1},$$

and consequently $\lim_{\nu\to\infty} \log\det(1 - A_\nu)$ exists. We then put

$$\det(1 - A) = \exp\left(\lim_{\nu\to\infty} \log\det(1 - A_\nu)\right). \tag{8.4.8}$$

Notice that

$$|\det(1 - A)| \leq \prod_j (1 + s_j(A)) \leq e^{\|A\|_{C_1}}. \tag{8.4.9}$$

Using approximation by finite-rank operators, we also see that

$$\det((1 - A)(1 - B)) = \det(1 - A)\det(1 - B). \tag{8.4.10}$$

By the same argument, we can extend (8.4.7) to general trace-class operators for which $1 - (tA_1 + (1 - t)A_0)$ is invertible.

We now add a complex variable $z \in \mathbf{C}$ and consider the function $z \mapsto \det(1 - zA)$. If A is of finite rank, then this is an entire function of z. If $A \in C_1$, let $A_\nu \to A$ be a sequence of finite-rank operators. Then $|\det(1 - zA_\nu)| \le e^{|z|\|A_\nu\|_{C_1}}$. If $z\lambda_j(A) \neq 1, \forall j$, then $\det(1 - zA_\nu) \to \det(1 - zA)$ with locally uniform convergence in $\mathbf{C}\setminus\bigcup_j\{1/\lambda_j\}$, and it follows that $\det(1 - zA)$ is a holomorphic function on this set, that verifies

$$|\det(1 - zA)| \le e^{\|A\|_{C_1}|z|}. \tag{8.4.11}$$

Therefore, $\det(1 - zA_\nu)$ converges to an entire function $f(z)$ locally uniformly on $\mathbf{C}$. If $z = 1/\lambda_j(A)$, where $\lambda_j(A)$ is of multiplicity m, then exactly m eigenvalues of A_ν will converge to $\lambda_j(A)$, while the others will stay away from a neighborhood of this point (when ν is large enough). Considering the argument variation (Rouché), we conclude that $f(z)$ vanishes to the order m at $1/\lambda_j(A)$, and in particular we have $f(z) = \det(1 - zA)$ also at that point. In conclusion, we have

Proposition 8.4.2 *Let $A \in C_1$. Then $D_A(z) := \det(1 - zA)$ is an entire function whose zeros, counted with multiplicity, coincide with the values $1/\lambda_1(A), 1/\lambda_2(A), \ldots$ counted with the multiplicities of $\lambda_1(1), \lambda_2(A), \ldots$*

Observe that

$$D_A(0) = 1 \tag{8.4.12}$$

Also observe that $D_A(z)$ is of subexponential growth, in the sense that for every $\epsilon > 0$ there exists a constant $C_\epsilon > 0$ such that

$$|D_A(z)| \le C_\epsilon e^{\epsilon|z|}. \tag{8.4.13}$$

In fact, by a limiting argument, we have

$$|D_A(z)| \le \prod_{j=1}^{\infty}(1 + |z|s_j(A)) \le \prod_{j=1}^{N}(1 + |z|s_j(A))e^{\sum_{N+1}^{\infty} s_j(A))|z|}.$$

Here the prefactor is of polynomial growth for every fixed N and for a given $\epsilon > 0$, we can always choose $N > 0$ so large that the exponent is $\le \epsilon|z|$.

The next observation is that $W(z) := \prod_1^\infty(1 - z\lambda_j(A))$ is also an entire function of subexponential growth. This follows by the same argument, if we recall that the series $\sum_1^\infty |\lambda_j(A)|$ is convergent. We now use a special case of a theorem of Hadamard (see [4, Ch. 3, Section 3.2]): Since $D_A(z)$ and $W(z)$ have the same zeros (counted with multiplicity), we have

$$D_A(z) = W(z)e^{g(z)},$$

where $g(z)$ is an entire function. It is also clear that we can choose g with $g(0) = 0$. The function

$$\Re g(z) := \log |D_A(z)| - \sum_{j=1}^{\infty} \log |1 - \lambda_j z| \tag{8.4.14}$$

is harmonic. Let $R \geq 2$. For $|z| \leq R/2$, we have

$$\Re g(z) = \int_{|w|=R} P_R(z, w) \Re g(w) |dw|, \tag{8.4.15}$$

where $P_R(z, w) = R^{-1} P_1(z/R, w/R)$ is the Poisson kernel for the disc of radius R and $|dw|$ denotes the length element on the boundary of this disc. It is easy to see that

$$\frac{1}{CR} \leq P_R(z, w) \leq \frac{C}{R}, \text{ when } |z| \leq \frac{R}{2}, \ |w| = R, \tag{8.4.16}$$

where $C > 0$ is independent of R. Using the subexponential growth of $D_A(z)$, we get

$$\int_{|w|=R} P_R(z, w) \log |D_A(w)||dw| \leq \frac{C}{R} \epsilon R R \leq C \epsilon R, \tag{8.4.17}$$

for $R \geq R_\epsilon$ large enough. On the other hand,

$$\left| \int_{|w|=R} P_R(z, w) \sum_{1}^{\infty} \log |1 - \lambda_j w||dw| \right| \leq \sum_{1}^{\infty} \frac{1}{R} \int_{|w|=R} |\log |1 - \lambda_j w|||dw| \tag{8.4.18}$$

If $|\lambda_j| R \leq \frac{1}{2}$, we write $|\log |1 - \lambda_j w|| \leq C |\lambda_j| R$ and the sum over the corresponding j in (8.4.18) can be bounded by

$$\sum_{|\lambda_j| \leq \frac{1}{2R}} \frac{C}{R} 2\pi |\lambda_j| R R = 2\pi C R \sum_{|\lambda_j| \leq \frac{1}{2R}} |\lambda_j| = o(R), \ R \to \infty. \tag{8.4.19}$$

If $\frac{1}{2} \leq |\lambda_j| R \leq T$, where $T \gg 1$ is independent of R, then by straightforward estimates,

$$\frac{1}{R} \int_{|w|=R} |\log |1 - \lambda_j w|||dw| \leq C_T.$$

Let us estimate the number of λ_j in this case:

$$\sum_{\frac{1}{2}\le|\lambda_j|R\le T} 1 \le 2R \sum_{\frac{1}{2R}\le|\lambda_j|\le\frac{T}{R}} |\lambda_j| = o_T(R).$$

Hence

$$\sum_{\frac{1}{2}\le|\lambda_j|R\le T} \frac{1}{R}\int_{|w|=R} |\log|1-\lambda_j w|||dw| = o_T(R),\ R\to\infty. \tag{8.4.20}$$

It remains to consider the case $|\lambda_j|R \ge T$. Here $\log|1-\lambda_j R| \sim \log(|\lambda_j|R)$. Hence, with constants C and C_δ that are independent of T:

$$\begin{aligned}&\sum_{|\lambda_j|R\ge T} \frac{1}{R}\int_{|w|=R} |\log|1-\lambda_j w|||dw| \le C \sum_{|\lambda_j|R\ge T} \log(|\lambda_j|R)\\ &\le C_\delta \sum_{|\lambda_j|\ge\frac{T}{R}} |\lambda_j|^\delta R^\delta = C_\delta R^\delta \sum_{|\lambda_j|\ge\frac{T}{R}} |\lambda_j|^\delta.\end{aligned}$$

Put $\delta = 1/p$, $1 < p < \infty$, and let q be the conjugate index. Then, if N denotes the number of λ_j with $|\lambda_j| \ge T/R$, we get from Hölder's inequality

$$\sum_{|\lambda_j|\ge\frac{T}{R}} |\lambda_j|^{\frac{1}{p}} \le N^{\frac{1}{q}} \left(\sum_{|\lambda_j|\ge\frac{T}{R}} |\lambda_j|\right)^{\frac{1}{p}}. \tag{8.4.21}$$

Here $NT/R \le \sum|\lambda_j| \le C$, so $N \le CR/T$ and the expression (8.4.21) is bounded by $CR^{1/q}/T^{1/q}$. Hence

$$\sum_{|\lambda_j|\ge\frac{T}{R}} \frac{1}{R}\int_{|w|=R} |\log|1-\lambda_j w|||dw| \le \frac{CR^{\frac{1}{p}+\frac{1}{q}}}{T^{\frac{1}{q}}} = \frac{CR}{T^{\frac{1}{q}}}. \tag{8.4.22}$$

Combining the three cases, we find

$$\left|\int_{|w|=R} P_R(z,w)(\sum_1^\infty \log|1-\lambda_j w|)|dw|\right| \le o_T(R) + \frac{CR}{T^{\frac{1}{q}}} = o(R),\ R\to\infty \tag{8.4.23}$$

Combining this with (8.4.14), (8.4.15) and (8.4.17), we get

$$\Re g(z) \le o(R), \text{ on } |z| \le \frac{R}{2}.$$

Now we can apply Harnack's inequality (see e.g., [4, Section 6.3]) to the function $\Re g - o(R)$, which is ≤ 0 on the disc $|z| \le R/2$ and $\ge -o(R)$ at 0, and conclude that

$$\Re g \ge -o(R) \text{ on the disc } |z| \le \frac{R}{4}.$$

Since g is harmonic, it follows from the last two estimates that $\Re g = 0$. Hence g is constant and since we have chosen g with $g(0) = 0$, we get $g(z) = 0$ for all $z \in \mathbf{C}$. Thus, we have established

Theorem 8.4.3 *Let $A : \mathcal{H} \to \mathcal{H}$ be a trace-class operator with exactly N non-vanishing eigenvalues $\lambda_1(A), \lambda_2(A), \dots$, $0 \le N \le \infty$, repeated according to multiplicity, $0 \le N \le +\infty$. Then $D_A(z) = \det(1 - zA)$ satisfies*

$$D_A(z) = \prod_{j=1}^{N} (1 - \lambda_j z), \; z \in \mathbf{C}, \tag{8.4.24}$$

where the product is defined to be $= 1$ when $N = 0$.

From this we get the important Lidskii's theorem as a corollary (see [48, p. 101]):

Corollary 8.4.4 *If $A \in C_1$ we have*

$$tr\, A = \sum_{j=1}^{N} \lambda_j(A), \tag{8.4.25}$$

where $\lambda_j(A)$ are the non-vanishing eigenvalues as in Theorem 8.4.3.

Proof We know the result when A is of finite rank. In this case, we also know that

$$\frac{D_A'}{D_A} = \frac{\partial}{\partial z} \log \det(1 - zA) = -\mathrm{tr}\,((1 - zA)^{-1} A),$$

away from the zeros of $z \mapsto \det(1 - zA)$. In particular,

$$\frac{D_A'(0)}{D_A(0)} = -\mathrm{tr}\, A, \tag{8.4.26}$$

when A is of finite rank.

When $A \in C_1$, let A_ν be a sequence of finite-rank operators converging to A in the C_1 norm. Then $D_{A_\nu}(z) \to D_A(z)$, $D'_{A_\nu}(z) \to D'_A(z)$, when $\nu \to \infty$, uniformly for z in a neighborhood of 0. Since $D_A(z) \neq 0$ in such a neighborhood, we have also

$$\frac{D'_{A_\nu}(0)}{D_{A_\nu}(0)} \to \frac{D'_A(0)}{D_A(0)}.$$

By (8.4.26), we know that the right-hand side of the last relation is equal to $-\mathrm{tr}\, A_\nu$, and we also know that this quantity converges to $-\mathrm{tr}\, A$. Consequently (8.4.26) remains valid for general trace-class operators. In view of Theorem 8.4.3, we know on the other hand that

$$\frac{D'_A(0)}{D_A(0)} = -\sum_{j=1}^{N} \lambda_j(A),$$

and the corollary follows. □

The last proof also shows that

$$\frac{D'_A(z)}{D_A(z)} = \frac{\partial}{\partial z} \log D_A(z) = -\mathrm{tr}\,(1 - zA)^{-1} A = -\sum_{j=1}^{N} \frac{\lambda_j(A)}{1 - z\lambda_j(A)}, \tag{8.4.27}$$

for all z with $1 - z\lambda_j(A) \neq 0$, $\forall j$.

We end this section by recalling Jensen's formula and the standard application to getting bounds on the number of zeros of holomorphic functions. Let $f(z)$ be a holomorphic function on the open disc $D(0, R)$ with a continuous extension to the corresponding closed disc. Assume that $f(0) \neq 0$ and $f(z) \neq 0$ for $|z| = R$.

Assume first that $f(z)$ has no zeros at all. Then $\log|f(z)| = \Re \log f(z)$ is a harmonic function in the open disc which is continuous up to the boundary, and the mean value property of harmonic functions tells us that

$$\log|f(0)| = \frac{1}{2\pi} \int_0^{2\pi} \log|f(Re^{i\theta})| d\theta.$$

We now allow f to vanish and let $z_1, \dots, z_N$ be the zeros repeated according to their multiplicity. Since f is not allowed to vanish at 0 or at the boundary of $D(0, R)$, we have $0 < |z_j| < R$. Then

$$F(z) := f(z) \prod_{j=1}^{N} \frac{R^2 - \overline{z_j} z}{R(z - z_j)}$$

is holomorphic in the open disc, continuous up to the boundary and has no zeros in the closed disc of radius R. Moreover, $|F(z)| = |f(z)|$ when $|z| = R$, so according to the preceding paragraph, we have

$$\log |F(0)| = \frac{1}{2\pi} \int_0^{2\pi} \log |f(Re^{i\theta})| d\theta.$$

Expanding the left-hand side, we get Jensen's formula:

$$\log |f(0)| + \sum_{j=1}^{N} \log \frac{R}{|z_j|} = \frac{1}{2\pi} \int_0^{2\pi} \log |f(Re^{i\theta})| d\theta. \tag{8.4.28}$$

A standard application of this formula is to notice that if $N(R/2)$ is the number of zeros z_j of f with $|z_j| \leq R/2$, then we get

$$N(R/2) \log 2 \leq \frac{1}{2\pi} \int_0^{2\pi} \log |f(Re^{i\theta})| d\theta - \log |f(0)|. \tag{8.4.29}$$

Notes The presentation in Sect. 8.1 is very close to the one in Appendix A in [60]; further developments can be found in [105] and [148]. Sections 8.2–8.4 give a brief account of the classical theory of non-self-adjoint operators, following closely the monograph [48] (as in the lecture notes [136]). The proof of Lidskii's theorem (Corollary 8.4.4) is slightly simplified in comparison to [48].

Part II

Some General Results

9 Quasi-Modes in Higher Dimension

Recall that if $a(x, \xi)$ and $b(x, \xi)$ are two C^1-functions defined on some domain in $\mathbf{R}^{2n}_{x,\xi}$, then we can define the Poisson bracket to be the C^0-function on the same domain given by

$$\{a, b\} = a'_\xi \cdot b'_x - a'_x \cdot b'_\xi = H_a(b).$$

Here $H_a = a'_\xi \cdot \partial_x - a'_x \cdot \partial_\xi$ denotes the Hamilton vector field of a. The following result is due to Zworski [161], who obtained it via a semi-classical reduction from the above mentioned result of Hörmander. A direct proof was given in [40] and here we give a variant. We will assume some familiarity with symplectic geometry. (See e.g. [51, Ch. 5, 9] for real symplectic geometry, and [102] for corresponding notions in the complex domain.)

Theorem 9.0.1 *Let*

$$P(x, hD_x) = \sum_{|\alpha| \le m} a_\alpha(x)(hD_x)^\alpha, \; D_x = \frac{1}{i}\frac{\partial}{\partial x}, \tag{9.0.1}$$

have smooth coefficients in the open set $\Omega \subset \mathbf{R}^n$. Put $p(x, \xi) = \sum_{|\alpha| \le m} a_\alpha(x)\xi^\alpha$. Assume $z = p(x_0, \xi_0)$ with $\frac{1}{i}\{p, \overline{p}\}(x_0, \xi_0) > 0$. Then there exists $u = u_h \in C_0^\infty(\Omega)$, with $\|u\| = 1$, $\|(P - z)u\| = \mathcal{O}(h^\infty)$, when $h \to 0$. Moreover, u is concentrated to x_0 in the sense that if $W \subset \Omega$ is any fixed neighborhood of x_0, then $\|u\|_{L^2(\Omega \setminus W)} = \mathcal{O}(h^\infty)$.

J. Sjöstrand, *Non-Self-Adjoint Differential Operators, Spectral Asymptotics and Random Perturbations*, Pseudo-Differential Operators 14,
https://doi.org/10.1007/978-3-030-10819-9_9

More generally, the result remains valid if a_α depend on h in such a way that

$$a_\alpha = a_\alpha(x; h) \sim \sum_0^\infty a_{\alpha,k}(x)h^k \text{ in } C^\infty(\Omega),$$

now with $p(x, \xi) = \sum_{|\alpha|\leq m} a_{\alpha,0}(x)\xi^\alpha$.

In the case when the coefficients are all analytic near x_0 we can replace "h^∞" by "$e^{-1/Ch}$ for some $C > 0$". See [74, 132].

This implies that if P has an extension from $C_0^\infty(\Omega)$ to a closed densely defined operator $\widetilde{P} : L^2(\widetilde{\Omega}) \to L^2(\widetilde{\Omega})$, $\widetilde{\Omega} \supset \Omega$, and the resolvent $(\widetilde{P} - z)^{-1}$ exists for the value z above, then its norm is greater than any negative power of h when $h \to 0$ (and even exponentially large in the analytic case).

In the case $n \geq 2$, it was noticed in [104] that if $z = p(\rho)$ and $\Re p$, $\Im p$ are independent at ρ, then $\frac{1}{i}\{p, \overline{p}\}$ times the natural (non-vanishing) Liouville differential form (of maximal degree $2n - 2$) on $p^{-1}(z)$ is equal to a constant times the restriction of the closed form σ^{n-1} to $p^{-1}(z)$. Here

$$\sigma = \sum_1^n d\xi_j \wedge dx_j$$

is the symplectic 2-form on $T^*\Omega$.

If Γ is a compact connected component of $p^{-1}(z)$ on which $d\Re p$ and $d\Im p$ are pointwise independent, it follows that the average of $\frac{1}{i}\{p, \overline{p}\}$ over Γ with respect to the Liouville form has to vanish. Hence, if there is a point on Γ where the Poisson bracket is $\neq 0$, then there is also point where it is positive. In the case $n = 1$ we have a similar phenomenon, explained in the remarks after Proposition 4.2.6.

Example 9.0.2 Take $P = -h^2\Delta + V(x)$, $p(x, \xi) = \xi^2 + V(x)$, $\frac{1}{i}\{p, \overline{p}\} = -4\xi \cdot \Im V'(x)$.

Proof of Theorem 9.0.1 We will first treat the case when p is analytic in a neighborhood of (x_0, ξ_0) and use the same notation for analytic functions defined in a real neighborhood of some point, and their holomorphic extensions to a complex neighborhood of the same point. If ϕ is analytic near x_0 such that $\phi'(x_0) = \xi_0 \in \mathbf{R}^n$, and

$$\Im\phi''(x_0) > 0, \tag{9.0.2}$$

then we can define the complex Lagrangian manifold

$$\Lambda_\phi := \{(x, \phi'(x));\ x \in \text{neigh}\,(x_0, \mathbf{C}^n)\}. \tag{9.0.3}$$

As observed by Hörmander [82], the positivity assumption (9.0.2) can be formulated equivalently by saying that

$$\frac{1}{i}\sigma(t,\Gamma(t))>0,\ 0\neq t\in T_{(x_0,\xi_0)}(\Lambda_\phi). \tag{9.0.4}$$

Here:

- $T_{(x_0,\xi_0)}(\Lambda_\phi)$ denotes the tangent space (bundle) of Λ_ϕ as a smooth real submanifold of $\mathbf{C}^{2n}$. The corresponding tangent vectors $t=t_x\cdot\partial_x+\overline{t}_x\cdot\partial_{\overline{x}}+t_\xi\cdot\partial_\xi+\overline{t}_\xi\cdot\partial_{\overline{\xi}}$ split into a "holomorphic" component $t_{\mathrm{hol}}=t_x\cdot\partial_x+t_\xi\cdot\partial_\xi$ and an "antiholomorphic" one, $t_{\mathrm{ahol}}=\overline{t_{\mathrm{hol}}}=\overline{t}_x\cdot\partial_{\overline{x}}+\overline{t}_\xi\cdot\partial_{\overline{\xi}}$. Here, $\partial_x=(\partial_{x_1},\dots,\partial_{x_n})$, $\partial_{x_j}=(1/2)(\partial_{\Re x_j}+i^{-1}\partial_{\Im x_j})$, $\partial_{\overline{x}}=(\partial_{\overline{x}_1},\dots,\partial_{\overline{x}_n})$, $\partial_{\overline{x}_j}=\overline{\partial_{x_j}}$.
- $\Gamma:T_{(x_0,\xi_0)}\mathbf{C}^n\to T_{(x_0,\xi_0)}\mathbf{C}^n$ is the unique antilinear map which is equal to the identity on $T_{(x_0,\xi_0)}\mathbf{R}^{2n}$.
- σ denotes the symplectic 2-form $\sum_1^n d\xi_j\wedge dx_j$ on $\mathbf{C}^{2n}$ (now a differential form of type $(2,0)$), here viewed as a bilinear form on the complexified tangent space of the cotangent space at (x_0,ξ_0). In other terms, we have the identification and identity

  $$\sigma(t,s)=\langle\sigma|t\wedge s\rangle=\langle\sigma|t_{\mathrm{hol}}\wedge s_{\mathrm{hol}}\rangle$$

  and more explicitly,

  $$\sigma(t,s)=t_\xi\cdot s_x-t_x\cdot s_\xi,$$

  when s is represented similarly to t above. In practice we identify tangent vectors with their holomorphic parts, and define $zt=zt_{\mathrm{hol}}+\overline{z}t_{\mathrm{ahol}}$.
- Writing $t_{\mathrm{hol}}=t_x\cdot\partial_x+t_\xi\cdot\partial_\xi$ as above, we have $(\Gamma t)_{\mathrm{hol}}=\overline{t}_x\cdot\partial_x+\overline{t}_\xi\cdot\partial_\xi$. *From now on we identify tangent vectors and vector fields in the complex domain with their holomorphic parts, and by (very small) abuse of notation we write* $\overline{t}=\Gamma t$.
- More generally, let $\Lambda\subset\mathbf{C}^{2n}$ be a complex Lagrangian manifold, i.e., a complex manifold such that $\sigma|_\Lambda=0$, of maximal complex dimension with this property: $\dim(\Lambda)=n$. Assume that $(x_0,\xi_0)\in\Lambda$ and that (9.0.4) holds with Λ_ϕ replaced by Λ. Then $T_{(x_0,\xi_0)}\Lambda$ cannot contain any non-zero tangent vector (with holomorphic part equal to) $t_\xi\cdot\partial_\xi$, and it follows that $\Lambda=\Lambda_\phi$ with ϕ as above. If $t=t_x\cdot\partial_x+t_\xi\cdot\partial_\xi$ is (the holomorphic part of) a tangent vector to Λ_ϕ at (x_0,ξ_0), then $t_\xi=\phi''(x_0)t_x$ and

  $$\frac{1}{2i}\langle\sigma,t\wedge\overline{t}\rangle=\frac{1}{2i}(\phi''t_x\cdot\overline{t}_x-\overline{\phi}''t_x\cdot\overline{t}_x)=\Im(\phi''t_x\cdot\overline{t}_x),$$

  so $\Im\phi''(x_0)>0$.

Let z, p, (x_0, ξ_0) be as in the theorem. Then we observe that

$$\frac{1}{i}\sigma(H_p, \overline{H}_p)\left(= \frac{1}{i}\sigma(H_p, \Gamma(H_p))\right) = \frac{1}{i}\{p, \overline{p}\} > 0.$$

Moreover, the real set $\Sigma := p^{-1}(z)$ is a smooth symplectic manifold near (x_0, ξ_0). Indeed, the orthogonal space to $T_{x_0,\xi_0}\Sigma$ with respect to the real symplectic form is generated by $H_{\Re p}$, $H_{\Im p}$ and is transverse to $T_{x_0,\xi_0}\Sigma$, so the restriction to Σ of the symplectic 2-form is nondegenerate near (x_0, ξ_0). Using the Darboux theorem, we can identify it locally with $\mathbf{R}^{2(n-1)}$ and hence find a Lagrangian submanifold Λ' in its compexification, passing through (x_0, ξ_0), that satisfies the positivity condition (9.0.4). Viewing the complexification of Σ as a submanifold of $\mathbf{C}^{2n}$, we can define $\Lambda = \{\exp sH_p(\rho);\ s \in \mathrm{neigh}\,(0, \mathbf{C}),\ \rho \in \mathrm{neigh}\,((x_0, \xi_0), \Lambda')\}$. Using that H_p is symplectically orthogonal to the tangent space of Σ, it is then quite easy to verify that Λ is a complex Lagrangian manifold, contained in the complex characteristic hypersurface $\{\rho \in \mathrm{neigh}\,((x_0, \xi_0), \mathbf{C}^{2n});\ p(\rho) = 0\}$ and satisfying the positivity condition (9.0.4). Hence, Λ is of the form Λ_ϕ for an analytic function ϕ as in (9.0.2), (9.0.3), which also fulfills the eikonal equation

$$p(x, \phi'(x)) - z = 0. \tag{9.0.5}$$

We normalize ϕ by requiring that $\phi(x_0) = 0$. Then the function $e^{i\phi(x)/h}$ is rapidly decreasing with all its derivatives away from any neighborhood of x_0. By a complex version of the standard WKB-construction we can construct an elliptic symbol $a(x; h) \sim a_0(x) + ha_1(x) + \cdots$, by solving the suitable transport equations to infinite order at x_0, such that if $\chi \in C_0^\infty(\mathrm{neigh}\,(x_0, \mathbf{R}^n))$ is equal to 1 near x_0, then $u(x; h) = \chi(x)h^{-n/4}a(x; h)e^{i\phi(x)/h}$ has the required properties.

Now, we drop the analyticity assumption and assume merely that p is smooth, as in the statement of the theorem. Let $p^{(1)}$ be the 1st order Taylor polynomial of p at (x_0, ξ_0), so that

$$p(x, \xi) = p^{(1)}(x, \xi) + \mathcal{O}((x - x_0, \xi - \xi_0)^2),\ (x, \xi) \to (x_0, \xi_0).$$

Then the arguments above apply to $p^{(1)}$ and noticing that $\Sigma^{(1)} := (p^{(1)})^{-1}(z)$ is affine linear, we can find Λ' in the complexification of $\Sigma^{(1)}$ as above, also affine. Then $\Lambda^{(1)}$, defined similarly to Λ above, is an affine linear Lagrangian space of the form $\Lambda^{(2)} = \Lambda_{\phi^{(2)}}$, where $\phi^{(2)}$ is a 2nd order polynomial, with

$$\phi(x_0) = 0,\ \partial_x\phi^{(2)}(x_0) = \xi_0,\ \Im\partial_x^2\phi^{(2)} > 0,\ p^{(1)}(x, \partial_x\phi^{(2)}(x)) - z = 0,$$

so

$$p(x, \partial_x\phi^{(2)}(x)) - z = r,\ \text{where } r = \mathcal{O}((x - x_0)^2).$$

Look for $\psi^{(3)} = \mathcal{O}((x - x_0)^3)$ such that

$$p(x, \partial_x(\phi^{(2)} + \psi^{(3)})) - z = \mathcal{O}((x - x_0)^3).$$

This leads to a linear transport-type equation,

$$p'_\xi(x, \partial_x \phi^{(2)}) \cdot \partial_x \psi^{(3)} = -r + \mathcal{O}((x - x_0)^3),$$

which is easy to solve, using that $p'_\xi(x_0, \xi_0) \neq 0$. Iterating the construction, we can find $\psi^{(j)} = \mathcal{O}((x - x_0)^j)$, $j = 4, 5, \ldots$ such that if $\phi = \phi^{(2)} + \psi^{(3)} + \psi^{(4)} + \cdots$ in the sense of formal Taylor series, then in the same sense,

$$p(x, \partial_x \phi) = 0.$$

Again we can find a WKB solution $u = u_h$ with the required properties, noticing that $e^{i\phi(x)/h} f(x; h) = \mathcal{O}(h^\infty)$ in $L^2(\mathrm{neigh}\,(x_0, \mathbf{R}^n))$ if $f = \mathcal{O}(|x - x_0|^\infty + h^\infty)$. □

Notes The background is the same as in Sect. 4.2. Davies [33] showed that for the one-dimensional Schrödinger operator we can construct quasimodes for values of the spectral parameter that may be quite far from the spectrum of the operator. Zworski [161] observed that this result can be viewed as a special case of a more general result of Hörmander [78, 79] in the context of linear PDE.

Pravda-Starov [111] has generalized Theorem 9.0.1 by adapting a more refined quasimode construction of R. Moyer (in two dimensions) and Hörmander [83] for adjoints of operators that do not satisfy the Nirenberg-Trèves condition (Ψ) for local solvability.

Resolvent Estimates Near the Boundary of the Range of the Symbol

10

10.1 Introduction and Outline

In this chapter, which closely follows [140], we study bounds on the resolvent of a non-self-adjoint h-pseudodifferential operator P with (semi-classical) principal symbol p when $h \to 0$, when the spectral parameter is in a neighborhood of certain points on the boundary of the range of p. In Chap. 6 we have already described a very precise result of W. Bordeaux Montrieux in dimension 1. Here we consider a more general situation; the dimension can be arbitrary and we allow for more degenerate behaviour. The results will not be quite as precise as in the one-dimensional case.

In [40] we obtained resolvent estimates at certain boundary points in the following two cases:

(1) under a non-trapping condition,
(2) under a stronger "subellipticity condition".

The case (1) was studied in [40] with general and simple arguments related to the propagation of regularity; the treatment in case (2) was based on Hörmander's work on subellipticity for operators of principal type [83]. In this case the resolvent extends and has temperate growth in $1/h$ in discs of radius $\mathcal{O}(h \ln 1/h)$, centered at the appropriate boundary points. In case (2) we have an extension up to distance $\mathcal{O}(h^{k/(k+1)})$, where the integer $k \geq 2$ is determined by a condition of "subellipticity type".

In this chapter we concentrate on points of type (2) and obtain resolvent estimates by studying an associated semigroup as a Fourier integral operator with complex phase in the spirit of Maslov [99], Kucherenko [89], Melin and Sjöstrand [103]. (See also Menikoff and Sjöstrand [106], Matte [100].) It turned out to be more convenient to use Bargmann-

J. Sjöstrand, *Non-Self-Adjoint Differential Operators, Spectral Asymptotics and Random Perturbations*, Pseudo-Differential Operators 14,
https://doi.org/10.1007/978-3-030-10819-9_10

FBI transforms in the spirit of [132] and [60]. The semigroup method leads to a stronger result: The resolvent can be extended to a disc of radius $\mathcal{O}((h\ln 1/h)^{k/(k+1)})$ around the appropriate boundary points, as in dimension 1 when $k = 2$ in the result of Bordeaux Montrieux [16], cf. Theorem 6.1.4. In that case Bordeaux Montrieux also constructed quasimodes for values of the spectral parameter that are close to the boundary points.

We next state the results.

Let X denote either $\mathbf{R}^n$ or a compact smooth manifold of dimension n.

In the first case, let $m \in C^\infty(\mathbf{R}^{2n}; [1, +\infty[\,)$ be an order function (cf. [41, Def. 7.4], Sect. 5.1) so that for some $C_0, N_0 > 0$,

$$m(\rho) \le C_0 \langle \rho - \mu \rangle^{N_0} m(\mu), \quad \rho, \mu \in \mathbf{R}^{2n}, \tag{10.1.1}$$

where $\langle \rho - \mu \rangle = (1 + |\rho - \mu|^2)^{1/2}$. Let $P = P(x, \xi; h) \in S(m)$, in the sense that P is smooth in x, ξ and satisfies

$$|\partial_{x,\xi}^\alpha P(x, \xi; h)| \le C_\alpha m(x, \xi), \quad (x, \xi) \in \mathbf{R}^{2n}, \ \alpha \in \mathbf{N}^{2n}, \tag{10.1.2}$$

where C_α is independent of h. We also assume that

$$P(x, \xi; h) \sim p_0(x, \xi) + h p_1(x, \xi) + \cdots, \ \text{ in } S(m), \tag{10.1.3}$$

and write $p = p_0$ for the principal symbol. We adopt the ellipticity assumption

$$\exists w \in \mathbf{C}, \ C > 0, \ \text{ such that } |p(\rho) - w| \ge m(\rho)/C, \ \forall \rho \in \mathbf{R}^{2n}. \tag{10.1.4}$$

As in (5.1.3), let

$$P = P^{\mathrm{w}}(x, hD_x; h) = \mathrm{Op}(P(x, h\xi; h)) \tag{10.1.5}$$

be the Weyl quantization of the symbol $P(x, h\xi; h)$ that we can view as a closed unbounded operator on $L^2(\mathbf{R}^n)$.

In the second (compact manifold) case, we let $P \in S^m_{1,0}(T^*X)$ (the classical Hörmander symbol space) of order $m > 0$, meaning that

$$|\partial_x^\alpha \partial_\xi^\beta P(x, \xi; h)| \le C_{\alpha,\beta} \langle \xi \rangle^{m - |\beta|}, \quad (x, \xi) \in T^*X, \tag{10.1.6}$$

where $C_{\alpha,\beta}$ are independent of h. Similarly to (10.1.3), we assume that there exist $p_j \in S^{m-j}_{1,0}(T^*X)$ such that

$$P(x, \xi; h) - \sum_0^{N-1} h^j p_j(x, \xi) \in h^N S^{m-N}_{1,0}(T^*X), \ N = 1, 2, \ldots, \tag{10.1.7}$$

and we quantize the symbol $P(x, h\xi; h)$ in a standard (non-unique) way, by doing it for various local coordinates and gluing the quantizations together by means of a partition of unity. When $m > 0$ we make the ellipticity assumption

$$\exists C > 0, \text{ such that } |p(x,\xi)| \geq \frac{\langle \xi \rangle^m}{C}, \quad |\xi| \geq C. \tag{10.1.8}$$

Let $\Sigma(p) = \overline{p(T^*X)}$ (the closure of $p(T^*X)$) and let $\Sigma_\infty(p)$ be the set of accumulation points of $p(\rho_j)$ for all sequences $\rho_j \in T^*X$, $j = 1, 2, 3, \ldots$ that tend to infinity. The main result of this chapter is taken from [40, 140]:

Theorem 10.1.1 *We adopt the general assumptions above. Let $z_0 \in \partial\Sigma(p) \setminus \Sigma_\infty(p)$ and assume that $dp \neq 0$ at every point of $p^{-1}(z_0)$. Then for every such point ρ there exists $\theta \in \mathbf{R}$ (unique up to a multiple of π) such that $d(e^{-i\theta}(p - z_0))$ is real at ρ. We write $\theta = \theta(\rho)$. Consider the following two cases:*

(1) *For every $\rho \in p^{-1}(z_0)$, the maximal integral curve of $H_{\Re(e^{-i\theta(\rho)}p)}$ through the point ρ is not contained in $p^{-1}(z_0)$.*
(2) *There exists an integer $k \geq 1$ such that for every $\rho \in p^{-1}(z_0)$, there exists $j \in \{1, 2, \ldots, k\}$ such that*

$$H_p^j(\overline{p})(\rho)/(j!) \neq 0.$$

Here $H_p = p'_\xi \cdot \partial_x - p'_x \cdot \partial_\xi$ is the Hamiltonian field, viewed as a differential operator.

In case (1), there exists a constant $C_0 > 0$ such that for every constant $C_1 > 0$ there is a constant $C_2 > 0$ such that the resolvent $(z - P)^{-1}$ is well-defined for $|z - z_0| < C_1 h \ln \frac{1}{h}$, $h < \frac{1}{C_2}$, and satisfies the estimate

$$\|(z - P)^{-1}\| \leq \frac{C_0}{h} \exp\left(\frac{C_0}{h}|z - z_0|\right). \tag{10.1.9}$$

In case (2), there exists a constant $C_0 > 0$ such that for every constant $C_1 > 0$ there is a constant $C_2 > 0$ such that the resolvent $(z - P)^{-1}$ is well-defined for $|z - z_0| < C_1(h \ln \frac{1}{h})^{k/(k+1)}$, $h < \frac{1}{C_2}$ and satisfies the estimate

$$\|(z - P)^{-1}\| \leq \frac{C_0}{h^{\frac{k}{k+1}}} \exp\left(\frac{C_0}{h}|z - z_0|^{\frac{k+1}{k}}\right). \tag{10.1.10}$$

In [40] the bound $\mathcal{O}(1/h)$ in (10.1.9) was obtained for $|z - z_0| \leq h/\mathcal{O}(1)$, as a well as a less precise polynomial bound for $|z - z_0| < C_1 h \ln \frac{1}{h}$. The condition in case (2) was formulated a little differently, but the two formulations lead to the same microlocal models and are therefore equivalent.

Let us now consider a special situation of interest for evolution equations, namely the case when

$$z_0 \in i\mathbf{R}, \tag{10.1.11}$$

$$\Re p(\rho) \geq 0 \text{ in neigh}\left(p^{-1}(z_0), T^*X\right). \tag{10.1.12}$$

Theorem 10.1.2 *We adopt the general assumptions above. Let $z_0 \in \partial\Sigma(p) \setminus \Sigma_\infty(p)$ and assume* (10.1.11), (10.1.12). *Also assume that $dp \neq 0$ on $p^{-1}(z_0)$, so that $d\Im p \neq 0$, $d\Re p = 0$ on that set. Consider the two cases:*

(1) *For every $\rho \in p^{-1}(z_0)$, the maximal integral curve of $H_{\Im p}$ through the point ρ contains a point where $\Re p > 0$.*
(2) *There exists an integer $k \geq 1$ such that for every $\rho \in p^{-1}(z_0)$, we have $H^j_{\Im p}\Re p(\rho) \neq 0$ for some $j \in \{1, 2, \dots, k\}$.*

Then, in case (1)*, there exists a constant $C_0 > 0$ such that for every constant $C_1 > 0$ there is a constant $C_2 > 0$ such that the resolvent $(z - P)^{-1}$ is well-defined for*

$$|\Im(z - z_0)| < \frac{1}{C_0}, \quad \frac{-1}{C_0} < \Re z < C_1 h \ln\frac{1}{h}, \quad h < \frac{1}{C_2},$$

and satisfies the estimate

$$\|(z - P)^{-1}\| \leq \begin{cases} \frac{C_0}{|\Re z|}, & \text{if } \Re z \leq -h, \\ \frac{C_0}{h}\exp(\frac{C_0}{h}\Re z), & \text{if } \Re z \geq -h. \end{cases} \tag{10.1.13}$$

In case (2)*, there exists a constant $C_0 > 0$ such that for every constant $C_1 > 0$ there is a constant $C_2 > 0$ such that the resolvent $(z - P)^{-1}$ is well-defined for*

$$|\Im(z - z_0)| < \frac{1}{C_0}, \quad \frac{-1}{C_0} < \Re z < C_1\left(h \ln\frac{1}{h}\right)^{\frac{k}{k+1}}, \quad h < \frac{1}{C_2}, \tag{10.1.14}$$

and satisfies the estimate

$$\|(z - P)^{-1}\| \leq \begin{cases} \frac{C_0}{|\Re z|}, & \text{if } \Re z \leq -h^{\frac{k}{k+1}}, \\ \frac{C_0}{h^{\frac{k}{k+1}}}\exp(\frac{C_0}{h}(\Re z)_+^{\frac{k+1}{k}}), & \text{if } \Re z \geq -h^{\frac{k}{k+1}}. \end{cases} \tag{10.1.15}$$

As stated in [140], the proof in case (1) follows from an inspection of the proof in [40] and we shall not elaborate more on that. From now on we concentrate on the case (2).

Away from the set $p^{-1}(z_0)$ we can use ellipticity, and we then need to obtain microlocal estimates near a point $\rho \in p^{-1}(z_0)$. The proof of Theorem 10.1.1 can be reduced to that of Theorem 10.1 by means of a factorization and will not be given below. See [140].

The main idea of the proof of Theorem 10.1.2 is to study $\exp(-tP/h)$ microlocally[1] for $0 \le t \ll 1$ and to show that in this case

$$\| \exp\left(-\frac{tP}{h}\right) \| \le C \exp\left(-\frac{t^{k+1}}{Ch}\right), \tag{10.1.16}$$

for some constant $C > 0$. This implies that $\| \exp\left(-\frac{tP}{h}\right) \| = \mathcal{O}(h^\infty)$ for $t \ge h^\delta$ when $\delta(k+1) < 1$, and the formula

$$(z - P)^{-1} = -\frac{1}{h} \int_0^\infty \exp\left(\frac{t(z-P)}{h}\right) dt, \tag{10.1.17}$$

leads to (10.1.15).

The most direct way of studying $\exp(-tP/h)$, or rather a microlocal version of that operator, is to view it as a Fourier integral operator with complex phase [89, 99, 100, 103] of the form

$$U(t)u(x) = \frac{1}{(2\pi h)^n} \iint e^{\frac{i}{h}(\phi(t,x,\eta) - y\cdot\eta)} a(t, x, \eta; h) u(y) dy d\eta, \tag{10.1.18}$$

where the phase ϕ should have a non-negative imaginary part and satisfy the Hamilton–Jacobi equation:

$$i\partial_t \phi + p(x, \partial_x \phi) = \mathcal{O}((\Im \phi)^\infty), \text{ locally uniformly}, \tag{10.1.19}$$

with the initial condition

$$\phi(0, x, \eta) = x \cdot \eta. \tag{10.1.20}$$

This can be carried out in a classical manner, for instance by adapting the method of [103] to the case of non-homogeneous symbols following a reduction used in [100, 106].

[1] Roughly, we say that two distributions u_1 and u_2 of temperate growth, also with respect to h, are equal microlocally near a point in phase space, if there exists an h-pseudodifferential operator χ (used very much as a cutoff function in ordinary distribution theory) with symbol equal to 1 near that point, such that $\chi(u_1 - u_2)$ is $\mathcal{O}(h^\infty)$ in L^2. Roughly, the similar notion for operators is obtained by using cutoffs to the right and to the left. We are here at the level of ideas only and avoid formal definitions.

The following variant seems more practical: Let

$$Tu(x) = Ch^{-\frac{3n}{4}} \int e^{\frac{i}{h}\phi(x,y)} u(y) dy,$$

be an FBI—or (generalized) Bargmann–Segal—transform that we treat in the spirit of Fourier integral operators with complex phase as in [132]. Here ϕ is holomorphic in a neighborhood of $(x_0, y_0) \in \mathbf{C}^n \times \mathbf{R}^n$, and $-\phi'_y(x_0, y_0) = \eta_0 \in \mathbf{R}^n$, $\Im \phi''_{y,y}(x_0, y_0) > 0$, $\det \phi''_{x,y}(x_0, y_0) \neq 0$. Let κ_T be the associated canonical transformation. Then, T is bounded $L^2 \to H_{\Phi_0} := \mathrm{Hol}(\Omega) \cap L^2(\Omega, e^{-2\Phi_0/h} L(dx))$ and has (microlocally) a bounded inverse, where Ω is a small complex neighborhood of x_0 in $\mathbf{C}^n$. The weight Φ_0 is smooth and strictly pluri-subharmonic. If $\Lambda_{\Phi_0} := \{(x, \frac{2}{i}\frac{\partial \Phi_0}{\partial x}); x \in \mathrm{neigh}(x_0)\}$, then (locally) $\Lambda_{\Phi_0} = \kappa_T(T^*X)$.

The operator $\widetilde{P} = TPT^{-1}$ can be defined locally modulo $\mathcal{O}(h^\infty)$ (cf. [91]) as a bounded operator from $H_\Phi \to H_\Phi$ when the weight Φ is smooth and satisfies $\Phi' - \Phi'_0 = \mathcal{O}(h^\delta)$ for some $\delta > 0$. (In the analytic framework it suffices that $\Phi' - \Phi'_0$ is small.) Egorov's theorem applies in this situation, so the principal symbol $\widetilde{p}$ of $\widetilde{P}$ is given by $\widetilde{p} \circ \kappa_T = p$. Thus (under the assumptions of Theorem 10.1.2), we have $\Re \widetilde{p}|_{\Lambda_{\Phi_0}} \geq 0$. This in turn can be used to see that for $0 \leq t \leq h^\delta$, we have $e^{-t\widetilde{P}/h} = \mathcal{O}(1)$: $H_{\Phi_0} \to H_{\Phi_t}$, where $\Phi_t \leq \Phi_0$ is determined by the real Hamilton–Jacobi problem

$$\frac{\partial \Phi_t}{\partial t} + \Re \widetilde{p}(x, \frac{2}{i}\frac{\partial \Phi_t}{\partial x}) = 0, \;\; \Phi_{t=0} = \Phi_0. \tag{10.1.21}$$

Now the bound (10.1.16) follows from the estimate

$$\Phi_t \leq \Phi_0 - \frac{t^{k+1}}{C}, \tag{10.1.22}$$

where $C > 0$. In [140] we obtained this estimate by studying certain I-Lagrangian deformations of the real phase space prior to the FBI-transformation. Here we shall use a shorter and more direct approach.

In Sect. 10.4 we discuss some examples.

There are many works that develop similar ideas and results, see [11,68–70,76,77,153, 154].

10.2 Evolution Equations on the Transform Side

Let $\widetilde{P}(x, \xi; h)$ be a smooth symbol defined in $\mathrm{neigh}((x_0, \xi_0), \Lambda_{\Phi_0})$, with an asymptotic expansion

$$\widetilde{P}(x, \xi; h) \sim \widetilde{p}(x, \xi) + h\widetilde{p}_1(x, \xi) + \ldots \text{ in } C^\infty(\mathrm{neigh}((x_0, \xi_0), \Lambda_{\Phi_0})).$$

Let $\widetilde{P}$ also denote an almost holomorphic extension (see e.g., [103] and Appendix 6.A) to a complex neighborhood of (x_0, ξ_0):

$$\widetilde{P}(x,\xi;h) \sim \widetilde{p}(x,\xi) + h\widetilde{p}_1(x,\xi) + \cdots \quad \text{in } C^\infty(\text{neigh}((x_0,\xi_0), \mathbf{C}^{2n})),$$

where $\widetilde{p}$, $\widetilde{p}_j$ are smooth extensions such that

$$\overline{\partial}\widetilde{p},\ \overline{\partial}\widetilde{p}_j = \mathcal{O}(\text{dist}((x,\xi), \Lambda_{\Phi_0})^\infty).$$

Then, as shown in [91] and later in [104], if $u = u_h$ is holomorphic in a neighborhood V of x_0 and belonging to $H_{\Phi_0}(V)$ in the sense that $\|u\|_{L^2(V, e^{-2\Phi_0/h}L(dx))}$ is finite and of temperate growth in $1/h$ when h tends to zero, then $\widetilde{P}u = \widetilde{P}(x, hD_x; h)u$ can be defined in any smaller neighborhood $W \Subset V$ by the formula

$$\widetilde{P}u(x) = \frac{1}{(2\pi h)^n} \iint_{\Gamma(x)} e^{\frac{i}{h}(x-y)\cdot\theta} \widetilde{P}((x+y)/2, \theta; h)u(y)dyd\theta, \tag{10.2.1}$$

where $\Gamma(x)$ is a good contour (in the sense of [132]) of the form $\theta = \frac{2}{i}\frac{\partial\Phi_0}{\partial x}(\frac{x+y}{2}) + \frac{i}{C_1}(\overline{x-y})$, $|x-y| \leq 1/C_2$, $C_1, C_2 > 0$. Then $\overline{\partial}\widetilde{P}$ is negligible, i.e., of norm $\mathcal{O}(h^\infty)$: $H_{\Phi_0}(V) \to L^2_{\Phi_0}(W)$, and modulo such negligible operators, $\widetilde{P}$ is independent of the choice of a good contour. By solving a $\overline{\partial}$-problem (assuming, as we may, that our neighborhoods are pseudoconvex) we can always correct $\widetilde{P}$ with a negligible operator such that (after an arbitrarily small decrease of W) $\widetilde{P} = \mathcal{O}(1) : H_{\Phi_0}(V) \to H_{\Phi_0}(W)$. Also, if $\Phi = \Phi_0 + \mathcal{O}(h \ln \frac{1}{h})$ in C^2, then $\widetilde{P} = \mathcal{O}(h^{-N_0}) : H_\Phi(V) \to H_\Phi(W)$, for some N_0. By means of Stokes' formula, we can show that $\widetilde{P}$ will change only by a negligible term if we replace Φ_0 by Φ in the definition of $\Gamma(x)$, and then it follows that $\widetilde{P} = \mathcal{O}(1) : H_\Phi(V) \to H_\Phi(W)$.

Recall [104] (see also [121]) that the identity operator $H_{\Phi_0}(V) \to H_{\Phi_0}(W)$ is, up to a negligible operator (i.e., an operator of norm $\mathcal{O}(h^\infty)$) of the form

$$Iu(x) = h^{-n} \iint e^{\frac{2}{h}\Psi_0(x,\overline{y})} a(x,\overline{y};h)u(y)e^{-\frac{2}{h}\Phi_0(y)}dyd\overline{y}, \tag{10.2.2}$$

where $\Psi_0(x,y)$, $a(x,y;h)$ are almost holomorphic on the antidiagonal $y = \overline{x}$ with $\Psi_0(x,\overline{x}) = \Phi_0(x)$, $a(x,y;h) \sim a_0(x,y) + ha_1(x,y) + \cdots$, $a_0(x,\overline{x}) \neq 0$. More generally a pseudodifferential operator like $\widetilde{P}$ takes the form

$$\begin{aligned} \widetilde{P}u(x) &= h^{-n} \iint e^{\frac{2}{h}\Psi_0(x,\overline{y})} q(x,\overline{y};h)u(y)e^{-\frac{2}{h}\Phi_0(y)}dyd\overline{y} \\ q_0(x,\overline{x}) &= \widetilde{p}\left(x, \frac{2}{i}\frac{\partial\Phi_0}{\partial x}(x)\right) a_0(x,\overline{x}), \end{aligned} \tag{10.2.3}$$

and where q_0 denotes the first term in the asymptotic expansion of the symbol q. In this discussion, Φ_0 can be replaced by any other smooth exponent Φ which is $\mathcal{O}(h^\delta)$ close to Φ_0 in C^∞ and we make the corresponding replacement of Ψ_0. A well-known consequence of the strict pluri-subharmonicity of Φ is that

$$2\Re\Psi(x,\overline{y}) - \Phi(x) - \Phi(y) \asymp -|x-y|^2, \tag{10.2.4}$$

so the uniform boundedness $H_\Phi \to H_\Phi$ follows from the domination of the modulus of the kernel of $e^{-\Phi/h} \circ \widetilde{P} \circ e^{\Phi/h}$ by a Gaussian convolution kernel.

Now, we study the evolution problem

$$(h\partial_t + \widetilde{P})\widetilde{U}(t) = 0,\ \widetilde{U}(0) = 1, \tag{10.2.5}$$

where t is restricted to the interval $[0, h^\delta]$ for some arbitrarily small but fixed $\delta > 0$. We review the approximate solution of this problem by a geometrical optics construction: Look for $\widetilde{U}(t)$ of the form

$$\widetilde{U}(t)u(x) = h^{-n} \iint e^{\frac{2}{h}\Psi_t(x,\overline{y})} a_t(x,\overline{y};h)u(y)e^{-2\Phi_0(y)/h} dy d\overline{y}, \tag{10.2.6}$$

where Ψ_t, a_t depend smoothly on all the variables and $\Psi_{t=0} = \Psi_0$, $a_{t=0} = a_0$ in (10.2.3), so $\widetilde{U}(0) = 1$ up to a negligible operator.

Formally $\widetilde{U}(t)$ is the Fourier integral operator

$$\widetilde{U}(t)u(x) = h^{-n} \iint e^{\frac{2}{h}(\Psi_t(x,\theta) - \Psi_0(y,\theta))} a_t(x,\theta;h)u(y) dy d\theta, \tag{10.2.7}$$

where we choose the integration contour $\theta = \overline{y}$. Writing $2\Psi_t(x,\theta) = i\phi_t(x,\theta)$ leads to more standard notation and we impose the eikonal equation

$$i\partial_t\phi + \widetilde{p}(x, \phi'_x(x,\theta)) = 0. \tag{10.2.8}$$

We cannot hope to solve the eikonal equation exactly, but we can do so to infinite order at $t = 0$, $x = \overline{y} = \theta$. If we put

$$\Lambda_{\phi_t(\cdot,\theta)} = \{(x, \phi'_x(t,x,\theta))\}, \tag{10.2.9}$$

then

$$\Lambda_{\phi_t(\cdot,\theta)} = \exp\left(t\widehat{H}_{\frac{1}{i}\widetilde{p}}\right)(\Lambda_{\phi_0(\cdot,\theta)}), \tag{10.2.10}$$

to infinite order at $t = 0$, $\theta = x$. Here $\widehat{H}_{\widetilde{p}} = H_{\widetilde{p}} + \overline{H_{\widetilde{p}}}$ denotes the real vector field associated to the (1,0)-field $H_{\widetilde{p}}$, and similarly for $\widehat{H}_{\frac{1}{i}\widetilde{p}}$. (As in Chap. 9 we sometimes drop the distinction between a vector field of type (1,0) and the corresponding real vector field $\widehat{\nu} = \nu + \overline{\nu}$.) At a point where $\overline{\partial}\widetilde{p} = 0$, we have

$$\widehat{H}_{\widetilde{p}} = H^{\Re\sigma}_{\Re\widetilde{p}} = H^{\Im\sigma}_{\Im\widetilde{p}}, \quad \widehat{H}_{i\widetilde{p}} = -H^{\Re\sigma}_{\Im\widetilde{p}} = H^{\Im\sigma}_{\Re\widetilde{p}}, \tag{10.2.11}$$

where the other fields are the Hamiltonian fields of $\Re\widetilde{p}$, $\Im\widetilde{p}$ with respect to the real symplectic forms $\Re\sigma$ and $\Im\sigma$, respectively. See [104, 132]. Thus (10.2.10) can be written as

$$\Lambda_{\phi_t(\cdot,\theta)} = \exp(tH^{-\Im\sigma}_{\Re\widetilde{p}})(\Lambda_{\phi_0(\cdot,\theta)}). \tag{10.2.12}$$

A complex Lagrangian manifold is also I-Lagrangian (i.e., a Lagrangian manifold for $\Im\sigma$) and (10.2.12) can be viewed as a relation between I-Lagrangian manifolds. It defines the I-Lagrangian manifold $\Lambda_{\phi_t(\cdot,\theta)}$ unambiguously, once we have fixed an almost holomorphic extension of $\widetilde{p}$. Locally, a smooth I-Lagrangian manifold Λ, for which the x-space projection $\Lambda \ni (x,\xi) \mapsto x \in \mathbf{C}^n$ is a local diffeomorphism, takes the form $\Lambda = \Lambda_\Phi$, where Φ is real and smooth and

$$\Lambda_\Phi := \{(x, \frac{2}{i}\frac{\partial\Phi}{\partial x});\ x \in \Omega\},\ \Omega \subset \mathbf{C}^n \text{ open.}$$

With a slight abuse of notation, we can therefore identify the $\mathbf{C}$-Lagrangian manifold Λ_{ϕ_0} with the I-Lagrangian manifold $\Lambda_{-\Im\phi_0}$, since $\frac{\partial\phi_0}{\partial x} = \frac{2}{i}\frac{\partial(-\Im\phi_0)}{\partial x}$, at all points where $\overline{\partial}_x\phi_0 = 0$.

Relation (10.2.4) shows that

$$\Phi_0(x) + \Phi_0(\overline{\theta}) - (-\Im\phi_0(x,\theta)) \asymp |x - \overline{\theta}|^2. \tag{10.2.13}$$

Define

$$\Lambda_{\Phi_t} = \exp(tH^{-\Im\sigma}_{\Re\widetilde{p}})(\Lambda_{\Phi_0}), \tag{10.2.14}$$

and fix the t-dependent constant in this definition of Φ_t by imposing the real Hamilton–Jacobi equation,

$$\partial_t\Phi_t + \Re\widetilde{p}(x, \frac{2}{i}\frac{\partial\Phi_t}{\partial x}) = 0,\ \Phi_{t=0} = \Phi_0. \tag{10.2.15}$$

The real part of (10.2.8) gives a similar equation for $-\Im\phi_t$,

$$\partial_t(-\Im\phi) + \Re\widetilde{p}(x, \frac{2}{i}\frac{\partial}{\partial x}(-\Im\phi)) = 0. \tag{10.2.16}$$

Relation (10.2.13) implies that Λ_{Φ_0} and $\Lambda_{-\Im\phi_0(\cdot,\theta)}$ intersect transversally at $(x_0(\theta), \xi_0(\theta))$, where $(x_t(\overline{\theta}), \xi_t(\overline{\theta})) := \exp(tH_{\Re\widetilde{p}}^{-\Im\sigma})(\overline{\theta}, \frac{2}{i}\frac{\partial\Phi_0}{\partial x}(\overline{\theta}))$. Equations (10.2.14) and (10.2.12) then show that Λ_{Φ_t} and $\Lambda_{-\Im\phi_t(\cdot,\theta)}$ intersect transversally at $(x_t(\overline{\theta}, \xi_t(\overline{\theta}))$. Then (10.2.13), (10.2.15), (10.2.16) imply that $\Phi_t(x) - (-\Im\phi_t(x,\theta)) = \mathcal{O}(|x - x_t(\overline{\theta})|^2)$, and again by (10.2.13) and the transversal intersection, we get

$$\Phi_t(x) + \Phi_0(\overline{\theta}) - (-\Im\phi_t(x,\theta)) \asymp |x - x_t(\overline{\theta})|^2. \tag{10.2.17}$$

Determining a_t by solving a sequence of transport equations, we arrive at the following result:

Proposition 10.2.1 *The operator $\widetilde{U}(t)$ constructed above is $\mathcal{O}(1) : H_{\Phi_0}(V) \to H_{\Phi_t}(W)$ ($W \Subset V$ being small pseudoconvex neighborhoods of a fixed point x_0) uniformly for $0 \le t \le h^\delta$ and it solves the problem* (10.2.5) *up to negligible terms. This local statement makes sense, since by* (10.2.17) *we have*

$$2\Re\Psi_t(x,\overline{y}) - \Phi_t(x) - \Phi_0(y) \asymp -|x - x_t(y)|^2. \tag{10.2.18}$$

Using (10.2.5), we get up to negligible errors that

$$(h\partial_t + \widetilde{P})[\widetilde{P}, \widetilde{U}(t)] = 0, \ [\widetilde{P}, \widetilde{U}(0)] = 0,$$

and examining this evolution problem, we conclude that the Fourier integral operator $[\widetilde{P}, \widetilde{U}(t)]$ is negligible. In particular,

$$h\partial_t\widetilde{U}(t) + \widetilde{U}(t)\widetilde{P} = 0, \ 0 \le t \le h^\delta, \tag{10.2.19}$$

up to negligible errors.

Let us briefly recall an alternative approach, leading to the same weights Φ_t (cf. [104]). Consider formally

$$(e^{-t\widetilde{P}/h}u|e^{-t\widetilde{P}/h}u)_{H_{\Phi_t}} = (u_t|u_t)_{H_{\Phi_t}}, \ u \in H_{\Phi_0},$$

and look for Φ_t such that the time derivative of this expression vanishes to leading order. We get

$$\begin{aligned} 0 &\approx h\partial_t \int u_t\overline{u}_t e^{-2\Phi_t/h}L(dx) \\ &= -\left((\widetilde{P}u_t|u_t)_{H_{\Phi_t}} + (u_t|\widetilde{P}u_t)_{H_{\Phi_t}} + \int 2\frac{\partial\Phi_t}{\partial t}(x)|u|^2 e^{-2\Phi_t/h}L(dx)\right). \end{aligned}$$

Here

$$(\widetilde{P}u_t|u_t)_{H_{\Phi_t}} = \int (\widetilde{p}|_{\Lambda_{\Phi_t}} + \mathcal{O}(h))|u_t|^2 e^{-2\Phi_t/h} L(dx),$$

and similarly for $(u_t|\widetilde{P}u_t)_{H_{\Phi_t}}$, so we would like to have

$$0 \approx \int (2\frac{\partial \Phi_t}{\partial t} + 2\Re \widetilde{p}|_{\Lambda_{\Phi_t}} + \mathcal{O}(h))|u_t|^2 e^{-2\Phi_t/h} L(dx).$$

We choose Φ_t to be the solution of (10.2.15). Then the preceding discussion again shows that $e^{-t\widetilde{P}/h} = \mathcal{O}(1) : H_{\Phi_0} \to H_{\Phi_t}$.

$\Re \widetilde{p}$ is constant along the integral curves of $H_{\Re \widetilde{p}}^{-\Im\sigma}$. Therefore, the second term in (10.2.15) is ≥ 0, so

$$\Phi_t \leq \Phi_0,\ t \geq 0, \tag{10.2.20}$$

under the assumption that

$$\Re \widetilde{p}|_{\Lambda_{\Phi_0}} \geq 0. \tag{10.2.21}$$

Recall that we confine our discussion to the interval $0 \leq t \leq h^\delta$.

Write $\Phi_t = \Phi_0 - \Psi_t$,

$$\widetilde{p}\left(x, \frac{2}{i}\partial_x \Phi_t\right) = \widetilde{p}\left(x, \frac{2}{i}\partial_x \Phi_0\right) - \widetilde{p}'_\xi\left(x, \frac{2}{i}\partial_x \Phi_0\right) \cdot \frac{2}{i}\partial_x \Psi_t + \mathcal{O}\left((\partial_x \Psi_t)^2\right).$$

Using this is in (10.2.15), we get

$$-\partial_t \Psi_t + \Re \widetilde{p}\left(x, \frac{2}{i}\partial_x \Phi_0\right) - \Re\left(\widetilde{p}'_\xi\left(x, \frac{2}{i}\partial_x \Phi_0\right) \cdot \frac{2}{i}\partial_x \Psi_t\right) + \mathcal{O}\left((\partial_x \Psi_t)^2\right) = 0.$$

Define $a = \Re \widetilde{p}$, $b = \Im \widetilde{p}$ on Λ_{Φ_0} and extend these functions to a neighborhood by means of almost holomorphic extension, so that $\widetilde{p} = a + ib$ also in a neighborhood of Λ_{Φ_0}. On Λ_{Φ_0}, we have

$$\partial_t \Psi_t + \frac{1}{2}\left(\frac{2}{i}\left(a'_\xi + ib'_\xi\right)) \cdot \partial_x + \overline{\frac{2}{i}\left(a'_\xi + ib'_\xi\right)) \cdot \partial_x}\right)\Psi_t + \mathcal{O}\left((\partial_x \Psi_t)^2\right) = a. \tag{10.2.22}$$

If $\mu = \sum_1^n \mu_j \partial_{x_j}$, let $\widehat{\mu} = \mu + \overline{\mu} = \sum(\mu_j \partial_{x_j} + \overline{\mu_j \partial_{x_j}})$ be the associated real vector field with the same action as a differential operator on holomorphic functions. Then we get

$$\partial_t \Psi_t + \widehat{b'_\xi \cdot \partial_x} \Psi_t - i\widehat{a'_\xi \cdot \partial_x} \Psi_t + \mathcal{O}((\partial_x \Psi_t)^2) = a, \ \Psi_0 = 0, \tag{10.2.23}$$

where b'_ξ, a'_ξ and a are evaluated on Λ_{Φ_0}. Here we observe that $\nu := \widehat{b'_\xi \cdot \partial_x}$ is the x-space projection of the real Hamiltonian field of $b_{|\Lambda_{\Phi_0}}$ with respect to the real symplectic form $\sigma|_{\Lambda_{\Phi_0}}$.

Notice first that (10.2.23) can be written as $\partial_t \Psi_t + \widetilde{\nu}\Psi_t = a$, where $\widetilde{\nu}$ is some real t-dependent vector field on Λ_{Φ_0} (locally identified with $\mathbf{C}^n_x$). It follows again that $\Psi_t \geq 0$. Then by the standard inequality $\nabla f = \mathcal{O}(f^{1/2})$ for non-negative C^2 functions, we have $(\partial \Psi_t)^2 = \mathcal{O}(\Psi_t)$,

$$a'_\xi \cdot \partial_x \Psi_t = \mathcal{O}(|a'_\xi||\partial_x \Psi_t|) = \mathcal{O}(1) a^{1/2} \Psi_t^{1/2} = \mathcal{O}(1)(\epsilon a + \epsilon^{-1}\Psi_t),$$

uniformly with respect to $\epsilon > 0$. (Because of the almost holomorphy, we have $|a'_\xi| = \mathcal{O}(\nabla a_{|\Lambda_{\Phi_0}})$.) Using this in (10.2.23), we get

$$\partial_t \Psi_t + \nu \Psi_t + \mathcal{O}(1+\epsilon^{-1})\Psi_t = (1+\mathcal{O}(\epsilon))a. \tag{10.2.24}$$

This is a differential inequality along the integral curves of $\nu = H_{\Im \widetilde{p}}$ on Λ_{Φ_0} and fixing $\epsilon > 0$ small enough, we get

$$\Psi_t(\exp t H_{\Im \widetilde{p}}(\rho)) \asymp \widetilde{J}(t,\rho), \tag{10.2.25}$$

for $\rho \in \operatorname{neigh}((x_0,\xi_0),\Lambda_{\Phi_0})$, $0 < t \ll 1$, where

$$\widetilde{J}(t,\rho) = \int_0^t \Re \widetilde{p}(\exp s H_{\Im \widetilde{p}}(\rho)) ds. \tag{10.2.26}$$

Notice that

$$\begin{aligned} &\widetilde{J}(t,\kappa_T(\rho)) = \int_0^t \Re p(\exp s H_{\Im p}(\rho)) ds =: J(t,\rho), \\ &0 \leq J \in C^\infty(\operatorname{neigh}((0,\rho_0),[0,+\infty[\times \mathbf{R}^{2n})). \end{aligned} \tag{10.2.27}$$

Now, introduce the following assumption corresponding to the case (2) in Theorem 10.1.2,

$$H^j_{\Im p}(\Re p)(\rho_0) \begin{cases} = 0, & \text{if } j \leq k-1, \\ > 0, & \text{if } j = k, \end{cases} \tag{10.2.28}$$

where k has to be even since $\Re p \geq 0$. We will work in a sufficiently small neighborhood of ρ_0. Then

$$\partial_t^{j+1} J(0, \rho_0) = H^j_{\Im p}(\Re p)(\rho_0) \begin{cases} = 0, & \text{if } j \leq k-1, \\ > 0, & \text{if } j = k, \end{cases} \tag{10.2.29}$$

Proposition 10.2.2 *Let* $0 \leq J(t, \rho) \in C^\infty(\mathrm{neigh}\,(0, \rho_0), [0, +\infty[\, \times \mathbf{R}^{2n})$ *satisfy* (10.2.29) *for some* $k \geq 1$. *Then there is a constant* $C > 0$ *such that*

$$J(t, \rho) \geq \frac{t^{k+1}}{C}, \quad (t, \rho) \in \mathrm{neigh}\,((0, \rho_0), [0, +\infty[\, \times \mathbf{R}^{2n}). \tag{10.2.30}$$

Proof Assume that (10.2.30) does not hold. Then there is a sequence $(t_\nu, \rho_\nu) \in [0, +\infty[\, \times \mathbf{R}^{2n}$ converging to $(0, \rho_0)$ such that

$$\frac{J(t_\nu, \rho_\nu)}{t_\nu^{k+1}} \to 0,$$

and since $J(t, \rho)$ is an increasing function of t, we get

$$\sup_{0 \leq t \leq t_\nu} \frac{J(t, \rho_\nu)}{t_\nu^{k+1}} \to 0. \tag{10.2.31}$$

Introduce the Taylor expansion,

$$J(t, \rho_\nu) = a_\nu^{(0)} + a_\nu^{(1)} t + \cdots + a_\nu^{(k+1)} t^{k+1} + \mathcal{O}(t^{k+2}),$$

and define

$$u_\nu(s) = \frac{J(t_\nu s, \rho_\nu)}{t_\nu^{k+1}}, \ 0 \leq s \leq 1.$$

Then, by (10.2.31),

$$\sup_{0 \leq s \leq 1} u_\nu(s) \to 0, \ \nu \to \infty.$$

On the other hand,

$$u_\nu(s) = \underbrace{\frac{a_\nu^{(0)}}{t_\nu^{k+1}} + \frac{a_\nu^{(1)}}{t_\nu^k} s + \cdots + a_\nu^{(k+1)} s^{k+1}}_{=: p_\nu(s)} + \mathcal{O}(t_\nu s^{k+2}),$$

so

$$\sup_{0\le s\le 1} |p_\nu(s)| \to 0,\ \nu \to \infty.$$

The corresponding coefficients of p_ν have to tend to 0, and, in particular,

$$a_\nu^{(k+1)} = \frac{1}{(k+1)!}(\partial_t^{k+1} J(0,\rho_\nu) \to 0$$

which is in contradiction with (10.2.29). □

From (10.2.26), (10.2.27), (10.2.30), we get

Proposition 10.2.3 *Under the assumption* (10.2.28), *we have*

$$\Phi_t(x) \le \Phi_0(x) - \frac{t^{k+1}}{C},\ (t,x) \in neigh\,((0,x_0), [0,\infty[\,\times \mathbf{C}^n). \tag{10.2.32}$$

10.3 Resolvent Estimates

Let P be an h-pseudodifferential operator satisfying the general assumptions of the introduction.

Let $z_0 \in (\partial\Sigma(p)) \setminus \Sigma_\infty(p)$. We will treat the case (2) of Theorem 10.1.2, so that

$$z_0 \in i\mathbf{R}, \tag{10.3.1}$$

$$\Re p(\rho) \ge 0 \text{ in neigh}\,(p^{-1}(z_0), T^*X), \tag{10.3.2}$$

$$\forall \rho \in p^{-1}(z_0),\ \exists j \le k,\ \text{such that } H_{\Im p}^j \Re p(\rho) > 0. \tag{10.3.3}$$

Proposition 10.3.1 *$\exists C_0 > 0$ such that $\forall C_1 > 0$, $\exists C_2 > 0$ such that for z, h as in* (10.1.14) *and $u \in C_0^\infty(X)$ it holds that*

$$\begin{aligned} &|\Re z|\|u\| \le C_0\|(z-P)u\|, \text{ when } \Re z \le -h^{\frac{k}{k+1}}, \\ &h^{\frac{k}{k+1}}\|u\| \le C_0 \exp\left(\frac{C_0}{h}(\Re z)_+^{\frac{k+1}{k}}\right)\|(z-P)u\|, \\ &\text{when } \Re z \ge -h^{\frac{k}{k+1}} \le \Re z \le \mathcal{O}(1)\left(h\ln\frac{1}{h}\right)^{\frac{k}{k+1}}. \end{aligned} \tag{10.3.4}$$

Proof The required estimate is easy to obtain microlocally in the region where $P - z_0$ is elliptic, in the sense that if $\chi \in C_0^\infty(T^*X)$ and $\operatorname{supp}\chi \cap p^{-1}(z_0) = \emptyset$ and writing χ also for a corresponding h-quantization, then by standard calculus [41], we have $\|\chi u\| \leq \mathcal{O}(1)\|(z - P)u\| + \mathcal{O}(h^\infty)\|u\|$ for small values of $|z - z_0|$. It then suffices to show the following statement:

For every $\rho_0 \in p^{-1}(z_0)$, there exists $\chi \in C_0^\infty(T^*X)$, equal to 1 near ρ_0, such that for z, h as in (10.1.14) and letting χ also denote a corresponding h-pseudodifferential operator, we have

$$
\begin{aligned}
&|\Re z|\|\chi u\| \leq C_0\|(z - P)u\| + C_N h^N\|u\|, \text{ when } \Re z \leq -h^{\frac{k}{k+1}},\\
&h^{\frac{k}{k+1}}\|\chi u\| \leq C_0 \exp\left(\frac{C_0}{h}(\Re z)_+^{\frac{k+1}{k}}\right)\|(z - P)u\| + C_N h^N\|u\|,\\
&\text{when } -h^{\frac{k}{k+1}} \leq \Re z \leq \mathcal{O}(1)\left(h\ln\frac{1}{h}\right)^{\frac{k}{k+1}},
\end{aligned}
\tag{10.3.5}
$$

where $N \in \mathbf{N}$ is arbitrary.

When $\Re z \leq -h^{k/(k+1)}$ this is an easy consequence of the semi-classical sharp Gårding inequality (see for instance [41]), so from now on we assume that $\Re z \geq -h^{k/(k+1)}$.

Let $T : L^2 \to H_{\Phi_0}(\operatorname{neigh}(x_0, \mathbf{C}^n))$ be an FBI transform as in Sect. 10.1, with $(y_0, \eta_0) = \rho_0$. We then have a conjugated operator (cf. [132]) $\widetilde{P}$ as in (10.2.1), with $(x_0, \xi_0) = \kappa_T(y_0, \eta_0) \in \Lambda_{\Phi_0}$, and $\Lambda_{\Phi_0} = \kappa_T(T^*X)$, $\widetilde{p} = p \circ \kappa_T^{-1}$ near (x_0, ξ_0), with $\widetilde{p}$ the principal symbol of $\widetilde{P}$, such that

$$
\|TPu - \widetilde{P}Tu\|_{H_{\Phi_0}(V)} \leq \mathcal{O}(h^\infty)\|u\|_{L^2}, \tag{10.3.6}
$$

where V is a small neighborhood of x_0. When $\operatorname{supp}\chi$ is small enough, we also have

$$
\|\chi u\| \leq \mathcal{O}(1)\|Tu\|_{H_{\Phi_0}(V)} + \mathcal{O}(h^\infty)\|u\|_{L^2}. \tag{10.3.7}
$$

It suffices to show that

$$
\begin{aligned}
\|u\|_{H_{\Phi_0}(V_1)} \leq\ & h^{-\frac{k}{k+1}} C_0 \exp\left(\frac{C_0}{h}(\Re z)_+^{\frac{k+1}{k}}\right)\|(\widetilde{P} - z)u\|_{H_{\Phi_0}(V_2)}\\
&+ \mathcal{O}(h^\infty)\|u\|_{H_{\Phi_0}(V_3)},
\end{aligned}
\tag{10.3.8}
$$

$u \in H_{\Phi_0}(V_3)$, where $V_1 \Subset V_2 \Subset V_3$ are neighborhoods of x_0, given by $(x_0, \xi_0) = \kappa_T(\rho_0) \in \Lambda_{\Phi_0}$.

From Proposition 10.2.3 and the fact that $\widetilde{U}(t) : H_{\Phi_0}(V_2) \to H_{\Phi_t}(V_1)$, we see that

$$
\|\widetilde{U}(t)u\|_{H_{\Phi_0}(V_1)} \leq Ce^{-t^{k+1}/(Ch)}\|u\|_{H_{\Phi_0}(V_2)}. \tag{10.3.9}
$$

Choose $\delta > 0$ small enough so that $\delta(k+1) < 1$ and put

$$\widetilde{R}(z) = \frac{1}{h}\int_0^{h^\delta} e^{\frac{tz}{h}}\widetilde{U}(t)dt. \tag{10.3.10}$$

Before verifying that $\widetilde{R}$ is an approximate left inverse to $\widetilde{P} - z$, we study the norm of this operator in H_{Φ_0}. We have in $\mathcal{L}(H_{\Phi_0}(V_2), H_{\Phi_0}(V_1))$:

$$\|e^{\frac{tz}{h}}\widetilde{U}(t)\| \leq C\exp\left(\frac{1}{h}\left(t\Re z - \frac{t^{k+1}}{C}\right)\right). \tag{10.3.11}$$

The right-hand side is $\mathcal{O}(h^\infty)$ for $t = h^\delta$, since $\delta(k+1) < 1$ and $\Re z \leq \mathcal{O}(1)\,(h\ln(1/h))^{k/(k+1)}$.

We get

$$\|\widetilde{R}(z)\| \leq \frac{C}{h}\int_0^{+\infty}\exp\left(\frac{1}{h}\left(t\Re z - \frac{t^{k+1}}{C}\right)\right)dt = \frac{C^{\frac{k+2}{k+1}}}{h^{\frac{k}{k+1}}} I\left(\frac{C^{\frac{1}{k+1}}}{h^{\frac{k}{k+1}}}\Re z\right), \tag{10.3.12}$$

where

$$I(s) = \int_0^\infty e^{st - t^{k+1}}dt. \tag{10.3.13}$$

Lemma 10.3.2 *We have*

$$I(s) = \mathcal{O}(1), \text{ when } |s| \leq 1, \tag{10.3.14}$$

$$I(s) = \frac{\mathcal{O}(1)}{|s|}, \text{ when } s \leq -1, \tag{10.3.15}$$

$$I(s) \leq \mathcal{O}(1)s^{-\frac{k-1}{2k}}\exp\left(\frac{k}{(k+1)^{\frac{k+1}{k}}}s^{\frac{k+1}{k}}\right), \text{ when } s \geq 1. \tag{10.3.16}$$

Proof The first two estimates are straightforward, so we concentrate on the last one, where we may also assume that $s \gg 1$. A computation shows that the exponent $f_s(t) = st - t^{k+1}$ on $[0, +\infty[$ has a unique critical point $t = t(s) = (s/(k+1))^{1/k}$, which is a nondegenerate maximum, since

$$f_s''(t(s)) = -k(k+1)^{\frac{1}{k}}s^{\frac{k-1}{k}},$$

with critical value

$$f_s(t(s)) = \frac{k}{(k+1)^{\frac{k+1}{k}}} s^{\frac{k+1}{k}}.$$

Now

$$f_s''(t) = -(k+1)kt^{k-1} \lesssim f_s''(t(s)), \text{ for } \frac{t(s)}{2} \le t < +\infty,$$

so $\int_{t(s)/2}^{\infty} e^{st-t^{k+1}} dt$ satisfies the required upper bound. Here we write $a \lesssim b$ when $a, b \ge 0$ and $a \le \mathcal{O}(b)$.

On the other hand,

$$f_s(t(s)) - f_s(t) \ge \frac{s^{\frac{k+1}{k}}}{C}, \text{ for } 0 \le t \le \frac{t(s)}{2},\ s \gg 1,$$

so

$$\int_0^{\frac{t(s)}{2}} e^{st-t^{k+1}} dt \le \mathcal{O}(1) s^{\frac{1}{k}} \exp\left(f_s(t(s)) - \frac{s^{\frac{k+1}{k}}}{C}\right),$$

and (10.3.16) follows. □

Applying the lemma to (10.3.12), we get

Proposition 10.3.3 *We have*

$$\|\widetilde{R}(z)\| \le \frac{C}{h^{\frac{k}{k+1}}}, \quad \text{if } |\Re z| \le \mathcal{O}(1) h^{\frac{k}{k+1}}, \tag{10.3.17}$$

$$\|\widetilde{R}(z)\| \le \frac{C}{|\Re z|}, \quad \text{if } -1 \ll \Re z \le -h^{\frac{k}{k+1}}, \tag{10.3.18}$$

$$\|\widetilde{R}(z)\| \le \frac{C}{h^{\frac{k}{k+1}}} \exp\left(C_k \frac{(\Re z)^{\frac{k+1}{k}}}{h}\right), \quad h^{\frac{k}{k+1}} \le \Re z \le C\left(h \ln \frac{1}{h}\right)^{\frac{k}{k+1}}. \tag{10.3.19}$$

From the beginning of the proof of Lemma 10.3.2, or more directly from (10.3.11), we see that

$$\|e^{\frac{tz}{h}} \widetilde{U}(t)\| \le C \exp \frac{C_k}{h} (\Re z)_+^{\frac{k+1}{k}},$$

which is bounded by some negative power of h, since $\Re z \leq \mathcal{O}(1)(h \ln \frac{1}{h})^{\frac{k}{k+1}}$. Working locally, we then see that modulo a negligible operator,

$$\widetilde{R}(z)(\widetilde{P} - z) \equiv \frac{1}{h} \int_0^{h^\delta} e^{\frac{tz}{h}} (-h\partial_t - z)\widetilde{U}(t)dt \equiv 1,$$

where the last equivalence follows from an integration by parts and the fact that $e^{tz/h}\widetilde{U}(t)$ is negligible for $t = h^\delta$. Combining this with Proposition 10.3.3, we get (10.3.8), and then (10.3.5) by means of (10.3.6), which completes the proof of Proposition 10.3.1. □

We can now complete the

Proof of Theorem 10.1.2 Using standard pseudodifferential machinery (see for instance [41]) we first notice that P has discrete spectrum in a neighborhood of z_0 and that $P - z$ is a Fredholm operator of index 0 from $\mathcal{D}(P)$ to L^2 when z varies in a small neighborhood of z_0. On the other hand, Proposition 10.3.1 implies that $P - z$ is injective and hence bijective for $\Re z \leq \mathcal{O}(h \ln \frac{1}{h})^{\frac{k}{k+1}}$ and we also get the corresponding bounds on the resolvent. □

As for the proof of Theorem 10.1.1, we refer to [140]: It is based on a factorization with the help of Malgrange's preparation theorem which allows for a reduction to a situation close to that of Theorem 10.1.2.

10.4 Examples

Consider

$$P = -h^2\Delta + iV(x), \;\; V \in C^\infty(X; \mathbf{R}), \tag{10.4.1}$$

where X is either a smooth compact manifold of dimension n or $X = \mathbf{R}^n$. In the second case we assume that $p = \xi^2 + iV(x)$ belongs to a symbol space $S(m)$, where $m \geq 1$ is an order function. If $V \in C_b^\infty(\mathbf{R}^2)$, then we can take $m = 1 + \xi^2$, and if $\partial^\alpha V(x) = \mathcal{O}((1 + |x|)^2)$ for all $\alpha \in \mathbf{N}^n$ and satisfies the ellipticity condition $|V(x)| \geq C^{-1}|x|^2$ for $|x| \geq C$, for some constant $C > 0$, then we can take $m = 1 + \xi^2 + x^2$.

We have $\Sigma(p) = [0, \infty[+ i\overline{V(X)}$. When X is compact $\Sigma_\infty(p)$ is empty, and when $X = \mathbf{R}^n$, we have $\Sigma_\infty(p) = [0, \infty[+ i\Sigma_\infty(V)$, where $\Sigma_\infty(V)$ is the set of accumulation points at infinity of V.

Let $z_0 = x_0 + iy_0 \in \partial\Sigma(p) \setminus \Sigma_\infty(p)$.

- In the case $x_0 = 0$ we see that Theorem 10.1.2 (2) is applicable with $k = 2$, provided that y_0 is not a critical value of V.

- When $x_0 > 0$, then y_0 is either the maximum or the minimum value of V. Assume that $V^{-1}(y_0)$ is finite and that each element of that set is a non-degenerate maximum or minimum. Then Theorem 10.1.2 (2) is applicable to $\pm iP$ with $k = 2$. By allowing a more complicated behaviour of V near its extreme points, we can produce examples where the same result applies with $k > 2$.

Next, consider the non-self-adjoint harmonic oscillator

$$Q = -\frac{d^2}{dy^2} + iy^2 \tag{10.4.2}$$

on the real line, studied by Boulton [22] and Davies [34]. Introduce a large spectral parameter $E = i\lambda + \mu$, where $\lambda \gg 1$ and $|\mu| \ll \lambda$. The change of variables $y = \sqrt{\lambda}x$ permits us to identify Q with $Q = \lambda P$, where $P = -h^2\frac{d^2}{dx^2} + ix^2$ and $h = 1/\lambda \to 0$. Hence $Q - E = \lambda(P - (1 + i\frac{\mu}{\lambda}))$ and Theorem 10.1.2 (2) is applicable with $k = 2$. We conclude that $(Q - E)^{-1}$ is well-defined and of polynomial growth in λ (which can be specified further), respectively $\mathcal{O}(\lambda^{-1})$ when

$$\frac{\mu}{\lambda} \le C_1(\lambda^{-1}\ln\lambda)^{\frac{2}{3}} \quad \text{and} \quad \frac{\mu}{\lambda} \le C_1 \quad \text{respectively,}$$

for any fixed $C_1 > 0$, i.e., when

$$\mu \le C_1\lambda^{\frac{1}{3}}(\ln\lambda)^{\frac{2}{3}} \quad \text{and} \quad \mu \le C_1\lambda^{\frac{1}{3}} \quad \text{respectively.} \tag{10.4.3}$$

Actually, we are here in dimension 1 with $k = 2$, so Theorem 6.1.4 gives an even more precise result.

We refer to [137] for some comments about the Kramers–Fokker–Planck operator

$$P = hy \cdot \partial_x - V'(x) \cdot h\partial_y + \frac{1}{2}(y - h\partial_y) \cdot (y + h\partial_y) \tag{10.4.4}$$

on $\mathbf{R}^{2n} = \mathbf{R}^n_x \times \mathbf{R}^n_y$, where V is smooth and real-valued.

Notes This chapter gives bounds on the resolvent of an h-pseudodifferential operator P near the boundary of the range of the semi-classical principal symbol. We follow [140] quite closely; the results are the same but one essential step of the proof has been simplified. They partially extend those of [40]. They are more general than those in Chap. 6: the dimension can be arbitrary and we allow a higher order of degeneration. In the case $\{p, \{p, \overline{p}\}\} \neq 0$ it would be interesting to know if one can get the same precision as in the one-dimensional case in Chap. 6.

From Resolvent Estimates to Semigroup Bounds

11

11.1 Introduction

In Chap. 10 we saw a concrete example of how to get resolvent bounds from semigroup bounds. Naturally, one can go in the opposite direction and in this chapter we discuss some abstract results of that type, including the Hille–Yoshida and Gearhardt–Prüss–Hwang–Greiner theorems. As for the latter, we also give a result of Helffer and the author [62] that provides a more precise bound on the semigroup. We refer to [62] for the discussion of some examples.

11.2 General Results

We start by recalling some general results and here we follow [43]. Let $\mathcal{B}$ be a complex Banach space.

Definition 11.2.1 A map $[0, +\infty[\ni t \mapsto T(y) \in \mathcal{L}(\mathcal{B}, \mathcal{B})$ is a strongly continuous semigroup if

$$\begin{cases} T(t+s) = T(t)T(s), \ t, s \geq 0, \\ T(0) = 1, \end{cases} \tag{11.2.1}$$

and the orbit maps

$$[0, +\infty[\ni t \mapsto T(t)x \in \mathcal{B} \tag{11.2.2}$$

are continuous for every $x \in \mathcal{B}$.

J. Sjöstrand, *Non-Self-Adjoint Differential Operators, Spectral Asymptotics and Random Perturbations*, Pseudo-Differential Operators 14,
https://doi.org/10.1007/978-3-030-10819-9_11

Example 11.2.2 Let $A \in \mathrm{Mat}\,(n, n)$ (the space of complex $n \times n$ matrices) and put $T(t) = \exp(tA) : \mathbf{C}^n \to \mathbf{C}^n$. This is a uniformly continuous semigroup: $\|T(t) - T(s)\| \to 0$, $t \to s$ for every $s \geq 0$.

Example 11.2.3 Translation semigroups: Let $\mathcal{B} = L^p([0, +\infty[\,)$, $1 \leq p < +\infty$ and define $T(t) : \mathcal{B} \to \mathcal{B}$ by $T(t)u(x) = u(x + t)$. Then $T(t)$, $0 \leq t < +\infty$ is a strongly continuous semigroup which is not uniformly continuous.

Example 11.2.4 Let $\mathcal{B} = L^2(\mathbf{R}^n)$ and let $T(t)u = U(t, x) \in C^1([0, +\infty[); H^0(\mathbf{R}^n))$, be the solution of the heat equation

$$\partial_t U(t, x) = \Delta_x U(t, x),\ U(0, x) = u(x).$$

Then $T(t)$ is a strongly continuous semigroup and also a contraction semigroup in the sense that $\|T(t)\| \leq 1$ for all $t \geq 0$.

Let $(T(t))_{0 \leq t < \infty}$ be a strongly continuous semi-group. The Banach-Steinhaus theorem implies that there exists $M_0 \geq 1$ such that $\|T(t)\| \leq M_0$ for $0 \leq t \leq 1$.

Proposition 11.2.5 *If* $(T(t))_{0 \leq t < \infty}$ *is a strongly continuous semigroup, then there exist* $M \geq 1$, $\omega \in \mathbf{R}$ *such that we have have the following property* (P(M, ω))*:*

$$\|T(t)\| \leq Me^{\omega t},\ t \geq 0. \tag{11.2.3}$$

Proof If $t \geq 0$, let $[t] \in \mathbf{N}$ be the integer part of t, so that $t = [t] + s$, with $0 \leq s < 1$. Then with M_0 as above, we get

$$\|T(t)\| = \|(T(1))^{[t]}T(s)\| \leq \|T(1)\|^{[t]}\|T(s)\| \leq M_0^{[t]+1} \leq M_0 \cdot M_0^t.$$

Thus we have (11.2.3) with $M = M_0$ and $\omega = \ln M_0$. □

Definition 11.2.6 Let $(T(t))_{0 \leq t < \infty}$ be a strongly continuous semigroup. The growth bound $\omega_0 \in [-\infty, +\infty[$ is

$$\omega_0 := \inf\{\omega \in \mathbf{R};\ \exists\, M = M_\omega < +\infty \text{ such that } \|T(t)\| \leq Me^{\omega t},\ \forall t \in [0, +\infty[\,\}.$$

If $T(t) = \exp(tA) : \mathbf{C}^n \to \mathbf{C}^n$, $A \in \mathrm{Mat}\,(n, n)$, then $\omega_0 = \max_{\lambda \in \sigma(A)} \Re\lambda$.

The Generator

Definition 11.2.7 Let $T(t),\ t \geq 0$ be a strongly continuous semigroup. We define the generator $A : \mathcal{B} \subset \mathcal{D}(A) \to \mathcal{B}$ to be the linear operator with domain

$$\mathcal{D}(A) = \{x \in \mathcal{B};\ \lim_{\epsilon \to 0} \epsilon^{-1}(T(\epsilon) - 1)x \text{ exists in } \mathcal{B}\},$$

acting on $x \in \mathcal{D}(A)$ as $Ax = \lim_{\epsilon \to 0} \epsilon^{-1}(T(\epsilon) - 1)x$.

Theorem 11.2.8 *The generator of a strongly continuous semigroup is closed and densely defined.*

One can show that if $x \in \mathcal{D}(A)$, then for every $t \geq 0$, $T(t)(\mathcal{D}(A)) \subset \mathcal{D}(A)$ and $\partial_t T(t)x = AT(t)x = T(t)Ax$. A general problem is to characterize the generators of semigroups. The following result gives some necessary conditions.

Theorem 11.2.9 *Let $T(t),\ t \geq 0$ be a strongly continuous semigroup with generator A and let $M \geq 1$, $\omega \in \mathbf{R}$ be such that* (P(M, ω)) *holds.*

If $\lambda \in \mathbf{C}$ and

$$\widetilde{R}(\lambda)x := \int_0^\infty e^{-\lambda s} T(s)x ds = \lim_{T \to +\infty} \int_0^T e^{-\lambda s} T(s)x ds$$

exists for all $x \in \mathcal{B}$ (as a limit of vector-valued Riemann integrals), then $\lambda \in \rho(A)$ and $\widetilde{R}(\lambda) = (\lambda - A)^{-1}$. In particular, if $\Re\lambda > \omega$, then $\lambda \in \rho(A)$ and

$$\|(\lambda - A)^{-1}\| \leq \frac{M}{\Re\lambda - \omega}.$$

Corollary 11.2.10 *Under the assumptions of Theorem 11.2.9, let ω_0 be the growth bound of $T(t)$. Then the spectrum of A is contained in the half-plane $\Re\lambda \leq \omega_0$.*

The Hille–Yoshida theorem (see e.g. [43, Theorem II.3.5]) characterizes generators of contraction semigroups (when taking $\omega = 0$):

Theorem 11.2.11 *Let $\omega \in \mathbf{R}$ and let $A : \mathcal{B} \supset \mathcal{D}(A) \to \mathcal{B}$ be a linear operator. The following properties are equivalent:*

(a) *A generates a strongly continuous semigroup $T(t)$, $t \geq 0$, which satisfies* (P($1, \omega$)).
(b) *A is closed, densely defined. For every $\lambda > \omega$ we have $\lambda \in \rho(A)$ and*

$$\|(\lambda - A)^{-1}\| \leq \frac{1}{\lambda - \omega}.$$

(c) *A is closed, densely defined. For every* $\lambda \in \mathbf{C}$ *with* $\Re\lambda > \omega$ *we have* $\lambda \in \rho(A)$ *and*

$$\|(\lambda - A)^{-1}\| \leq \frac{1}{\Re\lambda - \omega}.$$

Example 11.2.12 Let $-A = -(h\partial_x)^2 + ix^2 : L^2 \to L^2$ with domain B^2 defined in the beginning of Sect. 2.5. The numerical range of A is contained in the third quadrant and in particular in the half-plane $\Re\lambda < 0$, so we know that $\|(\lambda - A)^{-1}\| \leq 1/\Re\lambda$ when $\Re\lambda > 0$. Consequently, A generates a contraction semigroup.

The following theorem of Feller, Miyadera and Phillips (see e.g. [43, Theorem II.3.8]) characterizes generators of general strongly continuous semigroups:

Theorem 11.2.13 *Let* $\omega \in \mathbf{R}$, $M \geq 1$ *and let* $A : \mathcal{B} \supset \mathcal{D}(A) \to \mathcal{B}$ *be a linear operator. The following properties are equivalent:*

(a) *A generates a strongly continuous semigroup* $T(t)$, $t \geq 0$ *which satisfies* (P(M, ω)).
(b) *A is closed, densely defined. For every* $\lambda > \omega$ *we have* $\lambda \in \rho(A)$ *and*

$$\|(\lambda - A)^{-n}\| \leq \frac{M}{\lambda - \omega}, \quad \forall n \in \mathbf{N} \setminus \{0\}.$$

(c) *A is closed, densely defined. For every* $\lambda \in \mathbf{C}$ *with* $\Re\lambda > \omega$ *we have* $\lambda \in \rho(A)$ *and*

$$\|(\lambda - A)^{-n}\| \leq \frac{M}{\Re\lambda - \omega}, \quad \forall n \in \mathbf{N} \setminus \{0\}.$$

A drawback with the Hille–Yoshida theorem is that it is essentially limited to the special case of contraction semigroups. The Feller–Miyadera–Phillips theorem gives a general characterization of generators of strongly continuous semigroups, but it may be less clear how to verify the conditions of that theorem in concrete applications.

11.3 The Gearhardt–Prüss–Hwang–Greiner Theorem

In this section we first recall the Gearhardt–Prüss–Hwang–Greiner theorem (see [43], Theorem V.I.11, [152], Theorem 19.1 as well as [158], [37]) and then give a variant with explicit bounds on the norm of the semigroup due to Helffer and the author [62], also published in [57]. The GPHG-theorem reads:

Theorem 11.3.1

(a) *Assume that* $\mathcal{B} = \mathcal{H}$ *is a Hilbert space and that* $\|(z - A)^{-1}\|$ *is uniformly bounded in the half-plane* $\Re z \geq \omega$. *Then there exists a constant* $M > 0$ *such that* P(M, ω) *holds.*

(b) *If* $\mathrm{P}(M,\omega)$ *holds, then for every* $\alpha > \omega$, $\|(z-A)^{-1}\|$ *is uniformly bounded in the half-plane* $\Re z \geq \alpha$.

The part (b) follows from the more precise statement in Theorem 11.2.9.

An idea in the proof is to use that the resolvent and the inhomogeneous equation $(\partial_t - A)u = w$ in exponentially weighted spaces are related via Fourier–Laplace transform and we can use Plancherel's formula. This is why we need to work in a Hilbert space. Variants of this simple idea have also been used in more concrete situations. See [25, 46, 66, 122]. Before stating the more explicit version, we make some simple remarks.

We can improve slightly the conclusion of (a). If the assumption in (a) is true for some ω, then it is automatically true for some $\omega' < \omega$:

Lemma 11.3.2 *If for some* $r(\omega) > 0$, $\|(z-A)^{-1}\| \leq \frac{1}{r(\omega)}$ *for* $\Re z > \omega$, *then for every* $\omega' \in]\omega - r(\omega), \omega]$ *we have*

$$\|(z-A)^{-1}\| \leq \frac{1}{r(\omega) - (\omega - \omega')}, \quad \Re z > \omega'.$$

Proof If $z \in \mathbb{C}$ and $\Re z > \omega'$, we take a sequence of numbers $z_0 \in \mathbb{C}$ such that $\Im z_0 = \Im z$ and $\Re z_0 \searrow \omega$, and the lemma follows from Proposition 2.1.6. □

Remark 11.3.3 Let

$$\omega_1 = \inf\{\omega \in \mathbf{R};\ \{z \in \mathbb{C};\ \Re z > \omega\} \subset \rho(A) \text{ and } \sup_{\Re z > \omega} \|(z-A)^{-1}\| < \infty\}.$$

For $\omega > \omega_1$, we define $r(\omega)$ by

$$\frac{1}{r(\omega)} = \sup_{\Re z > \omega} \|(z-A)^{-1}\|.$$

Then $r(\omega)$ is an increasing function of ω; for every $\omega \in]\omega_1, \infty[$, we have $\omega - r(\omega) \geq \omega_1$ and for $\omega' \in [\omega - r(\omega), \omega]$ we have

$$r(\omega') \geq r(\omega) - (\omega - \omega').$$

We may state all this more elegantly by saying that r is a Lipschitz function on $]\omega_1, +\infty[$ satisfying

$$0 \leq \frac{dr}{d\omega} \leq 1\,.$$

Moreover, if $\omega_1 > -\infty$, then $r(\omega) \to 0$ when $\omega \searrow \omega_1$.

Remark 11.3.4 By Theorem 11.2.9, we already know that $\|(z-A)^{-1}\|$ is uniformly bounded in the half-plane $\Re z \geq \beta$ if $\beta > \omega_0$, where ω_0 is the growth bound for $(T(t))_{0\leq t<\infty}$. If $\alpha \leq \omega_0$, we see that $\|(z-A)^{-1}\|$ is uniformly bounded in the half-plane $\Re z \geq \alpha$, provided that

- we have this uniform boundedness on the line $\Re z = \alpha$,
- A has no spectrum in the half-plane $\Re z \geq \alpha$,
- $\|(z-A)^{-1}\|$ does not grow too wildly in the strip $\alpha \leq \Re z \leq \beta$, namely, $\|(z-A)^{-1}\| \leq \mathcal{O}(1)\exp(\mathcal{O}(1)\exp(k|\Im z|))$, where $k < \pi/(\beta-\alpha)$.

We then also have

$$\sup_{\Re z\geq\alpha} \|(z-A)^{-1}\| = \sup_{\Re z=\alpha} \|(z-A)^{-1}\|. \tag{11.3.1}$$

This follows from the subharmonicity of $\ln\|(z-A)^{-1}\|$, Hadamard's theorem (or the Phragmén–Lindelöf theorem in exponential coordinates) and the maximum principle.

The following result of B. Helffer and the author [62] (see also [57]) gives explicit bounds on the semigroup in the GPHG-theorem.

Theorem 11.3.5 *We make the assumptions of Theorem* 11.3.1, (a) *and define* $r(\omega) > 0$ *by*

$$\frac{1}{r(\omega)} = \sup_{\Re z\geq\omega} \|(z-A)^{-1}\|.$$

Let $m(t) \geq \|T(t)\|$ *be a continuous positive function. Then for all* $t, a, \widetilde{a} > 0$, *such that* $t = a + \widetilde{a}$, *we have*

$$\|T(t)\| \leq \frac{e^{\omega t}}{r(\omega)\|\frac{1}{m}\|_{e^{-\omega\cdot}L^2([0,a])}\|\frac{1}{m}\|_{e^{-\omega\cdot}L^2([0,\widetilde{a}])}}. \tag{11.3.2}$$

Here the norms are always the natural ones obtained from $\mathcal{H}$, L^2, thus for instance $\|T(t)\| = \|T(t)\|_{\mathcal{L}(\mathcal{H},\mathcal{H})}$, if u is a function on $\mathbf{R}$ with values in $\mathbf{C}$ or in $\mathcal{H}$, $\|u\|$ denotes the natural L^2 norm; when the norm is taken over a subset J of $\mathbf{R}$, this is indicated with a "$L^2(J)$". In (11.3.2) we also have the natural norm in the exponentially weighted space $e^{-\omega\cdot}L^2([0,a])$, and similarly with $\widetilde{a}$ instead of a; $\|f\|_{e^{-\omega\cdot}L^2([0,a])} = \|e^{\omega\cdot}f(\cdot)\|_{L^2([0,a])}$.

Notice that we only need the bound $m(t)$ for small t, say $0 \leq t \leq 1$ and that we have (11.3.2) with $m(t)$ replaced by $m(t)1_{[0,1]}(t) + \infty 1_{]1,+\infty[}$. Thus from the given bound $m(t)$ for small times and the resolvent bound, we get a global bound $\widetilde{m}(t)$ on $\|T(t)\|$. We can then replace $m(t)$ by $\min\{m(t), \widetilde{m}(t)\}$ and reapply the theorem. It is an interesting problem to understand what would be the optimal bound that we can get from such an iteration. Some steps in that direction were taken in [62].

The following variant of the main result, obtained with Helffer in [62], could be useful in problems of return to equilibrium.

Theorem 11.3.6 *We make the assumptions of Theorem* 11.3.5, *so that* (11.3.2) *holds. Let* $\widetilde{\omega} < \omega$ *and assume that* A *has no spectrum on the line* $\Re z = \widetilde{\omega}$ *and that the spectrum of* A *in the half-plane* $\Re z > \widetilde{\omega}$ *is compact (and included in the strip* $\widetilde{\omega} < \Re z < \omega$*). Assume that* $\|(z-A)^{-1}\|$ *is uniformly bounded on* $\{z \in \mathbf{C};\ \Re z \geq \widetilde{\omega}\} \setminus U$*, where* U *is any neighborhood of* $\sigma_+(A) := \{z \in \sigma(A);\ \Re z > \widetilde{\omega}\}$*, and define* $r(\widetilde{\omega})$ *by*

$$\frac{1}{r(\widetilde{\omega})} = \sup_{\Re z = \widetilde{\omega}} \|(z-A)^{-1}\|.$$

Then for every $t > 0$*,*

$$T(t) = T(t)\Pi_+ + R(t) = T(t)\Pi_+ + T(t)(1-\Pi_+),$$

where for all $a, \widetilde{a} > 0$ *with* $a + \widetilde{a} = t$*,*

$$\|R(t)\| \leq \frac{e^{\widetilde{\omega} t}}{r(\widetilde{\omega}) \|1/m\|_{e^{-\widetilde{\omega}\cdot} L^2([0,a])} \|1/m\|_{e^{-\widetilde{\omega}\cdot} L^2([0,\widetilde{a}])}} \|I - \Pi_+\|. \tag{11.3.3}$$

Here Π_+ *denotes the spectral projection associated to* $\sigma_+(A)$*:*

$$\Pi_+ = \frac{1}{2\pi i} \int_{\partial V} (z-A)^{-1} dz,$$

where V *is any compact neighborhood of* $\sigma_+(A)$ *with* C^1 *boundary, disjoint from* $\sigma(A) \setminus \sigma_+(A)$*.*

We refer to [62] for the proofs of the last two theorems, which are based on the study of the inhomogeneous equation

$$(\partial_t - A)u = w \text{ on } \mathbf{R} \tag{11.3.4}$$

and exponentially weighted L^2 spaces on the real line, as well as Laplace transforms.

Notes In this chapter we have collected some results about semigroup bounds that follow from estimates on the resolvent in certain half-planes.

Counting Zeros of Holomorphic Functions 12

12.1 Introduction

In this chapter we will generalize Proposition 3.4.6 of Hager about counting the zeros of holomorphic functions of exponential growth. In [56] we obtained such a generalization, by weakening the regularity assumptions on the functions ϕ. However, due to some logarithmic losses, we were not quite able to recover Hager's original result, and we still had a fixed domain Γ with smooth boundary.

In many spectral problems, the domain should be allowed to depend on h and other parameters, it could be a long thin rectangle, and the boundary regularity should be relaxed. In this chapter, based on [141] we revisit systematically the proof of the counting proposition in [56] and obtain a general and quite natural result allowing an h-dependent exponent ϕ to be merely continuous and the h-dependent domain Γ to have Lipschitz boundary. The result generalizes the two earlier ones. By allowing suitable small changes of the points z_j we also get rid of the logarithmic losses. Compared with the results in [141] we relax a subharmonicity assumption about the exponent in the exponential bounds.

We next formulate the results. Let $\Gamma \Subset \mathbf{C}$ be an open set and $\gamma = \partial\Gamma$ be the boundary of Γ. Let $r : \gamma \to]0, \infty[$ be a Lipschitz function of Lipschitz modulus $\le 1/2$:

$$|r(x) - r(y)| \le \frac{1}{2}|x - y|, \quad x, y \in \gamma. \tag{12.1.1}$$

We further assume that γ is Lipschitz in the following precise sense, where r enters:

There exists a constant C_0 such that for every $x \in \gamma$ there exist new affine coordinates $\widetilde{y} = (\widetilde{y}_1, \widetilde{y}_2)$ of the form $\widetilde{y} = U(y - x)$, $y \in \mathbf{C} \simeq \mathbf{R}^2$ being the old coordinates, where $U = U_x$ is orthogonal, such that the intersection of Γ and the rectangle $R_x := \{y \in$

J. Sjöstrand, *Non-Self-Adjoint Differential Operators, Spectral Asymptotics and Random Perturbations*, Pseudo-Differential Operators 14,
https://doi.org/10.1007/978-3-030-10819-9_12

$\mathbf{C};\ |\widetilde{y}_1| < r(x),\ |\widetilde{y}_2| < C_0 r(x)\}$ takes the form

$$\{y \in R_x;\ \widetilde{y}_2 > f_x(\widetilde{y}_1),\ |\widetilde{y}_1| < r(x)\}, \tag{12.1.2}$$

where $f_x(\widetilde{y}_1)$ is Lipschitz on $[-r(x), r(x)]$, with Lipschitz modulus $\le C_0$ (i.e., $|f_x(t) - f_y(s)| \le C_0|t-s|$). Notice that our assumption (12.1.2) remains valid if we decrease r.

It will be convenient to extend the function $r(x)$ to all of $\mathbf{C}$, by putting

$$r(x) = \inf_{y\in\gamma}(r(y) + \frac{1}{2}|x-y|). \tag{12.1.3}$$

The extended function is also Lipschitz with modulus $\le \frac{1}{2}$:

$$|r(x) - r(y)| \le \frac{1}{2}|x-y|,\ x, y \in \mathbf{C}.$$

Notice that

$$r(x) \ge \frac{1}{2}\mathrm{dist}\,(x, \gamma), \tag{12.1.4}$$

where $\mathrm{dist}\,(x, y) = \inf_{y\in\gamma}|x-y|$ denotes the distance from x to γ, and that

$$|y-x| \le r(x) \Longrightarrow \frac{r(x)}{2} \le r(y) \le \frac{3r(x)}{2}. \tag{12.1.5}$$

For convenience, we shall also assume that Γ is simply connected.

Before stating the general version of zero counting result, we review some facts about Radon measures. If Ω is open in $\mathbf{R}^n$ for some $n \ge 1$, we say that μ is a (Radon) measure on Ω if it is a distribution on Ω of order 0. We will only consider real-valued measures i.e., $\langle\mu, \psi\rangle \in \mathbf{R}$ for every $\psi \in C_0(\Omega)$. Thus for every $K \Subset \Omega$ there exists $C_K > 0$ such that

$$|\langle\mu, \psi\rangle| \le C_K\|\psi\|_{L^\infty}, \tag{12.1.6}$$

for all $\psi \in C_0(\Omega)$ with support in K. If we have can find a constant $C = C_K$ which is independent of K, then we say that μ is a finite (Radon) measure. Notice that if μ is a measure, then its restriction to any open set $\Omega' \Subset \Omega$ is finite. When μ is finite we let $|\mu|(\Omega)$ denote the smallest K-independent constant in (12.1.6). We say that μ is positive and write $\mu \ge 0$ if $\langle\mu, \psi\rangle \ge 0$ for all $0 \le \psi \in C_0(\Omega)$. We have the Jordan decomposition $\mu = \mu_+ - \mu_-$ of μ into the difference of two positive Radon measures which is minimal in the sense that if $\mu = \nu_+ - \nu_-$ with $\nu_\pm \ge 0$, then $\mu_\pm \le \nu_\pm$. See [45, Corollary 2]. We write $|\mu| = \mu_+ + \mu_-$. When $\nu \ge 0$ is a finite Radon measure we can define $\nu(\Omega) = \lim\langle\nu, \psi_j\rangle$, where $C_0(\Omega; [0,1]) \ni \psi_j \nearrow 1_\Omega$, $j \to \infty$. This is in agreement with the earlier definition of $|\mu|(\Omega)$. Also when μ is a finite Radon measure on Ω, we define

$$\mu(\Omega) = \lim_{C_0(\Omega;[0,1])\ni\psi_j\nearrow 1_\Omega} \langle \mu, \psi_j \rangle. \tag{12.1.7}$$

The main result of this chapter is:

Theorem 12.1.1 *Let* $\Gamma \Subset \mathbf{C}$ *be open, simply connected and have Lipschitz boundary* γ *with an associated Lipschitz weight* r *as in* (12.1.1)–(12.1.3). *Put* $\widetilde{\gamma}_{\alpha r} = \bigcup_{x\in\gamma} D(x, \alpha r(x))$ *for any constant* $\alpha > 0$. *Let* $z_j^0 \in \gamma$, $j \in \mathbf{Z}/N\mathbf{Z}$ *be distributed along the boundary in the positively oriented sense such that*[1]

$$r(z_j^0)/4 \le |z_{j+1}^0 - z_j^0| \le r(z_j^0)/2.$$

Then there exists a positive constant C_1^0, *depending only on the constant* C_0 *in the assumption around* (12.1.2), *such that for every constant* $C_1 \ge C_1^0$, *there exists a constant* $C_2 > 0$ *such that we have the following for any* $z_j \in D(z_j^0, r(z_j^0)/(2C_1))$:

Let $0 < h \le 1$ *and let* ϕ *be a continuous function defined on some open neighborhood of the closure of* $\widetilde{\gamma}_r$ *such that* $\mu := \Delta\phi L(dz)$ *is a finite Radon measure there, and denote by the same symbol a distribution extension to* $\Gamma \cup \widetilde{\gamma}_r$. *If* u *is a holomorphic function on* $\Gamma \cup \widetilde{\gamma}_r$ *satisfying*

$$h \ln |u| \le \phi(z) \text{ on } \widetilde{\gamma}_r, \tag{12.1.8}$$

$$h \ln |u(z_j)| \ge \phi(z_j) - \epsilon_j, \text{ for } j = 1, 2, \dots, N, \tag{12.1.9}$$

where $\epsilon_j > 0$, *then the number of zeros of* u *in* Γ *satisfies*

$$\begin{aligned} &\left| \#(u^{-1}(0) \cap \Gamma) - \frac{1}{2\pi h}\mu(\Gamma) \right| \le \\ &\frac{C_2}{h} \left(|\mu|(\widetilde{\gamma}_r) + \sum_{j=1}^{N} \left(\epsilon_j + \int_{D(z_j, \frac{r(z_j)}{4C_1})} \left| \ln \frac{|w - z_j|}{r(z_j)} \right| \mu_+(dw) \right) \right). \end{aligned} \tag{12.1.10}$$

Here $\mu(\Gamma)$ *and* $|\mu|(\widetilde{\gamma}_r)$ *are well defined as in* (12.1.7) *(with smooth test functions equal to* 1 *away from a neighborhood of* $\widetilde{\gamma}_r$ *in the case of* μ*).* μ_+ *and* $|\mu|$ *are defined from the Jordan decomposition.*

[1] Here "4" can be replaced by any fixed constant >2.

By observing that the average of $\left|\ln \frac{|w-z_j|}{r(z_j)}\right|$ with respect to the Lebesgue measure $L(dz_j)$ over $D(z_j^0, \frac{r(z_j^0)}{2C_1})$ is $\mathcal{O}(1)$, we can get rid of the logarithmic terms in Theorem 12.1.1, by making a suitable choice of $z_j = \widetilde{z}_j$, and we get:

Theorem 12.1.2 *Let $\Gamma \Subset \mathbf{C}$ be simply connected and have Lipschitz boundary γ with an associated Lipschitz weight r as in* (12.1.1)–(12.1.3). *Let $z_j^0 \in \gamma$, $j \in \mathbf{Z}/N\mathbf{Z}$, be distributed along the boundary in the positively oriented sense such that (see Footnote 1)*

$$r(z_j^0)/4 \le |z_{j+1}^0 - z_j^0| \le r(z_j^0)/2.$$

Then there exists a positive constant C_1^0, depending only on the constant C_0 in the assumption around (12.1.2), *such that for every constant $C_1 \ge C_1^0$, there exists a constant $C_2 > 0$ such that we have the following:*

Let $0 < h \le 1$ and let ϕ be a continuous function defined on some neighborhood of the closure of $\widetilde{\gamma}_r$ such that $\mu := \Delta\phi L(dz)$ is a finite Radon measure there and denote by the same symbol a distribution extension to $\Gamma \cup \widetilde{\gamma}_r$. Then there exist $\widetilde{z}_j \in D(z_j^0, \frac{r(z_j^0)}{2C_1})$ such that if u is a holomorphic function on $\Gamma \cup \widetilde{\gamma}_r$ satisfying

$$h \ln |u| \le \phi(z) \text{ on } \widetilde{\gamma}_r, \tag{12.1.11}$$

and

$$h \ln |u(\widetilde{z}_j)| \ge \phi(\widetilde{z}_j) - \epsilon_j,\ \ j = 1, 2, \ldots, N, \tag{12.1.12}$$

instead of (12.1.9), *then*

$$\left|\#(u^{-1}(0) \cap \Gamma) - \frac{1}{2\pi h}\mu(\Gamma)\right| \le \frac{C_2}{h}\left(|\mu|(\widetilde{\gamma}_r) + \sum \epsilon_j\right). \tag{12.1.13}$$

Of course, if we already know (in the subharmonic case) that

$$\int_{D(z_j, \frac{r(z_j)}{4C_1})} \left|\ln \frac{|w - z_j|}{r(z_j)}\right| \mu_+(dw) = \mathcal{O}(1)\mu_+\left(D(z_j, \frac{r(z_j)}{4C_1})\right), \tag{12.1.14}$$

then we can keep $\widetilde{z}_j = z_j$ in (12.1.10) and get (12.1.13). This is the case if we assume that μ_+ is equivalent to the Lebesgue measure $L(d\omega)$ in the following sense:

$$\frac{\mu_+(dw)}{\mu_+(D(z_j, \frac{r(z_j)}{4C_1}))} \asymp \frac{L(dw)}{L(D(z_j, \frac{r(z_j)}{4C_1}))} \text{ on } D(z_j, \frac{r(z_j)}{4C_1}), \tag{12.1.15}$$
$$\text{uniformly for } j = 1, 2, \ldots, N.$$

Then we get,

Theorem 12.1.3 *Under the conditions of Theorem* 12.1.1 *as well as the assumption* (12.1.14) *or the stronger assumption* (12.1.15), *conditions (12.1.8) and (12.1.9), imply* (12.1.13).

In particular, we recover Proposition 3.4.6, where Γ, ϕ are independent of h, γ of class C^∞ and $\phi \in C^2(\mathrm{neigh}\,(\gamma))$. Then $|\mu| \le \mathcal{O}(1)L(dz)$ and it suffices to choose $r = \sqrt{\epsilon}$, $\epsilon_j = \epsilon$ and to notice that we can replace ϕ by $\phi + \epsilon$. The counting proposition in [56] can also be recovered. (In the subharmonic case we have $\mu = \mu_+ = |\mu|$ near $\widetilde{\gamma}_r$.)

A simple application will be given in Sect. 12.4. Other (for us) more important applications will be given in Chaps. 13, 14, 17.

The outline of the chapter is the following:

In Sect. 12.2 we consider thin neighborhoods of the boundary where the width is variable and determined by the function r. We verify that we can find such neighborhoods with smooth boundary and estimate the derivatives of the function that defines the boundary. Then we develop some exponentially weighted estimates for the Laplacian in such domains in the spirit of what can be done for the Schrödinger equation [59] and in a large number of works in thin domains, see for instance [21,49]. From that we also deduce pointwise estimates on the corresponding Green kernel.

In Sect. 12.3 we prove the main results by following the general strategy of the proof of the corresponding result in [56] and carry out the averaging argument that leads to the elimination of the logarithms.

In Sect. 12.4 we consider as a simple illustration the zeros of sums of exponentials of holomorphic functions. These results can also be obtained with more direct methods, cf. [15,36,72].

This chapter is a slightly improved version of [141], where a last section—not included here—establishes a connection with classical results on zeros of entire functions.

12.2 Thin Neighborhoods of the Boundary and Weighted Estimates

Let Γ, $\gamma = \partial\Gamma$, r be as in the introduction.

We can find a smooth function $\widetilde{r}(x)$ such that

$$\frac{1}{C}r(x) \le \widetilde{r}(x) \le r(x), \quad |\nabla\widetilde{r}(x)| \le \frac{1}{2}, \quad \partial^\alpha\widetilde{r}(x) = \mathcal{O}(\widetilde{r}^{1-|\alpha|}), \tag{12.2.1}$$

where $C > 0$ is a universal constant. Indeed, let $0 \le \chi \in C_0^\infty(D(0,1))$ satisfy $\int \chi(z)L(dz) = 1$ and put, for $\epsilon > 0$ small enough and $C_1 > 0$ large enough,

$$\widetilde{r}(x) = C_1^{-1}\int \chi\left(\frac{x-y}{\epsilon r(y)}\right)(\epsilon r(y))^{-2})r(y)L(dy).$$

From now on, we replace $r(x)$ by $\widetilde{r}(x)$ and drop the tilde. Then (12.1.1), (12.1.2), (12.1.5) remain valid and (12.1.4) remains valid in the weakened form

$$r(x) \geq \frac{1}{C}\mathrm{dist}\,(x, \gamma), \tag{12.2.2}$$

where $C > 0$ is a new constant.

Consider the signed distance to γ:

$$g(x) = \begin{cases} \mathrm{dist}\,(x, \gamma), \text{ if } x \in \Gamma \\ -\mathrm{dist}\,(x, \gamma), \text{ if } x \in \mathbf{C} \setminus \Gamma \end{cases} \tag{12.2.3}$$

Possibly after multiplying r by a small positive constant, we deduce from (12.1.2) that for every $x \in \gamma$ there exists a normalized constant real vector field $\nu = \nu_x$ (namely $\partial_{\widetilde{y}_1}$, cf. (12.1.2)) such that

$$\nu(g) \geq \frac{1}{C} \text{ in } R_x, \quad C = (1 + C_0^2)^{-1/2}. \tag{12.2.4}$$

In the set $\bigcup_{x \in \gamma} R_x$, we consider the regularized function

$$g_\epsilon(x) = \int \frac{1}{(\epsilon r(x))^2} \chi\left(\frac{x-y}{\epsilon r(x)}\right) g(y) L(dy), \tag{12.2.5}$$

where $0 \leq \chi \in C_0^\infty(D(0,1))$, $\int \chi(x) L(dx) = 1$. Here $\epsilon > 0$ is small and we notice that $r(x) \asymp r(y)$, $g(y) = \mathcal{O}(r(y))$, when $\chi((x-y)/(\epsilon r(x)) \neq 0$. It follows that $g_\epsilon(x) = \mathcal{O}(r(x))$ and more precisely, since g is Lipschitz, that

$$g_\epsilon(x) - g(x) = \mathcal{O}(\epsilon r(x)). \tag{12.2.6}$$

Differentiating (12.2.5), we get

$$\begin{aligned} \nabla_x g_\epsilon(x) = (\nabla_x g)_\epsilon + 2 \int \frac{-\nabla r(x)}{\epsilon^2 r(x)^2} \chi\left(\frac{x-y}{\epsilon r(x)}\right) \frac{g(y)}{r(x)} L(dy) \\ + \int \frac{1}{\epsilon^2 r(x)^2} \chi'\left(\frac{x-y}{\epsilon r(x)}\right) \cdot \frac{x-y}{\epsilon r(x)} (-\nabla r(x)) \frac{g(y)}{r(x)} L(dy), \end{aligned} \tag{12.2.7}$$

where $(\nabla_x g)_\epsilon$ is defined as in (12.2.5) with g replaced by $\nabla_x g$. It follows that

$$\nabla_x g_\epsilon(x) - (\nabla g)_\epsilon(x) = \mathcal{O}(1) \sup_{y \in D(x, \epsilon r(x))} \frac{|g(y)|}{r(x)}. \tag{12.2.8}$$

In particular, $\nabla_x g_\epsilon = \mathcal{O}(1)$ and with $\nu = \nu_x$:

$$\nu(g_\epsilon)(y) \geq \frac{1}{2C}, \text{ when } y \in R_x, \text{ and } \sup_{|z-y|\leq \epsilon r(y)} |g(z)| \ll r(y). \tag{12.2.9}$$

Differentiating (12.2.7) further, we get

$$\partial^\alpha g_\epsilon(x) = \mathcal{O}_\alpha((\epsilon r(x))^{1-|\alpha|}), \ |\alpha| \geq 1. \tag{12.2.10}$$

Let $C > 0$ be large enough but independent of ϵ. Put

$$\widehat{\gamma}_{\epsilon, C\epsilon r} = \{x \in \bigcup_{y\in\gamma} R_y; \ |g_\epsilon(x)| < C\epsilon r(x)\}. \tag{12.2.11}$$

If $C > 0$ is sufficiently large, then in the coordinates associated to (12.1.2), $\widehat{\gamma}_{\epsilon, C\epsilon r}$ takes the form

$$f_x^-(\widetilde{y}_1) < \widetilde{y}_2 < f_x^+(\widetilde{y}_1), \ |\widetilde{y}_1| < r(x), \tag{12.2.12}$$

where $f_x^\pm$ are smooth on $[-r(x), r(x)]$ and satisfy

$$\partial_{\widetilde{y}_1}^k f_x^\pm = \mathcal{O}_k((\epsilon r(x))^{1-k}), \ k \geq 1, \tag{12.2.13}$$

$$0 < f_x^+ - f_x, \ f_x - f_x^- \asymp C\epsilon r(x). \tag{12.2.14}$$

Later, we will fix $\epsilon > 0$ small enough and write

$$\gamma_r = \widehat{\gamma}_{\epsilon, C\epsilon r} \text{ and more generally, } \gamma_{\alpha r} = \widehat{\gamma}_{\epsilon, C\epsilon\alpha r}. \tag{12.2.15}$$

We shall next establish an exponentially weighted estimate for the Dirichlet Laplacian in γ_r:

Proposition 12.2.1 *Let $C > 0$ be sufficiently large and $\epsilon > 0$ sufficiently small in the definition of γ_r in (12.2.15). Then there exists a new constant $C_{\mathrm{new}} > 0$ such that if $\psi \in C^2(\overline{\gamma}_r)$ and*

$$|\psi_x'| \leq \frac{1}{C_{\mathrm{new}} r}, \tag{12.2.16}$$

we have

$$\|e^\psi Du\| + \frac{1}{C_{\mathrm{new}}}\|\frac{1}{r}e^\psi u\| \leq C_{\mathrm{new}}\|re^\psi \Delta u\|, \ u \in (H_0^1 \cap H^2)(\gamma_r), \tag{12.2.17}$$

where $\|w\|$ denotes the L^2 norm when the function w is scalar and we write

$$(v|w) = \int \sum v_j(x)\overline{w}_j(x)L(dx),\ \ \|v\| = \sqrt{(v|v)},$$

for $\mathbf{C}^n$-valued functions with components in L^2. H_0^1 and H^2 are the standard Sobolev spaces.[2]

Proof Let $\psi \in C^2(\overline{\gamma}_r; \mathbf{R})$ and put

$$-\Delta_\psi = e^\psi \circ (-\Delta) \circ e^{-\psi} = D_x^2 - (\psi_x')^2 + i(\psi_x' \circ D_x + D_x \circ \psi_x'),$$

where the last term is formally anti-self-adjoint. Then for every $u \in (H^2 \cap H_0^1)(\gamma_r)$,

$$\Re(-\Delta_\psi u|u) = \|D_x u\|^2 - ((\psi_x')^2 u|u). \tag{12.2.18}$$

We need an a priori estimate for D_x. Let $v : \overline{\gamma}_r \to \mathbf{R}^n$ be sufficiently smooth. We sometimes consider v as a vector field. Then for $u \in (H^2 \cap H_0^1)(\gamma_r)$:

$$(Du|uv) - (uv|Du) = i(\operatorname{div}(v)u|u).$$

Assume $\operatorname{div}(v) > 0$. If $v = \nabla w$, then $\operatorname{div}(v) = \Delta w$, and our assumption is then that w is strictly subharmonic. Then

$$\int \operatorname{div}(v)|u|^2 dx \le 2\|uv\|\|Du\| \le \|Du\|^2 + \|uv\|^2,$$

which we write

$$\int (\operatorname{div}(v) - |v|^2)|u|^2 dx \le \|Du\|^2.$$

Using this in (12.2.18), we get

$$\frac{1}{2}\|Du\|^2 + \int \left(\frac{1}{2}(\operatorname{div}(v) - |v|^2) - (\psi_x')^2\right)|u|^2 dx$$
$$\le \|\frac{1}{k}(-\Delta_\psi)u\|\|ku\| \le \frac{1}{2}\|\frac{1}{k}(-\Delta_\psi)u\|^2 + \frac{1}{2}\|ku\|^2,$$

[2] In the following we simply write C instead of C_{new}, following the general rule that constants change from time to time.

where k is any positive continuous function on $\overline{\gamma}_r$. We write this as

$$\frac{1}{2}\|Du\|^2 + \int \left(\frac{1}{2}(\mathrm{div}\,(v) - |v|^2 - k^2) - (\psi_x')^2\right)|u|^2 dx \le \frac{1}{2}\|\frac{1}{k}(-\Delta_\psi)u\|^2. \tag{12.2.19}$$

We shall see that we can choose v so that

$$\mathrm{div}\,(v) \ge r^{-2}, \quad |v| \le \mathcal{O}(r^{-1}). \tag{12.2.20}$$

After replacing v by $C^{-1}v$ for a sufficiently large constant C, we achieve that

$$\mathrm{div}\,(v) - |v|^2 \asymp r^{-2}. \tag{12.2.21}$$

Before continuing, let us establish (12.2.20). Let $g = g_\epsilon$ be the function in the definition of $\gamma_r = \widehat{\gamma}_{\epsilon,C\epsilon r}$ in (12.2.11), so that $C^{-1} \le |\nabla g| \le 1$ (with the new C independent of ϵ, C in (12.2.11)), $\partial^\alpha g = \mathcal{O}_\epsilon(r(x)^{1-|\alpha|})$. Put

$$v = \nabla(e^{\lambda g/r}), \tag{12.2.22}$$

where $\lambda > 0$ will be sufficiently large. Notice that

$$\nabla\left(\frac{g}{r}\right) = \frac{\nabla g}{r} - \frac{g\nabla r}{r^2},$$

where

$$\left|\frac{\nabla g}{r}\right| \asymp \frac{1}{r}$$

uniformly with respect to ϵ, and

$$\left|\frac{g\nabla r}{r^2}\right| = \mathcal{O}(1)\frac{g}{r}\frac{1}{r} = \mathcal{O}(\epsilon)\frac{1}{r},$$

in γ_r, so if we fix $\epsilon > 0$ sufficiently small, then

$$\left|\nabla\left(\frac{g}{r}\right)\right| \asymp \frac{1}{r}.$$

We have

$$v = e^{\frac{\lambda g}{r}}\lambda\nabla\left(\frac{g}{r}\right), \quad |v| \asymp e^{\lambda\mathcal{O}(\epsilon)}\frac{\lambda}{r},$$

so the second part of (12.2.20) holds for every fixed value of λ. Further,

$$\operatorname{div}(v) = e^{\frac{\lambda g}{r}}\left(\lambda^2\left|\nabla\left(\frac{g}{r}\right)\right|^2 + \lambda\Delta\left(\frac{g}{r}\right)\right).$$

Here,

$$\left|\nabla\left(\frac{g}{r}\right)\right|^2 \asymp \frac{1}{r^2},\quad \Delta\left(\frac{g}{r}\right) = \mathcal{O}\left(\frac{1}{r^2}\right),$$

so if we fix λ large enough, we also get the first part of (12.2.20).

If we choose $k = (Cr)^{-1}$ for a sufficiently large constant C, we get from (12.2.21) and (12.2.16) that

$$\frac{1}{2}(\operatorname{div}(v) - |v|^2 - k^2) - (\psi_x')^2 \asymp r^{-2}.$$

Thus, with a new sufficiently large constant C, (12.2.19) yields

$$\|Du\|^2 + \frac{1}{C}\int_{\gamma_r}\frac{1}{r^2}|u|^2dx \leq C\|r(-\Delta_\psi)u\|^2, \tag{12.2.23}$$

which we can also write as

$$\|Du\| + \frac{1}{C}\|\frac{1}{r}u\| \leq C\|r(-\Delta_\psi)u\|. \tag{12.2.24}$$

Thus,

$$\|De^\psi u\| + \frac{1}{C}\|\frac{1}{r}e^\psi u\| \leq \|re^\psi \Delta u\|,$$

and by (12.2.16),

$$\|e^\psi Du\| \leq \|De^\psi u\| + \|\mathcal{O}(1/r)e^\psi u\| \leq \mathcal{O}(1)\|re^\psi\Delta u\|,$$

so we get (12.2.17) with a new constant C. □

If $\Omega \Subset \mathbf{C}$ has smooth boundary, let G_Ω, P_Ω denote the Green and the Poisson kernels of Ω, so that the Dirichlet problem,

$$\Delta u = v,\ u|_{\partial\Omega} = f,\quad u, v \in C^\infty(\overline{\Omega}),\ f \in C^\infty(\partial\Omega),$$

has the unique solution

$$u(x) = \int_{\Omega} G_{\Omega}(x, y)v(y)L(dy) + \int_{\partial\Omega} P_{\Omega}(x, y)f(y)|dy|.$$

Recall that $-G_{\Omega} \geq 0$, $P_{\Omega} \geq 0$. It is also clear that

$$-G_{\Omega}(x, y) \leq C - \frac{1}{2\pi} \ln |x - y|, \tag{12.2.25}$$

where $C > 0$ only depends on the diameter of Ω. Indeed, let $-G_0(x, y)$ denote the right-hand side of (12.2.25) and choose $C > 0$ large enough so that $-G_0 \geq 0$ on $\Omega \times \Omega$. Then on the operator level,

$$G_{\Omega}v = G_0 v - P_{\Omega}(G_0 v|_{\partial\Omega}),$$

so that

$$G_{\Omega}(x, y) = G_0(x, y) - \int_{\partial\Omega} P_{\Omega}(x, z)G_0(z, y)|dz|,$$

and hence $G_{\Omega} \geq G_0$, $-G_{\Omega} \leq -G_0$. The same argument (replacing G_0 by $G_{\widetilde{\Omega}}$ with $\widetilde{\Omega} \supset \Omega$) shows that $-G_{\Omega}$ is an increasing function of Ω:

$$\Omega_1 \subset \Omega_2 \Longrightarrow -G_{\Omega_1} \leq -G_{\Omega_2} \text{ on } \Omega_1 \times \Omega_1.$$

We will also use the elementary scaling property:

$$G_{\Omega}(x/t, y/t) = G_{t\Omega}(x, y),\ x, y \in t\Omega, t > 0. \tag{12.2.26}$$

Proposition 12.2.2 *Under the same assumptions as in Proposition* 12.2.1 *there exists a (new) constant $C > 0$ such that we have*

$$-G_{\gamma_r}(x, y) \leq C - \frac{1}{2\pi} \ln \frac{|x - y|}{r(y)},\ \textit{when } |x - y| \leq \frac{r(y)}{C}, \tag{12.2.27}$$

$$-G_{\gamma_r}(x, y) \leq C \exp(-\frac{1}{C} \int_{\pi_\gamma(y)}^{\pi_\gamma(x)} \frac{1}{r(t)} |dt|),\ \textit{when } |x - y| \geq \frac{r(y)}{C}, \tag{12.2.28}$$

where it is understood that the integral is evaluated along γ from $\pi_\gamma(y) \in \gamma$ to $\pi_\gamma(x) \in \gamma$, where $\pi_\gamma(y)$, $\pi_\gamma(x)$ denote points in γ with $|x - \pi_\gamma(x)| = \operatorname{dist}(x, \gamma)$, $|y - \pi_\gamma(y)| = \operatorname{dist}(y, \gamma)$, and we choose these two points (when they are not uniquely defined) and the intermediate segment in such a way that the integral is as small as possible.

Proof Let $y \in \gamma_r$, and put $t = r(y)$. Then we can find $\Omega \Subset \mathbf{C}$ uniformly bounded (with respect to y) whose boundary is uniformly bounded in the C^∞ sense,[3] such that γ_r coincides with $y + t\Omega =: \Omega_y$ in $D(y, 2r(y)/C)$, $\Omega_y \subset D(y, 4r(y)/C)$ and $r \asymp r(y)$ in that disc. In view of (12.2.25) and (12.2.26) we see that $-G_{\Omega_y}(x, y)$ satisfies the upper bound in (12.2.27).

Let $\chi = \chi(\frac{x-y}{r(y)})$ be a standard cutoff equal to 1 on $D(y, \frac{r(y)}{C})$ with $\operatorname{supp}\chi(\frac{\cdot-y}{r(y)}) \subset D(y, 2r(y)/C)$, and write the identity:

$$G_{\gamma_r}(\cdot, y) = \chi\left(\frac{\cdot - y}{r(y)}\right) G_{\Omega_y}(\cdot, y) - G_{\gamma_r}[\Delta, \chi\left(\frac{\cdot - y}{r(y)}\right)]G_{\Omega_y}(\cdot, y). \tag{12.2.29}$$

Using that the non-negative function $-G_{\Omega_y}$ satisfies (12.2.27), we see that the L^2-norm of $G_{\Omega_y}(\cdot, y)$ over the cutoff region (i.e., the support of the x-gradient of the cutoff) is $\mathcal{O}(r(y))$. G_{Ω_y} is harmonic with boundary value 0 in a neighborhood of $\operatorname{supp}\nabla\chi$. From standard estimates for elliptic boundary value problems, we conclude after scaling that the L^2-norm of $\nabla_x G_{\Omega_y}(x, y)$ over the same region is $\mathcal{O}(1)$. It follows that

$$\|[\Delta, \chi(\frac{\cdot - y}{r(y)})]G_{\Omega_y}(\cdot, y)\| = \mathcal{O}(\frac{1}{r(y)}),$$

and hence, by applying (12.2.17) with $\psi = 0$ to

$$u = G_{\gamma_r}[\Delta, \chi(\frac{\cdot - y}{r(y)})]G_{\Omega_y}(\cdot, y),$$

we get

$$\frac{1}{r}G_{\gamma_r}[\Delta, \chi(\frac{\cdot - y}{r(y)})]G_{\Omega_y}(\cdot, y) = \mathcal{O}(1), \text{ in } L^2(\gamma_r).$$

Away from $\operatorname{supp}[\Delta, \chi(\frac{\cdot-y}{r(y)})]$ the function $G_{\gamma_r}[\Delta, \chi(\frac{\cdot-y}{r(y)})]G_{\Omega_y}(\cdot, y)$ is harmonic on γ_r with boundary value zero and, appealing as above to apriori estimates for elliptic boundary value problems and scaling, we conclude that inside the region where $\chi(\frac{\cdot-y}{r(y)}) = 1$, it is $\mathcal{O}(1)$. From (12.2.29) we then get the estimate (12.2.27).

To get (12.2.28), we apply (12.2.17) and (12.2.16) to the second term in (12.2.29) with $\psi(x) = \psi(x, y)$ equal to the distance from y to x with respect to the Lithner–Agmon distance $(\widetilde{C}r)^{-2}|dx|^2$ with $\widetilde{C} > 0$ large enough. Then (12.2.16) holds, so (12.2.17)

[3] That is, given by an equation $f(z) = 0$, where f belongs to a bounded family of smooth real functions with the property that $f(z) = 0 \Longrightarrow |\nabla f(z)| \geq 1/\mathcal{O}(1)$ uniformly for all f in the family.

applies with

$$u = G_{\gamma_r}[\Delta, \chi((\cdot - y)/r(y))]\Psi_{\Omega_y}(\cdot, y), \;\; \Delta u = [\Delta, \chi((\cdot - y)/r(y))]\Psi_{\Omega_y}(\cdot, y).$$

Noticing that

$$\psi(x) \geq \mathcal{O}(1)^{-1} \int_{\pi_\gamma(y)}^{\pi_\gamma(x)} \frac{1}{r(t)} |dt|,$$

we get for $|x - y| \geq r(y)/\mathcal{O}(1)$:

$$\| r^{-1} \exp\left(\mathcal{O}(1)^{-1} \int_{\pi_\gamma(y)}^{\pi_\gamma(\cdot)} \frac{1}{r(t)} |dt| \right) G_{\gamma_r(\cdot, y)} \| \leq \mathcal{O}(1).$$

Using again that $|u| \leq \mathcal{O}(r(y))\|u\|$ in a domain Ω_y, where u is harmonic, near a part of the boundary where $u = 0$, we get (12.2.28) □

We will also need a lower bound on G_{γ_r} on suitable subsets of γ_r. For $\epsilon > 0$ fixed and sufficiently small, we say that $M \Subset \gamma_r$ is an elementary piece of γ_r if

- $M \subset \gamma_{(1-\frac{1}{C})r}$, cf. (12.2.15),
- $\frac{1}{C} \leq \frac{r(x)}{r(y)} \leq C,\ x, y \in M$,
- there exists $y \in M$ such that $M = y + r(y)\widetilde{M}$, where $\widetilde{M}$ belongs to a bounded set of relatively compact subsets of $\mathbf{C}$ with smooth boundary.

In the following, it will be tacitly understood that we choose our elementary pieces with some uniform control (C fixed and uniform control on the $\widetilde{M}$).

Proposition 12.2.3 *If M is an elementary piece in γ_r, then*

$$-G_{\gamma_r}(x, y) \asymp 1 + |\ln \frac{|x - y|}{r(y)}|,\ x, y \in M. \tag{12.2.30}$$

Proof We just outline the argument. First, by using arguments from the proof of Proposition 12.2.2 (without any exponential weights), we see that

$$-G_{\gamma_r}(x, y) \asymp -\ln \frac{|x - y|}{r(y)}, \text{ when } x, y \in M,\ \frac{|x - y|}{r(y)} \ll 1. \tag{12.2.31}$$

Next, if M' is a slightly larger elementary piece of the form $y + (1 + \frac{1}{C})r(y)\widetilde{M}$, then from Harnack's inequality for the positive harmonic function $-G_{\gamma_r}(\cdot, y)$ on $M' \setminus D(y, \frac{1}{2C}r(y))$, we see that $-G_{\gamma_r}(x, y) \asymp 1$ in $M \setminus D(y, \frac{1}{C}r(y))$, which together with (12.2.31) gives (12.2.30). □

12.3 Distribution of Zeros

Let ϕ be a real C^2 function defined in some neighborhood of $\overline{\gamma_r}$. Put

$$\mu = \mu_\phi = \Delta\phi L(dz). \tag{12.3.1}$$

Let u be a holomorphic function defined in a neighborhood of $\Gamma \cup \overline{\gamma_r}$. We assume that

$$h \ln |u(z)| \leq \phi(z), \; z \in \overline{\gamma_r}. \tag{12.3.2}$$

Lemma 12.3.1 *Let $z_0 \in M$, where M is an elementary piece, such that*

$$h \ln |u(z_0)| \geq \phi(z_0) - \epsilon, \; 0 < \epsilon \ll 1. \tag{12.3.3}$$

Then the number of zeros of u in M is

$$\leq \frac{C}{h}\left(\epsilon + \int_{\gamma_r} -G_{\gamma_r}(z_0, w)\mu(dw)\right). \tag{12.3.4}$$

Here C does not depend on ϕ, ϵ.

Since ϕ is of class C^2, we have $\mu(dw) = \Delta\phi(w)L(dw)$, $\Delta\phi \in C^0$ and

$$\int_{\gamma_r} G_{\gamma_r}(z_0, w)\mu(dw) = -\phi(z_0) + \int_{\partial\gamma_r} P_{\gamma_r}(z_0, w)\phi(w)|dw|, \tag{12.3.5}$$

where P is the Poisson kernel.

Proof of Lemma 12.3.1 Let

$$n_u(dz) = \sum 2\pi\delta(z - z_j),$$

where z_j are the zeros of u counted with their multiplicity. We may assume that no z_j is situated on $\partial\gamma_r$. Then, since $\Delta \ln |u| = n_u$,

$$\begin{aligned} h \ln |u(z)| = & \\ & \int_{\gamma_r} G_{\gamma_r}(z, w)hn_u(dw) + \int_{\partial\gamma_r} P_{\gamma_r}(z, w)h \ln |u(w)||dw| \\ & \leq \int_{\gamma_r} G_{\gamma_r}(z, w)hn_u(dw) + \int_{\partial\gamma_r} P_{\gamma_r}(z, w)\phi(w)|dw| \\ & = \int_{\gamma_r} G_{\gamma_r}(z, w)hn_u(dw) + \phi(z) - \int_{\gamma_r} G_{\gamma_r}(z, w)\mu(dw). \end{aligned} \tag{12.3.6}$$

Putting $z = z_0$ in (12.3.6) and using (12.3.3), we get

$$\int_{\gamma_r} -G_{\gamma_r}(z_0, w) h n_u(dw) \le \epsilon + \int_{\gamma_r} -G_{\gamma_r}(z_0, w)\mu(dw).$$

Now by (12.2.30),

$$-G_{\gamma_r}(z_0, w) \ge \frac{1}{C},\ \ w \in M,$$

and we get (12.3.4). □

Notice that this argument is basically the same as when using Jensen's formula to estimate the number of zeros of a holomorphic function in a disc and was used in the proof of Proposition 3.4.6.

Let z_j^0, z_j be as in Theorem 12.1.1 and keep the assumption that ϕ is of class C^2 near $\overline{\gamma}_r$. We may arrange so that $\widetilde{\gamma}_{r/C_1} \subset \gamma_r \subset \widetilde{\gamma}_r$. In particular, the assumptions of Theorem 12.1.1 imply (12.3.2). Now we sharpen the assumption (12.3.3) and assume, as in Theorem 12.1.1, that

$$h \ln |u(z_j)| \ge \phi(z_j) - \epsilon_j. \tag{12.3.7}$$

Let $M_j \subset \gamma_r$ be elementary pieces such that

$$\begin{aligned} z_j \in M_j,\ \operatorname{dist}(z_j, M_k) \ge \frac{r(z_j)}{C} \text{ when } k \neq j,\\ \gamma_{\tilde{r}} \subset \bigcup_j M_j,\ \tilde{r} = (1 - \frac{1}{\widetilde{C}})r, \end{aligned} \tag{12.3.8}$$

where $\widetilde{C} \gg 1$. Recall that $\gamma_r = \widehat{\gamma}_{\epsilon, C\epsilon r}$, where C, ϵ are now fixed (cf. (12.2.11)), and that $\gamma_{\alpha r} = \widehat{\gamma}_{\epsilon, \alpha C\epsilon r}$.

According to Lemma 12.3.1, we have

$$\#(u^{-1}(0) \cap M_j) \le \frac{C_3}{h}\left(\epsilon_j + \int_{\gamma_r} -G_{\gamma_r}(z_j, w)\mu(dw)\right). \tag{12.3.9}$$

Consider the harmonic functions on $\gamma_{\tilde{r}}$,

$$\Psi(z) = h\left(\ln |u(z)| + \int_{\gamma_{\tilde{r}}} -G_{\gamma_{\tilde{r}}}(z, w) n_u(dw)\right), \tag{12.3.10}$$

$$\Phi(z) = \phi(z) + \int_{\gamma_{\tilde{r}}} -G_{\gamma_{\tilde{r}}}(z, w)\mu(dw). \tag{12.3.11}$$

Then $\Phi(z) = \phi(z)$ on $\partial\gamma_{\widetilde{r}}$. Similarly, $\Psi(z) \geq h \ln |u(z)|$ on $\gamma_{\widetilde{r}}$, with equality on $\partial\gamma_{\widetilde{r}}$.

Consider the harmonic function

$$H(z) = \Phi(z) - \Psi(z), \; z \in \gamma_{\widetilde{r}}. \tag{12.3.12}$$

Then on $\partial\gamma_{\widetilde{r}}$, we have, by (12.3.2), that

$$H(z) = \phi(z) - h \ln |u(z)| \geq 0,$$

so by the maximum principle,

$$H(z) \geq 0, \;\; \text{on } \gamma_{\widetilde{r}}. \tag{12.3.13}$$

By (12.3.7), we have

$$\begin{aligned} H(z_j) =& \Phi(z_j) - \Psi(z_j) \\ =& \phi(z_j) - h \ln |u(z_j)| \\ &+ \int_{\gamma_{\widetilde{r}}} -G_{\gamma_{\widetilde{r}}}(z_j, w)\mu(dw) - \int_{\gamma_{\widetilde{r}}} -G_{\gamma_{\widetilde{r}}}(z_j, w)hn_u(dw) \\ \leq& \epsilon_j + \int_{\gamma_{\widetilde{r}}} -G_{\gamma_{\widetilde{r}}}(z_j, w)\mu(dw). \end{aligned} \tag{12.3.14}$$

Harnack's inequality[4] implies that

$$H(z) \leq \mathcal{O}(1)(\epsilon_j + \int -G_{\gamma_{\widetilde{r}}}(z_j, w)\mu(dw)) \text{ on } M_j \cap \gamma_{\widehat{r}}, \; \widehat{r} = (1 - \frac{1}{\widetilde{C}})\widetilde{r}. \tag{12.3.15}$$

Now assume that u extends to a holomorphic function in a neighborhood of $\Gamma \cup \overline{\gamma_r}$. We want to evaluate the number of zeros of u in Γ. Using (12.3.9), we first have

$$\#(u^{-1}(0) \cap \gamma_{\widetilde{r}}) \leq \frac{C}{h} \sum_{j=1}^{N} \left(\epsilon_j + \int_{\gamma_r} -G_{\gamma_r}(z_j, w)\mu(dw) \right). \tag{12.3.16}$$

[4] Which says that if $\Omega \Subset \mathbf{C}$ is a connected open set with smooth boundary and $K \subset \Omega$ a compact subset, then there exists a constant $C = C_{\Omega,K} > 0$ such that $u(z_1) \leq Cu(z_2)$ for all $z_1, z_2 \in K$ and every non-negative harmonic function u on Ω, uniformly if Ω varies in a family of uniformly bounded subsets of $\mathbf{C}$ and dist $(K, \partial\Omega) \geq 1/\mathcal{O}(1)$ uniformly, see e.g., [4, Section 6.3].

Let $\chi \in C_0^\infty(\Gamma \cup \gamma_{\widetilde{r}}; [0, 1])$ be equal to 1 on Γ. Of course, χ will have to depend on r, but we may assume that for all $k \in \mathbf{N}$,

$$\nabla^k \chi = \mathcal{O}(r^{-k}). \tag{12.3.17}$$

We are interested in

$$\int \chi(z) h n_u(dz) = \int_{\gamma_{\widetilde{r}}} h \ln |u(z)| \Delta \chi(z) L(dz). \tag{12.3.18}$$

On $\gamma_{\widetilde{r}}$ we have

$$\begin{aligned} h \ln |u(z)| &= \Psi(z) - \int_{\gamma_{\widetilde{r}}} -G_{\gamma_{\widetilde{r}}}(z, w) h n_u(dw) \\ &= \Phi(z) - H(z) - \int_{\gamma_{\widetilde{r}}} -G_{\gamma_{\widetilde{r}}}(z, w) h n_u(dw) \\ &= \phi(z) + \int_{\gamma_{\widetilde{r}}} -G_{\gamma_{\widetilde{r}}}(z, w) \mu(dw) - H(z) - \int_{\gamma_{\widetilde{r}}} -G_{\gamma_{\widetilde{r}}}(z, w) h n_u(dw) \\ &=: \phi(z) + R(z), \end{aligned} \tag{12.3.19}$$

where the last equality defines $R(z)$.

Inserting this in (12.3.18), we get

$$\int \chi(z) h n_u(dz) = \int \chi(z) \mu(dz) + \int R(z) \Delta \chi(z) L(dz). \tag{12.3.20}$$

(Here we also used some extension of ϕ to Γ with $\mu = \Delta\phi$.) The task is now to estimate $R(z)$ and the corresponding integral in (12.3.20). Put

$$\mu_j = \mu_+(M_j \cap \gamma_{\widetilde{r}}), \tag{12.3.21}$$

where $\mu = \mu_+ - \mu_-$ is the Radon decomposition of μ, here with $\mu_\pm = m_\pm L(dz)$, $0 \le m_\pm \in C^0$, $m_+ m_- = 0$. Using the exponential decay property (12.2.28) (equally valid for $G_{\gamma_{\widetilde{r}}}$) we get for $z \in M_j \cap \gamma_{\widetilde{r}}$, $\mathrm{dist}\,(z, \partial M_j) \ge r(z_j)/\mathcal{O}(1)$:

$$\left| \int_{\gamma_{\widetilde{r}}} -G_{\gamma_{\widetilde{r}}}(z, w) \mu(dw) \right| \le \int_{M_j \cap \gamma_{\widetilde{r}}} -G_{\gamma_{\widetilde{r}}}(z, w) \mu_+(dw) + \mathcal{O}(1) \sum_{k \ne j} \mu_k e^{-\frac{1}{C_0}|j-k|}, \tag{12.3.22}$$

where $|j-k|$ denotes the natural distance from j to k in $\mathbf{Z}/N\mathbf{Z}$ and $C_0 > 0$. Similarly, from (12.3.15) we get

$$H(z) \le \mathcal{O}(1)\left(\epsilon_j + \int_{M_j\cap\gamma_{\widetilde{r}}} -G_{\gamma_{\widetilde{r}}}(z_j, w)\mu_+(dw) + \sum_{k\neq j} e^{-\frac{1}{C_0}|j-k|}\mu_k\right), \tag{12.3.23}$$

for $z \in M_j \cap \gamma_{\widetilde{r}}$.

This gives the following estimate on the contribution from the first two terms in $R(z)$ to the last integral in (12.3.20), where we also use that $\Delta\chi = \mathcal{O}(r^{-2})$:

$$\begin{aligned} &\int_{\gamma_{\widetilde{r}}}\left(\int_{\gamma_{\widetilde{r}}} -G_{\gamma_{\widetilde{r}}}(z,w)\mu(dw) - H(z)\right)\Delta\chi(z)L(dz) \\ &= \mathcal{O}(1)\sum_j(\epsilon_j + \int_{M_j\cap\gamma_{\widetilde{r}}} -G_{\gamma_{\widetilde{r}}}(z_j,w)\mu_+(dw)) + \sum_{k\neq j} e^{-\frac{1}{C_0}|j-k|}\mu_k) \\ &+ \mathcal{O}(1)\sum_j \int_{M_j\cap\gamma_{\widetilde{r}}}\int_{M_j\cap\gamma_{\widetilde{r}}} -G_{\gamma_{\widetilde{r}}}(z,w)\mu_+(dw)|\Delta\chi(z)|L(dz). \end{aligned} \tag{12.3.24}$$

Here by (12.2.30),

$$\int_{M_j\cap\gamma_{\widetilde{r}}} -G_{\gamma_{\widetilde{r}}}(z,w)|\Delta\chi(z)|L(dz) = \mathcal{O}(1), \tag{12.3.25}$$

so (12.3.24) leads to

$$\begin{aligned} &\int_{\gamma_{\widetilde{r}}}\left(\int_{\gamma_{\widetilde{r}}} -G_{\gamma_{\widetilde{r}}}(z,w)\mu(dw) - H(z)\right)\Delta\chi(z)L(dz) \\ &= \mathcal{O}(1)\left(\mu_+(\gamma_{\widetilde{r}}) + \sum_j \epsilon_j + \sum_j \int_{M_j\cap\gamma_{\widetilde{r}}} -G_{\gamma_{\widetilde{r}}}(z_j,w)\mu_+(dw)\right). \end{aligned} \tag{12.3.26}$$

The contribution from the last term in $R(z)$ (in (12.3.19)) to the last integral in (12.3.20) is

$$\int_{z\in\gamma_{\widetilde{r}}}\int_{w\in\gamma_{\widetilde{r}}} G_{\gamma_{\widetilde{r}}}(z,w)hn_u(dw)\Delta\chi(z)L(dz). \tag{12.3.27}$$

Here, by using an estimate similar to (12.3.22) with $\mu(dw)$ replaced by $L(dz)$, together with (12.3.25), we get

$$\int_{z\in\gamma_{\widetilde{r}}} G_{\gamma_{\widetilde{r}}}(z,w)(\Delta\chi)(z)L(dz) = \mathcal{O}(1),$$

so the expression (12.3.27) is by (12.3.16) equal to

$$
\begin{aligned}
&\mathcal{O}(h)\#(u^{-1}(0)\cap\gamma_{\tilde r})\\
&=\mathcal{O}(1)\sum_{j=1}^{N}\Big(\epsilon_j+\int_{\gamma_r}-G_{\gamma_r}(z_j,w)\mu_+(dw)\Big)\\
&=\mathcal{O}(1)\left(\mu_+(\gamma_r)+\sum_{j=1}^{N}\left(\epsilon_j+\int_{M_j}-G_{\gamma_r}(z_j,w)\mu_+(dw)\right)\right).
\end{aligned}
\tag{12.3.28}
$$

This is quite similar to (12.3.26). Using Proposition 12.2.2, we have

$$
\begin{aligned}
&\int_{M_j\cap\gamma_{\tilde r}}-G_{\gamma_{\tilde r}}(z_j,w)\mu_+(dw)\le\\
&\mathcal{O}(1)\left(\int_{|w-z_j|\le\frac{r(z_j)}{C}}\left|\ln\left(\frac{|z_j-w|}{r(z_j)}\right)\right|\mu_+(dw)+\mu_+(M_j\cap\gamma_{\tilde r})\right)
\end{aligned}
$$

and similarly for the last integral in (12.3.28). Using all this in (12.3.20), we get

$$
\begin{aligned}
&\int\chi(z)hn_u(dz)=\int\chi(z)\mu(dz)\\
&+\mathcal{O}(1)\left(\mu_+(\gamma_r)+\sum_j\left(\epsilon_j+\int_{|w-z_j|\le r(z_j)/C}\left|\ln\left(\frac{|z_j-w|}{r(z_j)}\right)\right|\mu_+(dw)\right)\right).
\end{aligned}
\tag{12.3.29}
$$

Now we observe that

$$
\left|\#(u^{-1}(0)\cap\Gamma)-\frac{1}{2\pi h}\int\chi(z)hn_u(dz)\right|\le\#(u^{-1}(0)\cap\gamma_{\tilde r}),
$$

which can be estimated by means of (12.3.28). Combining this with (12.3.29) and the fact that

$$
\left|\mu(\Gamma)-\int\chi(z)\mu(dz)\right|\le|\mu|(\gamma_r),\ \ |\mu|=\mu_++\mu_-,
$$

we get

$$
\begin{aligned}
&|\#(u^{-1}(0)\cap\Gamma)-\frac{1}{2\pi h}\mu(\Gamma)|\le\\
&\frac{\mathcal{O}(1)}{h}\left(|\mu|(\gamma_r)+\sum_j(\epsilon_j+\int_{|w-z_j|\le\frac{r(z_j)}{C}}\left|\ln\left(\frac{|z_j-w|}{r(z_j)}\right)\right|\mu_+(dw))\right).
\end{aligned}
\tag{12.3.30}
$$

This completes the proof of Theorem 12.1.1 in the case when ϕ is of class C^2 near $\overline{\gamma}_r$.

Now we remove the C^2 assumption and take ϕ continuous as in Theorem 12.1.1 As there, we let $\mu = \Delta\phi L(dz)$. Let $0 \le \chi \in C_0^\infty(D(0,1))$ be a radial function with $\int \chi(z)L(dz) = 1$, and put $\chi_\epsilon = \epsilon^{-2}\chi(\cdot/\epsilon)$ for $0 < \epsilon \ll 1$. Then $\phi_\epsilon := \chi_\epsilon * \phi$ (convolution of χ_ϵ and ϕ) and $\mu_\epsilon := \chi_\epsilon * \mu = (\Delta\phi_\epsilon)L(dz)$, are well defined in slightly shrinked domains, and hence the C^2 special case of the theorem is applicable. If u is as in the theorem, we have (cf. (12.1.8) and (12.1.9)):

$$h\ln|u| \le \phi_\epsilon(z) + \min \epsilon_j \quad \text{on } \widetilde{\gamma}_r,$$
$$h\ln|u(z_j)| \ge \phi_\epsilon(z_j) - 2\epsilon_j,\ \ j = 1, 2, \dots, N,$$

when $\epsilon > 0$ is small enough, since $\phi_\epsilon \to \phi$ uniformly, when $\epsilon \to 0$. We can then apply the C^2-version of the theorem, and replace r with $r/2$ in the conclusion (12.1.10), and get

$$\begin{aligned} &|\#(u^{-1}(0)\cap\Gamma) - \frac{1}{2\pi h}\mu_\epsilon(\Gamma)| \le \\ &\frac{C_2}{h}\left(|\mu_\epsilon|(\widetilde{\gamma}_{r/2}) + \sum_1^N\left(\epsilon_j + \int_{D(z_j,\frac{r(z_j)}{8C_1})}\left|\ln\left(\frac{|w-z_j|}{r(z_j)}\right)\right|\mu_{\epsilon,+}(dw)\right)\right). \end{aligned} \tag{12.3.31}$$

Since χ_ϵ is radial and hence even, we have

$$\mu_\epsilon(\Gamma) = \int 1_\Gamma * \chi_\epsilon(z)\mu(dz)$$

when ϵ is small enough. Also,

$$\mu(\Gamma) = \lim_{j\to\infty}\int \chi_j(z)\mu(dz),$$

where $0 \le \chi_j \in C_0^\infty(\Gamma)$ is an increasing sequence of functions, equal to 1 away from a small neighborhood of γ, that converges to 1_Γ. It follows that

$$|\mu(\Gamma) - \mu_\epsilon(\Gamma)| \le 2|\mu|(\widetilde{\gamma}_r), \tag{12.3.32}$$

when ϵ is small enough.

Let us next look at the right-hand side in (12.3.31). From the Jordan decomposition $\mu = \mu_+ - \mu_-$, we get $\mu_\epsilon = \chi_\epsilon * \mu_+ - \chi_\epsilon * \mu_-$, $\chi_\epsilon * \mu_\pm \ge 0$, in a slightly shrinked neighborhood of $\overline{\gamma}_r$. By the minimality of the Jordan decomposition, we have $\mu_{\epsilon,\pm} \le \chi_\epsilon * \mu_\pm$ and hence $|\mu_\epsilon| \le \chi_\epsilon * |\mu|$. When ϵ is small enough, we get

$$|\mu_\epsilon|(\widetilde{\gamma}_{r/2}) \le |\mu|(\widetilde{\gamma}_r)|.$$

The integral in the general term in the sum in (12.3.31) is

$$\begin{aligned}
&\leq \int_{D(z_j,r_j/(8C_1))} \left|\ln\left(\frac{|w-z_j|}{r_j}\right)\right| (\chi_\epsilon * \mu_+)(dw) \\
&= \int \chi_\epsilon * \left(1_{D(z_j,r_j/(8C_1))}(w)\left|\ln\left(\frac{|w-z_j|}{r_j}\right)\right|\right) \mu_+(dw) \qquad (12.3.33) \\
&\leq \int_{D(z_j,r_j/(4C_1))} \chi_\epsilon * \left|\left(\ln\frac{|\cdot - z_j|}{r_j}\right)\right| (w)\mu_+(dw),
\end{aligned}$$

where $r_j := r(z_j)$ and we assume that ϵ is small enough.

Notice that here we work in a region where $\ln(|w-z_j|/r_j) < 0$. By the spherical mean-value property for harmonic functions, we have

$$\chi_\epsilon * \ln(|\cdot - z_j|/r_j)(w) = \ln|w-z_j|/r_j$$

when $|w-z_j| \geq \epsilon$. Comparing

$$\Delta_w \ln(|w-z_j|/r_j) = 2\pi\delta_{z_j},$$

$$\Delta_w(\chi_\epsilon * \ln(|\cdot - z_j|/r_j)) = 2\pi\chi_\epsilon(\cdot - z_j),$$

in log-radial coordinates, $t = \ln|\cdot - z_j|$, we see that the equality above extends to an inequality inside $D(z_j,\epsilon)$[5]:

$$0 \geq \chi_\epsilon * \ln(|\cdot - z_j|/r_j)(w) \geq \ln(|w-z_j|/r_j),$$

so the last integral in (12.3.33) can be estimated from above by

$$\int_{D(z_j,r_j/(4C_1))} |\ln(|w-z_j|/r_j)|\mu_+(dw). \qquad (12.3.34)$$

Combining (12.3.32)–(12.3.34) with (12.3.31), we get (12.1.7) which completes the proof of Theorem 12.1.1. □

[5]This also follows more directly from the spherical mean value inequality for subharmonic functions.

We next discuss when the contribution from the logarithmic integrals in (12.1.10) can be eliminated or simplified. Let r, C_1, z_j^0 be as in Theorem 12.1.1. We get

$$\int_{D\left(z_j^0,\frac{r(z_j^0)}{2C_1}\right)}\int_{D\left(z,\frac{r(z)}{4C_1}\right)}\left|\ln\left(\frac{|w-z|}{r(z)}\right)\right|\mu_+(dw)\frac{L(dz)}{L\left(D\left(z_j^0,\frac{r(z_j^0)}{2C_1}\right)\right)}$$

$$\le\int_{D\left(z_j^0,\frac{r(z_j^0)}{2C_1}\right)}\int_{D\left(z_j^0,\frac{r(z_j^0)}{C_1}\right)}\left|\ln\left(\frac{|w-z|}{r(z)}\right)\right|\mu_+(dw)\frac{L(dz)}{L\left(D\left(z_j^0,\frac{r(z_j^0)}{2C_1}\right)\right)}$$

$$\le\mathcal{O}(1)\mu_+\left(D\left(z_j^0,\frac{r(z_j^0)}{C_1}\right)\right),$$

where we changed the order of integrations in the last step and also used that $r(z)\asymp r(z_j^0)$ in $D(z_j^0,\frac{r(z_j^0)}{C_1})$. We conclude that the mean value of

$$D\left(z_j^0,\frac{r(z_j^0)}{2C_1}\right)\ni z\mapsto\int_{D(z,\frac{r(z)}{4C_1})}\left|\ln\left(\frac{|w-z|}{r(z)}\right)\right|\mu_+(dw)$$

is $\mathcal{O}(1)\mu_+\left(D(z_j^0,\frac{r(z_j^0)}{C_1})\right)$. Thus we can find $\widetilde{z}_j\in D(z_j^0,\frac{r(z_j^0)}{2C_1})$ such that

$$\int_{D(\widetilde{z}_j,\frac{r(\widetilde{z}_j)}{4C_1})}\left|\ln\left(\frac{|w-\widetilde{z}_j|}{r(\widetilde{z}_j)}\right)\right|\mu_+(dw)=\mathcal{O}(1)\mu_+(D(z_j^0,r(z_j^0)/C_1)).\tag{12.3.35}$$

This establishes Theorem 12.1.2. □

Remark 12.3.2 Under the assumptions leading to (12.3.35), the set $\Omega_j(C)$ of all $\widetilde{z}\in D(z_j^0,r(z_j^0)/(2C_1))$ such that

$$\int_{D(\widetilde{z}_j,\frac{r(\widetilde{z}_j)}{4C_1})}\left|\left(\ln\frac{|w-\widetilde{z}_j|}{r(\widetilde{z}_j)}\right)\right|\mu_+(dw)\ge C\mu_+(D(z_j^0,r(z_j^0)/C_1)),$$

satisfies

$$\frac{L(\Omega_j(C))}{L(D(z_j^0,r(z_j^0/(2C_1))}\to 0,\ C\to+\infty.$$

This makes it possible to find $\tilde{z}_j \in D(z_j^0, r(z_j^0)/(2C_1))$ simultaneously satisfying several (finally many) estimates of the form (12.3.35), for instance if we have several measures μ_+^j, $1 \le N < \infty$.

12.4 Application to Sums of Exponential Functions

Consider the function

$$u(z;h) = \sum_1^N e^{\phi_j(z)/h}, \tag{12.4.1}$$

where N is finite and ϕ_j are holomorphic in the open set $\Omega \subset \mathbf{C}$ and independent of h for simplicity. Put

$$\psi_j(z) = \Re\phi_j(z), \tag{12.4.2}$$

let $\Gamma \Subset \Omega$ have C^∞ boundary γ and assume

$$\begin{aligned} &\forall x \in \gamma,\ \Psi(x) := \max_j \psi_j(x) \text{ is attained} \\ &\text{for at most two different values of } j. \end{aligned} \tag{12.4.3}$$

$$\begin{aligned} &\text{If } x \in \gamma,\ \Psi(x) = \psi_j(x) = \psi_k(x),\ j \neq k, \\ &\text{then } \nu(x, \partial_x)(\psi_j(x) - \psi_k(x)) \neq 0, \end{aligned} \tag{12.4.4}$$

where ν denotes the normalized tangent vector field (say positively oriented) to γ. We shall see that Theorem 12.1.2 allows us to determine the number of zeros of u in Γ up to $\mathcal{O}(1)$. This result can be further strengthened by using direct arguments (see for instance [72]), but the purpose of this section is simply to illustrate the results above. We also notice that the results will be valid if u is holomorphic in Ω, but with the representation (12.4.1) and the ϕ_j defined only in a neighborhood of γ.

Define Ψ in Ω or more generally in a neighborhood of γ in Ω as in (12.4.3). Define the curves $\gamma_{j,k} \subset \Omega$ near γ by

$$\gamma_{j,k} = \{x;\ \Psi(x) = \psi_j(x) = \psi_k(x)\},\ j \neq k.$$

We shall establish the following result (without any claim of novelty, see [72] as well as [15, 36]). For a closely related old result on entire functions, see [93], Chapter VI, Section 3, Theorem 9, attributed to Pfluger [110].

Proposition 12.4.1 *We have*

$$\left|\#(u^{-1}(0)\cap\Gamma)-\frac{1}{2\pi h}\int_{\Gamma}\Delta\Psi(z)L(dz)\right|=\mathcal{O}(1). \tag{12.4.5}$$

Here, in the case when ψ_j and Ψ are defined only in a neighborhood of γ, we take any distribution extension of Ψ to a neighborhood of Γ. Notice that near γ the function Ψ is subharmonic and $\Delta\Psi$ is supported by the union of the curves $\gamma_{j,k}$. On each such curve, $\Delta\Psi=\frac{\partial}{\partial n}(\psi_j-\psi_k)|dz|$, where n is the unit normal to $\gamma_{j,k}$, oriented so that $\frac{\partial}{\partial n}(\psi_j-\psi_k)>0$. This can be proved with the help of Green's formula, cf. the end of the proof of the main result in [146].

We shall prove Proposition 12.4.1 by means of Theorem 12.1.2.

Put

$$\Phi(z)=h\ln\left(\sum_{1}^{N}e^{\psi_j(z)/h}\right),\quad z\in\operatorname{neigh}(\gamma), \tag{12.4.6}$$

so that $\Phi(z)=hf(\frac{\psi_1}{h},..,\frac{\psi_N}{h})$, where

$$f(x)=\ln\left(\sum_{1}^{N}e^{x_j}\right). \tag{12.4.7}$$

If we define $\theta_j=e^{x_j}/\sum e^{x_k}$, then $\theta_j>0$, $\theta_1+\dots+\theta_N=1$, and

$$\partial_{x_j}f(x)=\theta_j, \tag{12.4.8}$$

$$f''(x)=\operatorname{diag}(\theta_j)-(\theta_j\theta_k)_{j,k}. \tag{12.4.9}$$

For $y\in\mathbf{R}^N$, we have

$$\langle f''(x)y|y\rangle=\sum\theta_j y_j^2-\left(\sum\theta_j y_j\right)^2,$$

which is ≥ 0, since the function $t\mapsto t^2$ is convex. Hence f is convex.

We apply this to $\Phi(z)$, now with $\theta_j=e^{\psi_j(z)/h}/\sum_k e^{\psi_k/h}$, and get

$$\partial_z\Phi(z)=\sum\theta_j\partial_z\psi_j,\quad \partial_z=\frac{1}{2}(\partial_{\Re z}+\frac{1}{i}\partial_{\Im z}) \tag{12.4.10}$$

$$\partial_{\bar z}\partial_z\Phi(z)=\frac{1}{h}\langle f''\partial_z\psi|\partial_{\bar z}\psi\rangle=\frac{1}{h}\left(\sum\theta_j|\partial_z\psi_j|^2-\left|\sum\theta_j\partial_z\psi_j\right|^2\right). \tag{12.4.11}$$

In the last calculation, we also used that ψ_j are harmonic. It follows that Φ is subharmonic near γ. Also notice that

$$\Delta\Phi(z) = \mathcal{O}(h^{-1}), \tag{12.4.12}$$

and that this estimate can be considerably improved away from the union of the $\gamma_{j,k}$: Assume for instance that in some open relatively compact subset, $\Psi(z) = \psi_1 \geq \max_{j\neq 1}\psi_j + \delta$, where $\delta > 0$. Then

$$\Phi = h\ln(e^{\psi_1/h}(1+\mathcal{O}(e^{-\delta/h}))) = \psi_1 + \mathcal{O}(he^{-\delta/h}). \tag{12.4.13}$$

Further,

$$\theta_1 = 1+\mathcal{O}(e^{-\delta/h}),\ \theta_j = \mathcal{O}(e^{-\delta/h}) \text{ for } j\neq 1, \tag{12.4.14}$$

so

$$\partial_z\Phi = \partial_z\psi_1 + \mathcal{O}(e^{-\delta/h}), \tag{12.4.15}$$

$$f''(e^{\psi_1/h},\dots,e^{\psi_N/h}) = \mathcal{O}(e^{-\delta/h}),$$

$$\partial_{\overline{z}}\partial_z\Phi = \mathcal{O}(\frac{1}{h}e^{-\delta/h}). \tag{12.4.16}$$

We will always be able to express the final result in terms of the simpler function Ψ:

Lemma 12.4.2 *We have*

$$\int_\Gamma \Delta\Phi L(dz) - \int_\Gamma \Delta\Psi L(dz) = \mathcal{O}(h). \tag{12.4.17}$$

Proof Using Green's formula, the left-hand side of (12.4.17) can be written

$$\int_\gamma \left(\frac{\partial\Phi}{\partial n} - \frac{\partial\Psi}{\partial n}\right)|dz|,$$

where n is the suitably oriented normal direction. It then suffices to apply (12.4.15), with ψ_1 replaced by Ψ, in the region where $d(z) := \mathrm{dist}\,(z, \bigcup_{j\neq k}\gamma_{j,k}) \gg h$ and use that the gradients of Φ, Ψ are $\mathcal{O}(1)$. □

We next notice that

$$h\ln|u(z;h)| \leq \Phi(z) \tag{12.4.18}$$

in a neighborhood of γ. On the other hand, for z near γ, $d(z) \gg h$, we have

$$h \ln |u(z; h)| \geq \Phi(z) - \mathcal{O}(h)e^{-d(z)/(Ch)}. \qquad (12.4.19)$$

We can now apply Theorem 12.1.2 with $r = \text{Const.}\, h, d(z_j^0) \geq Ch, \epsilon_j = \mathcal{O}(he^{-d(z_j^0)/(Ch)})$, $\phi = \Phi$. In view of (12.4.12), (12.4.16), we see that $\mu(\widetilde{\gamma}_r) = \mathcal{O}(h)$, $\sum \epsilon_j = \mathcal{O}(h)$, so

$$\#(u^{-1}(0) \cap \Gamma) - \frac{1}{2\pi h} \int_\Gamma \Delta\Phi L(dz) = \mathcal{O}(1),$$

and we obtain Proposition 12.4.1 from Lemma 12.4.2. □

Notes There has been a considerable activity in the study of the zero set of random holomorphic functions where the Edelman Kostlan formula has similarities with the above results (and the earlier ones by Hager and others mentioned above) and where further results have been obtained. See Sodin [149] Sodin and Tsirelson [150], Zrebiec [160], Shiffman et al. [125]. The two last papers deal with holomorphic functions of several variables and it would be of interest to see if our results also have extensions to that case. Our results are similar in spirit to classical results on zeros of entire functions, see Levin [93].

We have followed closely the paper [141] where ϕ was assumed to be subharmonic, which is certainly the main case of interest. In some applications however, the subharmonicity condition seems to be too restrictive and in this chapter we have made the effort to remove it.

Perturbations of Jordan Blocks

13

13.1 Introduction

In this chapter we shall study the spectrum of a random perturbation of the large Jordan block A_0, introduced in Sect. 2.4:

$$A_0 = \begin{pmatrix} 0\,1\,0\,0 \ldots 0 \\ 0\,0\,1\,0 \ldots 0 \\ 0\,0\,0\,1 \ldots 0 \\ \ldots\ldots\ldots \\ 0\,0\,0\,0 \ldots 1 \\ 0\,0\,0\,0 \ldots 0 \end{pmatrix} : \mathbf{C}^N \to \mathbf{C}^N.$$

- Zworski [162] noticed that for every $z \in D(0, 1)$, there are associated exponentially accurate quasimodes when $N \to \infty$. Hence the open unit disc is a region of spectral instability.
- We have spectral stability (a good resolvent estimate) in $\mathbf{C} \setminus \overline{D(0, 1)}$, since $\|A_0\| = 1$.
- $\sigma(A_0) = \{0\}$.

Thus, if $A_\delta = A_0 + \delta Q$ is a small (random) perturbation of A_0 we expect the eigenvalues to move inside a small neighborhood of $\overline{D(0, 1)}$. In the special case when $Qu = (u|e_1)e_N$, where $(e_j)_1^N$ is the canonical basis in $\mathbf{C}^N$, we have seen in Sect. 2.4 that the eigenvalues of A_δ are of the form

$$\delta^{1/N} e^{2\pi i k/N}, \ k \in \mathbf{Z}/N\mathbf{Z},$$

J. Sjöstrand, *Non-Self-Adjoint Differential Operators, Spectral Asymptotics and Random Perturbations*, Pseudo-Differential Operators 14,
https://doi.org/10.1007/978-3-030-10819-9_13

so if we fix $0 < \delta \ll 1$ and let $N \to \infty$, the spectrum "will converge to a uniform distribution on S^1".

Davies and Hager [38] studied random perturbations of A_0. They showed that with probability close to 1, most of the eigenvalues are close to a circle:

Theorem 13.1.1 *Let $A = A_0 + \delta Q$, $Q = (q_{j,k}(\omega))$, where $q_{j,k}$ are independent random variables $\sim \mathcal{N}_{\mathbf{C}}(0, 1)$. If $0 < \delta \le N^{-7}$, $R = \delta^{1/N}$, $\sigma > 0$, then with probability $\ge 1 - 2N^{-2}$, we have $\sigma(A_\delta) \subset D(0, RN^{3/N})$ and*

$$\#(\sigma(A_\delta) \cap D(0, Re^{-\sigma})) \le \frac{2}{\sigma} + \frac{4}{\sigma} \ln N.$$

The angular distribution was not treated in [38], except for more special perturbations, and the main purpose of this chapter is to do so following the general strategy of Hager's thesis that we have already followed in Chap. 3. A result by Guionnet et al. [52] implies that when δ is bounded from above by $N^{-\kappa-1/2}$ for some $\kappa > 0$ and from below by some negative power of N, then

$$\frac{1}{N} \sum_{\mu \in \sigma(A_\delta)} \delta(z - \mu) \to \text{the uniform measure on } S^1,$$

weakly in probability.

The main focus of the chapter is to get probabilistic statements about the distribution of eigenvalues near the critical radius $R = \delta^{1/N}$ when $Q = (q_{j,k})$, with $q_{j,k} \sim \mathcal{N}_{\mathbf{C}}(0, 1)$ independent. Then we know from Proposition 3.4.1 that $\|Q\|_{\mathrm{HS}} \le C_1 N$ with probability $\ge 1 - e^{-N^2}$. We distinguish between the cases of small and of larger perturbations.

- Small perturbations:

$$e^{-CN} \le \delta \le N^{-\frac{5}{2}-\epsilon_0}, \tag{13.1.1}$$

  where C, $\epsilon_0 > 0$ are fixed.
- Larger perturbations:

$$N^{\epsilon_0-\frac{5}{2}} \le \delta \ll N^{-\frac{3}{2}}.$$

The intermediate case can undoubtedly be treated along the lines of the case of larger perturbations.

As an outgrowth of this chapter, Vogel and the author studied in [147] the expectation density of eigenvalues inside the critical disc, adapting some methods from Vogel [155] devoted to Hager's operator, somewhat in the spirit of various works on zeros of random polynomials. This is a new and probably very rich area, but for the present book we finally preferred to limit the perspective to eigenvalue counting with probability. More recent results on large Toeplitz matrices can be found in [145, 146].

13.2 Main Results

Small Perturbations We now state the results and take $Q = (q_{j,k})$, with $q_{j,k} \sim \mathcal{N}_{\mathbf{C}}(0, 1)$ independent. Then we know from Proposition 3.4.1 (cf. Proposition 3.4.2) that $\|Q\|_{\mathrm{HS}} \le C_1 N$ with probability $\ge 1 - e^{-N^2}$. We distinguish between the cases of small and of larger perturbations. We first consider the case of small perturbations:

$$e^{-CN} \le \delta \le N^{-\frac{5}{2}-\epsilon_0}, \tag{13.2.1}$$

where $C,\ \epsilon_0 > 0$ are fixed. Define a "spectral radius" $R = R_\sigma > 0$ to be the unique solution in $]0, N/(N+1)[$ of

$$m(R) = 2C_1 N\delta, \ \text{where}\ m(R) := R^N(1-R). \tag{13.2.2}$$

In the beginning of Sect. 13.8, we will see that m is increasing on [0,N/(N+1)], decreasing on [N/(N+1),1] with a maximum $m_{\max} = m(N/(N+1)) = (1+o(1))/(eN)$, $N \to \infty$. Moreover, as we shall see in (13.8.3),

$$e^{\frac{1}{N}(\ln\delta+\ln N+\mathcal{O}(1))} \le R_\sigma \le e^{\frac{1}{N}(\ln\delta+2\ln N+\mathcal{O}(1))} \le e^{-\frac{1}{N}((\epsilon_0+\frac{1}{2})\ln N+\mathcal{O}(1))}, \tag{13.2.3}$$

and with probability $\ge 1 - e^{-N^2}$,

$$\sigma(A_\delta) \subset \overline{D(0, R_\sigma)}. \tag{13.2.4}$$

To see the last inclusion, we notice that by (2.4.4),

$$\|(z-A_0)^{-1}\| \le \frac{1-|z|^N}{|z|^N(1-|z|)} \le \frac{1}{m(|z|)}, \quad 0 < |z| \le \frac{N}{N+1},$$

and that

$$\|(z-A_0)^{-1}\| \le \frac{1}{m\left(\frac{N}{N+1}\right)}, \quad \frac{N}{N+1} \le |z| \le 1,$$

so when $\|Q\| \le C_1 N$ we get for $0 < |z| \le N/(N+1)$,

$$\delta\|Q\|\|(z-A_0)^{-1}\| \le C_1 N\delta/m(|z|),$$

which is $\leq 1/2$ for $|z| > R_\sigma$, since m is increasing on $[0, N/(N+1)]$. The same holds for $|z| \geq N/(N+1)$ by the maximum principle, and writing $z - A_\delta = (z - A_0)(1 + (z - A_0)^{-1}\delta Q)$ we get (13.2.4).

Theorem 13.2.1 *Let* $0 \leq \theta_1 < \theta_2 \leq 2\pi$, $\theta = \theta_2 - \theta_1$, $\Omega =]r_-, r_+[\, e^{i]\theta_1, \theta_2[}$, $0 < \epsilon_1 < 1$,

$$\frac{1}{\mathcal{O}(1)} \leq r_- \leq R_\sigma - \frac{1}{N}, \quad \text{and} \quad R_\sigma + \frac{1}{N} \leq r_+ \leq (1 + R_\sigma)/2.$$

With probability $\geq 1 - \mathcal{O}(N)e^{-N^{\epsilon_1}} - e^{-N^2}$, *we have*

$$\begin{aligned} &\left| \#(\sigma(A_\delta) \cap \Omega) - \frac{\theta}{2\pi} N \left(1 - \frac{r_-}{N(1-r_-)}\right) \right| \\ &\leq \mathcal{O}(\theta) N \left(\frac{N^{\epsilon_1 - 1}}{R_\sigma - r_-} + \frac{\delta N^{\frac{1}{2}}}{(1 - R_\sigma)^2} + N^{-\epsilon_0} \right) \\ &\quad + \mathcal{O}(1) N \left(\frac{N^{\epsilon_1} \ln N}{N} + \theta \left(\frac{R_\sigma}{r_-}\right)^{-\frac{N}{\mathcal{O}(1)}} + \theta \left(\frac{1}{R_\sigma}\right)^{-\frac{N}{\mathcal{O}(1)}} \right). \end{aligned} \tag{13.2.5}$$

In view of (13.2.4), the choice of $r_+ \geq R_\sigma$ is of no real interest (and the stronger lower bound for r_+, is merely a pointer to the proof, where we shall take $r_+ = (1 + R_\sigma)/2$), and we could take $r_+ = R_\sigma$. For (13.2.5) to be of interest when $\theta > 0$ is fixed, we would like the right-hand side to be $\ll N$. This is the case when $R_\sigma - r_- \gg N^{\epsilon_1 - 1}$, in particular when $R_\sigma - r_- \gg N^{\epsilon_2 - 1}$ for some $\epsilon_2 > \epsilon_1$. Then $r_-/(N(1 - r_-)) = \mathcal{O}(N^{-\epsilon_1})$ in the left-hand side. Thus for any such ϵ_2, we have a uniform angular distribution of the eigenvalues in a $N^{\epsilon_2 - 1}$-neighborhood of the circle $|z| = R_\sigma$. Choosing $\theta = 2\pi$, we see that most of the N eigenvalues belong to such a neighborhood.

Larger Perturbations Now assume,

$$N^{\epsilon_0 - \frac{5}{2}} \leq \delta \ll N^{-\frac{3}{2}}. \tag{13.2.6}$$

Then, when $\|Q\| \leq C_1 N$ (which holds with probability $\geq 1 - e^{-N^2}$) we have $\sigma(A_\delta) \subset D(0, 1 + C_1 \delta N)$ and we observe that $\delta N \ll \delta^{1/2} N^{1/4}$, $N^{-1} \ll \delta^{1/2} N^{1/4}$.

Theorem 13.2.2 *Let* $0 \le r_- \le 1 - 4(C_1\delta)^{\frac{1}{2}}N^{\frac{1}{4}}$, $r_+ \ge 1 + C_1\delta N$. *Let* $0 < \epsilon_1 < 1$. *Then with probability* $\ge 1 - \mathcal{O}(N)e^{-N^{\epsilon_1}}$, *we have*

$$\left|\#(\sigma(A_\delta)\cap\Omega) - \frac{\theta}{2\pi}\left(1 - \frac{r_-}{N(1-r_-)}\right)N\right| \le$$
$$\mathcal{O}(\theta)N\left(e^{-\frac{N^{\epsilon_0/2}}{\mathcal{O}(1)}} + \frac{\delta N^{\frac{1}{2}}}{(1-r_-)^2} + \frac{N^{\epsilon_1 - 1}}{1-r_-}\right)$$
$$+\mathcal{O}(1)N\left(N^{\epsilon_1-1}\ln N + \delta^{\frac{1}{2}}N^{\frac{1}{4}}\right). \qquad (13.2.7)$$

The right-hand side in (13.2.7) is $\ll N$ when $1-r_- \gg \delta^{1/2}N^{1/4}$ and $1-r_- \gg N^{\epsilon_1-1}$. Since

$$1 - r_- \ge \frac{1}{\mathcal{O}(1)}\delta^{\frac{1}{2}}N^{\frac{1}{4}} = \frac{1}{\mathcal{O}(1)}\delta^{\frac{1}{2}}N^{\frac{5}{4}}N^{-1} \ge \frac{1}{\mathcal{O}(1)}N^{\frac{\epsilon_0}{2}-1},$$

the last requirement on $1-r_-$ follows from the first, when $\epsilon_0/2 > \epsilon_1$, i.e., when ϵ_1 is chosen small enough. Again we have uniform angular distribution of the eigenvalues in $r_- \le |z| \le r_+$, since we also have $\frac{r_-}{N(1-r_-)} \ll 1$ for r_- in the range in the theorem. While in Theorem 13.2.1 we get uniform distribution in shells of thickness close to $1/N$ around the critical circle $|z| = R_\sigma$, we now need thicker shells that increase with δ and now contain the unit circle.

In the remainder of this chapter we prove the two theorems above. In Sect. 13.5, we also give an upper bound on the number of eigenvalues in discs inside the critical one.

13.3 Study in the Region $|z| > R$

Let F and R be as at the end of Sect. 2.4, satisfying (2.4.16).

Recall that $\det(z - A_0) = z^N$. We get the following upper bound:

$$\begin{aligned}|\det(z - A_\delta)| &= |\det(z-A_0)||\det(1-(z-A_0)^{-1}\delta Q)| \\ &\le |z|^N \exp\|(z-A_0)^{-1}\delta Q\|_{\mathrm{tr}} \\ &\le |z|^N \exp\left(F(|z|)\delta\|Q\|_{\mathrm{tr}}\right).\end{aligned}$$

For the first inequality above we refer to [48] or Sect. 8.4. Using that by (2.4.13),

$$\|(z-A_\delta)^{-1}\| \le F(|z|)(1 - F(|z|)\delta\|Q\|)^{-1},$$

when $F(|z|)\delta\|Q\| < 1$ (i.e., when $|z| > R$) we can permute the roles of A_0 and A_δ,

$$z - A_0 = (z - A_\delta)(1 + \delta(z - A_\delta)^{-1}Q),$$

and get a lower bound:

$$|z|^N = |\det(z - A_0)| \leq |\det(z - A_\delta)| \exp\left(\frac{F(|z|)\delta\|Q\|_{\mathrm{tr}}}{(1 - F(|z|)\delta\|Q\|)}\right).$$

In conclusion, under the assumption that $|z| > R$ we have

$$|z|^N \exp\left(-\frac{F(|z|)\delta\|Q\|_{\mathrm{tr}}}{1 - F(|z|)\delta\|Q\|}\right) \leq |\det(z - A_\delta)| \leq |z|^N \exp(F(|z|)\delta\|Q\|_{\mathrm{tr}}). \tag{13.3.1}$$

Later, we shall impose the condition (13.7.22), saying that $\|Q\|_{\mathrm{HS}} \leq C_1 N$ for a certain constant $C_1 > 0$. Then by the Cauchy–Schwarz inequality for the singular values of Q we know that $\|Q\|_{\mathrm{tr}} \leq C_1 N^{3/2}$ and (13.3.1) tells us that

$$|z|^N \exp\left(-\frac{F(|z|)\delta C_1 N^{3/2}}{1 - F(|z|)\delta\|Q\|}\right) \leq |\det(z - A_\delta)| \leq |z|^N \exp\left(F(|z|)\delta C_1 N^{3/2}\right). \tag{13.3.2}$$

Here and in (13.3.1), the second inequality does not require the assumption that $|z| > R$.

13.4 Study in the Region $|z| < 1$

Following [148], we introduce an auxiliary Grushin problem. Define $R_+ : \mathbf{C}^N \to \mathbf{C}$ by

$$R_+ u = u_1, \; u = (u_1 \, u_2 \, \ldots \, u_N)^{\mathrm{t}} \in \mathbf{C}^N. \tag{13.4.1}$$

Let $R_- : \mathbf{C} \to \mathbf{C}^N$ be defined by

$$R_- u_- = (0 \, 0 \, \ldots \, u_-)^{\mathrm{t}} \in \mathbf{C}^N. \tag{13.4.2}$$

Here, we identify vectors in $\mathbf{C}^N$ with column matrices. Then for $|z| < 1$, the operator

$$\mathcal{A}_0 = \begin{pmatrix} R_+ & 0 \\ A_0 - z & R_- \end{pmatrix} : \mathbf{C}^{N+1} \to \mathbf{C}^{1+N} \tag{13.4.3}$$

is bijective. In fact, its matrix is lower triangular with the entry 1 everywhere on the main diagonal, the entry $-z$ on the subdiagonal, and all other entries equal to 0. It has the inverse

$$\mathcal{E}_0 := \mathcal{A}_0^{-1} =: \begin{pmatrix} E_+^0 & E^0 \\ E_{-+}^0 & E_-^0 \end{pmatrix},$$

given by a lower triangular matrix with $(\mathcal{E}_0)_{j,k} = z^{j-k}$ for $1 \le k \le j \le N+1$. Then for $1 \le j \le N, 2 \le k \le N+1$,

$$E_{j,k}^0 = \begin{cases} 0, & \text{if } j < k, \\ z^{j-k}, & \text{if } k \le j \end{cases}. \tag{13.4.4}$$

Moreover,

$$E_+^0 = \begin{pmatrix} 1 \\ z \\ \vdots \\ z^{N-1} \end{pmatrix}, \quad E_-^0 = \begin{pmatrix} z^{N-1} & z^{N-2} & \dots & 1 \end{pmatrix}, \tag{13.4.5}$$

$$E_{-+}^0 = z^N. \tag{13.4.6}$$

Notice that this is equivalent to the more traditional approach: the inverse of $\begin{pmatrix} A_0 - z & R_- \\ R_+ & 0 \end{pmatrix}$ is $\begin{pmatrix} E^0 & E_+^0 \\ E_-^0 & E_{-+}^0 \end{pmatrix}$.

As in Sect. 2.4, we see that

$$\|E^0\| \le G(|z|), \ \|E_\pm^0\| \le G(|z|)^{\frac{1}{2}}, \ \|E_{-+}^0\| \le 1. \tag{13.4.7}$$

where $\|\cdot\|$ denote the natural operator norms and

$$G(|z|) := \min\left(N, \frac{1}{1-|z|}\right) \asymp 1 + |z| + |z|^2 + \dots + |z|^{N-1}. \tag{13.4.8}$$

The functions F (cf. (2.4.9)) and G are related by

$$F(R) = \frac{1}{R} G\left(\frac{1}{R}\right), \quad R \ge 1, \tag{13.4.9}$$

and we use this relation to extend the definition of G to $]0, +\infty[$.

Next, consider the natural Grushin problem for A_δ. If $\delta\|Q\|G(|z|) < 1$, we see that

$$\mathcal{A}_\delta = \begin{pmatrix} R_+ & 0 \\ A_\delta - z & R_- \end{pmatrix} \tag{13.4.10}$$

is bijective with inverse

$$\mathcal{E}_\delta = \begin{pmatrix} E_+^\delta & E^\delta \\ E_{-+}^\delta & E_-^\delta \end{pmatrix},$$

where

$$\begin{aligned} E^\delta &= E^0 - E^0\delta Q E^0 + E^0(\delta Q E^0)^2 - \cdots = E^0(1+\delta Q E^0)^{-1}, \\ E_+^\delta &= E_+^0 - E^0\delta Q E_+^0 + (E^0\delta Q)^2 E_+^0 - \cdots = (1+E^0\delta Q)^{-1}E_+^0, \\ E_-^\delta &= E_-^0 - E_-^0\delta Q E^0 + E_-^0(\delta Q E^0)^2 - \cdots = E_-^0(1+\delta Q E^0)^{-1}, \\ E_{-+}^\delta &= E_{-+}^0 - E_-^0\delta Q E_+^0 + E_-^0\delta Q E^0 \delta Q E_+^0 - \cdots \\ &= E_{-+}^0 - E_-^0\delta Q(1+E^0\delta Q)^{-1}E_+^0. \end{aligned} \tag{13.4.11}$$

We get

$$\begin{aligned} &\|E^\delta\| \le \frac{G(|z|)}{1-\delta\|Q\|G(|z|)}, \quad \|E_\pm^\delta\| \le \frac{G(|z|)^{\frac{1}{2}}}{1-\delta\|Q\|G(|z|)}, \\ &|E_{-+}^\delta - E_{-+}^0| \le \frac{\delta\|Q\|G(|z|)}{1-\delta\|Q\|G(|z|)}. \end{aligned} \tag{13.4.12}$$

Indicating derivatives with respect to δ with dots and omitting sometimes the super/subscript δ, we have

$$\dot{\mathcal{E}} = -\mathcal{E}\dot{\mathcal{A}}\mathcal{E} = -\begin{pmatrix} EQE_+ & EQE \\ E_-QE_+ & E_-QE \end{pmatrix}. \tag{13.4.13}$$

Integrating this from 0 to δ yields

$$\|E^\delta - E^0\| \le \frac{G(|z|)^2\delta\|Q\|}{(1-\delta\|Q\|G(|z|))^2}, \quad \|E_\pm^\delta - E_\pm^0\| \le \frac{G(|z|)^{\frac{3}{2}}\delta\|Q\|}{(1-\delta\|Q\|G(|z|))^2}. \tag{13.4.14}$$

Notice that $\det \mathcal{A}_0 = 1$. Combining (13.4.12) and (13.4.13), we get

$$|\partial_\delta \ln \det \mathcal{A}| = |\mathrm{tr}\,(\dot{\mathcal{A}}\mathcal{E})| = |\mathrm{tr}\, QE| \leq \|Q\|_{\mathrm{tr}}\|E\| \leq \|Q\|_{\mathrm{tr}}\frac{G(|z|)}{1-\delta\|Q\|G(|z|)},$$

and integration from 0 to δ yields

$$|\ln|\det \mathcal{E}_\delta|| = |\ln|\det \mathcal{A}_\delta|| \leq \frac{G(|z|)}{1-\delta\|Q\|G(|z|)}\delta\|Q\|_{\mathrm{tr}}. \tag{13.4.15}$$

We now sharpen the assumption that $\delta\|Q\|G(|z|) < 1$ to

$$\delta\|Q\|G(|z|) < 1/2. \tag{13.4.16}$$

Then

$$\begin{aligned} &\|E^\delta\| \leq 2G(|z|),\ \|E^\delta_\pm\| \leq 2G(|z|)^{\frac{1}{2}}, \\ &|E^\delta_{-+} - E^0_{-+}| \leq 2\delta\|Q\|G(|z|). \end{aligned} \tag{13.4.17}$$

Combining this with the identity $\dot{E}_{-+} = -E_- Q E_+$ (cf. (13.4.13)) and (13.4.14), we get

$$\|\dot{E}_{-+} + E^0_- Q E^0_+\| \leq 16G(|z|)^2\delta\|Q\|^2, \tag{13.4.18}$$

and after integration from 0 to δ,

$$E^\delta_{-+} = E^0_{-+} - \delta E^0_- Q E^0_+ + \mathcal{O}(1)G(|z|)^2(\delta\|Q\|)^2. \tag{13.4.19}$$

Using (13.4.5) and (13.4.6) we get, with $Q = (q_{j,k})$,

$$E^\delta_{-+} = z^N - \delta\sum_{j,k=1}^{N} q_{j,k} z^{N-j+k-1} + \mathcal{O}(1)G(|z|)^2(\delta\|Q\|)^2, \tag{13.4.20}$$

still under the assumption (13.4.16). More explicitly, the modulus of the remainder in (13.4.19) and (13.4.20) is bounded by $8G(|z|)^2\delta^2\|Q\|^2$.

13.5 Upper Bounds on the Number of Eigenvalues in the Interior

This section is somewhat independent of the main theme of the chapter, and we will use some estimates from Sects. 13.7 and 13.8.

Let $Q = (q_{j,k})_{1\le j,k\le N}$ be a random matrix where the entries are independent random variables $\sim \mathcal{N}_{\mathbf{C}}(0,1)$ and recall Proposition 3.4.1, which implies that

$$\|Q\|_{\mathrm{HS}} \le C_1 N, \text{ with probability } \ge 1 - e^{-N^2}. \tag{13.5.1}$$

We assume that

$$\delta C_1 N \le 1/2. \tag{13.5.2}$$

In this section we establish upper bounds on the number of eigenvalues in discs $D(0,r)$ with r not too large. We work in a disc $D(0,R_0)$, where $R_0 \le 1$ is the largest number such that

$$\delta C_1 N G(R_0) \le \frac{1}{2}. \tag{13.5.3}$$

Then with probability $\ge 1-e^{-N^2}$, we have (13.4.16) for every $z \in D(0,R_0)$. We can then apply (13.4.17) and see that with probability $\ge 1-e^{-N^2}$, we have

$$|E_{-+}^{\delta}(z)| \le |z|^N + 2\delta C_1 N G(|z|), \quad z \in D(0,R_0). \tag{13.5.4}$$

With the same probability, we have (13.4.20) with the explicit bound $8G(|z|)^2\delta^2(C_1N)^2$ on the modulus of the remainder term and we apply it for $z=0$:

$$|E_{-+}^{\delta}(0) + \delta q_{N,1}| \le 8\delta^2 (C_1 N)^2. \tag{13.5.5}$$

Since $q_{n,1} \sim \mathcal{N}_{\mathbf{C}}(0,1)$, we know that $|q_{n,1}| \ge \epsilon$ with probability $\ge e^{-\epsilon^2}$ for every fixed $\epsilon \ge 0$. Hence with probability $\ge e^{-\epsilon^2} - e^{-N^2}$, we have

$$|E_{-+}^{\delta}(0)| \ge \delta\epsilon - 8(C_1 N\delta)^2 = \delta(\epsilon - 8C_1^2\delta N^2). \tag{13.5.6}$$

For this to be of interest, we strengthen the assumption (13.5.2) (working in the limit $N \gg 1$) to

$$16 C_1^2 \delta N^2 < \epsilon \ll 1, \tag{13.5.7}$$

and in particular,

$$\delta < \frac{1}{16 C_1^2 N^2}. \tag{13.5.8}$$

Then (13.5.6) implies,

$$|E_{-+}^{\delta}(0)| \geq \frac{\delta\epsilon}{2}. \tag{13.5.9}$$

Recall that (still under the assumption that $\| Q\|_{\mathrm{HS}} \leq C_1 N$) the eigenvalues of $A_0+\delta Q$ in $D(0, R_0)$ coincide with the zeros of $E_{-+}^{\delta}(z)$ in the same set and the multiplicities agree (cf. Proposition 8.1.6). Let $0 < R < R_0$ and let $\lambda_1, \dots, \lambda_N$ be the zeros in $D(0, R)$ repeated according to their multiplicity. Then by Jensen's formula (8.4.28),

$$\ln |E_{-+}^{\delta}(0)| = \sum_{k=1}^{N} \ln \frac{|\lambda_k|}{R} + \frac{1}{2\pi R} \int_{|z|=R} \ln |E_{-+}^{\delta}(z)||dz|. \tag{13.5.10}$$

Equivalently,

$$\sum_{k=1}^{N} \ln \frac{R}{|\lambda_k|} = \ln \frac{1}{|E_{-+}^{\delta}(0)|} + \frac{1}{2\pi R} \int_{|z|=R} \ln |E_{-+}^{\delta}(z)||dz|.$$

For $0 < r < R$, let $M(r)$ denote the number of λ_k in $D(0, r)$. Then,

$$\sum_{k=1}^{N} \ln \frac{R}{|\lambda_k|} \geq M(r) \ln \frac{R}{r},$$

and in view of (13.5.4), (13.5.9) and (13.5.10), we get that with probability $\geq e^{-\epsilon^2} - e^{-N^2}$,

$$M(r) \ln \frac{R}{r} \leq \ln \frac{2}{\delta\epsilon} + \ln \left(R^N + 2C_1\delta N G(R)\right) = \ln \left(\frac{2(R^N + 2C_1\delta N G(R))}{\delta\epsilon}\right). \tag{13.5.11}$$

As we shall see in the beginning of Sect. 13.8, there is a unique $R = R_\sigma \in]0, 1 - 1/N]$ such that

$$R^N = 2C_1\delta N G(R),$$

provided that $m(1 - 1/N) \geq 2C_1\delta N$, where $m(t) = t^N(1 - t)$ for $0 \leq t \leq 1$, or more explicitly,

$$\left(1 - \frac{1}{N}\right)^N \frac{1}{N} \geq 2C_1\delta N,$$

$$\delta \leq \frac{1}{2C_1 N^2} \left(1 - \frac{1}{N}\right)^N = \frac{1}{2C_1 e N^2} \left(1 + \mathcal{O}\left(\frac{1}{N}\right)\right).$$

We also know from the beginning of Sect. 13.8 that

$$R^N \leq 2C_1\delta N G(R), \quad 0 \leq R \leq R_\sigma,$$

and that R_σ obeys the estimate (13.8.3).

Now assume,

$$0 < \delta < N^{-2} \min\left(\frac{1}{16C_1^2}, \frac{\left(1-\frac{1}{N}\right)^N}{2C_1}\right). \tag{13.5.12}$$

Then from (13.5.11) we see that for every $0 < r < R \leq R_\sigma$, $\epsilon > 0$ we have

$$M(r)\ln\frac{R}{r} \leq \ln\left(8C_1 N G(R)/\epsilon\right), \text{ with probability } \geq e^{-\epsilon^2} - e^{-N^2}. \tag{13.5.13}$$

Summing up, we have

Theorem 13.5.1 *Let $A_\delta = A_0 + \delta Q$, where $Q = (q_{j,k})_{1\leq j,k\leq N}$ and $q_{j,k}$ are independent random variables $\sim \mathcal{N}_{\mathbb{C}}(0,1)$. Let δ be in the range (13.5.12) and let $R = R_\delta$ be the unique solution in $]0, 1-1/N]$ of the equation $R^N = 2C_1\delta N G(R)$, so that (13.8.3) holds:*

$$N^{\frac{1}{N}}(2C_1\delta)^{\frac{1}{N}} \leq R_\sigma \leq \left(1 + \mathcal{O}\left(\frac{1}{N}\right)\right) N^{\frac{2}{N}}(2C_1\delta)^{\frac{1}{N}}.$$

Then for every (r, R, ϵ) with $0 < r < R \leq R_\sigma$, $\epsilon > 0$, we have (13.5.13), where $M(r)$ is the number of eigenvalues of A_δ in $D(0, r)$, counted with their multiplicities.

We shall next weaken the assumption (13.5.12) to

$$\delta \ll \frac{1}{N}. \tag{13.5.14}$$

Then Lemma 13.7.2 can be applied with $z = 0$ and we see that if $0 < \epsilon \leq C$ for any fixed $C > 0$, we have

$$|E_{-+}^\delta(0)| \geq \frac{\delta\epsilon}{2} \text{ and } |Q| \leq C_1 N, \text{ with probability } \geq 1 - \mathcal{O}(\epsilon^2) - e^{-N^2}. \tag{13.5.15}$$

Recall that $R_0 \leq 1$ is the largest number satisfying (13.5.3) and that (13.5.4) holds with probability $\geq 1 - e^{-N^2}$. Jensen's formula now gives (13.5.11) with probability $\geq 1 - \mathcal{O}(\epsilon^2) - e^{-N^2}$ for every fixed (r, R) with $0 < r < R < R_0$. Recall that we have defined $R = R_\sigma$ in $[0, 1-1/N]$ as the unique solution in that interval of the equation $R^N = 2C_1\delta NG(R)$ when $m(1-1/N) \geq 2C_1\delta N$ (asymptotically equivalent to $\delta \leq (2C_1 eN^2)^{-1}(1+\mathcal{O}(1/N))$. When $m(1-1/N) < 2C_1\delta N$, we define $R_\sigma = 1-1/N$ and notice that $R_\sigma < R_0$ also in this case. Again, we have $R^N < 2C_1\delta NG(R)$ when $R < R_\sigma$.

Theorem 13.5.2 *Instead of* (13.5.12), *we assume* (13.5.14) *and* $N \gg 1$. *When* $\delta > (1-1/N)^N/(2C_1N^2)$, *we extend the definition of* R_σ *by putting* $R_\sigma = 1-1/N$. *Let* $C > 0$ *be fixed. Then for* $0 < \epsilon \leq C$, $0 < r < R \leq R_\sigma$, *we have*

$$M(r)\ln\frac{R}{r} \leq \ln\left(8C_1NG(R)/\epsilon\right), \text{ with probability } \geq 1-\mathcal{O}(\epsilon^2)-e^{-N^2}. \tag{13.5.16}$$

13.6 Gaussian Elimination and Determinants

We review some standard material, see for instance [148] and Section 4 in [144]. Let $\mathcal{H}_j$, $\mathcal{G}_j$, $j = 1, 2$ be complex separable Hilbert spaces. Consider a bounded linear operator

$$\mathcal{A} = \begin{pmatrix} A_{11} & A_{12} \\ A_{21} & A_{22} \end{pmatrix} : \mathcal{H}_1 \times \mathcal{H}_2 \to \mathcal{G}_1 \times \mathcal{G}_2. \tag{13.6.1}$$

When $\mathcal{A}$ is bijective (with bounded inverse) we denote its inverse by

$$\mathcal{A}^{-1} = \mathcal{E} = \begin{pmatrix} E_{11} & E_{12} \\ E_{21} & E_{22} \end{pmatrix}. \tag{13.6.2}$$

Proposition 13.6.1

(1) *Assume that* A_{11} *is bijective. Then by Gaussian elimination we have the standard factorization into lower and upper triangular matrices:*

$$\mathcal{A} = \begin{pmatrix} A_{11} & 0 \\ A_{21} & 1 \end{pmatrix} \begin{pmatrix} 1 & A_{11}^{-1}A_{12} \\ 0 & A_{22} - A_{21}A_{11}^{-1}A_{12} \end{pmatrix}. \tag{13.6.3}$$

The first factor is bijective since A_{11} *is, so the bijectivity of* $\mathcal{A}$ *is equivalent to that of the second factor, which in turn is equivalent to that of* $A_{22} - A_{21}A_{11}^{-1}A_{12}$. *When* $\mathcal{A}$ *is bijective, we have the formula,*

$$\mathcal{A}^{-1} = \begin{pmatrix} 1 & a \\ 0 & (A_{22} - A_{21}A_{11}^{-1}A_{12})^{-1} \end{pmatrix} \begin{pmatrix} A_{11}^{-1} & 0 \\ b & 1 \end{pmatrix} =: \begin{pmatrix} E_{11} & E_{12} \\ E_{21} & E_{22} \end{pmatrix} =: \mathcal{E}, \tag{13.6.4}$$

where $a = -A_{11}^{-1}A_{12}(A_{22} - A_{21}A_{11}^{-1}A_{12})^{-1}$, $b = -A_{21}A_{11}^{-1}$ *and in particular,*

$$E_{22} = (A_{22} - A_{21}A_{11}^{-1}A_{12})^{-1}. \tag{13.6.5}$$

(2) *Now assume that* $\mathcal{A}$ *is bijective. Then* A_{11} *is bijective precisely when* E_{22} *is, and when that bijectivity holds we have*

$$\begin{aligned} E_{22}^{-1} &= A_{22} - A_{21}A_{11}^{-1}A_{12}, \\ A_{11}^{-1} &= E_{11} - E_{12}E_{22}^{-1}E_{21}. \end{aligned} \tag{13.6.6}$$

The first statement is clear. The second statement is also quite simple to verify, by solving for x_1 in the first equation in the system

$$\begin{cases} A_{11}x_1 + A_{12}x_2 = y_1, \\ A_{21}x_1 + A_{22}x_2 = y_2, \end{cases}$$

and substituting the result in the second equation.

Let now $\mathcal{H}_1 = \mathcal{G}_1$, $\mathcal{H}_2 = \mathcal{G}_2$ be of finite dimension and assume that $\mathcal{A}$ is bijective. From (13.6.4) and (13.6.5) we get

$$\det \mathcal{A}^{-1} = (\det E_{22}) \det A_{11}^{-1},$$

provided that A_{11} is bijective. This can be written

$$(\det \mathcal{A})(\det E_{22}) = \det A_{11}. \tag{13.6.7}$$

This identity extends to the case when A_{11} is not necessarily bijective, since $\det E_{22} = 0$ iff $\det A_{11} = 0$.

We apply this to $\mathcal{A} = \mathcal{A}_\delta$ in (13.4.10), or rather to $\begin{pmatrix} A_\delta - z & R_- \\ R_+ & 0 \end{pmatrix}$ with inverse $\begin{pmatrix} E^\delta & E_+^\delta \\ E_-^\delta & E_{-+}^\delta \end{pmatrix}$, under the assumption (13.4.16). Noticing that $E_{22} = E_{-+}^\delta$, $-A_{11} = z - A_\delta$ in this case and recalling (13.4.15), one obtains

$$\left| \ln |\det(z - A_\delta)| - \ln |E_{-+}^\delta| \right| \le 2G(|z|)\delta \|Q\|_{\mathrm{tr}}. \tag{13.6.8}$$

13.7 Random Perturbations and Determinants

Here to some extent we follow arguments of Vogel [155], see also [156].

We now let $Q = (q_{j,k}(\omega))_{j,k=1}^{N}$, where $q_{j,k}$ are independent random variables with the law $\mathcal{N}_{\mathbf{C}}(0,1)$. According to Proposition 3.4.1 we have

$$\mathbf{P}(\|Q\|_{\mathrm{HS}}^2 \geq x) \leq \exp\left(\frac{C_0}{2}N^2 - \frac{x}{2}\right)$$

and hence if $C_1 > 0$ is large enough,

$$\|Q\|_{\mathrm{HS}}^2 \leq C_1^2 N^2, \text{ with probability } \geq 1 - e^{-N^2}. \tag{13.7.1}$$

In particular (13.7.1) holds for the ordinary operator norm of Q. We continue to assume that $|z| < 1$ until further notice. We choose $\delta \geq 0$ so that

$$\delta C_1 N G(|z|) < \frac{1}{2}. \tag{13.7.2}$$

Then with probability $\geq 1 - e^{-N^2}$ (13.4.16) holds and so does (13.4.20), which gives

$$E_{-+}^{\delta} = z^N + \delta(Q|\overline{Z}) + \mathcal{O}(1)\,(G(|z|)\delta N)^2\,, \tag{13.7.3}$$

where

$$Z = \left(z^{N-j+k-1}\right)_{j,k=1}^{N}. \tag{13.7.4}$$

In the following, we often write $|\cdot|$ for the Hilbert–Schmidt norm $\|\cdot\|_{\mathrm{HS}}$. The scalar product in (13.7.3) is the corresponding one in the space of Hilbert–Schmidt operators $\simeq \mathbf{C}^{N^2}$. A straightforward calculation shows that

$$|Z| = \sum_0^{N-1} |z|^{2\nu} = \frac{1-|z|^{2N}}{1-|z|^2} = \frac{1-|z|^N}{1-|z|}\,\frac{1+|z|^N}{1+|z|}, \tag{13.7.5}$$

and in particular,

$$\frac{G_\Sigma(|z|)}{2} \leq |Z| \leq G_\Sigma(|z|), \tag{13.7.6}$$

where $G_\Sigma(|z|) = 1 + |z| + \cdots + |z|^{N-1} \asymp G(|z|)$ (cf. (13.4.8)). Then (13.7.3) shows that

$$|E_{-+}^{\delta} - z^N| \leq \delta N G_\Sigma(|z|)\,(C_1 + \mathcal{O}(1)G(|z|)\delta N)\,,$$

so

$$|E_{-+}^{\delta} - z^N| \leq \mathcal{O}(1)\delta N G(|z|). \tag{13.7.7}$$

From (13.7.3) and the Cauchy inequalities,[1] we get

$$d_Q E_{-+}^{\delta} = \delta |Z| (dQ|e_1) + \mathcal{O}\left(\frac{1}{N}\right) (\delta N G(|z|))^2 \tag{13.7.8}$$

in $\mathbf{C}^{N^2}$, where

$$e_1 = \frac{1}{|Z|}\overline{Z}. \tag{13.7.9}$$

Complete e_1 to an orthonormal basis $e_1, e_2, \dots, e_{N^2}$ in $\mathbf{C}^{N^2}$ and write

$$Q = Q' + Q_1 e_1, \quad Q' = \sum_2^{N^2} Q_k e_k \in (e_1)^{\perp}.$$

Then (13.7.3) and (13.7.8) give

$$E_{-+}^{\delta} = z^N + \delta |Z| Q_1 + \mathcal{O}\left((G(|z|)\delta N)^2\right), \tag{13.7.10}$$

$$d_Q E_{-+}^{\delta} = \delta |Z| d Q_1 + \mathcal{O}\left(N^{-1}(G(|z|)\delta N)^2\right), \tag{13.7.11}$$

for $|Q| \leq C_1 N$.

It will be convenient to extend $Q \mapsto E_{-+}^{\delta}(Q)$ to a smooth function $F : \mathbf{C}^{N^2} \to \mathbf{C}$, such that

$$\begin{aligned} F(Q) &= z^N + \delta |Z| Q_1 + \mathcal{O}\left((G(|z|)\delta N)^2\right) =: z^N + \delta |Z| f(Q), \\ d_Q F &= \delta |Z| d Q_1 + \mathcal{O}\left(N^{-1}(G(|z|)\delta N)^2\right), \end{aligned} \tag{13.7.12}$$

[1] If $u : B_{\mathbf{C}^M}(0, R) \to \mathbf{C}$ is a bounded holomorphic function, then the $\mathbf{C}^M$-norm $|du(z_0)|$ of the differential du at some point $z_0 \in B(0, R)$ is equal to the supremum of all $|u'_\omega(0)|$, where $u_\omega(w) = u(z_0 + w\omega)$, $w \in \mathrm{neigh}\,(0, \mathbf{C})$, when ω varies in the unit sphere $\partial B(0, 1)$ in $\mathbf{C}^M$. Applying the Cauchy inequality to all the functions u_ω and varying the point z_0 in $B(0, \theta R)$ for some fixed $0 < \theta < 1$, we get $\sup_{z \in B(0,\theta R)} |du(z)| \leq (C_\theta / R) \sup_{B(0,R)} |u|$.

and the remainders vanish outside $B_{\mathbf{C}^{N^2}}(0, 2C_1N)$. Indeed, we may assume that (13.7.10), (13.7.11) hold for $|Q| \le 2C_1N$ (adopting (13.7.2) with C_1 there replaced with $2C_1$) and take

$$F(Q) = z^N + \delta|Z|Q_1 + \chi\left(\frac{|Q|}{2C_1N}\right)\mathcal{O}\left((G(|z|)\delta N)^2\right),$$

where the $\mathcal{O}(\cdots)$ is the same quantity as in the first equation in (13.7.12) and

$$\chi \in C_0^\infty(]-1, 1[\,; [0, 1]),\ \chi = 1 \text{ on } [-1/2, 1/2].$$

Since F will in general not be holomorphic, the last estimate holds in the space of real-linear mappings: $\mathbf{C}^{N^2} \to \mathbf{C}$. The function f satisfies

$$f(Q) = Q_1 + \mathcal{O}(NG(|z|)\delta N), \tag{13.7.13}$$

$$d_Q f = dQ_1 + \mathcal{O}(G(|z|)\delta N). \tag{13.7.14}$$

We now strengthen the assumption (13.7.2) to

$$G(|z|)\delta N \ll 1. \tag{13.7.15}$$

Then the map $\mathbf{C} \ni Q_1 \mapsto f(Q_1, Q') \in \mathbf{C}$ is bijective for every Q' and has an inverse $g = g(\zeta, Q')$, satisfying

$$g(\zeta, Q') = \zeta + \mathcal{O}(NG(|z|)\delta N), \tag{13.7.16}$$

$$d_{\zeta,Q'} g(\zeta, Q') = d\zeta + \mathcal{O}(G(|z|)\delta N). \tag{13.7.17}$$

Let $\mu(d\zeta)$ be the direct image of the Gaussian measure $\pi^{-N^2}e^{-|Q|^2}L(dQ)$ under f. We study μ in $D(0, C)$ for any fixed $C > 0$. For $\varphi \in \mathcal{C}_0(D(0, C))$, we get

$$\begin{aligned}
\int \varphi(\zeta)\mu(d\zeta) &= \int \varphi(f(Q))\pi^{-N^2}e^{-|Q|^2}L(dQ) \\
&= \int_{\mathbb{C}^{N^2-1}} \pi^{1-N^2}e^{-|Q'|^2}\left(\int_{\mathbb{C}} \pi^{-1}e^{-|Q_1|^2}\varphi(f(Q))L(dQ_1)\right)L(dQ') \\
&= \int_{\mathbb{C}^{N^2-1}} \pi^{1-N^2}e^{-|Q'|^2}\left(\int_{\mathbb{C}} \pi^{-1}e^{-|g(\zeta,Q')|^2}\varphi(\zeta)L(d_\zeta g)\right)L(dQ'),
\end{aligned}$$

where

$$L(d_\zeta g) = L(dQ_1) = \det\left(\frac{\partial(Q_1, \overline{Q_1})}{\partial(\zeta, \overline{\zeta})}\right) L(d\zeta).$$

We get for $\varphi \in \mathcal{C}_0(D(0, C))$,

$$\int \varphi(\zeta)\mu(d\zeta)$$
$$= \int_{\mathbb{C}} \varphi(\zeta) \left(\int_{\mathbb{C}^{N^2-1}} \pi^{-1} e^{-|g(\zeta,Q')|^2} \pi^{1-N^2} e^{-|Q'|^2} \det\left(\frac{\partial(Q_1, \overline{Q_1})}{\partial(\zeta, \overline{\zeta})}\right) L(dQ') \right) L(d\zeta),$$

so that in $D(0, C)$

$$\mu(d\zeta) = \left(\int_{\mathbb{C}^{N^2-1}} (\pi^{-1} e^{-|g(\zeta,Q')|^2} \pi^{1-N^2} e^{-|Q'|^2} \det\left(\frac{\partial(Q_1, \overline{Q_1})}{\partial(\zeta, \overline{\zeta})}\right) L(dQ') \right) L(d\zeta). \tag{13.7.18}$$

We conclude that for $|\zeta_0|, r \le \mathcal{O}(1)$, the probability that $|Q| \le C_1 N$ and $f(Q) \in D(\zeta_0, r)$ is bounded from above by

$$\int_{\mathbb{C}^{N^2-1}} \int_{\zeta \in D(\zeta_0,r)} \pi^{-1} e^{-|g(\zeta,Q')|^2} L(d_\zeta g) \pi^{1-N^2} e^{-|Q'|^2} L(dQ'). \tag{13.7.19}$$

From (13.7.17) we infer that

$$g(\zeta, Q') - g(\zeta_0, Q') = (1 + \mathcal{O}(G\delta N))(\zeta - \zeta_0),$$

so

$$\{g(\zeta, Q');\ \zeta \in D(\zeta_0, r)\} \subset D(g(\zeta_0, Q'), \widetilde{r}),\ \ \widetilde{r} = (1 + \mathcal{O}(\delta N G))r$$

and the last integral is

$$\le \int_{\mathbb{C}^{N^2-1}} \int_{D(g(\zeta_0,Q'),\widetilde{r})} \pi^{-1} e^{-|\omega|^2} L(d\omega) \pi^{1-N^2} e^{-|Q'|^2} L(dQ').$$

Here the inner integral is

$$\le \int_{D(0,\widetilde{r})} \frac{1}{\pi} e^{-|\omega|^2} L(d\omega) = 1 - e^{-\widetilde{r}^2}. \tag{13.7.20}$$

Indeed, by rotation symmetry, we may assume that $g(\zeta_0, Q') = t \geq 0$, and by Fubini's theorem, we are reduced to show that $F(t) \leq F(0)$, where

$$F(t) = \int_{t-\widetilde{r}}^{t+\widetilde{r}} e^{-s^2} ds.$$

It then suffices to observe that $F'(t) \leq 0$.

Thus the integral in (13.7.19) is bounded by

$$1 - e^{-\widetilde{r}^2} \leq (1 + \mathcal{O}(G\delta N))(1 - e^{-r^2}).$$

In terms of E_{-+}^{δ}, we get under the assumption (13.7.15):

Lemma 13.7.1 *We recall* (13.7.6), (13.7.12). *For* $0 \leq t \leq C\delta G(|z|)$, $|z|^N \leq C\delta G(|z|)$, *the probability that* $|Q| \leq C_1 N$ *and that* $|E_{-+}^{\delta}| \leq t$ *is*

$$\leq (1 + \mathcal{O}(G(|z|)\delta N)) \left(1 - \exp\left(-\left(\frac{t}{\delta |Z|}\right)^2\right)\right).$$

Recall that $|Q| \leq C_1 N$ with probability $\geq 1 - e^{-N^2}$. It follows that for t, z as in the lemma,

$$\begin{aligned}
&\mathbf{P}(|Q| \leq C_1 N \text{ and } |E_{-+}^{\delta}| > t) \\
&= \mathbf{P}(|Q| \leq C_1 N) - \mathbf{P}(|Q| \leq C_1 N \text{ and } |E_{-+}^{\delta}| \leq t) \\
&\geq 1 - e^{-N^2} - (1 + \mathcal{O}(G(|z|)\delta N)) \left(1 - e^{-\left(\frac{t}{|Z|\delta}\right)^2}\right).
\end{aligned}$$

Assuming

$$|z|^N,\ t \leq CG(|z|)\delta, \tag{13.7.21}$$

we get

Lemma 13.7.2 *We work in the region* $|z| < 1$ *and assume* (13.7.15), (13.7.21). *Then*

$$\mathbf{P}(|Q| \leq C_1 N \textit{ and } |E_{-+}^{\delta}| > t) \geq 1 - \mathcal{O}(1) \left(\frac{t}{G(|z|)\delta}\right)^2 - e^{-N^2}.$$

From the bound

$$|Q| \leq C_1 N, \tag{13.7.22}$$

which we adopt from now on, and the Cauchy–Schwarz inequality for the singular values of Q, we know that $\|Q\|_{\mathrm{tr}} \le C_1 N^{3/2}$. It is of some interest to introduce the following strengthening of (13.7.15):

$$G(|z|)\delta N^{\frac{3}{2}} \le \mathcal{O}(1). \tag{13.7.23}$$

From (13.6.8) we get

$$\left|\ln|\det(z - A_\delta)| - \ln|E_{-+}^\delta|\right| \le 2G\delta C_1 N^{3/2}.$$

Thus,

$$|\det(z - A_\delta)| \ge e^{-2G\delta C_1 N^{3/2}}|E_{-+}^\delta|,\ N \to \infty$$

and Lemma 13.7.2 gives

Proposition 13.7.3 *Under the assumptions* (13.7.23), (13.7.21), *we have*

$$\mathbf{P}(|Q| \le C_1 N \text{ and } |\det(z - A_\delta)| > t) \ge 1 - \mathcal{O}(1)\left(\frac{t}{G(|z|)\delta}\right)^2 - e^{-N^2}.$$

If we replace (13.7.23) *by the weaker assumption* (13.7.15) *and keep* (13.7.21), *we have*

$$\mathbf{P}(|Q| \le C_1 N \text{ and } |\det(z - A_\delta)| > te^{-2C_1 G\delta N^{3/2}}) \ge 1 - \mathcal{O}(1)\left(\frac{t}{G(|z|)\delta}\right)^2 - e^{-N^2}.$$

We sum up the estimates obtained so far.

Theorem 13.7.4 *Consider $A_\delta = A_0 + \delta Q$, where the entries of Q are independent $\mathcal{N}_{\mathbf{C}}(0,1)$ random variables. Then with probability $\ge 1 - e^{-N^2}$ we have* (13.7.22)*: $\|Q\|_{\mathrm{HS}} \le C_1 N$ and we assume this estimate from now on. Let $G(r)$, $F(R)$ be defined in* (13.4.8) *and* (2.4.9), (2.4.10) *respectively, so that by* (13.4.9),

$$G(r) = \frac{1}{r}F\left(\frac{1}{r}\right),\ F(R) = \frac{1}{R}G\left(\frac{1}{R}\right).$$

Then we have

(A) *A general upper bound:*

$$|\det(z - A_\delta)| \le |z|^N \exp\left(F(|z|)C_1\delta N^{3/2}\right),\ z \ne 0. \tag{13.7.24}$$

(B) *An exterior lower bound: If* $F(|z|)C_1\delta N < 1/2$,

$$|\det(z - A_\delta)| \geq |z|^N \exp\left(-2F(|z|)C_1\delta N^{3/2}\right), \quad z \neq 0. \tag{13.7.25}$$

(C) *Interior upper and lower bounds: If* $|z| < 1$ *and* $G(|z|)C_1\delta N < 1/2$,

$$|\det(z - A_\delta)| \begin{cases} \leq \left(|z|^N + 2C_1\delta N G(|z|)\right) e^{2G(|z|)C_1\delta N^{3/2}}, \\ \geq \left(|z|^N - 2C_1\delta N G(|z|)\right)_+ e^{-2G(|z|)C_1\delta N^{3/2}}. \end{cases} \tag{13.7.26}$$

(D) *An interior probabilistic lower bound. Assume that* $|z| < 1$ *and that we have* (13.7.23), (13.7.21). *Then we have*

$$|\det(z - A_\delta)| > t \text{ with probability } \geq 1 - \mathcal{O}(1)\left(\frac{t}{G(|z|)\delta}\right)^2 - e^{-N^2}. \tag{13.7.27}$$

Replacing (13.7.23) *by* (13.7.15) *and keeping* (13.7.21), *leads to the weaker estimate,*

$$|\det(z - A_\delta)| > te^{-2C_1\delta G N^{3/2}} \text{ with probability } \geq 1 - \mathcal{O}(1)\left(\frac{t}{G(|z|)\delta}\right)^2 - e^{-N^2}. \tag{13.7.28}$$

Proof (A) and (B) follow from (13.3.2). (C) follows from the last part of (13.4.12), (13.6.8) and the fact that $\|Q\|_{\mathrm{tr}} \leq C_1 N^{3/2}$. (D) follows from Proposition 13.7.3. □

Remark 13.7.5 Notice that $F(1) = G(1) = N$ and recall that F is decreasing, while G is increasing. Let $r_\sigma, R_\sigma > 0$ be determined by the conditions (cf. (13.2.2))

$$G(r_\sigma) = \frac{1}{2C_1\delta N} = F(R_\sigma).$$

- If $N < 1/(2C_1\delta N)$, then $0 < R_\sigma < 1 < r_\sigma < +\infty$ and from (B), we conclude that $\sigma(A_\delta) \subset \overline{D(0, R_\sigma)} \subset D(0, 1)$. This is a case of small perturbations.
- If $N > 1/(2C_1\delta N)$, the spectrum of A_δ is not necessarily included in the closed unit disc. This is a case of "larger perturbations".
- In the limiting case $N = 1/(2C_1\delta N)$, we have $\sigma(A_\delta) \subset \overline{D(0, 1)}$.

In the next section we discuss the distribution of eigenvalues in a case of small perturbations.

13.8 Eigenvalue Distribution for Small Random Perturbations

In this section we study the eigenvalue distribution when the spectral radius R_σ in Remark 13.7.5 is smaller than 1 with some logarithmic margin. By (2.4.9), we have $F(R) = R^{-N}G(R)$ for $R \leq 1$ and when $R \leq 1 - 1/N$ we have

$$F(R) = \frac{1}{m(R)}, \text{ where } m(R) := R^N(1-R). \tag{13.8.1}$$

Thus, we are interested in the equation

$$m(R) = 2C_1\delta N, \; 0 \leq R \leq 1 - 1/N. \tag{13.8.2}$$

We have

$$m'(R) = (N+1)R^{N-1}\left(\frac{N}{N+1} - R\right),$$

and hence we have a unique critical point $R = R_{\max} = N/(N+1)$, which is the point of maximum of m. For N large, we get

$$m_{\max} := m(R_{\max}) = \frac{1 + \mathcal{O}(\frac{1}{N})}{eN}.$$

A necessary condition for (13.8.2) is of course that $m_{\max} > 2C_1\delta N$. We will establish simple upper and lower bounds on $R = R_\sigma$ in (13.8.2) under the apriori assumption that $R_\sigma \in [0, 1 - 1/N]$ and N is large.

Clearly $R_\sigma^N > 2C_1\delta N$, since $0 < 1 - R_\sigma < 1$, so

$$R_\sigma > R_1 := (2C_1\delta N)^{\frac{1}{N}} = e^{\frac{1}{N}(\ln 2C_1\delta + \ln N)}.$$

Similarly, $R_\sigma < R_{\max} = 1 - 1/(N+1)$, so $1 - R_\sigma > 1/(N+1)$ and hence

$$\begin{aligned} &R_\sigma^N/(N+1) < 2C_1\delta N, \\ &R_\sigma < R_2 := (2C_1\delta N(N+1))^{1/N} = e^{\frac{1}{N}(\ln 2C_1\delta + 2\ln N) + \mathcal{O}(N^{-1})}. \end{aligned}$$

We conclude that

$$e^{\frac{1}{N}(\ln 2C_1\delta + \ln N)} \leq R_\sigma \leq e^{\frac{1}{N}(\ln 2C_1\delta + 2\ln N) + \mathcal{O}(N^{-1})}. \tag{13.8.3}$$

The smallness condition on the perturbation that we first adopt in this section (to be sharpened later) is that $\ln 2C_1\delta + 2\ln N \le -\epsilon_0 \ln N$ for some fixed $\epsilon_0 > 0$, i.e.,

$$2C_1\delta \le N^{-2-\epsilon_0}. \tag{13.8.4}$$

It is slightly stronger than the one for small perturbations in Remark 13.7.5 and it implies, as we have just seen, that

$$R_\sigma \le 1 - \frac{\epsilon_0 \ln N + \mathcal{O}(1)}{N}, \tag{13.8.5}$$

and in particular the apriori assumption $R_\sigma \le 1 - 1/N$ is satisfied.

In the remainder of this section, we restrict the attention to the disc $D(0, 1 - 1/N)$ where $F(|z|) = |z|^{-N}(1-|z|)^{-1} = m(z)^{-1}$ and $G(|z|) = (1-|z|)^{-1}$. The upper bounds in (A) and (C) in Theorem 13.7.4 become respectively,

$$\ln|\det(z - A_\delta)| \le N \ln|z| + \frac{C_1\delta N^{\frac{3}{2}}}{|z|^N(1-|z|)},$$

$$\ln|\det(z - A_\delta)| \le \ln\left(|z|^N + \frac{2C_1\delta N}{1-|z|}\right) + \frac{2C_1\delta N^{\frac{3}{2}}}{1-|z|}.$$

We will also assume, for simplicity, that δ is at most exponentially decaying: there exists $C_0 > 0$ such that

$$e^{-C_0 N} \le \delta. \tag{13.8.6}$$

As we shall see more precisely below, we may neglect the term $2C_1\delta N/(1-|z|)$ in the argument of the last logarithm as long as it is $\ll |z|^N$. Now the two terms are equal precisely when $m(|z|) = 2C_1\delta N$, i.e., when $|z| = R_\sigma$, and therefore $|z|^N$ will be the dominant term in the region $|z| \ge R_\sigma + \epsilon \ln N/N$ for any fixed $\epsilon > 0$. So in that region, in order to compare the upper bounds in (A) and (C), it suffices to compare

$$\text{(a)}\ \frac{C_1\delta N^{\frac{3}{2}}}{|z|^N(1-|z|)} \quad \text{and} \quad \text{(c)}\ \frac{2C_1\delta N^{\frac{3}{2}}}{1-|z|}.$$

These two terms are of the same order of magnitude when $|z| = 1 - 1/N$, but when $|z| = 1 - \epsilon(\ln N)/N$ the first one dominates the second one with a factor which is roughly N^ϵ. Though this is not a rigorous proof that the bound in (C) is sharper than the one in (A) in all parameter ranges, we choose to use only (C), (D) in $D(0, 1 - 1/N)$.

Let us first review some general facts about radial subharmonic functions and about functions of the form $\ln(u + v)$.

In polar coordinates $z = re^{i\theta}$, $r > 0$, $\theta \in S^1$, we have

$$r^2\Delta = (r\partial_r)^2 + \partial_\theta^2.$$

From this we see that a real continuous radial function $\phi = \phi(r)$, defined on an annulus centered at 0, is subharmonic iff $(r\partial_r)^2\phi \geq 0$, or equivalently iff $\phi(e^t)$ is convex.

We also recall:

Lemma 13.8.1 *Let $\Omega \subset \mathbf{C}$ be open and let $u, v \in C^2(\Omega; \mathbf{R})$ be subharmonic functions. Then $w := \ln(e^u + e^v)$ is subharmonic. The statement also holds after replacing "subharmonic" with "convex".*

Proof Write $\partial = \frac{\partial}{\partial z}$, $\overline{\partial} = \frac{\partial}{\partial \overline{z}}$. Differentiating the relation $e^w = e^u + e^v$, we get $e^w\partial w = e^u\partial u + e^v\partial v$. Applying $\overline{\partial}$ to this, we get after some computations

$$\overline{\partial}\partial w = e^{u-w}\overline{\partial}\partial u + e^{v-w}\overline{\partial}\partial v + e^{u+v-2w}(|\partial u|^2 + |\partial v|^2 - 2\Re \partial u \overline{\partial v})$$

which is ≥ 0, since each of the three terms in the right-hand side is ≥ 0.

The proof in the convex case is basically the same. □

Consider again $w = \ln(e^u + e^v)$, with $u, v \in C^2(\Omega; \mathbf{R})$. We have

$$\max(u, v) \leq w \leq \max(u, v) + \ln 2.$$

Further, at a point where $u \geq v$, we can write $e^u + e^v = e^u(1 + e^{-(u-v)})$ to see that

$$w = u + \ln(1 + e^{-(u-v)}) = u + e^{-(u-v)} + \mathcal{O}(e^{-2(u-v)}).$$

Similarly,

$$dw = \frac{e^u}{e^u + e^v}du + \frac{e^v}{e^u + e^v}dv = (1 - \mathcal{O}(e^{-(u-v)}))du + \mathcal{O}(e^{-(u-v)})dv. \tag{13.8.7}$$

Now, consider the function $\ln\big(r^N + 2C_1\delta N/(1-r)\big)$, $r = |z| \in]0, 1[$, which is of the form $w = \ln(e^u + e^v)$ with $u = N\ln r$, $v = \ln\frac{2C_1\delta N}{1-r}$, where $r = |z|$. u is harmonic and we check that v is subharmonic, i.e. $v = \ln\frac{2C_1\delta N}{1-e^t}$ is convex for $t < 0$. Thus w is subharmonic by the lemma. For later use we also notice that $1/(r-1)$ is subharmonic for $r > 1$.

We consider the left-most point $r = r_0$ where $u(r) = v(r)$. This equation is equivalent to $m(r) = 2C_1\delta N$, already studied, so $r_0 = R_\sigma$.

It is clear that $u(r) \le v(r)$ for $0 < r \le R_\sigma$, with equality precisely for $r = R_\sigma$ and we look for a quantitative statement. We have

$$\partial_r(u-v) = \frac{N}{r} - \frac{1}{1-r}, \quad u - v = N \ln \frac{r}{R_\sigma} + \ln \frac{1-r}{1-R_\sigma}, \tag{13.8.8}$$

and for $r \le R_\sigma$, $\frac{1}{1-r} \le \frac{1}{1-R_\sigma}$. Here, $R_\sigma = e^{-s}$, $s \asymp \ln(1/\delta)/N$, so

$$1 - R_\sigma \ge \frac{\ln 1/\delta}{\mathcal{O}(1)N},$$

$$\frac{1}{1-R_\sigma} \le \frac{\mathcal{O}(N)}{\ln 1/\delta} \ll \frac{N}{r}, \text{ for } r \le R_\sigma.$$

Hence, in the large N limit,

$$\partial_r(u-v) = (1+o(1))\frac{N}{r}, \; 0 < r \le R_\sigma. \tag{13.8.9}$$

In particular,

$$u - v = (1+o(1))N(\ln r - \ln R_\sigma), \; 0 < r \le R_\sigma. \tag{13.8.10}$$

We now restrict the attention to a region

$$r \le 1 - \frac{\ln N}{\mathcal{O}(N)}. \tag{13.8.11}$$

In this region we still have

$$\frac{1}{1-r} \le \frac{\mathcal{O}(N)}{\ln N} \ll \frac{N}{r}$$

and (13.8.9), (13.8.10) remain valid.

Lemma 13.8.2 *When* (13.8.11) *holds, we have in the limit of large N,*

$$\partial_r(u-v) = (1+o(1))\frac{N}{r}, \; u - v = (1+o(1))N\left(\ln r - \ln R_\sigma\right). \tag{13.8.12}$$

Here,

$$u(r) = N \ln r, \; v(r) = \ln \frac{2C_1\delta N}{1-r}. \tag{13.8.13}$$

Recalling that $G(r) = 1/(1-r)$ when $r \leq 1 - 1/N$, we define $h(r)$ by

$$Nh(r) = \frac{2C_1\delta N^{\frac{3}{2}}}{1-r} + \ln\left(r^N + \frac{2C_1\delta N}{1-r}\right) = N^{\frac{1}{2}}e^{v(r)} + \ln\left(e^u + e^v\right). \qquad (13.8.14)$$

Also, put

$$U(z) = \ln|\det(z - A_\delta)|, \qquad (13.8.15)$$

so that $U(z) \leq Nh(|z|)$ for $|z| \leq 1 - (\ln N)/\mathcal{O}(N)$ by (13.7.26). We have seen that $\ln(e^u + e^v)$ is subharmonic and it follows from that discussion, or by direct computation, that e^v is subharmonic, so we have

Lemma 13.8.3 *$h(|z|)$ is subharmonic away from* 0 *on* $D(0, 1 - 1/N)$.

Comparison of the Bounds (13.7.26) **When $R_\sigma < |z| \leq 1 - (\ln N)/\mathcal{O}(N)$** We write the lower bound in (13.7.26) as

$$U(z) \geq N(h(|z|) - \epsilon(|z|)),\ N\epsilon(r) = 2N^{\frac{1}{2}}e^{v(r)} + \ln\left(e^u + e^v\right) - \ln\left(e^u - e^v\right). \qquad (13.8.16)$$

We saw prior to (13.8.7) that

$$\ln\left(e^u + e^v\right) = u + e^{-(u-v)} + \mathcal{O}\left(e^{-2(u-v)}\right),$$

when $u \geq v$ (i.e., when $r \geq R_\sigma$ for our special functions u, v). Now restrict r by imposing that

$$r \geq R_\sigma + \frac{1}{N}, \qquad (13.8.17)$$

so that $u(r) \geq v(r) + 1 + o(1)$ by (13.8.12). Then $e^u \geq e^{1+o(1)}e^v$ and as prior to (13.8.7), we get

$$\ln\left(e^u - e^v\right) = u - e^{-(u-v)} + \mathcal{O}\left(e^{-2(u-v)}\right).$$

Hence, for $R_\sigma + 1/N \leq r \leq 1 - \ln N/\mathcal{O}(N)$,

$$N\epsilon(r) = \frac{4C_1\delta N^{\frac{3}{2}}}{1-r} + 2e^{-(u-v)} + \mathcal{O}\left(e^{-2(u-v)}\right). \qquad (13.8.18)$$

From (13.8.12) we have

$$e^{u-v} = \left(\frac{r}{R_\sigma}\right)^{N(1+o(1))},$$

and (13.8.18) gives,

$$N\epsilon(r) \leq \frac{4C_1\delta N^{\frac{3}{2}}}{1-r} + 3\left(\frac{r}{R_\sigma}\right)^{-N(1+o(1))}. \tag{13.8.19}$$

Comparison of the Lower Bound (13.7.27) and the Upper Bound in (13.7.26) We now work in the region $r < R_\sigma$. Write $t = e^s$. Then (13.7.27) tells us that if $|z|^N$, $t \leq CG(|z|)\delta$, we have

$$U(z) > s \text{ with probability } \geq 1 - (1-|z|)^2 e^{2(s-\ln\delta)} - e^{-N^2}. \tag{13.8.20}$$

Notice that $|z|^N \leq 2C_1G(|z|)\delta$, for $|z| \leq R_\sigma$. Here we have to reconciliate the wishes to have a large lower bound and a probability very close to 1. Recalling the upper bound (13.8.4) on δ with the corresponding fixed parameter $\epsilon_0 > 0$, we fix a new parameter ϵ_1 with $0 < \epsilon_1 \ll 1$ and choose

$$s = \ln\delta - N^{\epsilon_1}. \tag{13.8.21}$$

We get from (13.8.20) that

$$U(z) \geq \ln\delta - N^{\epsilon_1} \text{ with probability } \geq 1 - e^{-2N^{\epsilon_1}} - e^{-N^2}. \tag{13.8.22}$$

This estimate can be written

$$U(z) \geq N\left(h(|z|) - \epsilon(|z|)\right),$$

where

$$\epsilon(|z|) = \frac{2C_1\delta N^{\frac{1}{2}}}{1-r} + \frac{1}{N}\ln\left(\frac{2C_1N}{1-r} + \frac{r^N}{\delta}\right) + N^{\epsilon_1-1}.$$

Now, restrict r to the region

$$r \leq R_\sigma - \frac{1}{N}. \tag{13.8.23}$$

Then $2C_1N/(1-r)$ dominates over r^N/δ by a factor close to e or larger, so

$$\ln\left(\frac{2C_1N}{1-r}+\frac{r^N}{\delta}\right)\asymp \ln\frac{N}{1-r}\le \mathcal{O}(\ln N),$$

and we get

$$\epsilon(r)\le \frac{2C_1\delta N^{\frac{1}{2}}}{1-r}+\frac{\mathcal{O}(N^{\epsilon_1})}{N}. \tag{13.8.24}$$

Summing up, we have

Proposition 13.8.4 *For each $z\in D(0,R_\sigma-1/N)$, we have*

$$U(z)\ge N(h(|z|)-\epsilon(r)),\ \textit{with probability}\ \ge 1-e^{-2N^{\epsilon_1}}-e^{-N^2}, \tag{13.8.25}$$

where $\epsilon(r)>0$ is a function satisfying (13.8.24).

We also want to study $\Delta(h(|z|))=(r^{-2}(r\partial_r)^2h)(|z|)$ and more precisely its integral over annuli centered at $z=0$. We have

$$\int_{r_1<|z|<r_2}\Delta(h(|z|))L(dz)=2\pi\int_{r_1}^{r_2}\partial_r r\partial_r h(r)dr=2\pi\left[r\partial_r h(r)\right]_{r_1}^{r_2}, \tag{13.8.26}$$

where $h(r)$ is given in (13.8.14), so we need to study

$$r\partial_r\ln\left(r^N+\frac{2C_1\delta N}{1-r}\right)$$

on each side of $r=R_\sigma$. We will use (13.8.7) and (13.8.12), with u, v as in (13.8.13).

For $R_\sigma\le r\le 1-(\ln N)/\mathcal{O}(N)$ we get

$$r\partial_r\ln\left(r^N+\frac{2C_1\delta N}{1-r}\right)=(1-\mathcal{O}(e^{-(u-v)}))N+\mathcal{O}(e^{-(u-v)})\frac{r}{1-r},$$

where by (13.8.12),

$$u-v=N(1+o(1))\ln\frac{r}{R_\sigma}.$$

Since $\frac{r}{1-r}=\mathcal{O}(N)$, we get

$$\frac{1}{N}r\partial_r\ln\left(r^N+\frac{2C_1\delta N}{1-r}\right)=\left(1+\mathcal{O}(1)\left(\frac{r}{R_\sigma}\right)^{-(1+o(1))N}\right). \tag{13.8.27}$$

For $r < R_\sigma$ we also use (13.8.7) with a permutation of the roles of u and v:

$$r\partial_r \ln\left(r^N + \frac{2C_1\delta N}{1-r}\right) = (1 - \mathcal{O}(e^{-(v-u)}))\frac{r}{1-r} + \mathcal{O}(e^{-(v-u)})N.$$

By (13.8.12), we get as before,

$$\frac{1}{N}r\partial_r \ln\left(r^N + \frac{2C_1\delta N}{1-r}\right) = \frac{r}{N(1-r)} + \mathcal{O}(1)\left(\frac{r}{R_\sigma}\right)^{(1+o(1))N}. \tag{13.8.28}$$

So far, we have treated the contributions to $r\partial_r h(r)$ from the second term in (13.8.14). The contribution from the first term is

$$r\partial_r \frac{2C_1\delta N^{\frac{1}{2}}}{1-r} = \frac{2C_1\delta N^{\frac{1}{2}}}{1-r}\frac{r}{1-r}. \tag{13.8.29}$$

By (13.8.4), we have

$$r\partial_r \frac{2C_1\delta N^{\frac{1}{2}}}{1-r} = \mathcal{O}(1)N^{-\epsilon_0-\frac{1}{2}}\frac{r}{1-r}. \tag{13.8.30}$$

Rather than distinguishing further between different regions in r, we now strengthen (13.8.4) to

$$2C_1\delta \le N^{-\frac{5}{2}-\epsilon_0^{\text{new}}}, \tag{13.8.31}$$

for some $\epsilon_0^{\text{new}}(= \epsilon_0 - \frac{1}{2}) > 0$. Then (13.8.30) becomes

$$r\partial_r \frac{2C_1\delta N^{\frac{1}{2}}}{1-r} = \mathcal{O}(1)N^{-\epsilon_0^{\text{new}}}\frac{r}{N(1-r)}, \tag{13.8.32}$$

which is small compared to the leading terms, 1 and $\frac{r}{N(1-r)}$ in (13.8.27) and (13.8.28), respectively

Proposition 13.8.5 *Under the assumption* (13.8.31), *we have*

$$\begin{aligned} r\partial_r h(r) = &\underbrace{2C_1\delta N^{\frac{1}{2}}\frac{r}{(1-r)^2}}_{\mathcal{O}\left(N^{-\epsilon_0^{\text{new}}}\frac{r}{N(1-r)}\right)} \\ &+\begin{cases} 1 + \mathcal{O}(1)(r/R_\sigma)^{-(1+o(1))N}, & R_\sigma \le r \le 1 - \frac{\ln N}{\mathcal{O}(1)N}, \\ \frac{r}{N(1-r)} + \mathcal{O}(1)(R_\sigma/r)^{-(1+o(1))N}, & 0 < r \le R_\sigma. \end{cases} \end{aligned} \tag{13.8.33}$$

The derivative of the first term to the right is $\mathcal{O}\left(\delta N^{\frac{1}{2}}(1-r)^{-3}\right)$.

Estimates of the Subharmonic Measure of Small Discs We next estimate the integral of the function $\Delta h(z)$ over small discs contained in $D(0, 1-(\ln N)/(\mathcal{O}(1)N))$. As a preparation (and this will also be used directly), we shall study the integral (13.8.26) over the annulus $D(0, r_1, r_2) = \{z \in \mathbf{C};\ r_1 < |z| < r_2\}$. We shall assume for simplicity that

$$r_2 \geq \frac{1}{\mathcal{O}(1)}. \tag{13.8.34}$$

From (13.8.26) and Proposition 13.8.5 it follows directly that,

Proposition 13.8.6 *Let $I(r_1, r_2)$ be the integral in* (13.8.26) *and assume* (13.8.31), (13.8.34).

(a) *If $r_2 \leq R_\sigma$, we have*

$$\begin{aligned} I(r_1, r_2) = &\frac{2\pi}{N} \frac{r_2 - r_1}{(1-r_1)(1-r_2)} \\ &+ \mathcal{O}(1)\left(\left(\frac{R_\sigma}{r_2}\right)^{-(1+o(1))N} + \delta N^{\frac{1}{2}} \frac{r_2 - r_1}{(1-r_1)(1-r_2)^2}\right), \end{aligned}$$

where the last term in the remainder is $\leq \delta N^{\frac{1}{2}+3} = \mathcal{O}(N^{1-\epsilon_0^{\mathrm{new}}})$.

(b) *If $R_\sigma \leq r_1 < r_2 \leq 1 - (\ln N)/(\mathcal{O}(1)N)$, we have*

$$I(r_1, r_2) = \mathcal{O}(1)\left(\delta N^{\frac{1}{2}} \frac{r_2 - r_1}{(1-r_1)(1-r_2)^2} + \left(\frac{r_1}{R_\sigma}\right)^{-(1+o(1))N}\right).$$

(c) *If $r_1 \leq R_\sigma \leq r_2$, we have*

$$\begin{aligned} I(r_1, r_2) = 2\pi &- \frac{2\pi r_1}{N(1-r_1)} \\ + \mathcal{O}(1)&\left(\delta N^{\frac{1}{2}} \frac{r_2 - r_1}{(1-r_1)(1-r_2)^2} + \left(\frac{r_2}{R_\sigma}\right)^{-(1+o(1))N} + \left(\frac{R_\sigma}{r_1}\right)^{-(1+o(1))N}\right). \end{aligned}$$

Consider open discs $D(z, \rho)$ with $|z| = r$, $r + \rho = r_2$, $r - \rho = r_1$, so that $\rho = (r_2 - r_1)/2$. Then, since h is a radial function,

$$\int_{D(z,\rho)} \Delta h(z) L(dz) \leq \mathcal{O}(1) I(r_1, r_2)\rho \tag{13.8.35}$$

We first concentrate on the case (a) in the last proposition and assume, in addition, that

$$\rho \leq \theta_0 (R_\sigma - r) \tag{13.8.36}$$

for some fixed $\theta_0 \in]0, 1[$, so that $R_\sigma - r$, $R_\sigma - r_1$, $R_\sigma - r_2$ are all of the same order of magnitude. The same holds if we replace R_σ by 1.

Then we get

$$I(r_1, r_2) = \mathcal{O}(1) \left(\frac{1}{N} \frac{\rho}{(1-r)^2} + (R_\sigma / r)^{-N/\mathcal{O}(1)} \right),$$

$$\int_{D(z,\rho)} \Delta h(z) L(dz) = \mathcal{O}(\rho) \left(\frac{1}{N} \frac{\rho}{(1-r)^2} + (R_\sigma / r)^{-N/\mathcal{O}(1)} \right). \tag{13.8.37}$$

If $\widetilde{\gamma}$ is a "band" of the type, $\{z \in \mathbf{C};\ \mathrm{dist}\,(z, \gamma) \leq \rho\}$, where γ is a Lipschitz curve passing through z, then we define the Δh-density along $\widetilde{\gamma}$ at $z \in \gamma$ by

$$\Delta h\text{-dens}\,(z, \rho) = \frac{1}{\rho} \int_{D(z,\rho)} \Delta h L(dz), \tag{13.8.38}$$

which under the present assumptions satisfies

$$\Delta h\text{-dens}\,(z, \rho) = \mathcal{O}(1) \left(\frac{1}{N} \frac{\rho}{(1-r)^2} + (R_\sigma / r)^{-N/\mathcal{O}(1)} \right). \tag{13.8.39}$$

Next, recall that under the same assumptions, $\epsilon(r)$ is given by (13.8.24):

$$\epsilon(r) \leq \frac{2 C_1 \delta N^{1/2}}{1-r} + \frac{\mathcal{O}(N^{\epsilon_1})}{N} \leq \mathcal{O}(1) \frac{N^{\epsilon_1}}{N}.$$

Here we also assume (13.8.23). In both cases (a) and (b), we define the ϵ-density by

$$\epsilon\text{-dens}\,(z, \rho) = \frac{\epsilon(|z|)}{\rho},$$

so in the present case (a), we have

$$\epsilon\text{-dens}\,(z, \rho) \leq \mathcal{O}(1) \frac{N^{\epsilon_1}}{N\rho}. \tag{13.8.40}$$

We want to choose ρ in (13.8.36) making $(\Delta h\text{-dens} + \epsilon\text{-dens}\,)(z, \rho)$ as small as possible. The minimum of $\rho \mapsto \frac{\rho}{N(1-r)^2} + \frac{N^{\epsilon_1}}{N\rho}$ over $\mathbf{R}_+$ is attained at $\rho = N^{\epsilon_1/2}(1-r)$ which is too large for (13.8.36) to hold, so we choose $\rho = \rho_{\min} = \theta_0(R_\sigma - r)$ and get the corresponding value,

$$(\epsilon\text{-dens} + \Delta h\text{-dens})_{\min}(z) = \mathcal{O}(1)\left(\frac{N^{\epsilon_1}}{N(R_\sigma - r)} + (R_\sigma/r)^{-N/\mathcal{O}(1)}\right). \tag{13.8.41}$$

We next do the same estimates in the case (b) of the proposition, now under the additional assumption

$$\rho \le \theta_0(r - R_\sigma),\ r_2 \le (1 + R_\sigma)/2, \tag{13.8.42}$$

with θ_0 as before. We get

$$\Delta h\text{-dens}(z, \rho) = \mathcal{O}(1)\left(\frac{\delta N^{1/2}\rho}{(1 - R_\sigma)^3} + (r/R_\sigma)^{-N/\mathcal{O}(1)}\right), \tag{13.8.43}$$

and by (13.8.19),

$$\epsilon\text{-dens}(z, \rho) = \frac{\mathcal{O}(1)}{\rho}\left(\frac{\delta N^{1/2}}{1 - R_\sigma} + \frac{1}{N}(r/R_\sigma)^{-N/\mathcal{O}(1)}\right). \tag{13.8.44}$$

Restrict ρ further by imposing

$$\rho \ge \frac{1}{N}. \tag{13.8.45}$$

Then

$$(\Delta\text{-dens} + \epsilon\text{-dens})(z, \rho) = \mathcal{O}(1)\left(\frac{\delta N^{1/2}\rho}{(1 - R_\sigma)^3} + \frac{\delta N^{1/2}}{\rho(1 - R_\sigma)} + (r/R_\sigma)^{-N/\mathcal{O}(1)}\right).$$

Without the condition (13.8.42) the optimal choice of ρ would be $\rho = 1 - R_\sigma$ which however is too large, so we settle for $\rho = \rho_{\min} = \theta(r - R_\sigma)$ and get

$$(\Delta h\text{-dens} + \epsilon\text{-dens})_{\min}(z) = \mathcal{O}(1)\left(\frac{\delta N^{1/2}}{(r - R_\sigma)(1 - R_\sigma)} + (r/R_\sigma)^{-N/\mathcal{O}(1)}\right). \tag{13.8.46}$$

We shall next combine our estimates with Theorem 12.1.2, with "$1/h$" there replaced by N and "ϕ" by "h". Let

$$\Omega = \{z \in \mathbf{C};\ \theta_1 < \arg z < \theta_2,\ r_- < |z| < r_+\}, \tag{13.8.47}$$

where $1/\mathcal{O}(1) \le r_- \le R_\sigma - 1/N$, $R_\sigma + 1/N \le r_+ \le (1+R_\sigma)/2$. Let $\theta = \theta_2 - \theta_1$. The main term in (12.1.13) in Theorem (12.1.2) is

$$\frac{N}{2\pi}\int_\Omega \Delta h(z) L(dz) = N\theta \left[r\partial_r h(r) \right]_{r_-}^{r_+},$$

which according to (13.8.33) is equal to

$$\begin{aligned} N\theta \Bigg(1 + \frac{2C_1\delta N^{\frac{1}{2}} r_+}{(1-r_+)^2} - \frac{2C_1\delta N^{\frac{1}{2}} r_-}{(1-r_-)^2} - \frac{r_-}{N(1-r_-)} \\ + \mathcal{O}\left(\left(\frac{r_+}{R_\sigma}\right)^{-\frac{N}{\mathcal{O}(1)}} + \left(\frac{R_\sigma}{r_-}\right)^{-\frac{N}{\mathcal{O}(1)}} \right) \Bigg) \\ = N\theta \left(1 - \frac{r_-}{N(1-r_-)} + \mathcal{O}(1)\left(N^{-\epsilon_0^{\mathrm{new}}} + \left(\frac{r_+}{R_\sigma}\right)^{-\frac{N}{\mathcal{O}(1)}} + \left(\frac{R_\sigma}{r_-}\right)^{-\frac{N}{\mathcal{O}(1)}} \right) \right). \end{aligned} \tag{13.8.48}$$

In Theorem 12.1.2 we choose the local radius $r = r(z)$, $z \in \partial\Omega$ of the form $r(z) = \max(\theta_0 ||z| - R_\sigma|, 2/N)$, where we recognize $\rho = \rho(|z|) = \theta_0||z| - R_\sigma|$. We let z_j^0 in that theorem avoid a $1/N$-neighborhood of the circle $|z| = R_\sigma$.

We now apply Theorem 12.1.2 to $u := \det(z - A_\delta)$, $U = \ln|u|$ with Ω, r as above, $z_j^0, \widetilde{z}_j$ as in the theorem, "ϕ" equal to h (and "h" in the theorem equal to $1/N$). Then with probability $\ge 1 - e^{-N^2}$ we have (cf. (C) in Theorem 13.7.4 and (13.8.14)),

$$U \le Nh(|z|) = \frac{2C_1 N^{3/2}}{1-|z|} + \ln\left(|z|^N + \frac{2C_1\delta N}{1-|z|} \right), \quad |z| \le 1 - \frac{1}{N}. \tag{13.8.49}$$

Moreover, for each j with $|z_j^0| < R_\sigma - 1/N$ we have with probability $\ge 1 - e^{-N^{\epsilon_1}} - e^{-N^2}$,

$$U(\widetilde{z}_j) \ge N(h(|\widetilde{z}_j|) - \epsilon_j), \ \epsilon_j = \epsilon(|\widetilde{z}_j|), \ \epsilon(r) \le \frac{2C_1\delta N^{1/2}}{1-r} + \frac{\mathcal{O}(N^{\epsilon_1})}{N}, \tag{13.8.50}$$

by Proposition 13.8.4. By (13.8.6), we have with probability $\ge 1 - e^{-N^2}$

$$U(\widetilde{z}_j) \ge N(h(|\widetilde{z}_j|) - \epsilon_j), \text{ when } |z_j^0| > R_\sigma, \ \epsilon_j = \epsilon(|\widetilde{z}_j|), \tag{13.8.51}$$

where now, by (13.8.19),

$$\epsilon(r) \le \frac{4C_1\delta N^{1/2}}{1-r} + \frac{3}{N}\left(\frac{r}{R_\sigma}\right)^{-N(1+o(1))}.$$

The number of points $\tilde{z}_j$ is $\le \mathcal{O}(N)$, hence inequalities (13.8.50) hold simultaneously with probability $\ge 1 - \mathcal{O}(N)e^{-N^{\epsilon_1}} - e^{-N^2}$, and Theorem 12.1.2 applies.

The remainder term in (12.1.13) is

$$\mathcal{O}(1)N\left(\int_{\{z\in\partial\Omega;\ ||z|-R_\sigma|\ge\frac{1}{N}\}} (\epsilon\text{-dens} + \Delta h\text{-dens})_{\min}(z)|dz| + \mathcal{O}\left(\frac{1}{N}\right)\right), \tag{13.8.52}$$

where the last term corresponds to $\{z \in \partial\Omega;\ ||z| - R_\sigma| < 1/N\}$, more precisely to the integral of Δh over a N^{-1}-neighborhood of that set. The boundary integral can be decomposed into 4 terms, $B_{\rm ia}$, $B_{\rm ir}$, $B_{\rm ea}$, $B_{\rm er}$, where i stands for interior, e for exterior, a for arc and r for radial. From (13.8.41), we get

$$B_{\rm ia} = \mathcal{O}(1)\theta\left(\frac{N^{\epsilon_1}}{N(R_\sigma - r_-)} + (R_\sigma/r_-)^{-\frac{N}{\mathcal{O}(1)}}\right),$$

$$\begin{aligned} B_{\rm ir} &= \mathcal{O}(1)\int_{r_-}^{R_\sigma-\frac{1}{N}}\left(\frac{N^{\epsilon_1}}{N(R_\sigma - r)} + (R_\sigma/r)^{-\frac{N}{\mathcal{O}(1)}}\right)dr \\ &= \mathcal{O}(1)\left(\frac{N^{\epsilon_1}\ln N}{N} + \frac{1}{N}\right) = \mathcal{O}(1)\frac{N^{\epsilon_1}\ln N}{N}. \end{aligned}$$

From (13.8.46) we get

$$B_{\rm ea} = \mathcal{O}(1)\theta\left(\frac{\delta N^{\frac{1}{2}}}{(r_+ - R_\sigma)(1 - R_\sigma)} + (r_+/R_\sigma)^{-\frac{N}{\mathcal{O}(1)}}\right),$$

$$\begin{aligned} B_{\rm er} &= \int_{R_\sigma+\frac{1}{N}}^{r_+}\left(\frac{C_1\delta N^{\frac{1}{2}}}{(r - R_\sigma)(1 - R_\sigma)} + (r/R_\sigma)^{-\frac{N}{\mathcal{O}(1)}}\right)dr \\ &= \mathcal{O}(1)\left(\frac{\delta N^{\frac{1}{2}}\ln N}{1 - R_\sigma} + \frac{1}{N}\right). \end{aligned}$$

Then the remainder (13.8.52) is $\le \mathcal{O}(N)$ times

$$\theta\frac{N^{\epsilon_1 - 1}}{R_\sigma - r_-} + \theta\frac{\delta N^{\frac{1}{2}}}{(r_+ - R_\sigma)(1 - R_\sigma)} + \frac{N^{\epsilon_1}\ln N}{N} + \theta\left(\frac{R_\sigma}{r_-}\right)^{-\frac{N}{\mathcal{O}(1)}} + \theta\left(\frac{r_+}{R_\sigma}\right)^{-\frac{N}{\mathcal{O}(1)}}. \tag{13.8.53}$$

Combining this with (13.8.48) and Theorem 12.1.2, we get

Theorem 13.8.7 *Let* $e^{-CN} \le \delta \le N^{-\frac{5}{2}-\epsilon_0^{\text{new}}}$ *for some* $\epsilon_0^{\text{new}} > 0$*, so that* R_σ *is bounded from above in* (13.8.5) *and bounded from below by* $1/\mathcal{O}(1)$ *in* (13.8.3)*. Let*

$$\frac{1}{\mathcal{O}(1)} \le r_- \le R_\sigma - \frac{1}{N},\quad R_\sigma + \frac{1}{N} \le r_+ \le (1+R_\sigma)/2.$$

Define Ω *by* (13.8.47)*,* $\theta := \theta_2 - \theta_1$ *and let* $\epsilon_1 > 0$ *be small and fixed. Then with probability*

$$\ge 1 - \mathcal{O}(N)e^{-N^{\epsilon_1}} - e^{-N^2}, \tag{13.8.54}$$

we have

$$\begin{aligned} &\left| \#(\sigma(A_\delta)\cap\Omega) - \frac{\theta}{2\pi}N\left(1 - \frac{r_-}{N(1-r_-)}\right)\right| \le \\ &\quad \mathcal{O}(1)N\left(\theta\frac{N^{\epsilon_1-1}}{R_\sigma - r_-} + \theta\frac{\delta N^{\frac{1}{2}}}{(r_+ - R_\sigma)(1-R_\sigma)} + \theta N^{-\epsilon_0^{\text{new}}}\right. \\ &\qquad \left. + \frac{N^{\epsilon_1}\ln N}{N} + \theta\left(\frac{R_\sigma}{r_-}\right)^{-\frac{N}{\mathcal{O}(1)}} + \theta\left(\frac{r_+}{R_\sigma}\right)^{-\frac{N}{\mathcal{O}(1)}}\right). \end{aligned} \tag{13.8.55}$$

The eigenvalues of A_δ sit in $\overline{D(0, R_\sigma)}$ (with probability $\ge 1 - e^{-N^2}$), so we are free to choose r_+ conveniently. We choose $r_+ = (1+R_\sigma)/2$. Then (13.8.55) gives:

$$\begin{aligned} &\left| \#(\sigma(A_\delta)\cap\Omega) - \frac{\theta}{2\pi}N\left(1 - \frac{r_-}{N(1-r_-)}\right)\right| \le \\ &\quad \mathcal{O}(1)N\left(\theta\frac{N^{\epsilon_1-1}}{R_\sigma - r_-} + \theta\frac{\delta N^{\frac{1}{2}}}{(1-R_\sigma)^2} + \theta N^{-\epsilon_0^{\text{new}}}\right. \\ &\qquad \left. + \frac{N^{\epsilon_1}\ln N}{N} + \theta\left(\frac{R_\sigma}{r_-}\right)^{-\frac{N}{\mathcal{O}(1)}} + \theta R_\sigma^{\frac{N}{\mathcal{O}(1)}}\right). \end{aligned} \tag{13.8.56}$$

From this we get Theorem 13.2.1 with ϵ_0 there equal to ϵ_0^{new}. □

13.9 Eigenvalue Distribution for Larger Random Perturbations

When δ is larger we cannot exclude eigenvalues outside the unit disc. In addition to the upper bound on $\ln|\det(z - A_\delta)|$ that follows from the first part of (13.7.26), we shall use the one given by (13.7.24). However, in order to have some simplifications, we sacrifice the maximal sharpness and consider these estimates only in the regions $|z| < 1$ and $|z| > 1$, respectively, and in order to simplify further, we replace G and F by their upper bounds $1/(1-|z|)$ and $1/(|z|-1)$, respectively.

Our first problem is then to study the largest radial subharmonic function $h(r) = h_1(r)$ on $]0, \infty[$ such that

$$h(r) \leq \begin{cases} \frac{1}{N}\ln(r^N + \frac{2C_1\delta N}{1-r}) + \frac{2C_1\delta N^{\frac{1}{2}}}{1-r} =: f(r), & \text{if } 0 < r < 1, \\ \ln r + \frac{2C_1\delta N^{1/2}}{r-1} =: g(r), & \text{if } r > 1. \end{cases} \tag{13.9.1}$$

Here we replaced the C_1 in (13.7.24) by $2C_1$ to get more symmetric expressions. To simplify the expressions, we will introduce $\delta_{\text{new}} = 2C_1\delta$ and drop the subscript "new". Putting $r = e^t$ and subtracting the linear function Nt, we get the equivalent problem of studying the largest convex function $h(t) = h_0(t)$ on $\mathbf{R}$ (slightly abusing notation, by denoting functions of r and of $t = \ln r$ with the same letters) such that

$$h(t) \leq \begin{cases} f_0(t) := \frac{1}{N}\ln(e^{Nt} + \frac{\delta N}{1-e^t}) + \frac{\delta N^{\frac{1}{2}}}{1-e^t} - t, & \text{if } t < 0, \\ g_0(t) := \frac{\delta N^{1/2}}{e^t - 1}, & \text{if } t > 0. \end{cases} \tag{13.9.2}$$

The functions $f_0, g_0 > 0$ are convex, as we have already seen.

The problem is invariant under addition of a linear function of t, so we can replace f_0, g_0 by

$$f_s(t) = \frac{1}{N}\ln\left(e^{Nt} + \frac{\delta N}{1-e^t}\right) - (1-s)t + \frac{\delta N^{1/2}}{1-e^t}, \quad \text{if } t < 0,$$

$$g_s(t) = st + \frac{\delta N^{1/2}}{e^t - 1}, \quad \text{if } t > 0,$$

where we let the real parameter s vary in a small fixed neighborhood of $1/2$. (Thus, $f = f_1$, $g = g_1$.)

Let $f_s^{\min}$ and $g_s^{\min}$ denote the minima of f_s, and g_s, attained at $t_{\min}(f_s) < 0$ and $t_{\min}(g_s) > 0$ respectively. If we can choose s so that $f_s^{\min} = g_s^{\min}$, then the desired largest convex function $h = h_s$, satisfying (13.9.2) with (f, g) replaced by (f_s, g_s), is given by

$$h_s(t) = \begin{cases} f_s(t), & \text{if } t < t_{\min}(f_s), \\ f_s^{\min}(= g_s^{\min}), & \text{if } t_{\min}(f_s) \leq t \leq t_{\min}(g_s), \\ g_s(t), & \text{if } t > t_{\min}(g_s). \end{cases} \tag{13.9.3}$$

The desired function $h = h_1$ is then given by

$$h(t) = \begin{cases} f(t), & \text{if } t < t_{\min}(f_s), \\ f(t_{\min}(f_s))\frac{t_{\min}(g_s)-t}{t_{\min}(g_s)-t_{\min}(f_s)} + g(t_{\min}(g_s))\frac{t-t_{\min}(f_s)}{t_{\min}(g_s)-t_{\min}(f_s)}, \\ \text{if } t_{\min}(f_s) \le t \le t_{\min}(g_s), \\ g(t), & \text{if } t > t_{\min}(g_s). \end{cases} \tag{13.9.4}$$

In the following, we assume that for some fixed $\epsilon_0 > 0$,

$$N^{\epsilon_0 - 5/2} \le \delta \ll N^{-3/2}. \tag{13.9.5}$$

We start with the study of g_s and see that $t = t_{\min}(g_s)$ is given by

$$0 = g_s'(t) = s - \frac{\delta N^{1/2} e^t}{(e^t - 1)^2} = 0.$$

This leads to a second-order equation for $r = e^t$ which has the solutions:

$$r = 1 + \frac{\delta N^{1/2}}{2s} \pm \left(\frac{\delta}{s} N^{1/2} + \frac{\delta^2 N}{4s^2} \right)^{1/2}.$$

We choose the solution which is >1, that is, the one with the plus sign. From the upper bound in (13.9.5) and (recalling that s varies in a small fixed neighborhood of $1/2$) we know that $\delta N^{1/2} = \mathcal{O}(N^{-1})$ and we get by Taylor expanding,

$$r = 1 + (\delta/s)^{1/2} N^{1/4} + \mathcal{O}(\delta N^{1/2}).$$

Consequently,

$$t_{\min}(g_s) = (\delta/s)^{1/2} N^{1/4} + \mathcal{O}(\delta N^{1/2}). \tag{13.9.6}$$

A direct calculation gives,

$$g_s^{\min} = (1 + \mathcal{O}(\delta^{1/2} N^{1/4})) 2 (\delta s)^{1/2} N^{1/4}. \tag{13.9.7}$$

Also,

$$\partial_s g_s^{\min} = t_{\min}(g_s) = (1 + \mathcal{O}(\delta^{1/2} N^{1/4}))(\delta/s)^{1/2} N^{1/4}. \tag{13.9.8}$$

Another direct calculation shows that

$$\partial_t^2 g_s = \frac{\delta N^{\frac{1}{2}} e^t (e^t + 1)}{(e^t - 1)^3},$$

which implies that for $t \asymp t_{\min}(g_s) \asymp \delta^{1/2} N^{1/4}$,

$$\partial_t^2 g_s \asymp \delta^{-1/2} N^{-1/4} \asymp 1/t_{\min}(g_s). \tag{13.9.9}$$

In order to do the same work for f_s, we consider two simplified functions:

(a) The much simplified function

$$\widetilde{f_s}(t) = \frac{\delta N^{1/2}}{1 - e^t} - (1 - s)t.$$

(b) The less simplified function

$$\widehat{f_s}(t) = \frac{1}{N} \ln\left(\frac{\delta N}{1 - e^t}\right) + \widetilde{f_s}(t).$$

As we shall see, $e^{Nt} \ll \delta N/(1 - e^t)$ near the critical points $t_{\min}(\widetilde{f_s})$ and $t_{\min}(\widehat{f_s})$, so $\widehat{f_s}$ is a good approximation to f_s there.

From the symmetry relation

$$\widetilde{f_s}(-t) = \delta N^{1/2} + g_{1-s}(t), \tag{13.9.10}$$

the calculations for g_s give corresponding results for $\widetilde{f_s}$:

$$t_{\min}(\widetilde{f_s}) = -t_{\min}(g_{1-s}) = -(1 + \mathcal{O}(\delta^{1/2} N^{1/4})) \left(\frac{\delta}{1 - s}\right)^{1/2} N^{1/4}, \tag{13.9.11}$$

$$\partial_s \widetilde{f}_s^{\min} = t_{\min}(f_s) = -(1 + \mathcal{O}(\delta^{1/2} N^{1/4})) \left(\frac{\delta}{1 - s}\right)^{1/2} N^{1/4} \ll -1/N, \tag{13.9.12}$$

$$\partial_t^2 \widetilde{f_s} \asymp \delta^{-1/2} N^{-1/4} \asymp 1/|t_{\min}(\widetilde{f_s})|, \text{ for } t \asymp t_{\min}(\widetilde{f_s}). \tag{13.9.13}$$

We next look at $\widehat{f_s}$ and notice that

$$\partial_t \frac{1}{N} \ln \frac{\delta N}{1 - e^t} = \frac{e^t}{N(1 - e^t)} \asymp \frac{1}{N|t_{\min}(\widetilde{f_s})|} \text{ when } t \asymp t_{\min}(\widetilde{f_s}). \tag{13.9.14}$$

In particular,

$$(\partial_t \widehat{f}_s)(t_{\min}(\widetilde{f}_s)) \asymp \frac{1}{N|t_{\min}(\widetilde{f}_s)|} \asymp \frac{1}{\delta^{1/2}N^{5/4}}, \tag{13.9.15}$$

cf. (13.9.5).

Further,

$$\partial_t^2 \frac{1}{N} \ln \frac{\delta N}{1-e^t} = \frac{1}{N}\frac{e^t}{(1-e^t)^2} \asymp \frac{1}{N|t_{\min}(\widetilde{f}_s)|^2}, \text{ when } t \asymp t_{\min}(\widetilde{f}_s).$$

We compare this with (13.9.13) and find for $t \asymp t_{\min}(\widetilde{f}_s)$:

$$\partial_t^2 \frac{1}{N} \ln \frac{\delta N}{1-e^t} \asymp \frac{\partial_t^2 \widetilde{f}_s}{\delta^{1/2}N^{5/4}}, \tag{13.9.16}$$

so

$$\partial_t^2 \widehat{f}_s = \left(1 + \frac{\mathcal{O}(1)}{\delta^{1/2}N^{5/4}}\right) \partial_t^2 \widetilde{f}_s, \text{ when } t \asymp t_{\min}(\widetilde{f}_s). \tag{13.9.17}$$

Since $(\partial_t \widehat{f}_s)(t_{\min}(\widetilde{f}_s)) > 0$, we know that $t_{\min}(\widehat{f}_s) < t_{\min}(\widetilde{f}_s)$. For $t \le t_{\min}(\widetilde{f}_s)$, $t \asymp t_{\min}(\widetilde{f}_s)$ we have by (13.9.15), (13.9.17) and (13.9.13),

$$\begin{aligned} \partial_t \widehat{f}_s(t) &\le \frac{C/N}{|t_{\min}(\widetilde{f}_s)|} - \frac{1}{C|t_{\min}(\widetilde{f}_s)|}\left(t_{\min}(\widetilde{f}_s) - t\right) \\ &= \frac{1}{|t_{\min}(\widetilde{f}_s)|}\left(\frac{C}{N} - \frac{1}{C}\left(t_{\min}(\widetilde{f}_s) - t\right)\right), \end{aligned} \tag{13.9.18}$$

and we conclude that

$$t_{\min}(\widetilde{f}_s) - t_{\min}(\widehat{f}_s) \le \frac{C^2}{N} \le \mathcal{O}(1)\frac{|t_{\min}(\widetilde{f}_s)|}{\delta^{1/2}N^{5/4}} \le \mathcal{O}(1)\frac{\delta^{1/2}N^{1/4}}{\delta^{1/2}N^{5/4}} = \mathcal{O}\left(\frac{1}{N}\right). \tag{13.9.19}$$

Combining this estimate with the bound $\partial_t \widetilde{f}_s = \mathcal{O}(1)$ in the interesting region, we get

$$\begin{aligned} \widehat{f}_s^{\min} - \widetilde{f}_s^{\min} &= (\widehat{f}_s - \widetilde{f}_s)(t_{\min}(\widehat{f}_s)) + \widetilde{f}_s(t_{\min}(\widehat{f}_s)) - \widetilde{f}_s(t_{\min}(\widetilde{f}_s)) \\ &= \frac{1}{N}\left(\ln \frac{\delta N}{1-e^{t_{\min}(\widehat{f})}} + \mathcal{O}(1)\right) \\ &= \frac{1}{N}\left(\mathcal{O}(1) + \frac{1}{2}\ln\left(\delta N^{3/2}\right)\right) \\ &= \mathcal{O}(1)\frac{\ln N}{N}. \end{aligned} \tag{13.9.20}$$

We are now ready to look at f_s. The main observation here is that e^{Nt} is much smaller than $\delta N/(1-e^t)$ for $t \asymp t_{\min}(\widetilde{f_s})$. Indeed, for such t, we have

$$\frac{\delta N}{1-e^t} \asymp \frac{\delta N}{\delta^{1/2}N^{1/4}} \asymp \delta^{1/2}N^{5/4}N^{-1/2} \geq N^{(\epsilon_0-1)/2}/\mathcal{O}(1),$$

while

$$e^{Nt} \leq \exp(-\delta^{1/2}N^{5/4}/\mathcal{O}(1)) \leq \exp(-N^{\epsilon_0/2}/\mathcal{O}(1)),$$

so

$$\frac{\delta N}{1-e^t} + e^{Nt} = (1+\mathcal{O}(e^{-N^{\epsilon_0/2}/\mathcal{O}(1)}))\frac{\delta N}{1-e^t},$$

and consequently, for $t \asymp t_{\min}(\widetilde{f_s})$,

$$f_s - \widehat{f_s} = \mathcal{O}(1/N)e^{-N^{\epsilon_0/2}/\mathcal{O}(1)}. \tag{13.9.21}$$

In order to treat also the derivatives of the difference, we write

$$f_s - \widehat{f_s} = \frac{1}{N}\ln\left(1+\frac{e^{Nt}}{\left(\frac{\delta N}{1-e^t}\right)}\right). \tag{13.9.22}$$

We have just seen that

$$\frac{e^{Nt}}{\left(\frac{\delta N}{1-e^t}\right)} = \mathcal{O}(1/N)e^{-N^{\epsilon_0/2}/\mathcal{O}(1)},$$

and this extends to a complex domain:

$$\Re t \asymp t_{\min}(\widetilde{f_s}),\ \Im t = \mathcal{O}(1)t_{\min}(\widetilde{f_s}).$$

Hence by the Cauchy inequalities, we get for real $t \asymp t_{\min}(\widetilde{f_s})$, and $k \in \mathbf{N}$,

$$\partial_t^k\left(\frac{e^{Nt}}{\left(\frac{\delta N}{1-e^t}\right)}\right) = \mathcal{O}_k(1)e^{-N^{\epsilon_0/2}/\mathcal{O}(1)},$$

Using this in (13.9.22), we get for every $k \in \mathbf{N}$ that for $t \asymp t_{\min}(\widetilde{f_s})$,

$$\partial_t^k(f_s - \widehat{f_s}) = \mathcal{O}_k(1)\left(e^{-N^{\epsilon_0/2}/\mathcal{O}(1)}\right). \tag{13.9.23}$$

It is now clear that f_s has a unique critical point $t_{\min}(f_s) = t_{\min}(\widehat{f_s}) + \mathcal{O}\big(e^{-N^{\epsilon_0/2}/\mathcal{O}(1)}\big)$ which is a nondegenerate minimum and that

$$f_s^{\min} = f_s(t_{\min}) = \widehat{f}_s^{\min} + \mathcal{O}\big(e^{-N^{\epsilon_0/2}/\mathcal{O}(1)}\big). \tag{13.9.24}$$

Combining this with (13.9.20), we get

$$f_s^{\min} - \widetilde{f}_s^{\min} = \frac{1}{2N}\left(\ln\left(\delta N^{3/2}\right) + \mathcal{O}(1/N)\right). \tag{13.9.25}$$

By the symmetry relation (13.9.10), we know that

$$t_{\min}(g_{1/2}) = -t_{\min}(\widetilde{f}_{1/2}), \quad \widetilde{f}_{1/2}^{\min} = \delta N^{1/2} + g_{1/2}^{\min},$$

and in view of (13.9.25),

$$f_{1/2}^{\min} = \frac{1}{2N}\ln\left(\delta N^{3/2}\right) + \mathcal{O}(1/N) + g_{1/2}^{\min}. \tag{13.9.26}$$

We look for $s \approx 1/2$ such that $g_s^{\min} = f_s^{\min}$. Consider $F(s) := g_s^{\min} - f_s^{\min}$, so that

$$F\left(\frac{1}{2}\right) = -\frac{1}{2N}\ln\left(\delta N^{3/2}\right) + \mathcal{O}(1/N). \tag{13.9.27}$$

We have for s in a neighborhood of $1/2$:

$$\partial_s F(s) = t_{\min}(g_s) - t_{\min}(f_s) \asymp \delta^{1/2} N^{1/4}. \tag{13.9.28}$$

This gives

Proposition 13.9.1 *There exists a unique point s_0 in a neighborhood of $1/2$ such that*

$$f_{s_0}^{\min} = g_{s_0}^{\min}. \tag{13.9.29}$$

Moreover,

$$s_0 = \frac{1}{2} + \mathcal{O}(1)\frac{|\ln(\delta N^{3/2})| + 1}{\delta^{1/2} N^{5/4}} = \frac{1}{2} + o(1).$$

From the calculations above (cf. (13.9.6), (13.9.11), (13.9.19) and the line after (13.9.23)) we also have

$$t_{\min}(g_{s_0}) = (1 + o(1))(2\delta)^{1/2} N^{1/4}, \quad t_{\min}(f_{s_0}) = -(1 + o(1))(2\delta)^{1/2} N^{1/4}. \tag{13.9.30}$$

We can now apply (13.9.4) and the corresponding discussion and get

Proposition 13.9.2 *Let* $h = h_1$ *be the largest convex function in* (13.9.1). *Then there exists* $s_0 = \frac{1}{2} + o(1)$ *such that* h *is given by* (13.9.4) *with* $s = s_0$. *We also have* (13.9.30).

Recall that we work under the assumption (13.9.5): $N^{\epsilon_0 - 5/2} \le \delta \ll N^{-3/2}$ for some fixed $\epsilon_0 > 0$. Also recall that since (13.9.2) we have simplified the notations by writing δ for $2C_1\delta$. We now reinstate the original δ. Let us review the estimates in Theorem 13.7.4 that will be used:

With probability $\ge 1 - e^{-N^2}$ we have the interior upper bound (13.7.26):

$$\ln|\det(z - A_\delta)| \le \ln\left(r^N + \frac{2C_1\delta N}{1-r}\right) + \frac{2C_1\delta N^{3/2}}{1-r} =: Nf(r), \tag{13.9.31}$$

when $r < 1$, $\frac{C_1\delta N}{1-r} < \frac{1}{2}$. Here and below, we frequently write $r = |z|$.

With probability $\ge 1 - e^{-N^2}$, we have the exterior upper bound (13.7.24):

$$\ln|\det(z - A_\delta)| \le N\ln r + \frac{2C_1\delta N^{3/2}}{r-1} =: Ng(r), \tag{13.9.32}$$

when $r > 1$. Recall that $f = f_1$, $g = g_1$ in the notation of the beginning of this section. Let $h(r) = h_1(r)$ be the largest subharmonic function (of z) such that

$$h(r) \le f(r),\ 0 < r < 1, \quad h(r) \le g(r),\ r > 1.$$

If we identify functions of r with functions of t via the substitution $r = e^t$, Proposition 13.9.2 gives,

$$h(r) = \begin{cases} f(r) & \text{for } r \le 1 - (1+o(1))(4C_1\delta)^{1/2}N^{1/4}, \\ g(r) & \text{for } r \ge 1 + (1+o(1))(4C_1\delta)^{1/2}N^{1/4}, \end{cases} \tag{13.9.33}$$

now with the original δ reinstated.

Concerning (13.9.31), we notice that if $r \le 1 - (1+o(1))(4C_1\delta)^{1/2}N^{1/4}$, then by (13.9.5)

$$\frac{C_1\delta N}{1-r} \le \frac{1+o(1)}{2}(C_1\delta)^{1/2}N^{3/4} < 1/2,$$

and hence (13.9.31) applies. Thus we can apply the maximum principle and conclude that with probability $\ge 1 - e^{-N^2}$, we have

$$\ln|\det(z - A_\delta)| \le Nh(r),\ z \in \mathbf{C}. \tag{13.9.34}$$

With probability $\geq 1 - e^{-N^2}$ the exterior lower bound in (13.7.25)

$$\ln|\det(z - A_\delta)| \geq Ng(r) - 6\frac{C_1\delta N^{3/2}}{r-1}, \tag{13.9.35}$$

holds for $r > 1$ and $\frac{C_1\delta N}{r-1} < \frac{1}{2}$. As before, the last condition on r is satisfied when $r \geq 1 + (1 + o(1))(4C_1\delta)^{1/2}N^{1/4}$.

We also recall the interior probabilistic lower bound (13.7.28), which says that if

$$|z| < 1 - \frac{1}{N},\ |z|^N \leq \mathcal{O}(1)\frac{\delta}{1-r},\quad t \leq \mathcal{O}(1)\frac{\delta}{1-r},\ \frac{\delta N}{1-r} \ll 1,$$

then

$$\ln|\det(z - A_\delta)| > \ln t - \mathcal{O}(1)\frac{\delta N^{3/2}}{1-r},\ \text{with probability } \geq 1 - \mathcal{O}(1)\left(\frac{t(1-r)}{\delta}\right)^2 - e^{-N^2}. \tag{13.9.36}$$

We choose $t = e^{-N^{-\epsilon_1}}$ for some small fixed ϵ_1. (This is almost the same choice as in the case of small random perturbations, the factor δ is now squeezed between two powers of N and is therefore superfluous.) Then we see that

$$\ln|\det(z - A_\delta)| > -N^{\epsilon_1} - \mathcal{O}(1)\frac{\delta N^{3/2}}{1-r} \text{ with probability } \geq 1 - \mathcal{O}(1)\frac{(1-r)^2}{\delta^2}e^{-2N^{\epsilon_1}}, \tag{13.9.37}$$

for every fixed z with $r < 1 - 1/N$, $r^N \leq \mathcal{O}(\delta)/(1-r)$, provided that $e^{-N^{\epsilon_1}} \leq \mathcal{O}(\delta)/(1-r)$, $\delta N/(1-r) \ll 1$. When $r \leq 1 - (1 + o(1))(4C_1\delta)^{1/2}N^{1/4}$, these conditions on r are fulfilled, so (13.9.37) applies.

Recall that $f = h$ precisely when $t \leq t_{\min}(f_{s_0})$. From (13.9.30) and (13.9.1), we get in that region, $|f| \leq \mathcal{O}(\delta N^{1/2}/(1-r) + N^{\epsilon_1 - 1})$, so (13.9.37) implies that

$$\ln|\det(z - A_\delta)| \geq N(f(r) - \epsilon(r)), \tag{13.9.38}$$

where, as in (13.8.24),

$$\epsilon(r) \leq \mathcal{O}\left(N^{\epsilon_1 - 1} + \frac{\delta N^{1/2}}{1-r}\right),\ r < 1, \tag{13.9.39}$$

and we get

Proposition 13.9.3 *For every* $z \in D(0, 1-(1+o(1))(4C_1\delta)^{1/2}N^{1/4})$,

$$\ln|\det(z-A_\delta)| \geq N(h(r)-\epsilon(r)), \tag{13.9.40}$$

with probability as in (13.9.37). *Here* $\epsilon(r)$ *satisfies* (13.9.39).

We shall combine this with the upper bound (13.9.34) and the exterior lower bound (13.9.35), which gives

$$\ln|\det(z-A_\delta)| \geq N(h(r)-\epsilon(r)), \tag{13.9.41}$$

for $r \geq 1+(1+o(1))(4C_1\delta)^{1/2}N^{1/4}$, where

$$\epsilon(r) = 6\frac{C_1\delta N^{1/2}}{r-1}, \quad r > 1. \tag{13.9.42}$$

As in Sect. 13.8, we will study $I(r_1, r_2) = \int_{r_1<|z|<r_2} \Delta h(|z|)L(dz)$ (cf. (13.8.26)), confining to the case $|r_j - 1| \geq (1+o(1))(4C_1\delta)^{1/2}N^{1/4}$.

When $r \leq 1-(1+o(1))(4C_1\delta)^{1/2}N^{1/4}$, we have $h(r) = f(r)$ and by (13.9.21),

$$f - \widehat{f} = \mathcal{O}(e^{-N^{\epsilon_0/2}/\mathcal{O}(1)}) \tag{13.9.43}$$

and similarly for $(r\partial_r)^k(f-\widehat{f})$. Here (see (b) after (13.9.9))

$$\widehat{f}(r) = \frac{1}{N}\ln\left(\frac{2C_1\delta N}{1-r}\right) + \frac{2C_1\delta N^{1/2}}{1-r}.$$

We have

$$r\partial_r\widehat{f}(r) = \frac{r}{N(1-r)} + \frac{2C_1\delta N^{1/2}r}{(1-r)^2}. \tag{13.9.44}$$

This is an increasing function of r which reaches a value $\asymp 1$ at the right end point $r = 1-(1+o(1))(4C_1\delta)^{1/2}N^{1/4}$.

When $r \geq 1+(1+o(1))(4C_1\delta)^{1/2}N^{1/4}$, we have $h(r) = g(r) = \ln r + \frac{2C_1\delta N^{1/2}}{r-1}$ and we see that

$$r\partial_r h(r) = 1 - \frac{2C_1\delta N^{1/2}r}{(r-1)^2}. \tag{13.9.45}$$

We get the following result, similar to Proposition 13.8.6:

Proposition 13.9.4

(a) *When* $0 \leq r_1 < r_2 \leq 1 - (1 + o(1))(4C_1\delta)^{1/2}N^{1/4}$, *we have*

$$I(r_1, r_2) = \frac{r_2 - r_1}{N(1 - r_1)(1 - r_2)} + \mathcal{O}(1)\left(e^{-N^{\epsilon_0/2}/\mathcal{O}(1)} + \frac{C_1\delta N^{1/2}(r_2 - r_1)}{(1 - r_2)^3}\right).$$

(b) *When* $1 + (1 + o(1))(4C_1\delta)^{1/2}N^{1/4} \leq r_1 \leq r_2 \leq 2$, *we have*

$$I(r_1, r_2) = \mathcal{O}(1)\frac{C_1\delta N^{1/2}(r_2 - r_1)}{(r_1 - 1)^3}.$$

(c) *When* $1 - r_1,\ r_2 - 1 \geq (1 + o(1))(4C_1\delta)^{1/2}N^{1/4}$, *we have*

$$I(r_1, r_2) = 1 - \frac{r_1}{N(1 - r_1)} + \mathcal{O}(1)\left(e^{-N^{\epsilon_0/2}/\mathcal{O}(1)} + \frac{C_1\delta N^{1/2}r_1}{(1 - r_1)^2} + \frac{C_1\delta N^{1/2}r_2}{(r_2 - 1)^2}\right),$$

where

$$\frac{r_1}{N(1 - r_1)} \leq \mathcal{O}(1)N^{-\epsilon_0/2}.$$

We next turn to the Δh-density (cf. (13.8.35) and the subsequent estimates). Consider open discs $D(z, \rho)$, $|z| = r$, $r + \rho = r_2$, $r - \rho = r_1$, with r_1, r_2 as in case (a) or (b) of the proposition. Again we have (13.8.35), where in case (a) we assume in addition that

$$\rho \leq \theta_0(1 - r),\ \ \rho \geq \exp\left(-o(1)N^{\epsilon_0/2}\right), \tag{13.9.46}$$

for some fixed $\theta_0 \in]0, 1[$, so that $1 - r$, $1 - r_1$, $1 - r_2$ are of the same order of magnitude. Then in case (a),

$$I(r_1, r_2) = \mathcal{O}(1)\left(\frac{\rho}{N(1 - r)^2} + \frac{\delta N^{1/2}\rho}{(1 - r)^3}\right).$$

Defining Δh-dens as in (13.8.38), we get

$$\Delta h\text{-dens}\,(z, \rho) = \mathcal{O}(1)\left(\frac{\rho}{N(1 - r)^2} + \frac{\delta N^{1/2}\rho}{(1 - r)^3}\right). \tag{13.9.47}$$

In case (b), we take $|z| = r$, $r + \rho = r_2$, $r - \rho = r_1$ where r_1, r_2 are as in case (b) and in addition,

$$\rho \leq \theta_0(r-1), \tag{13.9.48}$$

for some fixed $\theta_0 \in]0, 1[$. Then,

$$\Delta h\text{-dens}\,(z, \rho) = \mathcal{O}\left(\frac{\delta N^{1/2}\rho}{(r-1)^3}\right). \tag{13.9.49}$$

We next look at the ϵ-density, defined as after (13.8.39). In case (a) we get from (13.9.38) and (13.9.39) that

$$\epsilon\text{-dens}\,(z, \rho) = \mathcal{O}(1)\frac{1}{\rho}\left(N^{\epsilon_1 - 1} + \frac{\delta N^{1/2}}{1-r}\right), \tag{13.9.50}$$

and in case (b) (cf. (13.9.35)) that

$$\epsilon\text{-dens}\,(z, \rho) = \mathcal{O}(1)\frac{1}{\rho}\frac{\delta N^{1/2}}{r-1}. \tag{13.9.51}$$

Thus, with the restrictions on ρ to be respected, we have in case (a)

$$\begin{aligned}
&(\Delta h\text{-dens} + \epsilon\text{-dens})(z, \rho) \leq \\
&\mathcal{O}(1)\left(\rho\left(\frac{1}{N(1-r)^2} + \frac{\delta N^{1/2}}{(1-r)^3}\right) + \frac{1}{\rho}\left(N^{\epsilon_1-1} + \frac{\delta N^{1/2}}{1-r}\right)\right) \\
&\leq \mathcal{O}(1)\left(\frac{\rho}{(1-r)^2} + \frac{1}{\rho}\right)\left(N^{\epsilon_1-1} + \frac{\delta N^{1/2}}{1-r}\right),
\end{aligned} \tag{13.9.52}$$

and in case (b)

$$\begin{aligned}
(\Delta h\text{-dens} + \epsilon\text{-dens})(z, \rho) &\leq \mathcal{O}(1)\left(\rho\frac{\delta N^{1/2}}{(r-1)^3} + \frac{1}{\rho}\frac{\delta N^{1/2}}{r-1}\right) \\
&\leq \mathcal{O}(1)\left(\frac{\rho}{(r-1)^2} + \frac{1}{\rho}\right)\frac{\delta N^{1/2}}{r-1}.
\end{aligned} \tag{13.9.53}$$

In both cases, we strengthen the assumption on r to

$$|r-1| \geq 2(4C_1\delta)^{1/2}N^{1/4} \tag{13.9.54}$$

and choose

$$\rho = \rho_{\min} = \frac{1}{3}|r-1|. \tag{13.9.55}$$

Then we get

$$(\Delta\text{-dens} + \epsilon\text{-dens})(z, \rho_{\min}) \le \mathcal{O}(1) \begin{cases} \frac{1}{1-r}\left(N^{\epsilon_1 - 1} + \frac{\delta N^{1/2}}{1-r}\right), & \text{in case (a)}, \\ \frac{1}{r-1}\frac{\delta N^{1/2}}{r-1}, & \text{in case } b). \end{cases} \tag{13.9.56}$$

We now adapt the discussion from (13.8.46) to Theorem 13.8.7. Define Ω as in (13.8.47), now with

$$1-r_-,\ r_+ - 1 \ge 2(1+o(1))(4C_1\delta)^{1/2}N^{1/4}, \tag{13.9.57}$$

and apply Theorem 12.1.2 with "$1/h$" there replaced by N and with "ϕ" replaced by $h(z)$. As noted after (13.8.47), the main term in that theorem is

$$\frac{N}{2\pi}\int_\Omega \Delta h(z) L(dz) = N\theta\, [r\partial_r h(r)]_{r_-}^{r_+},$$

which now according to (c) in Proposition 13.9.4 is equal to

$$N\theta\left(1 - \frac{r_-}{N(1-r_-)} + \mathcal{O}(1)\left(e^{-N^{\epsilon_0/2}/\mathcal{O}(1)} + \frac{\delta N^{1/2} r_-}{(1-r_-)^2} + \frac{\delta N^{1/2} r_+}{(r_+-1)^2}\right)\right). \tag{13.9.58}$$

Before continuing, let us observe that by (2.4.16), the spectrum of A_δ is contained in $D(0, R_\sigma)$ when $\|Q\| \le C_1 N$, where $R_\sigma \le R_0$ and R_0 is given by

$$\frac{1}{R_0 - 1} = \frac{1}{C_1\delta N}, \text{ that is, } R_0 = 1 + C_1\delta N.$$

Here $\delta N \ll \delta^{1/2}N^{1/4}$, so the spectrum of A_δ in Ω does not depend on the choice of r_+, satisfying (13.9.57). For simplicity, we choose $r_+ = r_+(r_-)$ such that $1 - r_- = r_+ - 1$.

The remainder terms in the theorem can be written as

$$\mathcal{O}(N)\left(\int_{\partial\Omega_{\mathrm{ia}}\cup\partial\Omega_{\mathrm{ea}}} (\epsilon\text{-dens} + \Delta h\text{-dens})(z, \rho_{\min})|dz| + R\right), \tag{13.9.59}$$

where $\partial\Omega_{\mathrm{ia}} = \partial\Omega \cap \{r = r_-\}$ and $\partial\Omega_{\mathrm{ea}} = \partial\Omega \cap \{r = r_+\}$ are the interior and exterior arcs in the boundary of Ω, and R is the contribution from the two radial segments in $\partial\Omega$. By (13.9.56), the integral in (13.9.59) is bounded by

$$\mathcal{O}(\theta)\frac{1}{1-r_-}\left(N^{\epsilon_1 - 1} + \frac{\delta N^{1/2}}{1-r_-}\right). \tag{13.9.60}$$

We have

$$\begin{aligned} R =& \mathcal{O}(1)\int_{r_-}^{r_{\min}(f_{s0})} (\epsilon\text{-dens} + \Delta h\text{-dens})(r, \rho_{\min})dr \\ &+ \mathcal{O}(1)\int_{r_{\min}(g_{s0})}^{r_+} (\epsilon\text{-dens} + \Delta h\text{-dens})(r, \rho_{\min})dr \\ &+ \mathcal{O}(1)\left(\left(r_{\min}(g_{s0}) - r_{\min}(f_{s0})\right) I(r_{\min}(f_{s0}), r_{\min}(g_{s0}))\right) \\ =& \mathcal{O}(1)\left(\int_{r_-}^{r_{\min}(f_{s0})} \left(\frac{N^{\epsilon_1 - 1}}{1-r} + \frac{\delta N^{\frac{1}{2}}}{(1-r)^2}\right) dr + \int_{r_{\min}(g_{s0})}^{r_+} \frac{\delta N^{\frac{1}{2}}}{(r-1)^2} dr\right) \\ &+ \mathcal{O}(1)\delta^{\frac{1}{2}} N^{\frac{1}{4}} I(r_{\min}(f_{s0}), r_{\min}(g_{s0})). \end{aligned}$$

Here, $I(r_{\min}(f_{s0}), r_{\min}(g_{s0})) = \mathcal{O}(1)$ by (c) in Proposition 13.9.4 and the second integral will satisfy the same upper bounds as the first one (up to a factor $\mathcal{O}(1)$), since r_+, r_- are symmetrically placed around 1, and $r_{\min}(f_{s0})$, $r_{\min}(g_{s0})$ have approximately the same property.

Thus,

$$\begin{aligned} R =& \mathcal{O}(1)\left(\int_{r_-}^{r_{\min}(f_{s0})} \left(\frac{N^{\epsilon_1 - 1}}{1-r} + \frac{\delta N^{\frac{1}{2}}}{(1-r)^2}\right) dr + \delta^{\frac{1}{2}} N^{\frac{1}{4}}\right) \\ =& \mathcal{O}(1)\left(N^{\epsilon_1 - 1}|\ln(\delta^{\frac{1}{2}} N^{\frac{1}{4}})| + \frac{\delta N^{\frac{1}{2}}(r_{\min}(f_{s0}) - r_-)}{(1 - r_{\min}(f_{s0}))(1-r_-)} + \delta^{\frac{1}{2}} N^{\frac{1}{4}}\right). \end{aligned}$$

Here the second term in the last expression is $\mathcal{O}(\delta^{1/2}N^{1/4})$ and $\ln(\delta^{1/2}N^{1/4}) = \mathcal{O}(1)\ln N$, since $N^{-2+\epsilon_0} \leq \delta N^{1/2} \leq N^{-1}$. Thus,

$$R = \mathcal{O}(1)\left(N^{\epsilon_1 - 1}\ln(N) + \delta^{\frac{1}{2}} N^{\frac{1}{4}}\right). \tag{13.9.61}$$

Applying Theorem 12.1.2, (13.9.58)–(13.9.61), we get,

Theorem 13.9.5 *Let $A_\delta = A_0 + \delta Q$ be the $N \times N$ matrix in the beginning of this chapter, where the entries of Q are independent complex random variables, $\sim \mathcal{N}_{\mathbb{C}}(0, 1)$*

and $N^{-\frac{5}{2}+\epsilon_0} \leq \delta \ll N^{-\frac{3}{2}}$ *for a fixed* $\epsilon_0 > 0$. *Define* Ω *as in* (13.8.47) *and put* $\theta = \theta_2 - \theta_1$. *Let*

$$r_{\min}(g_{s_0}) = \exp t_{\min}(g_{s_0}) = 1 + (1 + o(1))(4C_1\delta)^{\frac{1}{2}} N^{\frac{1}{4}},$$

$$r_{\min}(f_{s_0}) = \exp t_{\min}(f_{s_0}) = 1 - (1 + o(1))(4C_1\delta)^{\frac{1}{2}} N^{\frac{1}{4}},$$

be as in Proposition 13.9.1 *and* (13.9.30) (*with "*δ*" there replaced by* $2C_1\delta$). *Fix* $0 < \epsilon_1 \leq 1$ *and choose* r_-, r_+ *in the definition of* Ω, *so that*

$$\frac{1}{C} \leq r_- \leq r_{\min}(f_{s_0}), \quad r_+ \geq 1 + C_1 \delta N.$$

Then with probability $\geq 1 - \mathcal{O}(N)\exp(-N^{\epsilon_1})$, *we have*

$$\begin{aligned} \left| \#\left(\sigma(A_\delta) \cap \Omega\right) - \frac{\theta}{2\pi} N \left(1 - \frac{r_-}{N(1-r_-)}\right) \right| \leq \\ N\mathcal{O}(\theta) \left(e^{-N^{\epsilon_0/2}/\mathcal{O}(1)} + \frac{\delta N^{\frac{1}{2}}}{(1-r_-)^2} + \frac{N^{\epsilon_1 - 1}}{1 - r_-} \right) \\ + N\mathcal{O}(1)\left(N^{\epsilon_1 - 1} \ln N + \delta^{\frac{1}{2}} N^{\frac{1}{4}}\right). \end{aligned}$$

This gives Theorem 13.2.2.

Notes The eigenvalues of large Jordan blocks with random perturbations tend to accumulate to a certain critical circle [38] and have uniform distribution there [52]. In this chapter we give more refined results of the same nature, where Theorem 12.1.2 is the main ingredient in the proofs. Parallel and later works treat large Toeplitz matrices and the expected density of eigenvalues. Cf. [20, 145–147].

Part III

Spectral Asymptotics for Differential Operators in Higher Dimension

14 Weyl Asymptotics for the Damped Wave Equation

14.1 Eigenvalues of Perturbations of Self-adjoint Operators

The damped wave equation is closely related to non-self-adjoint perturbations of a self-adjoint operator P of the form

$$P_\epsilon = P + i\epsilon Q. \tag{14.1.1}$$

Here, P is a semi-classical pseudodifferential operator of order 0 on $L^2(X)$, where we consider two cases:

- $X = \mathbf{R}^n$ and P has the symbol $P \sim p(x,\xi) + hp_1(x,\xi) + \cdots$. in $S(m)$, as in Sect. 6.1, where the description is valid also in the case $n > 1$. We assume for simplicity that the order function $m(x,\xi)$ tends to $+\infty$, when (x,ξ) tends to ∞. We also assume that P is formally self-adjoint. Then by elliptic theory (and the ellipticity assumption on P) we know that P is essentially self-adjoint with purely discrete spectrum.
- X is a compact smooth manifold with positive smooth volume form dx and P is a formally self-adjoint differential operator, which in local coordinates takes the form,

$$P = \sum_{|\alpha|\le m} a_\alpha(x;h)(hD_x)^\alpha,\ m > 0$$

where $a_\alpha(x;h) \sim \sum_{k=0}^{\infty} h^k a_{\alpha,k}(x)$ in C^∞ and the "classical" principal symbol

$$p_m(x,\xi) = \sum_{|\alpha|=m} a_{\alpha,0}(x)\xi^\alpha,$$

J. Sjöstrand, *Non-Self-Adjoint Differential Operators, Spectral Asymptotics and Random Perturbations*, Pseudo-Differential Operators 14,
https://doi.org/10.1007/978-3-030-10819-9_14

satisfies

$$0 \leq p_m(x,\xi) \asymp |\xi|^m,$$

so m has to be even. In this case the semi-classical principal symbol is given by

$$p(x,\xi) = \sum_{|\alpha|\leq m} a_{\alpha,0}(x)\xi^\alpha.$$

We assume that $Q : L^2(X) \to L^2(X)$ is bounded with norm ≤ 1. This operator may depend on ϵ and h. Let $I = [a,b]$, $-\infty < a < b < +\infty$. We assume that a, b are not critical values of p:

$$dp \neq 0 \text{ on } p^{-1}(\partial I). \tag{14.1.2}$$

Then it is a classical result (see [41] and further references given there) that P has discrete spectrum and that we have the Weyl asymptotics

$$\#(\sigma(P)\cap I) = \frac{1}{(2\pi h)^n}\left(\mathrm{vol}\, p^{-1}(I) + \mathcal{O}(h)\right), \tag{14.1.3}$$

and that

$$\#(\sigma(P)\cap(\partial I + [-\delta,\delta\,])) = \mathcal{O}(1)\delta h^{-n}, \tag{14.1.4}$$

uniformly for $0 < h \leq \delta \ll 1$.

The following result is due to Markus and Matseev [97], see also [96, 134].

Theorem 14.1.1 *For $0 \leq \epsilon \ll 1$, the spectrum of P_ϵ is purely discrete and contained in $\mathbf{R} + i[-\epsilon,\epsilon]$. We have*

$$\#(\sigma(P)\cap(I + i[-\epsilon,\epsilon\,])) = \frac{1}{(2\pi h)^n}\left(\mathrm{vol}\, p^{-1}(I) + \mathcal{O}(\max(\epsilon,h))\right), \tag{14.1.5}$$

where the eigenvalues are counted with their algebraic multiplicity.

The inclusion $\sigma(P_\epsilon) \subset \mathbf{R} + i[-\epsilon,\epsilon\,]$ follows from the fact that

$$|\Im(P_\epsilon u|u)| \leq \epsilon\|u\|^2,$$

and from this estimate we also deduce that

$$\|(P_\epsilon - z)^{-1}\| \leq \frac{1}{|\Im z| - \epsilon}, \quad \text{when } |\Im z| > \epsilon.$$

The assumptions about P, and in particular the ellipticity assumption near infinity, imply that $(P - z_0)^{-1}$, $Q(P - z_0)^{-1}$ and $(P - z_0)^{-1}Q$ are compact when $z_0 \notin \sigma(P)$.[1] For such a z_0 (take for instance z_0 away from the real axis) we write

$$\begin{aligned} z - P_\epsilon &= (z_0 - P)(1 + (z_0 - P)^{-1}(z - z_0 - i\epsilon Q)) \\ &= (1 + (z - z_0 - i\epsilon Q)(z_0 - P)^{-1})(z_0 - P). \end{aligned}$$

Here, $(z_0 - P)^{-1}(z - z_0 - i\epsilon Q)$, $(z - z_0 - i\epsilon Q)(z_0 - P)^{-1}$ are compact for $z \neq z_0$ and since we can vary z_0, it follows from Fredholm theory that the spectrum of P_ϵ is discrete.

Proof of (14.1.5) This is the main part of the theorem and we shall follow the general strategy of [97] (see also [96, 134]) which consists in introducing a self-adjoint finite-rank perturbation of P which has no spectrum near ∂I. The use of the results in Chap. 12 will allow us to shorten the proof slightly.

Without loss of generality we may assume that $h \leq \epsilon$, for if $\epsilon < h$, we can write $\epsilon Q = h((\epsilon/h)Q)$ and use h as a new perturbation parameter, since $\|(\epsilon/h)Q\| \leq 1$.

Let $\epsilon < \delta \ll 1$. We define a δ-perturbation of P by moving the eigenvalues in $\partial I +] - \delta, \delta[$ away from that set, by minimal displacements. Let λ_j, $j \in \mathbf{N}$ denote the eigenvalues of P repeated with their multiplicity and let e_j, $j \in \mathbf{N}$ be a corresponding orthonormal basis of eigenvectors. Define,

$$\widetilde{\lambda}_j = \begin{cases} \text{a point in } \partial I + \{-\delta, \delta\} \text{ with dist}\,(\lambda_j, \partial I + \{-\delta, \delta\}) = |\lambda_j - \widetilde{\lambda}_j| \\ \text{when } \lambda_j \in \partial I +] - \delta, \delta[, \\ \lambda_j \text{ when } \lambda_j \in \mathbf{R} \setminus (\partial I +] - \delta, \delta[). \end{cases}$$

We define our modified operator to be

$$P^\delta u = \sum \widetilde{\lambda}_j (u|e_j) e_j, \tag{14.1.6}$$

where the spectral resolution of P is

$$Pu = \sum \lambda_j (u|e_j) e_j.$$

[1] $(P - z_0)^{-1}Q$ is also compact: $\mathcal{D}(P) \to \mathcal{D}(P)$ since it is related to $Q(P - z_0)^{-1} : L^2 \to L^2$ by conjugation with $P - z_0$.

It follows that

$$\|P^\delta - P\| \le \delta, \tag{14.1.7}$$

$$\|P^\delta - P\|_{\mathrm{tr}} \le \delta \#(\sigma(P) \cap (\partial I +]-\delta, \delta[)) \le \mathcal{O}(1)\delta^2 h^{-n}, \tag{14.1.8}$$

$$\sigma(P^\delta) \cap (\partial I +]-\delta, \delta[) = \emptyset. \tag{14.1.9}$$

By construction and (14.1.4),

$$\#(\sigma(P^\delta) \cap I) = \#(\sigma(P) \cap I) + \mathcal{O}(\delta h^{-n}). \tag{14.1.10}$$

Recalling that $P_\epsilon = P + i\epsilon Q$, we put

$$P_\epsilon^\delta = P^\delta + i\epsilon Q. \tag{14.1.11}$$

Then P_ϵ^δ has a purely discrete spectrum contained in $\mathbf{R} + i[-\epsilon, \epsilon]$ and more precisely,

$$\sigma(P_\epsilon^\delta) \subset (\mathbf{R} \setminus (\partial I +]-\delta, \delta[)) + \overline{D(0, \epsilon)}. \tag{14.1.12}$$

In particular $(\partial I +]-\delta + \epsilon, \delta - \epsilon[) + i\mathbf{R}$ is disjoint from the spectrum of P_ϵ^δ. By a simple deformation argument, we see that

$$\#(\sigma(P_\epsilon^\delta) \cap (I + i\mathbf{R})) = \#(\sigma(P^\delta) \cap I), \tag{14.1.13}$$

to be combined with (14.1.10).

In the following, we choose

$$\delta = C_0\epsilon, \tag{14.1.14}$$

where $C_0 > 1$ is a large constant to be fixed later.

Let

$$\Omega = I + i[-1, 1], \tag{14.1.15}$$

and consider the following neighborhood of $\partial\Omega$:

$$\widetilde{W} = \bigcup_{z \in \partial\Omega} D(z, \widetilde{r}(z)), \tag{14.1.16}$$

where

$$\widetilde{r}(z) = 4\epsilon + |\Im z|/4. \tag{14.1.17}$$

If C_0 in (14.1.14) is sufficiently large, then

$$D(z, 2\widetilde{r}(z)) \cap ((\mathbf{R} \setminus (\partial I +] - \delta + \epsilon, \delta - \epsilon [)) + i[-\epsilon, \epsilon]) = \emptyset, \quad z \in \widetilde{W}. \tag{14.1.18}$$

It follows that P_ϵ^δ has no spectrum in $\widetilde{W}$ and that

$$\|(P_\epsilon^\delta - z)^{-1}\| \le \mathcal{O}(1/\widetilde{r}(z)), \quad z \in \widetilde{W}. \tag{14.1.19}$$

We shall view the eigenvalues of P_ϵ in Ω as the zeros of a relative determinant. Let $0 \le \chi \in C_0^\infty(T^*X)$ be equal to 2 on a sufficiently large compact subset of T^*X. Then, if $\widetilde{p} = p + i\chi$, we see that $(\widetilde{p} - z)^{-1}$ is a uniformly bounded function for z in a fixed neighborhood of the closure of $\Omega \cup \widetilde{W}$. We can then quantize $\widetilde{p} - p + P$ as an h-pseudodifferential operator and get an operator $\widetilde{P}$ such that (for $h > 0$ small enough)

- $\widetilde{P} - P$ is uniformly bounded in $L^2(X)$,
- $\|\widetilde{P} - P\|_{\mathrm{tr}} < +\infty$,
- $\widetilde{P}$ has no spectrum in $\widetilde{W} \cup \Omega$ and the resolvent $(\widetilde{P} - z)^{-1}$ is uniformly bounded on that set. The same holds for $\widetilde{P}_\epsilon := \widetilde{P} + i\epsilon Q$.

The eigenvalues of P_ϵ in $\Omega \cup \widetilde{W}$ are the zeros of the determinant of a trace class perturbation of the identity, namely,

$$D_\epsilon(z) := \det\left((P_\epsilon - z)(\widetilde{P}_\epsilon - z)^{-1}\right) = \det\left(1 - (\widetilde{P} - P)(\widetilde{P}_\epsilon - z)^{-1}\right). \tag{14.1.20}$$

Similarly the eigenvalues of P_ϵ^δ in $\Omega \cup \widetilde{W}$ are the zeros of

$$D_\epsilon^\delta(z) = \det\left((P_\epsilon^\delta - z)(\widetilde{P}_\epsilon - z)^{-1}\right) = \det\left(1 - (\widetilde{P} - P + P - P^\delta)(\widetilde{P}_\epsilon - z)^{-1}\right) \tag{14.1.21}$$

and we have seen that there are no such values in $\widetilde{W}$.

Remark 14.1.2 We can identify the algebraic multiplicities of the eigenvalues and the corresponding zeros of a determinant by expanding the discussion in Sect. 5.4 slightly: Let $P(z) : \mathcal{F} \to \mathcal{G}$, $z \in A$ be a holomorphic family of Fredholm operators of index zero, where $A \subset \mathbf{C}$ is open and connected. Assume that $P(z)$ is bijective for at least one value of z. Then the set $\Gamma \subset A$ of points where it is not bijective is discrete. Let $z_0 \in \Gamma$. We can find $1 \le N_0 \in \mathbf{N}$ and bounded operators $R_+(z) : \mathcal{F} \to \mathbf{C}^{N_0}$, $R_-(z) : \mathbf{C}^{N_0} \to \mathcal{G}$, depending holomorphically on $z \in \mathrm{neigh}\,(z_0, A)$, of maximal rank, such that

$$\mathcal{P}(z) = \begin{pmatrix} P(z) & R_-(z) \\ R_+(z) & 0 \end{pmatrix} : \mathcal{F} \times \mathbf{C}^{N_0} \to \mathcal{G} \times \mathbf{C}^{N_0} \tag{14.1.22}$$

is bijective for $z \in \mathrm{neigh}\,(z_0, A)$. Define the multiplicity

$$m(z_0, P) = \mathrm{tr}\, \frac{1}{2\pi i} \int_\gamma P(z)^{-1} \partial_z P(z) dz, \tag{14.1.23}$$

where γ is the oriented boundary of the disc $D(z_0, r)$ with $r > 0$ so small that $\overline{D(z_0, r)} \cap \Gamma = \{z_0\}$. We shall see that $m(z_0, P)$ is well defined in general. When $P(z) = z - P$, it is the rank of the spectral projection, hence the usual multiplicity.

Let

$$\mathcal{E}(z) = \begin{pmatrix} E(z) & E_+(z) \\ E_-(z) & E_{-+}(z) \end{pmatrix} : \mathcal{G} \times \mathbf{C}^{N_0} \to \mathcal{F} \times \mathbf{C}^{N_0} \tag{14.1.24}$$

be the inverse of $\mathcal{P}(z)$. Then (cf. Sect. 5.4), we have

$$P(z)^{-1} = E(z) - E_+(z) E_{-+}^{-1} E_-(z), \tag{14.1.25}$$

and since $E(z)\partial_z P(z)$ is holomorphic near $\overline{D(z_0, r)}$ when $r > 0$ is small enough, we get

$$\frac{1}{2\pi i} \int_\gamma P(z)^{-1} \partial_z P(z) dz = -\frac{1}{2\pi i} \int_\gamma E_+(z) E_{-+}(z)^{-1} E_-(z) \partial_z P(z) dz$$

which is of trace class. Hence, $m(z_0, P)$ is well defined and we can take the trace of the last integral, move the trace inside the integral, and get

$$\begin{aligned} m(z_0, P) &= -\frac{1}{2\pi i} \int_\gamma \mathrm{tr}\Big(E_+(z) E_{-+}(z)^{-1} E_-(z) \partial_z P(z) \Big)\, dz \\ &= -\frac{1}{2\pi i} \int_\gamma \mathrm{tr}\Big(E_{-+}^{-1} E_- (\partial_z P) E_+ \Big)\, dz, \end{aligned} \tag{14.1.26}$$

using the cyclicity of the trace for the last equality. Using that $\partial_z \mathcal{E} = -\mathcal{E}(\partial_z \mathcal{P})\mathcal{E}$, we get

$$-\partial_z E_{-+} = E_-(\partial_z P) E_+ + E_{-+}(\partial_z R_+) E_+ + E_-(\partial_z R_-) E_{-+}, \tag{14.1.27}$$

$$-E_-(\partial_z P) E_+ = \partial_z E_{-+} + E_{-+}(\partial_z R_+) E_+ + E_-(\partial_z R_-) E_{-+},$$

and substitution of the last identity in (14.1.26) gives

$$m(z_0, P) = \mathrm{I} + \mathrm{II} + \mathrm{III}, \tag{14.1.28}$$

where

$$\mathrm{I} = \frac{1}{2\pi i} \int_\gamma \mathrm{tr}\Big(E_{-+}^{-1} \partial_z E_{-+}\Big)\, dz = m(z_0, \det E_{-+}),$$

the multiplicity of z_0 as a zero of $\det E_{-+}$,

$$\mathrm{II} = \frac{1}{2\pi i} \int_\gamma \mathrm{tr}\Big(E_{-+}^{-1} E_{-+}(\partial_z R_+) E_+\Big)\, dz = \frac{1}{2\pi i} \int_\gamma \mathrm{tr}((\partial_z R_+) E_+)\, dz = 0,$$

the integrand of the last integral being holomorphic,

$$\mathrm{III} = \frac{1}{2\pi i} \int_\gamma \mathrm{tr}\Big(E_{-+}^{-1} E_-(\partial_z R_-) E_{-+}\Big)\, dz = \frac{1}{2\pi i} \int_\gamma \mathrm{tr}(E_-(\partial_z R_-))\, dz = 0.$$

Thus,

$$m(z_0, P) = m(z_0, \det E_{-+}). \tag{14.1.29}$$

Similarly,

$$m(z_0, P) = \mathrm{tr} \frac{1}{2\pi i} \int_\gamma (\partial_z P(z)) P(z)^{-1} dz.$$

If $P(z) = P_1(z) P_2(z)$, where $P_2(z) : \mathcal{F} \to \mathcal{H}$, $P_1(z) : \mathcal{H} \to \mathcal{G}$ are Fredholm operators as above, one of which is bijective at $z = z_0$, then

$$m(z_0, P) = \begin{cases} m(z_0, P_1), & \text{when } P_2(z_0) \text{ is bijective,} \\ m(z_0, P_2), & \text{when } P_1(z_0) \text{ is bijective.} \end{cases} \tag{14.1.30}$$

The two cases are very similar and we only treat here the second one, when $P_1(z_0)$ is bijective. Let

$$\mathcal{P}_2(z) = \begin{pmatrix} P_2(z) & R_-^2(z) \\ R_+^2(z) & 0 \end{pmatrix}$$

be associated to P_2 as in (14.1.22). Let

$$\mathcal{P}_2^{-1} = \mathcal{E}_2 = \begin{pmatrix} E^2 & E_+^2 \\ E_-^2 & E_{-+}^2 \end{pmatrix}.$$

Then,

$$\mathcal{P} = \begin{pmatrix} P & P_1 R_-^2 \\ R_+^2 & 0 \end{pmatrix} = \begin{pmatrix} P_1 & 0 \\ 0 & 1 \end{pmatrix} \mathcal{P}_2$$

is bijective. The inverse of $\mathcal{P}$ is

$$\mathcal{E} = \begin{pmatrix} E^2 & E_+^2 \\ E_-^2 & E_{-+}^2 \end{pmatrix} \begin{pmatrix} P_1^{-1} & 0 \\ 0 & 1 \end{pmatrix} = \begin{pmatrix} E^2 P_1^{-1} & E_+^2 \\ E_-^2 P_1^{-1} & E_{-+}^2 \end{pmatrix},$$

and hence of the form (14.1.24) with $E_{-+} = E_{-+}^2$. Therefore, we have (14.1.30) when $P_1(z_0)$ is bijective. The other case can be treated similarly.

Finally, let us recall that when $\mathcal{F} = \mathcal{G}$, $P = 1 + K(z)$, where K is holomorphic and of trace class, then $m(z_0, P)$ as defined above, is equal to $m(z_0, \det P)$; the multiplicity of z_0 as a zero of $\det P$ (defined in Sect. 8.4).

We have

$$\begin{aligned} \frac{D_\epsilon(z)}{D_\epsilon^\delta(z)} &= \det\left((P_\epsilon - z)(\widetilde{P}_\epsilon - z)^{-1}(\widetilde{P}_\epsilon - z)(P_\epsilon^\delta - z)^{-1}\right) \\ &= \det\left((P_\epsilon - z)(P_\epsilon^\delta - z)^{-1}\right) \\ &= \det\left(1 - (P^\delta - P)(P_\epsilon^\delta - z)^{-1}\right). \end{aligned} \tag{14.1.31}$$

Thus,

$$\left|D_\epsilon(z)/D_\epsilon^\delta(z)\right| \le \exp \|(P^\delta - P)(P_\epsilon^\delta - z)^{-1}\|_{\mathrm{tr}}. \tag{14.1.32}$$

Here,

$$\|(P^\delta - P)(P_\epsilon^\delta - z)^{-1}\|_{\mathrm{tr}} \le \|(P_\epsilon^\delta - z)^{-1}\| \|P^\delta - P\|_{\mathrm{tr}},$$

and by (14.1.19), (14.1.8), we get for $z \in \widetilde{W}$:

$$\|(P^\delta - P)(P_\epsilon^\delta - z)^{-1}\|_{\mathrm{tr}} \le \mathcal{O}(1)\frac{\delta^2}{\widetilde{r}(z)h^n}.$$

From (14.1.32), we conclude that

$$\ln|D_\epsilon(z)| - \ln|D_\epsilon^\delta(z)| \le \mathcal{O}(1)\frac{\delta^2}{\widetilde{r}(z)h^n},\ z \in \widetilde{W}. \tag{14.1.33}$$

For every fixed $\theta > 0$, we have

$$\|(P_\epsilon - z)^{-1}\| \le \frac{\mathcal{O}_\theta(1)}{\widetilde{r}(z)}, \; z \in \widetilde{W}, \; |\Im z| \ge (1+\theta)\epsilon. \tag{14.1.34}$$

Using this and a formula similar to (14.1.31) with P_ϵ^δ replaced by P_ϵ, we get the upper bound

$$\ln |D_\epsilon^\delta(z)| - \ln |D_\epsilon(z)| \le \frac{\mathcal{O}(1)\delta^2}{\widetilde{r}(z)h^n},$$

in the same region as in (14.1.34), so in addition to (14.1.33) we have the lower bound

$$\ln |D_\epsilon(z)| - \ln |D_\epsilon^\delta(z)| \ge -\frac{\mathcal{O}(\delta^2)}{\widetilde{r}(z)h^n}, \quad z \subset \widetilde{W}, \; |\Im z| \ge (1+0)\epsilon. \tag{14.1.35}$$

In $\widetilde{W}$ we have

$$\ln |D_\epsilon^\delta(z)| = \Phi_0(z)/h^n, \tag{14.1.36}$$

where Φ_0 depends on the various parameters and is harmonic in $\widetilde{W}$. Extend Φ_0 to a smooth function on $\Omega \cup \widetilde{W}$ and notice that we have the exact formula

$$\#(\sigma(P_\epsilon^\delta) \cap \Omega) = \frac{1}{2\pi h^n} \int_\Omega \Delta\Phi_0(z) L(dz). \tag{14.1.37}$$

Let $r(z) = \widetilde{r}(z)/(1+\theta)$ for some fixed θ with $0 < \theta \ll 1$, and define (cf. (14.1.16))

$$W = \bigcup_{z \in \partial\Omega} D(z, r(z)). \tag{14.1.38}$$

From (14.1.33), (14.1.17), we have

$$\ln |D_\epsilon(z)| \le \Phi(z)/h^n, \; z \in \widetilde{W}, \tag{14.1.39}$$

where

$$\Phi(z) = \Phi_0(z) + \frac{C\delta^2}{((\Im z)^2 + \delta^2)^{1/2}}, \tag{14.1.40}$$

provided that C is large enough.

We can find $\chi \in C_0^\infty(\widetilde{W}, [0, 1])$, equal to 1 on W, such that

$$\nabla^\alpha_{\Re z, \Im z} \chi = \mathcal{O}(1) r(z)^{-|\alpha|}, \text{ for all } \alpha \in \mathbf{N}^2. \tag{14.1.41}$$

We extend the definition of Φ to $\widetilde{W} \cup \Omega$, by putting

$$\Phi(z) = \Phi_0(z) + \frac{C\delta^2 \chi(z)}{((\Im z)^2 + \delta^2)^{1/2}}. \tag{14.1.42}$$

Then,

$$\Delta\Phi - \Delta\Phi_0 = \frac{\mathcal{O}(\delta^2)}{(|\Im z| + \delta)^3},$$

and it follows that

$$\begin{aligned} \int |\Delta\Phi - \Delta\Phi_0| L(dz) &= \mathcal{O}(\delta^2) + \mathcal{O}(1) \int_0^1 \frac{\delta^2}{(s+\delta)^2} ds \\ &= \mathcal{O}(\delta). \end{aligned} \tag{14.1.43}$$

In particular, since Φ_0 is harmonic on $\widetilde{W}$,

$$\int_{\widetilde{W}} |\Delta\Phi| L(dz) = \mathcal{O}(\delta). \tag{14.1.44}$$

In view of the choice of $\widetilde{r}(z)$ in (14.1.17), if we take $\theta > 0$ small enough in the definition of $r(z)$, we can find distinct points $z_j^0 \in \partial\Omega$, $j \in \mathbf{Z}/N\mathbf{Z}$ distributed in the positive sense so that $j \mapsto \arg(z_j^0 - (a+b)/2)$ is increasing, with the properties:

1. There are precisely four points z_j^0 that minimize the distance to $\mathbf{R}$, namely $a \pm i5\epsilon/4$, $b \pm i5\epsilon/4$,
2. $r(z_j^0)/4 \leq |z_{j+1}^0 - z_j^0| \leq r(z_j^0)/2$.

It follows that $N = \mathcal{O}(1)\ln(1/\epsilon)$ and that

$$\partial\Omega \subset \bigcup_j D(z_j, r(z_j)/2). \tag{14.1.45}$$

Let

$$\epsilon_j = \frac{C\delta^2}{|\Im z_j| + \delta}, \tag{14.1.46}$$

for $C > 0$ sufficiently large. Then by (14.1.35) and (14.1.42),

$$\ln|D_\epsilon(z)| \geq \frac{1}{h^n}(\Phi(z) - \epsilon_j),\ z \in D(z_j, r(z_j)/C), \tag{14.1.47}$$

and we recall that $\ln|D_\epsilon(z)| \le \Phi(z)/h^n$ on W, by (14.1.39). By construction,

$$\sum \epsilon_j = \mathcal{O}(\delta). \tag{14.1.48}$$

To sum up the discussion, we have, by (14.1.10) and (14.1.13),

$$\#(\sigma(P_\epsilon^\delta) \cap \Omega) = \#(\sigma(P^\delta) \cap I) = \#(\sigma(P) \cap I) + \mathcal{O}(\delta h^{-n}).$$

Here $\#(\sigma(P_\epsilon^\delta) \cap \Omega)$ is given by the integral formula (14.1.37), where by (14.1.43),

$$\frac{1}{2\pi h^n} \int_\Omega \Delta\Phi_0 L(dz) = \frac{1}{2\pi h^n} \int_\Omega \Delta\Phi L(dz) + \mathcal{O}(\delta h^{-n}).$$

Thus,

$$\#(\sigma(P) \cap I) = \frac{1}{2\pi h^n} \int_\Omega \Delta\Phi L(dz) + \mathcal{O}(\delta h^{-n}).$$

On the other hand, Theorem 12.1.2, (14.1.39), (14.1.47), (14.1.44) and (14.1.48) show that

$$\left| \#(\sigma(P_\epsilon) \cap \Omega) - \frac{1}{2\pi h^n} \int_\Omega \Delta\Phi L(dz) \right| \le \mathcal{O}(\delta h^{-n}),$$

so

$$|\#(\sigma(P_\epsilon) \cap \Omega) - \#(\sigma(P) \cap I)| \le \mathcal{O}(\delta h^{-n}),$$

which in view of (14.1.3) concludes the proof of (14.1.5). □

Remark 14.1.3 Notice that Theorem 14.1.1 is an application of a more general abstract theorem, which we do not formulate in detail. In particular, there are extensions of the theorem to the case of boundary value problems. See [96, 97].

14.2 The Damped Wave Equation

Let X be compact Riemannian manifold of dimension n. The damped wave equation is of the form

$$(\partial_t^2 - \Delta + 2a(x)\partial_t)v(t, x) = 0, \quad (t, x) \in \mathbf{R} \times X. \tag{14.2.1}$$

Here Δ denotes the Laplace–Beltrami operator on X and we assume that $a \in C^\infty(X; \mathbf{R})$. Because of the "damping term" $2a\partial_t v$, this evolution is no longer energy conserving (for a suitable energy) and we have to expect exponential growth or decay of the solutions when $|t| \to \infty$. Clearly, this is related to the eigenfrequencies of the corresponding stationary problem. Put $v(t,x) = e^{it\tau}u(x)$, $\tau \in \mathbf{C}$. Then v solves (14.2.1) iff

$$(-\Delta - \tau^2 + 2ia(x)\tau)u(x) = 0. \tag{14.2.2}$$

When (u,τ) is a non-trivial solution of (14.2.2), we call τ an eigenfrequency, or simply an eigenvalue, and call u "the" corresponding eigenfunction. It is easy to show that if τ is an eigenfrequency, then

$$\inf a \le \Im\tau \le \sup a, \text{ when } \Re\tau \ne 0, \tag{14.2.3}$$

$$2\min(\inf a, 0) \le \Im\tau \le 2\max(\sup a, 0), \text{ when } \Re\tau = 0. \tag{14.2.4}$$

In this chapter we establish the most basic result about the distribution of eigenfrequencies, namely that their real parts obey Weyl asymptotics [97]. Many other important results concern the distribution of imaginary parts and the growth-decay of solutions and they are closely related to the geometry and the interplay between the damping coefficient a and the classical trajectories ("rays"). We refer to [66, 92, 118], where more references can be found. Some very detailed results in two dimensions are applicable to the eigenfrequencies. (See [73] for a more recent work on that theme.)

When $a = 0$, the eigenfrequencies are real and symmetrically distributed around 0. In fact, they are the square roots (with both signs) of the eigenvalues of $-\Delta$. In this case (neglecting the case of $\tau = 0$ which has the multiplicity two, as we shall see), we define the multiplicity of an eigenfrequency τ to be equal to that of τ^2 as an eigenvalue of $-\Delta$. Applying the standard result on the Weyl asymptotics for the eigenvalues of $-\Delta$ (see for instance [51] and further references given there) we have

Proposition 14.2.1 *When $a = 0$ the eigenfrequencies are real and symmetric around 0. The number of eigenfrequencies τ in $[0,\lambda]$, counted with their multiplicities, is equal to*

$$\left(\frac{\lambda}{2\pi}\right)^n \left(\mathrm{vol}\left(p^{-1}([0,1])\right) + \mathcal{O}\left(\lambda^{-1}\right)\right), \tag{14.2.5}$$

when $\lambda \to +\infty$. Here $p(x,\xi)$ denotes the principal symbol of $-\Delta$ (equal to the dual Riemannian metric on T^*X).

Back to the general case, we notice that the set of eigenfrequencies is symmetric under reflection in the imaginary axis and can be identified with the set of eigenvalues of the unbounded operator

$$\mathcal{P} = \begin{pmatrix} 0 & 1 \\ -\Delta & 2ia(x) \end{pmatrix} : H^1(X) \times H^0(X) \to H^1(X) \times H^0(X), \tag{14.2.6}$$

with domain $H^2 \times H^1$. In fact, the relation between (14.2.2) and

$$(\mathcal{P} - \tau) \begin{pmatrix} u_0 \\ u_1 \end{pmatrix} = 0, \tag{14.2.7}$$

is given by $u_0 = u$, $u_1 = \tau u$.

Let $\mathcal{P}_0$ be the operator in (14.2.6) with $a = 0$. Let $0 = \lambda_0 < \lambda_1 \le \cdots \to +\infty$ be the eigenvalues of $-\Delta$ and let $e_0, e_1, \ldots$ be a corresponding orthonormal basis of eigenfunctions. For $k \neq 0$, we put

$$f_k^{\pm} = \frac{1}{\sqrt{2}} \begin{pmatrix} \lambda_k^{-1/2} e_k \\ \pm e_k \end{pmatrix} \in H^1 \times H^0,$$

and for $k = 0$,

$$f_0^0 = \begin{pmatrix} e_0 \\ 0 \end{pmatrix}, \ f_0^1 = \begin{pmatrix} 0 \\ e_0 \end{pmatrix}.$$

Let $(e_0)^{\perp}$ denote the subspace L^2-orthogonal to $\mathbf{C}e_0$. Then $\lambda_1^{-1/2} e_1, \lambda_2^{-1/2} e_2, \ldots$ is an orthonormal basis in $H^1 \cap (e_0)^{\perp}$, equipped with the scalar product $[u|v] = (-\Delta u|v)$. It follows that $f_1, f_2, \ldots$ is an orthonormal basis in $\mathcal{H} = (H^1 \cap (e_0)^{\perp}) \times (H^0 \cap (e_0)^{\perp})$, with the scalar product

$$\left(\begin{pmatrix} u_0 \\ u_1 \end{pmatrix} \middle| \begin{pmatrix} \widetilde{u}_0 \\ \widetilde{u}_1 \end{pmatrix} \right)_{\mathcal{H}} = \left(\begin{pmatrix} -\Delta & 0 \\ 0 & 1 \end{pmatrix} \begin{pmatrix} u_0 \\ u_1 \end{pmatrix} \middle| \begin{pmatrix} \widetilde{u}_0 \\ \widetilde{u}_1 \end{pmatrix} \right)_{L^2 \times L^2} = [u_0|\widetilde{u}_0] + (u_1|\widetilde{u}_1).$$

Also $\mathcal{P}_0 f_k^{\pm} = \pm \lambda_k^{1/2} f_k^{\pm}$, so the restriction of $\mathcal{P}_0$ to $\mathcal{H}$ is self-adjoint with $f_k^{\pm}$ as a corresponding orthonormal basis of eigenvectors.

From this we see that $\mathcal{P}_0$ has a purely discrete spectrum, given by $0, \pm\sqrt{\lambda_k}$, $k \ge 1$; 0 is an eigenvalue of algebraic multiplicity two and the full spectral decomposition of $H^1 \times H^0$ is

$$H^1 \times H^0 = \mathbf{C} f_0^0 \oplus \mathbf{C} f_0^1 \oplus \mathcal{H}.$$

The algebraic multiplicity of $\pm\sqrt{\lambda_k}$ is equal to the multiplicity of λ_k as an eigenvalue of $-\Delta$, when $k \geq 1$.

When $\tau^2 \neq \lambda_k$ for all $k \geq 0$, the resolvent $(\mathcal{P}_0 - \tau)^{-1}$ is of the form

$$(\mathcal{P}_0 - \tau)^{-1} = (-\Delta - \tau^2)^{-1}(\mathcal{P}_0 + \tau). \tag{14.2.8}$$

Let

$$Q = \begin{pmatrix} 0 & 0 \\ 0 & 2ia \end{pmatrix}, \text{ so that } \mathcal{P} = \mathcal{P}_0 + Q.$$

Then we see that $Q(\mathcal{P}_0 - \tau)^{-1}$, $(\mathcal{P}_0 - \tau)^{-1}Q$ are compact on $H^1 \times H^0$ and $H^2 \times H^1$, respectively. The norms of these operators tend to zero when $\tau \to \infty$ in closed sectors that are disjoint from $\mathbf{R}$ away from 0. It follows that $\mathcal{P}$ has purely discrete spectrum and we define the multiplicity of an eigenfrequency to be equal to its algebraic multiplicity as an eigenvalue of $\mathcal{P}$. Thus the eigenfrequencies form a discrete set, confined to the region defined by (14.2.3), (14.2.4) and the elements have a natural multiplicity as defined above. (In [134] two further equivalent definitions of the multiplicity are given.) The set of eigenfrequencies, counted with their multiplicities, is invariant under the map $\tau \mapsto -\overline{\tau}$ of reflexion in the imaginary axis. We can then state a theorem which is due to Markus and Matseev [97]:

Theorem 14.2.2 *Under the assumptions above, the number of eigenfrequencies of Eq. (14.2.1) with real part in* $[0, \lambda]$, *counted with their multiplicity, is equal to*

$$\left(\frac{\lambda}{2\pi}\right)^n \left(\mathrm{vol}\left(p^{-1}([0, 1])\right) + \mathcal{O}\left(\lambda^{-1}\right)\right), \ \lambda \to +\infty. \tag{14.2.9}$$

Proof We reformulate the problem semi-classically. Let $h \asymp 1/\lambda$ and put $\tau = z/h$, so that $|z| \asymp 1$, when $|\tau| \asymp \lambda$. Then (14.2.2) becomes

$$(-h^2\Delta - z^2 + 2iahz)u = 0, \tag{14.2.10}$$

and we define the semi-classical eigenfrequencies in the obvious way and define the multiplicity of the semi-classical eigenfrequency z to be that of the eigenfrequency $\tau = z/h$. We then have to show that the number of semi-classical eigenfrequencies (counted with their multiplicity) with real part in $[0, b]$, where $b \asymp 1$, is equal to

$$\frac{1}{(2\pi h)^n}\left(\mathrm{vol}\left(p^{-1}([0, b])\right) + \mathcal{O}(h)\right), \ h \to 0. \tag{14.2.11}$$

The set of semi-classical eigenfrequencies is symmetric with respect to the imaginary axis, and we have a uniform bound on the number of eigenfrequencies on the imaginary axis, so equivalently, we have to show that the number of semi-classical eigenfrequencies with real part in $[a, b]$, where $a := -b$, is equal to

$$\frac{2}{(2\pi h)^n}\left(\mathrm{vol}\left(p^{-1}([0,b])\right)+\mathcal{O}(h)\right),\ h\to 0. \tag{14.2.12}$$

As in the non-semi-classical case, the eigenfrequencies appear as eigenvalues of a matrix operator, namely

$$\mathcal{P}_h=\begin{pmatrix}0 & 1\\ -h^2\Delta & 2iah\end{pmatrix}=\mathcal{P}_0+ihQ:\ H_h^1\times H^0\to H_h^1\times H^0, \tag{14.2.13}$$

where

$$Q=\begin{pmatrix}0 & 0\\ 0 & 2a\end{pmatrix},\quad \mathcal{P}_0=\begin{pmatrix}0 & 1\\ -h^2\Delta & 0\end{pmatrix}, \tag{14.2.14}$$

and it is here convenient to endow $H^k = H_h^k$ with the semi-classical norm

$$\|u\|_{H_h^k}=\|\langle hD\rangle^k u\|_{L^2}=\|(1-h^2\Delta)^{k/2}u\|_{L^2}.$$

For $\Im z \neq 0$, the resolvent of $\mathcal{P}_0$ is of the form

$$(\mathcal{P}_0-z)^{-1}=(-h^2\Delta-z^2)^{-1}(\mathcal{P}_0+z), \tag{14.2.15}$$

and using the semi-classical Sobolev norms, we see that

$$(\mathcal{P}_0-z)^{-1}=\mathcal{O}(1/|\Im z|):\ H_h^1\times H^0\to H_h^2\times H_h^1, \tag{14.2.16}$$

uniformly when $|z|\asymp 1$.

Again, $\mathcal{P}_0$ is self-adjoint on $\mathcal{H}=(H_h^1\times H^0)\cap\left((e_0)^\perp\times(e_0)^\perp\right)$ and has the spectral decomposition $\mathbf{C}f_0^0\oplus\mathbf{C}f_0^1\oplus(H_h^1\times H^0)\cap\left((e_0)^\perp\times(e_0)^\perp\right)$. The number of eigenvalues of $\mathcal{P}_0$ in $b+[-\delta,\delta\,]$ and in $-b+[-\delta,\delta\,]$ is $\mathcal{O}(\delta h^{-n})$ when $h<\delta\ll 1$ (and $b\asymp 1$), and we can construct a perturbation $\mathcal{P}_0^\delta$ of $\mathcal{P}_0$ which satisfies (14.1.7)–(14.1.9) with $I=[a,b\,]=[-b,b\,]$. We are therefore basically in the same situation as in the proof of Theorem 14.1.1 (with ϵ there equal to h), and the remainder of the proof is then the same. □

Notes We have established the Weyl law for the asymptotic distribution of the real parts of the eigenfrequencies for the damped wave equation, see Theorem 14.2.2, which is due to [97]. The problem of the distribution of the imaginary parts inside the spectral band is very deep. See [66, 73, 92, 118].

Distribution of Eigenvalues for Semi-classical Elliptic Operators with Small Random Perturbations, Results and Outline

15

In this chapter we will state a result asserting that for elliptic semi-classical (pseudo-)differential operators the eigenvalues are distributed according to Weyl's law "most of the time" in a probabilistic sense. The first three sections are devoted to the formulation of the results and in the last section we give an outline of the proof that will be carried out in Chaps. 16 and 17.

15.1 The Unperturbed Operator

In this section we describe the class of unperturbed operators.

Let X be a smooth compact manifold of dimension n. It is also possible to treat the case when $X = \mathbf{R}^n$ (cf. [137]), but we will concentrate on the compact manifold case.

Let P be an h-differential operator on X which in local coordinates takes the form

$$P = \sum_{|\alpha| \le m} a_\alpha(x; h)(hD)^\alpha, \tag{15.1.1}$$

where we use standard multiindex notation[1] and let $D = D_x = \frac{1}{i}\frac{\partial}{\partial x}$. We assume that the coefficients a_α are uniformly bounded in C^∞ for $h \in]0, h_0]$, $0 < h_0 \ll 1$. (We will also

[1] $|\alpha| = \|\alpha\|_{\ell^1}$, $(hD)^\alpha = h^{|\alpha|} D_{x_1}^{\alpha_1} \cdots D_{x_n}^{\alpha_n}$.

J. Sjöstrand, *Non-Self-Adjoint Differential Operators, Spectral Asymptotics and Random Perturbations*, Pseudo-Differential Operators 14,
https://doi.org/10.1007/978-3-030-10819-9_15

discuss the case when we only have some Sobolev space control of $a_0(x)$.) Assume

$$
\begin{aligned}
&a_\alpha(x;h) = a_\alpha^0(x) + \mathcal{O}(h) \text{ in } C^\infty,\\
&a_\alpha(x;h) = a_\alpha(x) \text{ is independent of } h \text{ for } |\alpha| = m.
\end{aligned}
\tag{15.1.2}
$$

Notice that this assumption is invariant under changes of local coordinates. The second part of (15.1.2) is for convenience only.

Also assume that P is elliptic in the classical sense, uniformly with respect to h:

$$
|p_m(x,\xi)| \geq \frac{1}{C}|\xi|^m, \tag{15.1.3}
$$

for some positive constant C, where the classical principal symbol

$$
p_m(x,\xi) = \sum_{|\alpha|=m} a_\alpha(x)\xi^\alpha \tag{15.1.4}
$$

is invariantly defined as a function on T^*X. It follows that $p_m(T^*X)$ is a closed cone in $\mathbf{C}$ and we assume that

$$
p_m(T^*X) \neq \mathbf{C}. \tag{15.1.5}
$$

If $z_0 \in \mathbf{C} \setminus p_m(T^*X)$, we see that $\lambda z_0 \notin \Sigma(p)$ if $\lambda \geq 1$ is sufficiently large and fixed, where $\Sigma(p) := p(T^*X)$ and p is the semiclassical principal symbol

$$
p(x,\xi) = \sum_{|\alpha|\leq m} a_\alpha^0(x)\xi^\alpha. \tag{15.1.6}
$$

Actually, (15.1.5) can be replaced by the weaker condition that $\Sigma(p) \neq \mathbf{C}$.

Standard elliptic theory shows that $P - z : H^m(X) \to L^2(X)$ is a holomorphic family of Fredholm operators, where $H^m(X)$ denotes the usual Soboloev space of order m. In particular, the index is independent of $z \in \mathbf{C}$ and the assumption (15.1.5) implies that $P - z$ is bijective for at least one value of z. Hence the index is zero and $P - z$ is bijective for z away from a discrete set. (When we only assume that $\Sigma(p) \neq \mathbf{C}$, we need to use the assumption that $h > 0$ is small enough.) It follows that if we consider P as an unbounded operator $L^2(X) \to L^2(X)$ with domain $\mathcal{D}(P) = H^m(X)$, then P has purely discrete spectrum.

In the case of multiplicative random perturbations we will need the symmetry assumption

$$
P = \Gamma P^* \Gamma, \tag{15.1.7}
$$

where P^* denotes the formal complex adjoint of P in $L^2(X, dx)$, with dx denoting some fixed smooth positive density of integration and Γ is the antilinear operator of complex conjugation: $\Gamma u = \overline{u}$. Notice that the left-hand side in (15.1.7) is equal to the "real" transpose P^{t}, defined by

$$\int (Pu)v dx = \int u(P^{\mathrm{t}}v)dx, \quad u, v \in C^\infty(X).$$

and that the assumption implies that

$$p(x, -\xi) = p(x, \xi). \tag{15.1.8}$$

15.2 The Random Perturbation

We will perturb our operator with a small constant factor times a potential which is a random linear combination of oscillatory functions. For convenience, the latter are chosen to be eigenfunctions of an elliptic differential operator.

Let $\widetilde{R}$ be an elliptic differential operator on X of order 2, such that for any choice of smooth local coordinates,

$$h^2\widetilde{R} = \sum (hD_{x_j})^* r_{j,k}(x) hD_{x_k}, \tag{15.2.1}$$

where the star indicates that we take the adjoint with respect to some fixed positive smooth density on X. We assume that the $r_{j,k}$ are smooth and independent of h and also that $\widetilde{R} \geq 0$ in the operator sense. Then $h^2\widetilde{R}$ is essentially self-adjoint with domain $H^2(X)$ and it has an orthonormal basis of eigenfunctions, $\epsilon_j \in L^2(X)$, $j = 1, 2, \dots$, with corresponding eigenvalues $\mu_j^2 = (h\mu_j^0)^2$, where $0 \leq \mu_j^0 \nearrow +\infty$. By Weyl's law for the large eigenvalues of a positive elliptic operator of order 2, we know that

$$\#\{j;\ \mu_j^0 \leq \lambda\} = (1 + o(1))(2\pi)^{-n} C_{\mathrm{W}} \lambda^n, \quad \lambda \to +\infty, \tag{15.2.2}$$

where C_{W} is the symplectic volume of $\{(x, \xi) \in T^*X;\ r(x, \xi) \leq 1\}$. Here

$$r(x, \xi) = \sum_{j,k} r_{j,k}(x) \xi_j \xi_k \tag{15.2.3}$$

is the principal symbol of $\widetilde{R}$ and also the semi-classical principal symbol of $h^2\widetilde{R}$.

Our random perturbation will be of the form δq_ω, where $\delta > 0$ is a small parameter and

$$q_\omega(x) = \sum_{0 < \mu_k \leq L} \alpha_k(\omega) \epsilon_k(x), \quad |\alpha|_{\mathbf{C}^D} \leq R. \tag{15.2.4}$$

Here we choose $L = L(h)$, $R = R(h)$ in the interval

$$\begin{aligned} h^{\frac{-3n}{s-\frac{n}{2}-\epsilon}} \ll L \leq Ch^{-M}, \quad M \geq \frac{3n}{s-\frac{n}{2}-\epsilon}, \\ \frac{1}{C}h^{-(\frac{n}{2}+\epsilon)M-\frac{3n}{2}} \leq R \leq Ch^{-\widetilde{M}}, \quad \widetilde{M} \geq \frac{3n}{2}+(\frac{n}{2}+\epsilon)M, \end{aligned} \tag{15.2.5}$$

for some $\epsilon \in]0, s-\frac{n}{2}[$, $s > \frac{n}{2}$, so by Weyl's law for the large eigenvalues of elliptic self-adjoint operators, the number of terms D in (15.2.4) is of the order of magnitude $(L/h)^n$. The small parameter δ will be of the form $\delta = \tau_0 h^{N_2-n}$, $0 < \tau_0 \leq h^2$, where $N_2 \geq N_2(n, s, \epsilon, M, \widetilde{M})$ is a sufficiently large constant.

The $\alpha_j(\omega)$ are random variables with a joint probability distribution

$$\mathbf{P}(d\alpha) = C(h)e^{\Phi(\alpha;h)}L(d\alpha), \tag{15.2.6}$$

where for some $N_4 > 0$,

$$|\nabla_\alpha \Phi| = \mathcal{O}(h^{-N_4}), \tag{15.2.7}$$

and $L(d\alpha)$ is the Lebesgue measure on $\mathbf{C}^D$. ($C(h)$ is the normalizing constant, ensuring that the probability of $B_{\mathbf{C}^D}(0, R)$ is equal to 1.)

15.3 The Result

Let $P_\delta = P + \delta q_\omega$, $\delta = \tau_0 h^{N_2-n}$, where q_ω is as in (15.2.4)–(15.2.7). Let

$$N_3 = (M+1)n, \ N_5 = N_4 + \widetilde{M}, \ N_6 = \max(N_3, N_5).$$

Let $\Omega \Subset \mathbf{C}$ b a fixed open simply connected set with smooth boundary. In Sect. 16.4 below (see Proposition 16.4.2) we introduce a continuous subharmonic function ϕ on Ω satisfying (16.4.6):

$$\frac{\Delta\phi}{2\pi}L(dz) = p_*(dxd\xi), \tag{15.3.1}$$

where the right-hand side denotes the direct image under p of the symplectic volume element.

Let $\Gamma \Subset \Omega$ be a Lipschitz domain as in Chap. 12 of constant scale $r = \sqrt{h}$. Let

$$G(w, \Gamma) = \int_{\partial\Gamma} \frac{h}{h+|w-z|^2}\frac{|dz|}{\sqrt{h}}. \tag{15.3.2}$$

Theorem 15.3.1 *Let $\widetilde{\delta} > 0$. Then with probability*

$$\geq 1 - \mathcal{O}(1)h^{-N_6-n}\left(\ln\frac{1}{\tau_0}\left(\ln\frac{1}{h}\right)^2\int_\Omega G(w,\Gamma)\Delta\phi(w)L(dw) + |\partial\Gamma|h^{\frac{1}{2}}\right)e^{-\frac{h^{-\widetilde{\delta}}}{\mathcal{O}(1)}},$$

the number of eigenvalues of P_δ in Γ (counted with their algebraic multiplicities) satisfies

$$\begin{aligned}&\left|\#(\sigma(P_\delta)\cap\Gamma) - \frac{1}{(2\pi h)^n}\mathrm{vol}\,(p^{-1}(\Gamma))\right| \leq \\ &\frac{\mathcal{O}(1)}{h^n}\left(h^{-\widetilde{\delta}}\ln\frac{1}{\tau_0}\left(\ln\frac{1}{h}\right)^2\int_\Omega G(w,\Gamma)\Delta\phi(w)L(dw) + |\partial\Gamma|h^{\frac{1}{2}}\right).\end{aligned} \tag{15.3.3}$$

Remark 15.3.2 Actually, we shall prove the theorem for the slightly more general operators, obtained by replacing P by $P_0 = P + \delta_0(h^{\frac{n}{2}}q_1^0 + q_2^0)$, for

$$0 \leq \delta_0 \leq h^2, \quad \|q_1^0\|_{H_h^s}, \|q_2^0\|_{H^s} \leq 1, \tag{15.3.4}$$

where $\|\cdot\|_{H_h^s}$ is the natural semi-classical norm on H^s that we shall review in Sect. 16.2. If we weaken (15.3.4) to

$$0 \leq \delta_0 \ \leq h^{2-\vartheta}, \tag{15.3.5}$$

for some fixed $\vartheta \ \in]0, 1/2[$, then we get the slightly weaker statement in the theorem obtained by replacing $|\partial\Gamma|h^{\frac{1}{2}}$ with $|\partial\Gamma|h^{\frac{1}{2}-\vartheta}$ both in the lower bound on the probability and in the estimate (15.3.3).

As in [56], we also have a result valid simultaneously for a family $\mathcal{C}$ of domains $\Gamma \subset \Omega$ satisfying the assumptions of Theorem 15.3.1 uniformly in the natural sense: With a probability as in Theorem 15.3.1 and after replacing h^{-N_6-n} there by $h^{-N_6-n-1/2}$ (or after increasing N_6 by $1/2$) the estimates (15.3.3) hold simultaneously for all $\Gamma \in \mathcal{C}$.

Remark 15.3.3 When $\widetilde{R}$ has real coefficients, we may assume that the eigenfunctions ϵ_j are real. Then, as will follow from the proofs below, we may restrict α in (15.2.4) to be in $\mathbf{R}^D$ so that q_ω is real, still with $|\alpha| \leq R$, and change $C(h)$ in (15.2.6) so that P becomes a probability measure on $B_{\mathbf{R}^D}(0, R)$. Then Theorem 15.3.1 remains valid.

Remark 15.3.4 The assumption (15.1.7) cannot be completely eliminated. Indeed, let $P = hD_x + g(x)$ on $\mathbf{T} = \mathbf{R}/(2\pi\mathbf{Z})$, where g is smooth and complex valued. Then (cf. Hager [54] and Chap. 3) the spectrum of P is contained in the line $\Im z = \int_0^{2\pi}\Im g(x)dx/(2\pi)$. This line will vary only very little under small multiplicative perturbations of P, so Theorem 15.3.1 cannot hold in this case.

In order to have a better understanding of the statement in Theorem 15.3.1, we need to compare the Weyl term

$$W = \frac{1}{2\pi} \int_{\Gamma} \Delta\phi(z) L(dz) \tag{15.3.6}$$

and the main contribution to the remainder in (15.3.3), given by

$$R = \int_{\Omega} G(w, \Gamma) \Delta\phi(w) L(dw). \tag{15.3.7}$$

As we shall see in Sect. 17.5, if $\partial\Gamma$ satisfies the regularity assumption (17.5.25):

$$|\partial\Gamma \cap D(w, R)| \leq \mathcal{O}(1)\sqrt{h}\left(\frac{R}{\sqrt{h}}\right)^{1+\kappa}, \ R \geq \sqrt{h},$$

where $|\gamma|$ denotes the length of the curve γ and $0 \leq \kappa \leq 1$, then we have (17.5.26) as well as the improvement away from Γ in Remark 17.5.1, (17.5.27), (17.5.28):

$$\begin{aligned} &G(w, \Gamma; h) \\ &= \begin{cases} \mathcal{O}(1)\left(1 + \frac{d(w,\partial\Gamma)}{\sqrt{h}}\right)^{\kappa-1} & \text{in general,} \\ \frac{\operatorname{diam}(\Gamma)}{\sqrt{h}}\left(1 + \frac{d(w,\partial\Gamma)}{\sqrt{h}}\right)^{\kappa-2}, & \text{when } d(w, \partial\Gamma) \geq \operatorname{diam}(\Gamma). \end{cases} \end{aligned} \tag{15.3.8}$$

Here $d(w, \partial\Gamma)$ denotes the distance from w to $\partial\Gamma$. From this estimate we expect—unless Γ is very small or if a considerable fraction of $p_*(dxd\xi)$ is concentrated to a $\partial\Gamma$—that $R \ll W$ and (15.3.3) indeed gives the asymptotic behaviour of the number of eigenvalues in Γ for moderate values of $\widetilde{\delta}$ and τ_0.

In the remainder of this section we shall discuss three closely related examples when (17.5.25) holds with $\kappa = 0$, so that

$$\begin{aligned} &G(w, \Gamma; h) \\ &= \mathcal{O}(1) \begin{cases} \left(1 + \frac{d(w,\partial\Gamma)}{\sqrt{h}}\right)^{-1} & \text{in general,} \\ \frac{\operatorname{diam}(\Gamma)}{\sqrt{h}}\left(1 + \frac{d(w,\partial\Gamma)}{\sqrt{h}}\right)^{-2}, & \text{when } d(w, \partial\Gamma) \geq \operatorname{diam}(\Gamma). \end{cases} \end{aligned} \tag{15.3.9}$$

Example 15.3.5 Let Γ be a fixed Lipschitz domain as in Chap. 12, independent of h with scale 1 (in the sense of that chapter), so (15.3.9) holds. Assume that

$$\operatorname{vol} p^{-1}(\partial\Gamma + D(0, t)) = \mathcal{O}(t^{\frac{1}{N_0}}), \ t \to 0, \tag{15.3.10}$$

for some $N_0 \geq 1$ and notice that this holds with $N_0 = 1$ when

$$p(\rho) \in \partial\Gamma \Longrightarrow d\Re p(\rho),\ d\Im p(\rho) \text{ are linearly independent.}$$

Then it is easy to check (as we shall do prior to Proposition 18.4.2) that

$$\int_\Gamma G(w,\Gamma)\Delta\phi(w)L(dw) = \mathcal{O}\left(h^{1/(2N_0)}\right),$$

where for notational simplicity we have suppressed a factor $\ln(1/h)$ on the right, when $N_0 = 1$. Theorem 15.3.1 now tells us that with probability

$$\geq 1 - \mathcal{O}(1)h^{-N_6-n}\left(\ln\frac{1}{\tau_0}\right)\left(\ln\frac{1}{h}\right)^2 h^{\frac{1}{2N_0}} e^{-\frac{h^{-\widetilde{\delta}}}{\mathcal{O}(1)}}, \tag{15.3.11}$$

we have

$$\begin{aligned} &\left|\#(\sigma(P_\delta)\cap\Gamma) - \frac{1}{(2\pi h)^n}\mathrm{vol}\,(p^{-1}(\Gamma))\right| \leq \\ &\frac{\mathcal{O}(1)}{h^n}h^{-\widetilde{\delta}}\left(\ln\frac{1}{\tau_0}\right)\left(\ln\frac{1}{h}\right)^2 h^{\frac{1}{2N_0}}, \end{aligned} \tag{15.3.12}$$

which is a significant estimate if $\ln 1/\tau_0 = \mathcal{O}(h^{-\beta})$, where $0 < \beta < 1/(2N_0)$, and we choose $\widetilde{\delta}$ small enough so that $\beta + \widetilde{\delta} < 1/(2N_0)$.

Example 15.3.6 Let $\widetilde{\Omega} \Subset p(T^*X) \cap \Omega$ be open and with no critical value of p in its closure:

$$p(\rho) \in \overline{\widetilde{\Omega}} \Longrightarrow d\Re p(\rho),\ d\Im p(\rho) \text{ are linearly independent.}$$

Then $\Delta\phi L(dz) \asymp L(dz)$ uniformly on $\widetilde{\Omega}$. Let Γ_0 be a bounded closed Lipschitz domain as in Chap. 12, independent of h and with scale 1 (in the sense of that chapter). We fix some $z_0 \in \widetilde{\Omega}$ and put $\Gamma := z_0 + \alpha\Gamma_0 \subset \widetilde{\Omega}$, where $\sqrt{h} \leq \alpha \ll 1$. Clearly $\partial\Gamma$ satisfies (17.5.25) with $\kappa = 0$, so we have (15.3.9) uniformly with respect to α.

We see that

$$W \asymp \alpha^2. \tag{15.3.13}$$

In order to estimate R, assume for simplicity that $z_0 = 0$. Then

$$
\begin{aligned}
R &= \int_{\Omega} G(w,\Gamma)\Delta\phi(w)L(dw) \lesssim \frac{\alpha}{\sqrt{h}}h + \int_{\widetilde{\Omega}} G(w,\Gamma)\Delta\phi(w)L(dw) \\
&\lesssim \alpha\sqrt{h} + \int_{\widetilde{\Omega}\cap\{d(w,\alpha\partial\Gamma_0)\leq\alpha\}} \left(1+\frac{d(w,\alpha\partial\Gamma_0)}{\sqrt{h}}\right)^{-1} L(dw) \\
&\quad + \int_{\widetilde{\Omega}\cap\{d(w,\alpha\partial\Gamma_0)\geq\alpha\}} \frac{\alpha}{\sqrt{h}}\left(1+\frac{d(w,\alpha\Gamma_0)}{\sqrt{h}}\right)^{-2} L(dw).
\end{aligned}
\tag{15.3.14}
$$

Here "$\lesssim$" means "$\leq \mathcal{O}(1)$ times". The first integral in the last term is equal to

$$
\begin{aligned}
&\alpha^2 \int_{d(\widetilde{w},\partial\Gamma_0)\leq 1} \left(1+\frac{\alpha d(\widetilde{w},\partial\Gamma_0)}{\sqrt{h}}\right)^{-1} L(d\widetilde{w}) \\
&\lesssim \alpha^2 \int_{\{d(\widetilde{w},\partial\Gamma_0)\leq \frac{\sqrt{h}}{\alpha}\}} L(d\widetilde{w}) + \alpha\sqrt{h}\int_{\{1\geq d(\widetilde{w},\partial\Gamma_0)\geq\frac{\sqrt{h}}{\alpha}\}} \frac{1}{d(\widetilde{w},\partial\Gamma_0)} L(d\widetilde{w}) \\
&\lesssim \alpha\sqrt{h} + \alpha\sqrt{h}\ln\frac{\alpha}{\sqrt{h}} = \alpha\sqrt{h}\left(1+\ln\frac{\alpha}{\sqrt{h}}\right).
\end{aligned}
$$

The second integral in the last member of (15.3.14) is equal to

$$
\begin{aligned}
\frac{\alpha}{\sqrt{h}}\alpha^2 &\int_{\frac{1}{\alpha}\widetilde{\Omega}\cap\{d(\widetilde{w},\partial\Gamma_0)\geq 1\}} \left(1+\frac{\alpha d(\widetilde{w},\partial\Gamma_0)}{\sqrt{h}}\right)^{-2} L(d\widetilde{w}) \\
&\lesssim \frac{\alpha^3}{\sqrt{h}} \int_{\frac{1}{\alpha}\widetilde{\Omega}\cap\{d(\widetilde{w},\partial\Gamma_0)\geq 1\}} \frac{h}{\alpha^2 d(\widetilde{w},\partial\Gamma_0)^2} L(d\widetilde{w}) \\
&\qquad\qquad \lesssim \alpha\sqrt{h}\ln\frac{1}{\alpha}.
\end{aligned}
$$

Thus,

$$
R \leq \mathcal{O}(1)\alpha\sqrt{h}\left(1+\ln\frac{\alpha}{\sqrt{h}}+\ln\frac{1}{\alpha}\right) = \mathcal{O}(1)\alpha\sqrt{h}\left(1+\ln\frac{1}{\sqrt{h}}\right).
$$

In conclusion, we have

$$
R \lesssim h^{\delta} W \text{ when } \frac{\alpha}{\sqrt{h}} \geq h^{-\delta}(1+\ln\frac{1}{h}),
$$

and the estimate (15.3.3) gives spectral asymptotics, provided that $\widetilde{\delta} < \delta$ and τ_0 is not too small, so that $\ln 1/\tau_0 = \mathcal{O}(h^{-\widehat{\delta}})$ for some $\widehat{\delta} < \delta - \widetilde{\delta}$.

Example 15.3.7 Let $z_0 \in \partial\Sigma$ and assume for simplicity that $p^{-1}(z_0)$ consists of just one point ρ_0. Our final assumptions will imply that

$$\{p, \{p, \overline{p}\}\} \neq 0, \tag{15.3.15}$$

at ρ_0. $d\Re p$ and $d\Im p$ have to be collinear at the point ρ_0 (since otherwise z_0 would belong to the interior of Σ) and in particular $\{p, \overline{p}\}(\rho_0) = 0$. Without loss of generality, we may assume that

$$dp(\rho_0) = d\Re p(\rho_0) \neq 0. \tag{15.3.16}$$

We may also assume for notational reasons that $z_0 = 0$. Since 0 belongs to $\partial\Sigma$, we see that one of the following holds:

1) $\Im p(\rho) \geq 0$ for all $\rho \in (\Re p)^{-1}(0) \cap \mathrm{neigh}\,(\rho_0)$.
2) $\Im p(\rho) \leq 0$ for all $\rho \in (\Re p)^{-1}(0) \cap \mathrm{neigh}\,(\rho_0)$.

Suppose, in order to fix the ideas, that we are in the first case. We then make the following generic assumption (which implies (15.3.15) and is equivalent to that condition when $n = 1$).

$$\Im p|_{(\Re p)^{-1}(0)\cap \mathrm{neigh}\,(\rho_0)} \text{ has a nondegenerate minimum at } \rho_0. \tag{15.3.17}$$

Then for $t \in \mathrm{neigh}\,(0, \mathbf{R})$, $\Im p|_{(\Re p)^{-1}(t)\cap \mathrm{neigh}\,(\rho_0)}$ has a nondegenerate minimum at a point $\rho(t)$ depending smoothly on t with $\rho(0) = \rho_0$. We have

$$\Im p(\rho) - \Im p(\rho(\Re p(\rho))) \asymp |\rho - \rho(\Re p(\rho))|^2 \tag{15.3.18}$$

for $\rho \in \mathrm{neigh}\,(\rho_0, T^*X)$. If $g(t) := \Im p(\rho(t))$, then near $z = 0$, Σ is given by $\Im z \geq g(\Re z)$ (and from (15.3.16) we see that $g'(0) = 0$).

Near ρ_0 the hypersurfaces $(\Re p)^{-1}(t)$ carry a natural Liouville measure[2] and the direct image of this measure under the map

$$(\Re p)^{-1}(\Re z) \ni \rho \longmapsto \Im p(\rho)$$

is of the form

$$f(z)(\Im z - g(\Re z))_+^{n-\frac{3}{2}} d\Im z, \text{ where } f > 0 \text{ is continuous,}$$

[2]When viewed as an $2n-1$ form λ on $(\Re p)^{-1}(t)$, it is given by $\lambda \wedge d\Re p = dx_1 \wedge \cdots \wedge dx_n \wedge d\xi_1 \wedge \cdots \wedge d\xi_n =$ the symplectic volume form.

since the Liouville measure of the set $\Im p(\rho) - \Im p(\rho(\Re z)) \leq r$ in $(\Re p)^{-1}(\Re z)$ is $\asymp r^{n-1/2}$. It follows that

$$p_*(dxd\xi) = f(z)(\Im z - g(\Re z))_+^{n-\frac{3}{2}} L(dz) \tag{15.3.19}$$

near $z = 0$. Notice that when $n = 1$ the corresponding density tends to $+\infty$ at the boundary of Σ, while in the case $n \geq 2$ it tends to 0.

We will compare $R(\Gamma)$ and $W(\Gamma)$ for small domains centered at points of the boundary of Σ. We may assume that $g \equiv 0$. Let $\Gamma = \Gamma(\alpha)$ be the domain

$$\Gamma(\alpha) = \{z \in \mathbf{C};\ |\Re z| < 2\alpha,\ 0 \leq \Im z < \alpha\},\ \text{for } \sqrt{h} \leq \alpha \ll 1. \tag{15.3.20}$$

We have

$$W \asymp \iint_{\substack{-\alpha<x<\alpha \\ 0\leq y<\alpha}} y^{n-\frac{3}{2}} dxdy \asymp \alpha^{n+\frac{1}{2}}. \tag{15.3.21}$$

We next look at R. Since the direct image measure vanishes in the lower half-plane (near 0) we may restrict our attention to $w = x + iy$ with $y \geq 0$. Since our domains are symmetric around the imaginary axis, it suffices to consider the case $x \geq 0$. In the quarter-plane $x, y \geq 0$ we have, by (15.3.9),

$$G(w, \Gamma) \lesssim \left(1 + \frac{d(w, \partial\Gamma)}{\sqrt{h}}\right)^{-1},\ \text{when } |x| \leq 2\alpha,\ y \leq 2\alpha,$$

$$G(w, \Gamma) \lesssim \frac{\alpha}{\sqrt{h}}\left(1 + \frac{|y - \alpha|}{\sqrt{h}}\right)^{-2},\ \text{when } 2\alpha \leq y \geq x,$$

$$G(w, \Gamma) \lesssim \frac{\alpha}{\sqrt{h}}\left(1 + \frac{|x - \alpha|}{\sqrt{h}}\right)^{-2},\ \text{when } 2\alpha \leq x \geq y.$$

The contribution to R from the region inside Γ is bounded by a constant times the contribution from the exterior region, so from now on we shall only consider the contribution from the region $x \geq 0$, $y \geq 0$, $\max(x, y) \geq \alpha$. Further, when $n \geq 2$, the contribution from the region $y \geq \max(\alpha, x)$ dominates the one from $x \geq \max(\alpha, y)$, while for $n = 1$ the latter contribution is dominant. When saying this, we tacitly restrict our attention to a neighborhood $\widetilde{\Omega}$ of 0 and observe that the contribution from $\Omega \setminus \widetilde{\Omega}$ is

$$\mathcal{O}(1)\alpha\sqrt{h}.$$

Case 1: $n = 1$ In this case

$$R \lesssim \iint_{2\alpha \geq x \geq \max(\alpha, y)} \left(1 + \frac{x-\alpha}{\sqrt{h}}\right)^{-1} y^{-\frac{1}{2}} dx dy$$
$$+ \iint_{\max(2\alpha, y) \leq x \leq 1} \frac{\alpha}{\sqrt{h}} \left(1 + \frac{x-\alpha}{\sqrt{h}}\right)^{-2} y^{-\frac{1}{2}} dx dy.$$

Here the first integral can be estimated by

$$\iint_{\substack{0 \leq y \leq 2\alpha \\ \alpha \leq x \leq 2\alpha}} \left(1 + \frac{x-\alpha}{\sqrt{h}}\right)^{-1} y^{-\frac{1}{2}} dx dy$$
$$\lesssim \sqrt{\alpha} \int_{\alpha}^{2\alpha} \left(1 + \frac{x-\alpha}{\sqrt{h}}\right)^{-1} dx = \sqrt{\alpha h} \ln \left(1 + \frac{\alpha}{\sqrt{h}}\right).$$

The second integral is equal to

$$\iint_{0 \leq y \leq 2\alpha \leq x \leq 1} \frac{\alpha}{\sqrt{h}} \left(1 + \frac{x-\alpha}{\sqrt{h}}\right)^{-2} y^{-\frac{1}{2}} dx dy$$
$$+ \iint_{2\alpha \leq y \leq x \leq 1} \frac{\alpha}{\sqrt{h}} \left(1 + \frac{x-\alpha}{\sqrt{h}}\right)^{-2} y^{-\frac{1}{2}} dx dy$$
$$= \frac{\alpha}{\sqrt{h}} \left(C\sqrt{\alpha h} \left[-\left(1 + \frac{x-\alpha}{\sqrt{h}}\right)^{-1} \right]_{x=2\alpha}^{1} \right.$$
$$\left. + \int_{2\alpha}^{1} \sqrt{h} \left[-\left(1 + \frac{x-\alpha}{\sqrt{h}}\right)^{-1} \right]_{y}^{1} y^{-\frac{1}{2}} dy \right)$$
$$\lesssim \alpha^{\frac{3}{2}} \left(1 + \frac{\alpha}{\sqrt{h}}\right)^{-1} + \alpha \int_{2\alpha}^{1} \left(1 + \frac{y-\alpha}{\sqrt{h}}\right)^{-1} y^{-\frac{1}{2}} dy$$
$$\lesssim \sqrt{\alpha h} + \alpha\sqrt{h} \int_{2\alpha}^{1} y^{-\frac{3}{2}} dy \lesssim \sqrt{\alpha h}.$$

Thus

$$R \lesssim \sqrt{\alpha h} \left(1 + \ln \left(1 + \frac{\alpha}{\sqrt{h}}\right)\right),$$

while $W \asymp \alpha^{3/2}$.

We conclude in this case that $R \lesssim h^{\delta} (\ln \frac{1}{h}) W$ if $\alpha \geq h^{\frac{1}{2}-\delta}\sqrt{h}$. The scaling argument in Chap. 6 could lead to improvements.

Case 2: $n \geq 2$ In this case

$$R \lesssim \iint_{\substack{2\alpha \geq y \geq \max(\alpha,x) \\ x \geq 0}} \left(1 + \frac{y-\alpha}{\sqrt{h}}\right)^{-1} y^{n-\frac{3}{2}} dx dy$$
$$+ \iint_{\max(2\alpha,x) \leq y \leq 1} \frac{\alpha}{\sqrt{h}} \left(1 + \frac{y-\alpha}{\sqrt{h}}\right)^{-2} y^{n-\frac{3}{2}} dx dy.$$

In the domain of integration of the first integral, we have $y \asymp \alpha$, $x \leq 2\alpha$, so this integral is

$$\lesssim 2\alpha\alpha^{n-\frac{3}{2}} \int_{\alpha}^{2\alpha} \left(1 + \frac{y-\alpha}{\sqrt{h}}\right)^{-1} dy \lesssim \alpha^{n-\frac{1}{2}} \sqrt{h} \left[\ln\left(1 + \frac{y-\alpha}{\sqrt{h}}\right)\right]_{\alpha}^{2\alpha}$$
$$\lesssim \alpha^{n-\frac{1}{2}} \sqrt{h} \ln\left(1 + \frac{\alpha}{\sqrt{h}}\right) \lesssim \alpha\sqrt{h}\alpha^{\frac{1}{2}} \ln\left(1 + \frac{\alpha}{\sqrt{h}}\right),$$

where in the last estimate we used that $n \geq 2$.

In the domain of integration of the second integral, we have

$$1 + \frac{y-\alpha}{\sqrt{h}} \asymp \frac{y}{\sqrt{h}},$$

so this integral is

$$\lesssim \int_0^1 \int_{\max(2\alpha,x)}^1 \frac{\alpha}{\sqrt{h}} \frac{h}{y^2} y^{n-\frac{3}{2}} dy dx \lesssim \alpha\sqrt{h} \int_0^1 \int_{\max(2\alpha,x)}^1 y^{n-\frac{7}{2}} dy dx.$$

When $n \geq 3$ this expression is clearly $\lesssim \alpha\sqrt{h}$. When $n = 2$ it is

$$\lesssim \alpha\sqrt{h} \int_0^2 (\alpha + x)^{-\frac{1}{2}} \lesssim \alpha\sqrt{h} \text{ also.}$$

Thus,

$$R \lesssim \left(1 + \alpha^{\frac{1}{2}} \ln \frac{\alpha}{\sqrt{h}}\right) \alpha\sqrt{h} \lesssim \alpha h^{\frac{1}{2}} \ln \frac{1}{h},$$

and we compare this with (15.3.21). It follows that if

$$h^{\frac{1}{2}-\delta} \ln \frac{1}{h} \lesssim \alpha^{n-1/2}, \ 0 < \delta \ll 1, \tag{15.3.22}$$

then

$$R \lesssim h^{\delta} W,$$

so Theorem 15.3.1 gives an asymptotic formula for the number of eigenvalues in Γ when $\alpha \ll 1$ fulfills (15.3.22).

15.4 Comparison of Theorems 15.3.1 and 1.1 in [137] and [139]

In the two indicated papers we treated respectively the case of h-pseudodifferential operators on $\mathbf{R}^n$ and of h-differential operators on a smooth compact manifold. The second work is a simple adaptation of the first, so we refer to the first one even though we shall compare with the result on manifolds. (The formulations are almost identical.) An extra assumption in those papers is that

$$\int_{D(z,r)} \Delta\phi(z)L(dz) = \mathcal{O}(r^{2\kappa}),\ 0 \le r \ll 1,\ z \in \Omega, \tag{15.4.1}$$

for some $\kappa\ \in]0,1]$. The class of random perturbations is the same there as in this chapter, except for the fact that some parameter ranges are slightly wider in [137, 139] but depending on κ. We here restrict the attention to the random perturbations of the present chapter.

The most interesting is probably to compare the right-hand sides of (1.14) in [137] and (15.3.3). To make things simple, let us notice that for every $\delta_1 > 0$ there exists $\delta_2 > 0$ such that if $\delta \ge e^{-h^{-\delta_2}}$ is as in Theorem 15.3.1, then with probability $\ge 1 - \mathcal{O}(1)e^{-h^{-\delta_2}}$, we have

$$\begin{aligned}&\left|\#(\sigma(P_\delta)\cap\Gamma) - \frac{1}{(2\pi h)^n}\mathrm{vol}\,(p^{-1}(\Gamma))\right| \le \\ &\frac{\mathcal{O}(1)}{h^n}\left(h^{-\delta_1}\int_\Omega G(w,\Gamma)\Delta\phi(w)L(dw) + |\partial\Gamma|h^{\frac{1}{2}}\right).\end{aligned} \tag{15.4.2}$$

Similarly, in Theorem 1.1 in [137] (and where Γ is a fixed domain with smooth boundary), for every $\delta_1 > 0$, there exists $\delta_2 > 0$ such that if $\delta \ge e^{-h^{-\delta_2}}$ is as in that theorem, then with probability $\ge 1 - \mathcal{O}(e^{-h^{-\delta_2}}/r)$, we have

$$\begin{aligned}&\left|\#(\sigma(P_\delta)\cap\Gamma) - \frac{1}{(2\pi h)^n}\mathrm{vol}\,(p^{-1}(\Gamma))\right| \le \\ &\frac{\mathcal{O}(1)}{h^n}\left(\frac{h^{\kappa-\delta_1}}{r} + r + \ln\left(\frac{1}{r}\right)\underbrace{\int_{\partial\Gamma+D(0,r)}\Delta\phi(w)L(dw)}_{=:F(r)}\right),\end{aligned} \tag{15.4.3}$$

for any $0 < r \ll 1$.

We neglect the factors $h^{-\delta_1}$ and the ln factor in (15.4.2), (15.4.3). Then we are reduced to comparing

$$A := \int_{\Omega} G(w, \Gamma)\Delta\phi L(dw) + h^{\frac{1}{2}} \tag{15.4.4}$$

and

$$B := \inf_{0<r\ll 1} \left(\frac{h^{\kappa}}{r} + r + F(r) \right). \tag{15.4.5}$$

Since Γ is a fixed domain with smooth boundary, the regularity assumption after (15.3.7) holds with κ there equal to 0 (not to be confounded with the κ in (15.4.1), (15.4.5)!), and by (15.3.8),

$$G(w, \Gamma) = \mathcal{O}(1) \left(1 + \frac{d(w, \partial\Gamma)}{\sqrt{h}} \right)^{-1}.$$

Hence

$$A \leq \mathcal{O}(1) \left(\int_0^{r_0} \left(1 + \frac{t}{\sqrt{h}} \right)^{-1} dF(t) + h^{\frac{1}{2}} \right). \tag{15.4.6}$$

Covering $\partial\Gamma + D(0, r)$ with $\asymp 1/r$ discs of radius $2r$ and using (15.4.1), we get

$$F(r) = \mathcal{O}(r^{2\kappa - 1}). \tag{15.4.7}$$

We end the section with a small calculation that indicates that in general A is smaller than B (and hence that Theorem 15.3.1 is sharper than the corresponding results in [137, 139]). For that we assume in addition that

$$F(r) \asymp r^{\alpha}, \tag{15.4.8}$$

for some $0 < \alpha \geq 2\kappa - 1$. Then

$$B \asymp \begin{cases} h^{\frac{\kappa}{2}}, & \text{if } \alpha \geq 1, \\ h^{\frac{\alpha\kappa}{\alpha+1}}, & \text{if } \alpha < 1. \end{cases} \tag{15.4.9}$$

The integral in (15.4.6) is equal to

$$\begin{aligned}
\left[\left(1+\frac{t}{\sqrt{h}}\right)^{-1}F(t)\right]_0^{r_0} &+ \frac{1}{\sqrt{h}}\int_0^{r_0}\left(1+\frac{t}{\sqrt{h}}\right)^{-2}F(t)dt \\
&\le \mathcal{O}(h^{\frac{1}{2}}) + \frac{\mathcal{O}(1)}{\sqrt{h}}\int_0^{r_0}\left(1+\frac{t}{\sqrt{h}}\right)^{-2}t^{\alpha}dt \\
&= \mathcal{O}(h^{\frac{1}{2}}) + \mathcal{O}(h^{\frac{\alpha}{2}})\int_0^{r_0/\sqrt{h}}(1+s)^{-2}s^{\alpha}ds \\
&= \mathcal{O}(1)\begin{cases} h^{\frac{\alpha}{2}}, & \text{if } \alpha < 1 \\ h^{\frac{\alpha}{2}}\left(\frac{1}{\sqrt{h}}\right)^{\alpha-1} = h^{\frac{1}{2}}, & \text{if } \alpha \ge 1, \end{cases}
\end{aligned}$$

neglecting a factor $\ln(1/h)$ when $\alpha = 1$. Thus,

$$A \lesssim \begin{cases} h^{\frac{1}{2}}, & \text{if } \alpha \ge 1, \\ h^{\frac{\alpha}{2}}, & \text{if } \alpha < 1. \end{cases} \tag{15.4.10}$$

Since $\kappa \le 1$, we have $A \lesssim B$ when $\alpha \ge 1$. From the inequalities $2\kappa - 1 \le \alpha > 0$, we have

$$\frac{\alpha\kappa}{\alpha+1} \le \frac{\alpha}{2},$$

so $A \lesssim B$ also when $\alpha < 1$. (Recall that we have neglected a factor $\ln(1/h)$ when $\alpha = 1$.) This is a rather clear indication that Theorem 15.3.1 gives a better remainder estimate than the corresponding results in [137, 139].

Notes In this chapter we announce a general result on Weyl asymptotics for elliptic operators on a compact manifold in the semi-classical limit with a small random potential. It is close to the main result of [139] with some generalizations and improvements. In the next two chapters we give the proof, following closely [137, 139], with the improvements coming from the use of the improved zero-counting result of Chap. 12.

Proof I: Upper Bounds

16

In this chapter we study upper bounds on singular values and determinants of certain operators related to P_δ. The bounds are not probabilistic; they only depend on a certain smallness of the perturbation.

16.1 Review of Some Calculus for h-Pseudodifferential Operators

We recall some basic h-pseudodifferential calculus on compact manifolds, including some fractional powers in the spirit of Seeley [123]. Recall from [83] that if $X \subset \mathbf{R}^n$ is open, $0 < \rho \leq 1$, $m \in \mathbf{R}$, then $S^m_\rho(X \times \mathbf{R}^n) = S^m_{\rho,1-\rho}(X \times \mathbf{R}^n)$ is the space of all $a \in C^\infty(X \times \mathbf{R}^n)$ such that for any $K \Subset X$, $\alpha, \beta \in \mathbf{N}^n$, there exists a constant $C = C(K, \alpha, \beta)$, such that

$$|\partial_x^\alpha \partial_\xi^\beta a(x, \xi)| \leq C \langle \xi \rangle^{m-\rho|\beta|+(1-\rho)|\alpha|}, \ (x, \xi) \in K \times \mathbf{R}^n. \tag{16.1.1}$$

When $a(x, \xi) = a(x, \xi; h)$ depends on the additional parameter $h \in]0, h_0]$ for some $h_0 > 0$, we say that $a \in S^m_\rho(X \times \mathbf{R}^n)$, if (16.1.1) holds uniformly with respect to h. For h-dependent symbols, we introduce $S^{m,k}_\rho = h^{-k} S^m_\rho$. When $\rho = 1$ it is customary to suppress the subscript ρ. We say that $R = R_h : \mathcal{E}'(X) \to C^\infty(X)$ is negligible if $\|\phi \circ R_h \circ \psi\|_{\mathcal{L}(H_h^{-N}, H_h^N)} \leq C_{N,\phi,\psi} h^N$ for all $\phi, \psi \in C_0^\infty(X)$, $N \in \mathbf{N}$. R_h is negligible iff the distribution kernel $K_R(x, y)$ satisfies $\partial_x^\alpha \partial_y^\beta K_R(x, y) = \mathcal{O}(h^N)$ for all α, β, N, uniformly on every compact subset of $X \times X$. Here $H_h^s(\mathbf{R}^n)$ is the Sobolev space $H^s(\mathbf{R}^n)$, equipped with the semi-classical norm $\|\langle hD \rangle^s u\|_{L^2}$, $s \in \mathbf{R}$. See Sect. 16.2.

Let now X be a compact n-dimensional manifold. We say that $R = R_h : \mathcal{D}'(X) \to C^\infty(X)$ is negligible, and write $R \equiv 0$, if the distribution-kernel K_R satisfies $\partial_x^\alpha \partial_y^\beta K_R(x, y) = \mathcal{O}(h^\infty)$ for all $\alpha, \beta \in \mathbf{N}^n$ (when expressed in local coordinates).

J. Sjöstrand, *Non-Self-Adjoint Differential Operators, Spectral Asymptotics and Random Perturbations*, Pseudo-Differential Operators 14,
https://doi.org/10.1007/978-3-030-10819-9_16

We say that an operator $P = P_h : C^\infty(X) \to \mathcal{D}'(X)$ belongs to the space $L^{m,k}(X)$ if $\phi \circ P_h \circ \psi$ is negligible for all $\phi, \psi \in C^\infty(X)$ with disjoint supports and if for every choice of local coordinates $x_1, \dots, x_n$, defined on the open subset $\widetilde{X} \subset X$ (that we view as a subset of $\mathbf{R}^n$), we have on $\widetilde{X}$ for every $u \in C_0^\infty(\widetilde{X})$:

$$Pu(x) = \frac{1}{(2\pi h)^n} \iint e^{\frac{i}{h}(x-y)\cdot\theta} a(x,\theta;h)u(y)dyd\theta + Ku(x), \tag{16.1.2}$$

where $a \in S^{m,k}(\widetilde{X} \times \mathbf{R}^n)$ and $K : \mathcal{E}'(\widetilde{X}) \to C^\infty(\widetilde{X})$ is negligible.

The correspondence $P \mapsto a$ is not globally well defined, but the various local maps give rise to a bijection

$$L^{m,k}(X)/L^{m-1,k-1}(X) \longrightarrow S^{m,k}(T^*X)/S^{m-1,k-1}(T^*X), \tag{16.1.3}$$

where we notice that $S^{m,k}(T^*X)$ is well defined in the natural way. The image $\sigma_P(x,\xi)$ of $P \in L^{m,k}(X)$ is called the principal symbol.

Pseudodifferential operators in the above classes map C^∞ to C^∞ and extend to well defined operators $\mathcal{D}'(X) \to \mathcal{D}'(X)$. They can therefore be composed with each other: If $P_j \in L^{m_j,k_j}(X)$, for $j = 1, 2$, then $P_1 \circ P_2 \in L^{m_1+m_2,k_1+k_2}$. Moreover $\sigma_{P_1 \circ P_2}(x,\xi) = \sigma_{P_1}(x,\xi)\sigma_{P_2}(x,\xi)$.

We can invert elliptic operators: If $P_h \in L^{m,k}$ is elliptic in the sense that $|\sigma_P(x,\xi)| \geq \frac{1}{C}h^{-k}\langle\xi\rangle^m$, then P_h is invertible (either as a map on C^∞ or on $\mathcal{D}'$) for $h > 0$ small enough, and the inverse Q belongs to $L^{-m,-k}$. (If we assume invertibility in the full range $0 < h \leq h_0$, then the conclusion holds in that range.) Notice that $\sigma_Q(x,\xi) = 1/\sigma_P(x,\xi) \in S^{-m,-k}/S^{-m-1,-k-1}$.

The proof of these facts is a routine application of the method of stationary phase, following for instance the presentation in [51]. See also Sect. 16.3, where we will consider a degenerate calculus.

Let

$$h^2\widetilde{R} = \sum (hD_{x_j})^* r_{j,k}(x) hD_{x_k}, \text{ with } r_{k,j} = \overline{r}_{j,k}, \tag{16.1.4}$$

be a self-adjoint non-negative elliptic operator with smooth h-independent coefficients on X, where the star indicates that we take the adjoint with respect to some fixed positive smooth density on X. Let $r(x,\xi) = \sum r_{j,k}\xi_j\xi_k$ be the (real-valued) principal symbol of $\widetilde{R}$ in the classical sense, so that $r(x,\xi)$ is a homogeneous polynomial in ξ with $r(x,\xi) \asymp |\xi|^2$. Then $P := h^2\widetilde{R}$ belongs to $L^{2,0}(X)$ and $\sigma_{h^2\widetilde{R}} = r$. It is a self-adjoint operator: $L^2(X) \to L^2(X)$ with domain $H^2(X)$, and by standard functional calculus we can define the self-adjoint operators $(1 + h^2\widetilde{R})^s$, $s \in \mathbf{R}$.

Proposition 16.1.1 *For every $s \in \mathbf{R}$, we have $(1 + h^2\widetilde{R})^s \in L^{2s,0}$ and the principal symbol is given by $(1 + r(x,\xi))^s$.*

Proof It suffices to show this for s sufficiently large negative. In that case we have

$$(1+h^2\widetilde{R})^s = \frac{1}{2\pi i}\int_\gamma (1+z)^s(z-h^2\widetilde{R})^{-1}dz, \tag{16.1.5}$$

where γ is the oriented boundary of the sector $|\arg(z+\frac{1}{2})| < \pi/4$. For $z \in \gamma$, and more generally for z in the complement of the sector, we write

$$(z-h^2\widetilde{R}) = |z|\left(\frac{z}{|z|} - \widetilde{h}^2\widetilde{R}\right),\ \widetilde{h} = \frac{h}{|z|^{1/2}},$$

and notice that $(z/|z|) - \widetilde{h}^2\widetilde{R} \in L^{2,0}$ is elliptic when we regard $\widetilde{h}$ as the new semi-classical parameter. By self-adjointness and positivity, we know that this operator is invertible, so $((z/|z|) - \widetilde{h}^2\widetilde{R})^{-1} \in L^{-2,0}$, and for every system of local coordinates the symbol (in the sense of $\widetilde{h}$-pseudodifferential operators) is

$$\frac{1}{(z/|z|) - r(x,\xi)} + a, \quad a \in S^{-3,-1}. \tag{16.1.6}$$

The symbol of $(z-h^2\widetilde{R})^{-1}$ as an h-pseudodifferential operator is therefore

$$\frac{1}{|z|(\frac{z}{|z|} - r(x,\frac{\xi}{|z|^{1/2}}))} + \frac{1}{|z|}a\left(x,\frac{\xi}{|z|^{1/2}};\frac{h}{|z|^{1/2}}\right). \tag{16.1.7}$$

Here the first term simplifies to $(z-r(x,\xi))^{-1}$ and the corresponding contribution to (16.1.5) has the symbol $(1+r(x,\xi))^s$.

The contribution from the remainder in (16.1.7) to the symbol in (16.1.5) is

$$b(x,\xi) := \frac{1}{2\pi i}\int_\gamma \frac{(1+z)^s}{|z|}a\left(x,\frac{\xi}{|z|^{1/2}};\frac{h}{|z|^{1/2}}\right)dz,$$

where we will use the estimate

$$\begin{aligned}\partial_x^\alpha\partial_\xi^\beta\frac{1}{|z|}a(x,\frac{\xi}{|z|^{1/2}};\frac{h}{|z|^{1/2}}) &= \mathcal{O}\left(\frac{h}{|z|^{(3+|\beta|)/2}}\langle\frac{\xi}{|z|^{1/2}}\rangle^{-3-|\beta|}\right)\\ &= \mathcal{O}(h)(|z|+|\xi|^2)^{-\frac{1}{2}(3+|\beta|)}.\end{aligned} \tag{16.1.8}$$

Thus,

$$\partial_x^\alpha\partial_\xi^\beta b = \mathcal{O}(1)h\int_\gamma |z|^s(|z|+|\xi|^2)^{-\frac{1}{2}(3+|\beta|)}|dz|. \tag{16.1.9}$$

- In a region $|\xi| = \mathcal{O}(1)$, we get

$$\partial_x^\alpha \partial_\xi^\beta b = \mathcal{O}(h).$$

- In the region $|\xi| \gg 1$ shift the contour γ in (16.1.5) to the oriented boundary of the sector $\arg(z+\frac{1}{2}|\xi|^2) < \frac{\pi}{4}$. Then we get (16.1.9) for the shifted contour and the integral can now be estimated by

$$\mathcal{O}(h)\int_{|\xi|^2/C}^{\infty} t^{s-\frac{3}{2}-\frac{|\beta|}{2}}dt = \mathcal{O}(h|\xi|^{2s-1-|\beta|}).$$

- There is also the negligible part in $(z-h^2\widetilde{R})^{-1}$, which by the same scaling arguments satisfies $\partial_x^\alpha \partial_y^\beta K(x,y,z;h) = \mathcal{O}((h/|z|^{1/2})^\infty)$ on the level of distribution kernels. After integration of $(1+z)^s K$ along γ, this gives rise to a negligible operator.

The proposition follows. □

16.2 Sobolev Spaces and Multiplication

We let $H_h^s(\mathbf{R}^n) \subset \mathcal{S}'(\mathbf{R}^n)$, $s \in \mathbf{R}$, denote the semiclassical Sobolev space of order s equipped with the norm $\|\langle hD\rangle^s u\|$, where norms without subscripts are the ones in L^2, ℓ^2 or the corresponding operator norms if nothing else is indicated. Here $\langle hD\rangle = (1+(hD)^2)^{1/2}$. Let $\widehat{u}(\xi) = \int e^{-ix\cdot\xi}u(x)dx$ denote the Fourier transform of the tempered distribution u on $\mathbf{R}^n$.

Proposition 16.2.1 *Let $s > n/2$. Then there exists a constant $C = C(s)$ such that for all $u, v \in H_h^s(\mathbf{R}^n)$, we have $u \in L^\infty(\mathbf{R}^n)$, $uv \in H_h^s(\mathbf{R}^n)$ and*

$$\|u\|_{L^\infty} \le Ch^{-n/2}\|u\|_{H_h^s}, \tag{16.2.1}$$

$$\|uv\|_{H_h^s} \le Ch^{-n/2}\|u\|_{H_h^s}\|v\|_{H_h^s}. \tag{16.2.2}$$

Proof The fact that $u \in L^\infty$ and the estimate (16.2.1) follow from Fourier's inversion formula and the Cauchy–Schwarz inequality:

$$|u(x)| \le \frac{1}{(2\pi)^n}\int \langle h\xi\rangle^{-s}(\langle h\xi\rangle^s|\widehat{u}(\xi)|)d\xi \le \frac{1}{(2\pi)^{n/2}}\|\langle h\cdot\rangle^{-s}\|\|u\|_{H_h^s}.$$

It then suffices to use that $\|\langle h\cdot\rangle^{-s}\| = C(s)h^{-n/2}$.

In order to prove (16.2.2), we pass to the Fourier transform side, and see that it suffices to show that with $\widetilde{u} = \langle h\cdot\rangle^s |\widehat{u}|$, $\widetilde{v} = \langle h\cdot\rangle^s |\widehat{v}|$,

$$\int w(\xi)\langle h\xi\rangle^s (\langle h\cdot\rangle^{-s}\widetilde{u} * \langle h\cdot\rangle^{-s}\widetilde{v})(\xi)d\xi \le C(s)h^{-\frac{n}{2}}\|\widetilde{u}\|\|\widetilde{v}\|\|w\|, \tag{16.2.3}$$

for all non-negative $\widetilde{u}, \widetilde{v}, w \in L^2$, where $*$ denotes convolution. Here the left-hand side can be written as

$$\iint_{\eta+\zeta=\xi} \frac{\langle h\xi\rangle^s}{\langle h\eta\rangle^s \langle h\zeta\rangle^s} w(\xi)\widetilde{u}(\eta)\widetilde{v}(\zeta)d\xi d\zeta \le \mathrm{I} + \mathrm{II},$$

where I and II denote the corresponding integrals over the sets $\{|\eta| \ge |\xi|/2\}$ and $\{|\zeta| \ge |\xi|/2\}$, respectively. Here

$$\begin{aligned} \mathrm{I} &\le C(s)\int (\int w(\xi)\widetilde{u}(\xi-\zeta)d\xi)\frac{\widetilde{v}(\zeta)}{\langle h\zeta\rangle^s}d\zeta \\ &\le C(s)\|w\|\|\widetilde{u}\|\|\frac{\widetilde{v}}{\langle h\cdot\rangle^s}\|_{L^1}. \end{aligned}$$

As in the proof of (16.2.1), we see that $\|\frac{\widetilde{v}}{\langle h\cdot\rangle^s}\|_{L^1} \le C(s)h^{-\frac{n}{2}}\|\widetilde{v}\|$, so I is bounded by a constant times $h^{-\frac{n}{2}}\|\widetilde{u}\|\|\widetilde{v}\|\|w\|$. The same estimate holds for II and (16.2.3) follows. □

Let X be a compact n-dimensional manifold. We cover X by finitely many coordinate neighborhoods $M_1, \dots, M_p$ and for each M_j, we let $x_1, \dots, x_n$ denote the corresponding local coordinates on M_j. Let $0 \le \chi_j \in C_0^\infty(M_j)$ have the property that $\sum_1^p \chi_j > 0$ on X. Define $H_h^s(X)$ to be the space of all $u \in \mathcal{D}'(X)$ such that

$$\|u\|_{H_h^s}^2 := \sum_1^p \|\chi_j\langle hD\rangle^s\chi_j u\|^2 < \infty. \tag{16.2.4}$$

By straightforward arguments one can show that this definition does not depend on the choice of the coordinate neighborhoods or on χ_j, and that different choices of these objects give norms in (16.2.4) which are uniformly equivalent when $h \to 0$.

This also follows from the h-pseudodifferential calculus on manifolds with symbols in the Hörmander space $S_{1,0}^m$, given in Sect. 16.1. We saw in Sect. 16.1 that $(1 + h^2\widetilde{R})^{s/2}$ is an h-pseudodifferential operator with symbol in $S_{1,0}^s$ and semiclassical principal symbol given by $(1 + r(x,\xi))^{s/2}$, where $r(x,\xi) = \sum_{j,k} r_{j,k}(x)\xi_j\xi_k$ is the semiclassical principal symbol of $h^2\widetilde{R}$. From Sect. 16.1, in particular Proposition 16.1.1, we get the following result for every $s \in \mathbf{R}$, which provides us with an equivalent definition of $H_h^s(X)$:

Proposition 16.2.2 *$H_h^s(X)$ is the space of all $u \in \mathcal{D}'(X)$ such that $(1+h^2\widetilde{R})^{s/2}u \in L^2$ and the norm $\|u\|_{H_h^s}$ is equivalent to $\|(1+h^2\widetilde{R})^{s/2}u\|$, uniformly when $h \to 0$.*

Let $(\mu_k^0)^2$, $k = 1, 2, \dots$ with $\mu_k^0 > 0$ be the eigenvalues of $\widetilde{R}$ repeated according to their multiplicity and let $\epsilon_1, \epsilon_2, \dots$ be a corresponding orthonormal basis of eigenfunctions. It follows from the basic Weyl asymptotics for elliptic self-adjoint operators on compact manifolds (see for instance [51] and further references given there) that

$$\#\{k;\ \mu_k^0 \le \lambda\} \asymp \lambda^n,$$

uniformly for $0 < h \ll 1$, $\lambda \ge 1$. If $u \in \mathcal{D}'(X)$, we have

$$u = \sum \alpha_k \epsilon_k,\ \alpha_k = (u|\epsilon_k),$$

where the sequence of α_k is of temperate growth and the series converges in the sense of distributions. From Proposition 16.2.2 we see that $u \in H_h^s$ iff $\sum \langle \mu_k \rangle^{2s}$ is finite and

$$\|u\|_{H_h^s}^2 \asymp \sum \langle \mu_k \rangle^{2s} |\alpha_k|^2,$$

uniformly with respect to h. Here $\mu_k := h\mu_k^0$, so that μ_k^2 are the eigenvalues of $h^2\widetilde{R}$.

Remark 16.2.3 From the first definition of H_h^s we see that Proposition 16.2.1 remains valid if we replace $\mathbf{R}^n$ by a compact n-dimensional manifold X.

Of course, $H_h^s(X)$ coincides with the standard Sobolev space $H^s(X) = H_1^s(X)$ and the norms are equivalent for each fixed value of h. We have the following variant of Proposition 16.2.1:

Proposition 16.2.4 *Let $s > n/2$. Then there exists a constant $C = C_s > 0$ such that*

$$\|uv\|_{H_h^s} \le C\|u\|_{H^s}\|v\|_{H_h^s},\ \forall u \in H^s(\mathbf{R}^n),\ v \in H_h^s(\mathbf{R}^n). \tag{16.2.5}$$

The result remains valid if we replace $\mathbf{R}^n$ by X.

Proof The adaptation to the case of a compact manifold is immediate by working in local coordinates, so it is enough to prove (16.2.5) in the $\mathbf{R}^n$-case.

Let $\chi \in C_0^\infty(\mathbf{R}^n)$ be equal to 1 in a neighborhood of 0. Write $u = u_1 + u_2$ with $u_1 = \chi(hD)u$, $u_2 = (1-\chi(hD))u$. Then, with hats indicating Fourier transforms, we have

$$\langle h\xi\rangle^s \widehat{u_1 v}(\xi) = \frac{1}{(2\pi)^n}\int \frac{\langle h\xi\rangle^s}{\langle h\eta\rangle^s}\left(\chi(h(\xi-\eta))\widehat{u}(\xi-\eta)\right)\langle h\eta\rangle^s \widehat{v}(\eta)d\eta.$$

Here $\langle h\xi\rangle/\langle h\eta\rangle = \mathcal{O}(1)$ on the support of $(\xi,\eta)\mapsto \chi(h(\xi-\eta))$, so

$$\|u_1 v\|_{H_h^s} \le \mathcal{O}(1)\|\widehat{u}\|_{L^1}\|v\|_{H_h^s} \le \mathcal{O}(1)\|u\|_{H^s}\|v\|_{H_h^s},$$

where we also used that $s > n/2$ in the last estimate.

When $1-\chi(h\xi)\neq 0$, we have $\langle h\xi\rangle^s \le Ch^s\langle\xi\rangle^s$, so $\|u_2\|_{H_h^s} \le Ch^s\|u\|_{H^s}$. From Proposition 16.2.1, we get

$$\|u_2 v\|_{H_h^s} \le Ch^{-\frac{n}{2}}\|u_2\|_{H_h^s}\|v\|_{H_h^s} \le \widetilde{C}h^{s-\frac{n}{2}}\|u\|_{H^s}\|v\|_{H_h^s} \le \widetilde{C}\|u\|_{H^s}\|v\|_{H_h^s},$$

when $h\le 1$. □

16.3 Bounds on Small Singular Values and Determinants in the Unperturbed Case

Let X be a compact manifold and let P be a differential operator on X, satisfying the assumptions of the unperturbed operator in Theorem 15.3.1. Thus P is of the form (15.1.1), and satisfies (15.1.2), (15.1.3), (15.1.5) where the classical and semi-classical symbols p_m and p are defined in (15.1.4) and (15.1.6) respectively. (We will not yet use the symmetry assumption (15.1.7).)

Recall that the spectral parameter z varies in the open simply connected set $\Omega \Subset \mathbf{C}^n$ with smooth boundary and that Ω contains a point z_0 which is not in the image of p. We can (cf. the proof of Proposition 5.1.3) construct a symbol $\widetilde{p}\in S^m(T^*X)$ which is equal to p outside any fixed given neighborhood of $p^{-1}(\overline{\Omega})$, such that $\widetilde{p}(\rho)-z$ is non-vanishing for every $z\in\overline{\Omega}$. Indeed, there is a smooth map $\kappa:\mathbf{C}\setminus\{z_0\}\to\mathbf{C}\setminus\{z_0\}$ such that $\kappa(\mathbf{C}\setminus\{z_0\})\cap\overline{\Omega}=\emptyset$ and $\kappa(z)=z$ for $|z|$ large. It then suffices to put $\widetilde{p}=\kappa\circ p$. Let $\widetilde{P}=P+\mathrm{Op}_h(\widetilde{p}-p)$, where we use the h-quantization in Sect. 16.1.

Using the h-pseudodifferential calculus, we see that for $h>0$ small enough,

$$\widetilde{P}-z : H_h^m(X)\to H^0(X) \tag{16.3.1}$$

is bijective for all $z\in\overline{\Omega}$ with a uniformly bounded inverse. Put

$$P_z = (\widetilde{P}-z)^{-1}(P-z) = 1-(\widetilde{P}-z)^{-1}(\widetilde{P}-P). \tag{16.3.2}$$

Notice that $\widetilde{P}-P=\mathrm{Op}_h(\widetilde{p}-p)$ is a smoothing operator of trace class $C_1(L^2,L^2)$ with the corresponding trace-class norm $=\mathcal{O}(h^{-n})$ (see [41, 119]).

In this section we shall estimate the number of small singular values of $P-z$ and P_z and obtain closely related upper bounds on $\ln|\det P_z|$. We introduce the operator

$$S=(P-z)^*(P-z). \tag{16.3.3}$$

(Later on we shall use the same symbol for a closely related bounded operator.)

In order to obtain these estimates, we shall develop a slightly degenerate pseudodifferential calculus. Let $h \le \alpha \ll 1$. A basic weight function in our calculus will be

$$\Lambda := \left(\frac{\alpha + s}{1 + s}\right)^{\frac{1}{2}}, \text{ where } s(\rho) = |p(\rho) - z|^2, \tag{16.3.4}$$

and we see that $\sqrt{\alpha/2} \le \Lambda \le 1$.

Consider first symbol properties of $1 + \frac{s}{\alpha}$ and its powers.

Proposition 16.3.1 *For every choice of local coordinates x on X, let (x, ξ) denote the corresponding canonical coordinates on T^*X. Then for all $\ell \in \mathbf{R}$, $\widetilde{\alpha}, \beta \in \mathbf{N}^n$, we have uniformly in ξ and locally uniformly in x:*

$$\partial_x^{\widetilde{\alpha}} \partial_\xi^\beta \left(1 + \frac{s}{\alpha}\right)^\ell = \mathcal{O}(1) \left(1 + \frac{s}{\alpha}\right)^\ell \Lambda^{-|\widetilde{\alpha}| - |\beta|} \langle \xi \rangle^{-|\beta|}. \tag{16.3.5}$$

Proof In the region $|\xi| \gg 1$ we see that $(1 + \frac{s}{\alpha})^\ell$ is an elliptic element of the Hörmander symbol class

$$\alpha^{-\ell} S_{1,0}^{2\ell m} =: \alpha^{-\ell} S(\langle \xi \rangle^{2\ell m}),$$

and $\Lambda \asymp 1$ there, so (16.3.5) holds. In the region $|\xi| = \mathcal{O}(1)$, we start with the case $\ell = 1$. Since $s \ge 0$, we have $\nabla s = \mathcal{O}(s^{\frac{1}{2}})$, so

$$\left|\nabla \left(1 + \frac{s}{\alpha}\right)\right| = \mathcal{O}\left(\frac{s^{\frac{1}{2}}}{\alpha}\right) \le \mathcal{O}(1) \left(1 + \frac{s}{\alpha}\right) (\alpha + s)^{-\frac{1}{2}} = \mathcal{O}(1) \left(1 + \frac{s}{\alpha}\right) \Lambda^{-1}.$$

For $k \ge 2$, we have

$$\left|\nabla^k \left(1 + \frac{s}{\alpha}\right)\right| = \mathcal{O}\left(\frac{1}{\alpha}\right) = \mathcal{O}(1) \left(1 + \frac{s}{\alpha}\right) \Lambda^{-2} \le \mathcal{O}(1) \left(1 + \frac{s}{\alpha}\right) \Lambda^{-k},$$

and we get (16.3.5) when $\ell = 1$.

If $\ell \in \mathbf{R}$, then $\partial_x^{\widetilde{\alpha}} \partial_\xi^\beta (1 + \frac{s}{\alpha})^\ell$ is a finite linear combination of terms

$$\left(1 + \frac{s}{\alpha}\right)^{\ell - k} \left(\partial_x^{\widetilde{\alpha}_1} \partial_\xi^{\beta_1} \left(1 + \frac{s}{\alpha}\right)\right) \cdots \left(\partial_x^{\widetilde{\alpha}_k} \partial_\xi^{\beta_k} \left(1 + \frac{s}{\alpha}\right)\right),$$

with $\widetilde{\alpha} = \widetilde{\alpha}_1 + \cdots + \widetilde{\alpha}_k$, $\beta = \beta_1 + \cdots + \beta_k$, and we get (16.3.5) in general. □

We next notice that when w belongs to a bounded subset of $\mathbf{C}$,

$$\frac{|\Im w|}{C}\left(1+\frac{s}{\alpha}\right) \leq \left|w-\frac{s}{\alpha}\right| \leq C\left(1+\frac{s}{\alpha}\right). \tag{16.3.6}$$

In fact, the second inequality is obvious, and so is the first one, when $\frac{s}{\alpha} \gg 1$. When $\frac{s}{\alpha} \leq \mathcal{O}(1)$, it follows from the fact that

$$1+\frac{s}{\alpha}=\mathcal{O}(1), \quad \left|w-\frac{s}{\alpha}\right| \geq |\Im w|.$$

From (16.3.5), (16.3.6), we get

$$\left|\partial_x^{\widetilde{\alpha}}\partial_\xi^\beta\left(w-\frac{s}{\alpha}\right)\right| \leq \mathcal{O}(1)\left(w-\frac{s}{\alpha}\right)\Lambda^{-|\widetilde{\alpha}|-|\beta|}\langle\xi\rangle^{-|\beta|}|\Im w|^{-1}. \tag{16.3.7}$$

When passing to $(w-\frac{s}{\alpha})^\ell$ and applying the proof of Proposition 16.3.1, we loose more powers of $|\Im w|$ which can still be counted precisely, but we refrain from doing so and simply state the following result:

Proposition 16.3.2 *For all $\ell \in \mathbf{R}$, $\widetilde{\alpha}, \beta \in \mathbf{N}^n$, there exists $J \in \mathbf{N}$, such that*

$$\partial_x^{\widetilde{\alpha}}\partial_\xi^\beta\left(w-\frac{s}{\alpha}\right)^\ell = \mathcal{O}(1)\left(1+\frac{s}{\alpha}\right)^\ell\Lambda^{-|\widetilde{\alpha}|-|\beta|}\langle\xi\rangle^{-|\beta|}|\Im w|^{-J}, \tag{16.3.8}$$

uniformly in ξ and locally uniformly in (x, w). In this estimate we can replace $(1+s/\alpha)^\ell$ with $(1+s/\alpha)^{\ell_1}(w-s/\alpha)^{\ell_2}$ for every (ℓ_1, ℓ_2) with $\ell_1+\ell_2=\ell$.

We next introduce some new symbol spaces.

Definition 16.3.3 Let $\widetilde{m}(x,\xi)$ be a weight function of the form $\widetilde{m}(x,\xi) = \langle\xi\rangle^k\Lambda^\ell$. We say that the family $a = a_w \in C^\infty(T^*X)$, $w \in D(0, C)$, belongs to $S_\Lambda(\widetilde{m})$ if for all $\widetilde{\alpha}, \beta \in \mathbf{N}^n$ there exists $J \in \mathbf{N}$ such that

$$\partial_x^{\widetilde{\alpha}}\partial_\xi^\beta a = \mathcal{O}(1)\widetilde{m}(x,\xi)\Lambda^{-|\widetilde{\alpha}|-|\beta|}\langle\xi\rangle^{-|\beta|}|\Im w|^{-J}. \tag{16.3.9}$$

Here, as in Proposition 16.3.2, it is understood that the estimate is expressed in canonical coordinates and is locally uniform in (x, w) and uniform in ξ. Notice that the set of estimates (16.3.9) is invariant under changes of local coordinates in X.

Let $U \subset X$ be a coordinate neighborhood that we shall view as a subset of $\mathbf{R}^n$ in the natural way. Let $a \in S_\Lambda(T^*U, \widetilde{m})$ be a symbol as in Definition 16.3.3, so that (16.3.9) holds uniformly in ξ and locally uniformly in (x, w). For fixed values of α, w the symbol

a belongs to $S^k_{1,0}(T^*U)$, so the classical h-quantization

$$Au = \mathrm{Op}_h(a)u(x) = \frac{1}{(2\pi h)^n} \iint e^{\frac{i}{h}(x-y)\cdot\eta} a(x, \eta; h)u(y)dyd\eta \tag{16.3.10}$$

is a well-defined operator $C_0^\infty(U) \to C^\infty(U)$, $\mathcal{E}'(U) \to \mathcal{D}'(U)$. In order to develop our rudimentary calculus on X, we first establish a pseudolocal property for the distribution kernel $K_A(x, y)$:

Proposition 16.3.4 *For all* $\widetilde{\alpha}, \beta \in \mathbf{N}^n$, $N \in \mathbf{N}$, *there exists* $M \in \mathbf{N}$ *such that*

$$\partial_x^{\widetilde{\alpha}} \partial_y^\beta K_A(x, y) = \mathcal{O}(h^N |\Im w|^{-M}), \tag{16.3.11}$$

locally uniformly on $U \times U \setminus \mathrm{diag}(U \times U)$.

Proof If $\gamma \in \mathbf{N}^n$, then $(x-y)^\gamma K_A(x, y)$ is the distribution kernel of $\mathrm{Op}_h((-hD_\xi)^\gamma a)$ and $(-hD_\xi)^\gamma a \in S_\Lambda\left(\widetilde{m}\,(h/\Lambda\langle\xi\rangle)^{|\gamma|}\right)$. By the observation after (16.3.4), $h/\Lambda \le \sqrt{2}h/\alpha^{\frac{1}{2}} \le \sqrt{2}h^{\frac{1}{2}}$. Thus for any $N \in \mathbf{N}$, there exists $M \in \mathbf{N}$ such that,

$$(x-y)^\gamma K_A(x, y) = \mathcal{O}(h^N |\Im w|^{-M}),$$

if $|\gamma| \ge \gamma(N)$ is large enough. Equation (16.3.11) follows when $\widetilde{\alpha} = \beta = 0$. Now, $\partial_x^{\widetilde{\alpha}} \partial_y^\beta K_A$ can be viewed as the distribution kernel of a new pseudodifferential operator of the same kind, so we get (16.3.11) for all $\widetilde{\alpha}, \beta$. □

This means that if $\phi, \psi \in C_0^\infty(U)$ have disjoint supports, then for every $N \in \mathbf{N}$, there exists $M \in \mathbf{N}$ such that $\phi A \psi : H^{-N}(\mathbf{R}^n) \to H^N(\mathbf{R}^n)$ with norm $\mathcal{O}(h^N |\Im w|^{-M})$, and this leads to a simple way of introducing pseudodifferential operators on X: Let $U_1, \ldots, U_s$ be coordinate neighborhoods that cover X. Let $\chi_j \in C_0^\infty(U_j)$ form a partition of unity and let $\widetilde{\chi}_j \in C_0^\infty(U_j)$ satisfy $\chi_j \prec \widetilde{\chi}_j$ in the sense that $\widetilde{\chi}_j$ is equal to 1 near $\mathrm{supp}\,(\chi_j)$. Let $a = (a_1, \ldots, a_s)$, where $a_j \in S_\Lambda(\widetilde{m})$. Then we quantize a by the formula

$$A = \sum_1^s \widetilde{\chi}_j \circ \mathrm{Op}_h(a_j) \circ \chi_j. \tag{16.3.12}$$

This is not an invariant quantization procedure, but it will suffice for our purposes.

We next study the composition to the left with non-exotic pseudodifferential operators. Let $U = U_j$ be one of the above coordinate neighborhoods, viewed as an open set in $\mathbf{R}^n$, and take $A = \mathrm{Op}_h(a)$, $a \in S_{1,0}(m_1)$, $m_1 = \langle\xi\rangle^r$, $B = \mathrm{Op}_h(b)$, $b \in S_\Lambda(m_2)$ with $m_2 = \langle\xi\rangle^k \Lambda^\ell$ as in Definition 16.3.3. We will assume that $\mathrm{supp}\,(b) \subset K \times \mathbf{R}^n$, where $K \subset U$ is compact. We are interested in $C = A \circ B$.

The symbol c of this composition is given by

$$\begin{aligned} c(x,\xi;h) &= e^{-\frac{i}{h}x\cdot\xi}A(b(\cdot,\xi)e^{\frac{i}{h}(\cdot)\cdot\xi})(x) \\ &= \frac{1}{(2\pi h)^n}\iint a(x,\eta)b(y,\xi)e^{\frac{i}{h}(x-y)\cdot(\eta-\xi)}dyd\eta \end{aligned} \tag{16.3.13}$$

Let $\chi \in C_0^\infty(B(0,\frac{1}{2}))$ be equal to 1 near 0, and let $d(x,\xi;h)$ be obtained from (16.3.13) after inserting a factor $1-\chi\,(\eta-\xi)/\langle\xi\rangle)$ in the integral. Then $|\eta-\xi|\geq\frac{1}{C}\langle\xi\rangle$ in the support of the integrand, so we can make repeated integrations by parts in the y-variables and get

$$\begin{aligned} &\forall N\in\mathbf{N},\ \widetilde{\alpha},\beta\in\mathbf{N}^n,\ \exists M\in\mathbf{N},\ \forall K\Subset U,\ \exists C>0; \\ &|\partial_x^{\widetilde{\alpha}}\partial_\xi^\beta d(x,\xi;h)|\leq C\frac{h^N\langle\xi\rangle^{-N}}{|\Im w|^M},\ (x,\xi)\in K\times\mathbf{R}^n. \end{aligned} \tag{16.3.14}$$

Up to such a term d, we get

$$\begin{aligned} c(x,\xi;h) &\equiv \frac{1}{(2\pi h)^n}\iint a(x,\eta)b(y,\xi)\chi\left(\frac{\eta-\xi}{\langle\xi\rangle}\right)e^{\frac{i}{h}(x-y)\cdot(\eta-\xi)}dyd\eta \\ &= \left(\frac{\langle\xi\rangle}{2\pi h}\right)^n\iint a\left(x,\langle\xi\rangle(\eta+\frac{\xi}{\langle\xi\rangle})\right)b(x+y,\xi)\chi(\eta)e^{\frac{-i\langle\xi\rangle}{h}y\cdot\eta}dyd\eta. \end{aligned} \tag{16.3.15}$$

We shall apply the method of stationary phase (see for instance [41, 51]) and pause to recall some facts about the expansion of the integral

$$J(t,u)=\frac{1}{(2\pi t)^n}\iint u(y,\eta)e^{-\frac{i}{t}y\cdot\eta}dyd\eta,\ u\in\mathcal{S}(\mathbf{R}^{2n}).$$

The exponent is of the form

$$\frac{i}{2t}Q\begin{pmatrix}y\\ \eta\end{pmatrix}\cdot\begin{pmatrix}y\\ \eta\end{pmatrix},$$

where

$$Q=\begin{pmatrix}0&-1\\-1&0\end{pmatrix}.$$

Then (cf. [51, Proposition 2.3]) we know that

$$J(t,u)\longrightarrow u(0),\ t\longrightarrow 0.$$

We also know (see [51, (2.4)]) that if $\mathcal{F} : L^2(\mathbf{R}^{2n}) \to L^2(\mathbf{R}^{2n})$ denotes the Fourier transform,

$$\mathcal{F}\left(\frac{1}{(2\pi t)^n} e^{\frac{i}{t} y\cdot\eta}\right)(y^*, \eta^*) = e^{-ity^*\cdot\eta^*},$$

then we can apply Plancherel's formula and write

$$\begin{aligned} J(t,u) &= \frac{1}{(2\pi)^{2n}} \iint (\mathcal{F}u)(y^*, \eta^*) e^{ity^*\cdot\eta^*} dy^* d\eta^* \\ &= \exp(tD_y \cdot \partial_\eta)(u)(0,0). \end{aligned}$$

By Taylor's formula with integral remainder, we have

$$\begin{aligned} J(t,u) &= \sum_0^{N-1} \frac{t^k}{k!} (\partial_\eta \cdot D_y)^k u(0,0) + R_N \\ &= \sum_{|\beta|<N} \frac{t^{|\beta|}}{\beta!} \partial_\eta^\beta D_y^\beta u(0,0) + R_N, \end{aligned}$$

where

$$R_N = t^N \frac{1}{(N-1)!} \int_0^1 (1-s)^N J((\partial_\eta \cdot D_y)^N u)(st) ds.$$

Applying this to the last expression in (16.3.15) with $t = h/\langle\xi\rangle$, we get

$$c(x,\xi;h) = \sum_{|\beta|<N} \frac{h^{|\beta|}}{\beta!} \partial_\xi^\beta a D_x^\beta b + R_N. \tag{16.3.16}$$

Here,

$$\begin{aligned} R_N = \left(\frac{h}{\langle\xi\rangle}\right)^N \frac{1}{(N-1)!} \int_0^1 (1-t)^N \times \\ J\left(t\frac{h}{\langle\xi\rangle}, (\partial_\eta \cdot D_y)^N \left(a(x, \langle\xi\rangle(\eta + \frac{\xi}{\langle\xi\rangle}))b(x+y,\xi)\chi(\eta)\right)\right) dt. \end{aligned} \tag{16.3.17}$$

By Plancherel's formula,

$$|J(s,u)| \le C \sum_{|\widetilde{\alpha}|+|\beta|\le 2n+1} \|\partial_y^{\widetilde{\alpha}} \partial_\eta^\beta u\|_{L^1}.$$

Indeed, $|\mathcal{F}u(y^*, \eta^*)|$ is bounded by $\mathcal{O}(1)\langle(y^*, \eta^*)\rangle^{-(2n-1)}$ times the sum in the right-hand side. Thus, we see that there exist exponents N_2, N_3 independent of N, such that

$$|R_N| \le C\left(\frac{h}{\langle\xi\rangle}\right)^N m_1(\xi)\langle\xi\rangle^{N_2}\alpha^{N_3-\frac{N}{2}}|\Im w|^{-M(N)}.$$

Similar estimates hold for the derivatives and using that N can be chosen arbitrarily large and that $\alpha \ge h^{1/2}$, we conclude:

Proposition 16.3.5 *Let $A = \mathrm{Op}_h(a)$, $a \in S_{1,0}(m_1)$, $m_1 = \langle\xi\rangle^r$, $B = \mathrm{Op}_h(b)$, $b \in S_\Lambda(m_2)$, $m_2 = \langle\xi\rangle^k\Lambda^\ell$ and assume that b has uniformly compact support in x. Then $A \circ B = \mathrm{Op}_h(c)$, where c belongs to $S_\Lambda(m_1m_2)$ and has the asymptotic expansion*

$$c \sim \sum \frac{h^{|\beta|}}{\beta!}\partial_\xi^\beta a(x,\xi)D_x^\beta b(x,\xi),$$

in the sense that for every $N \in \mathbf{N}$,

$$c = \sum_{|\beta|<N} \frac{h^{|\beta|}}{\beta!}\partial_\xi^\beta a(x,\xi)D_x^\beta b(x,\xi) + r_N(x,\xi;h),$$

where $r_N \in S_\Lambda(\frac{m_1m_2}{(\Lambda\langle\xi\rangle)^N}h^N)$.

We next make a parametrix construction for $w - \frac{1}{\alpha}S$, still with S as in (16.3.3), and most of the work will take place in a coordinate neighborhood U, viewed as an open set in $\mathbf{R}^n$. The symbol of $w - \frac{1}{\alpha}S$ is of the form

$$F = F_0 + F_{-1}, \quad F_0 = w - \frac{1}{\alpha}s,\ F_{-1} = \frac{h}{\alpha}s_{-1} \in S(\frac{h}{\alpha}\langle\xi\rangle^{2m-1}). \tag{16.3.18}$$

Recall (16.3.6) and Proposition 16.3.2. Put

$$E_0 = \frac{1}{w - \frac{1}{\alpha}s} \in S_\Lambda(\frac{\alpha}{\Lambda^2\langle\xi\rangle^{2m}}). \tag{16.3.19}$$

With Proposition 16.3.5 in mind, we first consider the formal composition

$$\begin{aligned} F\sharp E_0 &\sim \sum \frac{h^{|\beta|}}{\beta!}(\partial_\xi^\beta F)(D_x^\beta E_0) \\ &\sim 1 + \sum_{|\beta|\ge 1} \frac{h^{|\beta|}}{\beta!}(\partial_\xi^\beta F_0)(D_x^\beta E_0) + F_{-1}\sharp E_0. \end{aligned} \tag{16.3.20}$$

Here,

$$F_{-1}\sharp E_0 \in S_\Lambda(\frac{h}{\alpha}\langle\xi\rangle^{2m-1}\frac{\alpha}{\Lambda^2\langle\xi\rangle^{2m}}) = S_\Lambda(\frac{h}{\Lambda^2\langle\xi\rangle}).$$

Since F_0 also belongs to $S_\Lambda(\frac{1}{\alpha}\Lambda^2\langle\xi\rangle^{2m})$, we see that for $|\beta| \geq 1$,

$$h^{|\beta|}(\partial_\xi^\beta F_0)(D_x^\beta E_0) \in S_\Lambda(\frac{h^{|\beta|}}{\Lambda^{2|\beta|}\langle\xi\rangle^{|\beta|}}) \subset S_\Lambda(\frac{h}{\Lambda^2\langle\xi\rangle}),$$

and this can be improved for $|\beta| \geq 2$, using that $F \in S_{1,0}(\frac{1}{\alpha}\langle\xi\rangle^{2m})$:

$$h^{|\beta|}(\partial_\xi^\beta F_0)(D_x^\beta E_0) \in S_\Lambda\left(\frac{h}{\Lambda^2\langle\xi\rangle}\frac{h^{\frac{|\beta|-1}{2}}}{\langle\xi\rangle^{|\beta|-1}}\right).$$

Hence,

$$F\sharp E_0 = 1 + r_1,\ r_1 \in S_\Lambda(\frac{h}{\Lambda^2\langle\xi\rangle}),$$

where r_1 is a formal asymptotic sum, viewed as a sequence of partial sums.

Now put

$$E_1 = E_0 - r_1/(w - s/\alpha) \equiv E_0 \bmod (S_\Lambda(\frac{h\alpha}{\Lambda^4\langle\xi\rangle^{2m+1}})). \tag{16.3.21}$$

Then by the same estimates with an extra power of $h\Lambda^{-2}\langle\xi\rangle^{-1}$, we get

$$F\sharp E_1 = 1 + r_2,\ r_2 \in S_\Lambda((\frac{h}{\Lambda^2\langle\xi\rangle})^2),$$

and iterating the procedure we get

$$E_N \equiv \frac{1}{w - \frac{s}{\alpha}} \bmod S_\Lambda(\frac{\alpha}{\Lambda^2\langle\xi\rangle^{2m}}\frac{h}{\Lambda^2\langle\xi\rangle}), \tag{16.3.22}$$

such that

$$F\sharp E_N = 1 + r_N,\ r_N \in S_\Lambda((\frac{h}{\Lambda^2\langle\xi\rangle})^{N+1}). \tag{16.3.23}$$

In this construction we can work with partial sums of asymptotic series and

$$E_N \text{ is a holomorphic function of } w, \text{ for } |\xi| \geq C, \tag{16.3.24}$$

where C is independent of N. In fact, put $z = \alpha w$ and write

$$(z - S)^{-1} = \frac{1}{\alpha}(w - \alpha^{-1}S)^{-1}.$$

Start with $\alpha^{-1}E_0 = (z - s)^{-1}$ on the level of symbols and consider

$$(z - S)\sharp \frac{1}{\alpha} E_0 = 1 - \frac{hs_{-1}}{z - s} - \sum_{|\beta|\geq 1} \frac{h^{|\beta|}}{\beta!} \partial_\xi^\beta (s + hs_{-1}) D_x^\beta \left(\frac{1}{z - s}\right). \tag{16.3.25}$$

The right-hand side can be written in the general form

$$\sum_\beta \sum_{\substack{k\geq 0\\ \ell\geq 0}} \sum_{\substack{\gamma_1+..+\gamma_k+\\ \delta_1+\cdots+\delta_\ell=(\beta,\beta)}} C_{\beta,\delta,\gamma} h^{|\beta|} \underbrace{\frac{\partial_{x,\xi}^{\gamma_1} s}{z - s} \cdots \frac{\partial_{x,\xi}^{\gamma_k} s}{z - s}}_{\Gamma} \underbrace{\frac{\partial_{x,\xi}^{\delta_1} hs_{-1}}{z - s} \cdots \frac{\partial_{x,\xi}^{\delta_\ell} hs_{-1}}{z - s}}_{\Delta}, \tag{16.3.26}$$

where

1) If $k = 0$ or $\ell = 0$, the corresponding factor Γ or Δ is equal to 1 by definition.
2) $\gamma_j \neq 0$ for $1 \leq j \leq k$.
3) The term with $\beta = 0$, $\ell = 0$ is equal to 1.
4) Only finitely many values of ℓ are present for each β (actually $\ell = 0, 1$, to start with).

Writing $\gamma_j = (\gamma_{j,x}, \gamma_{j,\xi})$ and similarly for δ_j, we see that the general term in (16.3.26) belongs to the symbol class

$$\begin{aligned} S_\Lambda &\left(h^{|\beta|} \frac{\Lambda^{-|\gamma_1|}}{\langle\xi\rangle^{|\gamma_{1,\xi}|}} \cdots \frac{\Lambda^{-|\gamma_k|}}{\langle\xi\rangle^{|\gamma_{k,\xi}|}} \frac{h\langle\xi\rangle^{-1-\delta_{1,\xi}}}{\Lambda^2} \cdots \frac{h\langle\xi\rangle^{-1-\delta_{\ell,\xi}}}{\Lambda^2} \right) \\ &= S_\Lambda \left(h^{|\beta|} (\frac{h}{\Lambda^2})^\ell \Lambda^{-|\gamma_1|-\cdots-|\gamma_k|} \langle\xi\rangle^{-|\gamma_{1,\xi}|-\cdots-|\gamma_{k,\xi}|-|\delta_{1,\xi}|-\cdots-|\delta_{\ell,\xi}|-\ell} \right) \\ &\qquad \subset S_\Lambda \left(\left(\frac{h}{\Lambda^2 \langle\xi\rangle} \right)^{|\beta|+\ell} \right), \end{aligned}$$

where we used that $|\gamma_1| + \cdots + |\gamma_k| \leq 2|\beta|$, $\gamma_{1,\xi} + \cdots + \gamma_{k,\xi} + \delta_{1,\xi} + \cdots + \delta_{\ell,\xi} = \beta$.

Now assume that for some $N \in \{1, 2, \ldots\}$ found a finite sum

$$\frac{1}{\alpha} E_N = \frac{1}{z - s} \sum_\beta \sum_{\substack{k\geq 0\\ \ell\geq 0}} \sum_{\substack{\gamma_1+\cdots+\gamma_k+\\ \delta_1+..\delta_\ell=(\beta,\beta)}} C_{\beta,\gamma,\delta} h^{|\beta|} \frac{\partial_{x,\xi}^{\gamma_1} s}{z - s} \cdots \frac{\partial_{x,\xi}^{\gamma_k} s}{z - s} \frac{\partial_{x,\xi}^{\delta_1} hs_{-1}}{z - s} \cdots \frac{\partial_{x,\xi}^{\delta_\ell} hs_{-1}}{z - s}, \tag{16.3.27}$$

satisfying 1)–4), such that

$$(w - \frac{1}{\alpha}S)\sharp E_N = (z - S)\sharp \frac{1}{\alpha} E_N = 1 + R_N, \tag{16.3.28}$$

where R_N is of the form (16.3.26) with the additional restriction that $|\beta| + \ell > N$. Let $R_{N,N+1}$ be the sum of the terms in R_N with $|\beta| + \ell = N + 1$ and put $\frac{1}{\alpha} E_{N+1} = \frac{1}{\alpha} E_N - \frac{1}{z-s} R_{N,N+1}$. Then we get

$$(z - S)\sharp \frac{1}{\alpha} E_{N+1} = 1 + R_{N+1}, \tag{16.3.29}$$

where R_{N+1} is of the form (16.3.26) with $|\beta| + \ell > N + 1$. The symbol r_N of R_N is therefore as in (16.3.23).

Without loss of generality, we may assume that E_N in (16.3.23) coincides with E_N in (16.3.28).

Now we return to the manifold situation and denote by $E_N^{(j)}$, $r_N^{(j)}$ the corresponding symbols on T^*U_j, constructed above. Denote the operators by the same symbols, and put on the operator level

$$E_N = \sum_{j=1}^{s} \widetilde{\chi}_j E_N^{(j)} \chi_j, \tag{16.3.30}$$

with χ_j, $\widetilde{\chi}_j$ as in (16.3.12). Then

$$\begin{aligned}(w - \frac{1}{\alpha}S)E_N &= 1 - \sum_{j=1}^{s} \frac{1}{\alpha}[S, \widetilde{\chi}_j]E_{N-1}^{(j)}\chi_j + \sum_{j=1}^{s} \widetilde{\chi}_j r_N^{(j)} \chi_j \\ &=: 1 + R_N^{(1)} + R_N^{(2)} \\ &=: 1 + R_N.\end{aligned} \tag{16.3.31}$$

Proposition 16.3.4 implies that for every $\widetilde{N}$, there exists an $\widetilde{M}$ such that the trace-class norm of $R_N^{(1)}$ satisfies

$$\|R_N^{(1)}\|_{\mathrm{tr}} \le \mathcal{O}(h^{\widetilde{N}} |\Im w|^{-\widetilde{M}}). \tag{16.3.32}$$

As for the trace-class norm of $R_N^{(2)}$, we review next some facts about such norms for pseudodifferential operators:

If $A = a(x, D)$ is a pseudodifferential operator on $\mathbf{R}^n$, either in the Weyl or in the classical quantization, then A is of trace class and we have

$$\|A\|_{\mathrm{tr}} \le C \iint \sum_{|\beta|\le 2n+1} |\partial_{x,\xi}^{\beta} a| dx d\xi,$$

provided the integral is finite. In that case we also know that

$$\mathrm{tr}\,(A) = \frac{1}{(2\pi)^n} \iint a(x,\xi) dx d\xi.$$

See Robert [119], and also [41] for a sharper statement. Now an h-pseudodifferential operator $A = a(x, hD)$ is unitarily equivalent to $\widetilde{A} = a(h^{\frac{1}{2}}\widetilde{x}, h^{\frac{1}{2}} D_{\widetilde{x}})$, so

$$\begin{aligned}\|A\|_{\mathrm{tr}} &\le C \iint \sum_{|\beta|\le 2n+1} \partial_{\widetilde{x},\widetilde{\xi}}^{\beta} \left(a(h^{\frac{1}{2}}(\widetilde{x},\widetilde{\xi}))\right) d\widetilde{x} d\widetilde{\xi} \\ &= \frac{C}{h^n} \iint \sum_{|\beta|\le 2n+1} \left|(h^{\frac{1}{2}} \partial_{x,\xi})^{\beta} a\right| dx d\xi.\end{aligned}$$

Now, let $a \in S_{\Lambda}(m)$ be a symbol on T^*U with uniformly compact support in x. Then for $|\beta| \le 2n + 1$ we have

$$h^{\frac{|\beta|}{2}} \partial_{x,\xi}^{\beta} a = \mathcal{O}(1) m \left(\frac{h}{\alpha}\right)^{\frac{|\beta|}{2}} |\Im w|^{-M(\beta)}.$$

Thus, there exists $M \ge 0$ such that $a(x, hD_x)$ is of trace class and

$$\|a(x, hD)\|_{\mathrm{tr}} \le C h^{-n} \iint_{U\times\mathbf{R}^n} m(x,\xi) dx d\xi \, |\Im w|^{-M}, \tag{16.3.33}$$

provided the integral converges.

From (16.3.31), (16.3.23), we now get

$$\|R_N^{(2)}\|_{\mathrm{tr}} \le C h^{-n} |\Im w|^{-M(N)} \iint \left(\frac{h}{\Lambda^2\langle\xi\rangle}\right)^N dx d\xi,$$

and (16.3.32) then shows that we have the same estimate for R_N:

$$\|R_N\|_{\mathrm{tr}} \le C h^{-n} |\Im w|^{-M(N)} \iint \left(\frac{h}{\Lambda^2\langle\xi\rangle}\right)^N dx d\xi. \tag{16.3.34}$$

The contribution to this expression from the region where $\Lambda \geq 1/C$ is $\mathcal{O}(h^{N-n})$ $|\Im w|^{-M(N)}$.

For $0 \leq t \ll 1$, we put

$$V(t) = \iint_{s(x,\xi)\leq t} dx d\xi, \quad 0 \leq t \ll 1. \tag{16.3.35}$$

This is an increasing function of t and p is polynomial in ξ with $|p| \geq |\xi|^m$ for large ξ,

$$V(0) = 0. \tag{16.3.36}$$

For $N > 0$, put

$$V_N(\alpha) = \int_0^1 \left(1 + \frac{t}{\alpha}\right)^{-N} dV(t) = \int_0^{1/\alpha} (1+\tau)^{-N} dV(\alpha\tau) \tag{16.3.37}$$

The contribution to the integral in (16.3.34) from the region $0 \leq s \leq 1$ is equal to

$$\mathcal{O}(1) \int_0^1 \left(\frac{h}{\alpha + t}\right)^N dV(t) = \mathcal{O}(1) \left(\frac{h}{\alpha}\right)^N V_N(\alpha) \tag{16.3.38}$$

Remark 16.3.6 The choice of 1 as the upper bound of integration was somewhat arbitrary. If we replace it by a fixed constant $\theta \in]0, 1[$, then V_N changes by a quantity which is $\mathcal{O}(1)\alpha^N (V(1) - V(\theta))$.

The contribution to (16.3.34) from the region $s(x, \xi) > 1$ is $\mathcal{O}(h^{N_2})$ times some negative power of $|\Im w|$ for every N_2. In conclusion, we have

Proposition 16.3.7 *For all* $N, \widetilde{N} > 0$,

$$\|R_N\|_{\mathrm{tr}} \leq \mathcal{O}(1) h^{-n} \left(\left(\frac{h}{\alpha}\right)^N V_N(\alpha) + h^{\widetilde{N}} \right) |\Im w|^{-M(N)}. \tag{16.3.39}$$

From (16.3.31), we get on the level of operators

$$(w - \frac{1}{\alpha} S)^{-1} = E_N - (w - \frac{1}{\alpha} S)^{-1} R_N.$$

Using the functional Cauchy–Riemann formula (sometimes named after Dynkin and/or Helffer and Sjöstrand, cf. [41]) we have

$$\chi(\frac{1}{\alpha} S) = -\frac{1}{\pi} \int \frac{\partial \widetilde{\chi}}{\partial \overline{w}} (w - \frac{1}{\alpha} S)^{-1} L(dw), \tag{16.3.40}$$

where $\chi \in C_0^\infty(\mathbf{R})$ and $\widetilde{\chi} \in C_0^\infty(\mathbf{C})$ is an almost holomorphic extension. We get

$$\chi(\frac{1}{\alpha}S) = -\frac{1}{\pi}\int \frac{\partial \widetilde{\chi}}{\partial \overline{w}} E_N L(dw) + \frac{1}{\pi}\int \frac{\partial \widetilde{\chi}}{\partial \overline{w}} \left(w - \frac{1}{\alpha}S\right)^{-1} R_N L(dw) =: \mathrm{I} + \mathrm{II}. \tag{16.3.41}$$

From (16.3.39), (16.3.41) and the fact that $\partial\widetilde{\chi}/\partial\overline{w}$ is in C_0^∞ and vanishes to infinite order on the real axis, we get

$$\|\mathrm{II}\|_{\mathrm{tr}} = \mathcal{O}(1)h^{-n}\left(\left(\frac{h}{\alpha}\right)^N V_N(\alpha) + h^{\widetilde{N}}\right).$$

It remains to study the term I and for that we return to the expression (16.3.27) for (the symbol of) E_N (in local coordinates). Here $z = \alpha w$ and we see that E_N is equal to

$$\frac{1}{w-\frac{s}{\alpha}} \sum_\beta \sum_{k,\ell\geq 0} \sum_{\substack{\gamma_1+\dots\gamma_k+ \\ \delta_1+\dots+\delta_\ell=(\beta,\beta)}} C_{\beta,\gamma,\delta} h^{|\beta|} \frac{\partial^{\gamma_1}(\frac{s}{\alpha})}{w-\frac{s}{\alpha}} \cdots \frac{\partial^{\gamma_k}(\frac{s}{\alpha})}{w-\frac{s}{\alpha}} \frac{\partial^{\delta_1}(\frac{h}{\alpha}s_{-1})}{w-\frac{s}{\alpha}} \cdots \frac{\partial^{\delta_\ell}(\frac{h}{\alpha}s_{-1})}{w-\frac{s}{\alpha}}$$

The contribution from the general term to the symbol of I is

$$\begin{aligned} &-\frac{C_{\beta,\gamma,\delta}}{\pi}\int \frac{\partial\widetilde{\chi}(w)}{\partial\overline{w}} \frac{1}{(w-\frac{s}{\alpha})^{1+k+\ell}} L(dw) h^{|\beta|} \\ &\partial^{\gamma_1}(\frac{s}{\alpha})\cdots\partial^{\gamma_k}(\frac{s}{\alpha})\partial^{\delta_1}(\frac{h}{\alpha}s_{-1})\cdots\partial^{\delta_\ell}(\frac{h}{\alpha}s_{-1}) \\ &= \widetilde{C}_{\beta,\gamma,\delta}\chi^{(k+\ell)}(\frac{s}{\alpha})h^{|\beta|}\partial^{\gamma_1}(\frac{s}{\alpha})\cdots\partial^{\gamma_k}(\frac{s}{\alpha})\partial^{\delta_1}(\frac{h}{\alpha}s_{-1})\cdots\partial^{\delta_\ell}(\frac{h}{\alpha}s_{-1}). \end{aligned} \tag{16.3.42}$$

When $\beta = 0$, $\ell = 0$, we have $C_{\beta,\gamma,\delta} = 1$ and we get $\chi(\frac{s}{\alpha})$. For every $N \in \mathbf{N}$, the general term belongs to the symbol class

$$S_\Lambda\left(\left(\frac{\Lambda^2}{\alpha}\right)^{-N+k} \frac{h^{|\beta|}}{\Lambda^{|\gamma_1|+\dots+|\gamma_k|}}\left(\frac{h}{\alpha}\right)^\ell\right).$$

Here, we notice that $|\gamma_1| + \cdots + |\gamma_k| \leq 2|\beta|$, and since $h \leq \alpha$, $\Lambda \geq \alpha^{1/2}$ we conclude that the general term (16.3.42) belongs to

$$\left(\frac{h}{\alpha}\right)^{|\beta|+\ell} S_\Lambda\left(\left(\frac{\Lambda^2}{\alpha}\right)^{-N+k}\right).$$

The trace-class norm of the contribution from (16.3.42) to I is bounded by

$$\begin{aligned} \mathcal{O}(1)h^{-n}\left(\frac{h}{\alpha}\right)^{|\beta|+\ell}\int_0^1\left(1+\frac{t}{\alpha}\right)^{-(N-k)} dV(t) \\ \le \mathcal{O}(1)h^{-n}\left(\frac{h}{\alpha}\right)^{|\beta|+\ell} V_{N/2}(\alpha), \end{aligned}$$

where we used that N can be chosen $\geq 2k$.

Summing up, we have proved that for all $N, \widetilde{N} > 0$, the following holds.

Proposition 16.3.8 *Let $\chi \in C_0^\infty(\mathbf{R})$. For $0 < h \leq \alpha \leq 1/2$, we have*

$$\|\chi(\frac{1}{\alpha}S)\|_{\mathrm{tr}} = \mathcal{O}(1)h^{-n}(V_N(\alpha) + h^{\widetilde{N}}), \tag{16.3.43}$$

$$\mathrm{tr}\, \chi(\frac{1}{\alpha}S) = \frac{1}{(2\pi h)^n}\iint \chi(\frac{s(x,\xi)}{\alpha})dxd\xi + \mathcal{O}(h^{-n})\frac{h}{\alpha}(V_N(\alpha) + h^{\widetilde{N}}). \tag{16.3.44}$$

Remark 16.3.9 Recall that P_z is defined in (16.3.2). Using simple h-pseudo-differential calculus (for instance as in the appendix of [139]), we see that if we redefine S as follows

$$S = P_z^* P_z, \tag{16.3.45}$$

then in each local coordinate chart, $S = \mathrm{Op}_h(S)$, where $S \equiv s \bmod S_{1,0}(h\langle\xi\rangle^{-1})$ and s is now redefined as

$$s(x,\xi) = \left(\frac{|p(x,\xi) - z|}{|\widetilde{p}(x,\xi) - z|}\right)^2. \tag{16.3.46}$$

The discussion goes through without any changes (now with $m = 0$) and we still have Proposition 16.3.8 with the new choice of S, s and the corresponding redefinition of $V(t)$, now restricted to $0 \leq t \leq 1/2$. Notice here that if V_{new} denotes this new function, then for some $C \geq 1$, we have

$$\frac{1}{C}V(\frac{t}{C}) \leq V_{\mathrm{new}}(t) \leq CV(Ct),\ 0 \leq t \ll 1.$$

In the remainder of this section, we choose S, s as in (16.3.45), (16.3.46). In this case S is a trace-class perturbation of the identity, whose symbol is $1+\mathcal{O}(h^\infty/\langle\xi\rangle^\infty)$ and similarly for all its derivatives, in a region $|\xi| \geq \mathrm{Const}$.

Let $0 \leq \chi \in C_0^\infty([0,\infty[)$ with $\chi(0) > 0$ and let $\alpha_0 > 0$ be small and fixed. From standard pseudodifferential calculus in the spirit of [104, Section 2], we get

$$\ln \det \left(S + \alpha_0 \chi(\frac{1}{\alpha_0} S) \right) = \frac{1}{(2\pi h)^n} \left(\iint \ln \left(s + \alpha_0 \chi(\frac{1}{\alpha_0} s) \right) dx d\xi + \mathcal{O}(h) \right). \tag{16.3.47}$$

Here is an outline of the

Proof of (16.3.47) $P = S + \alpha_0 \chi(S/\alpha_0)$ is an h-pseudodifferential operator with symbol in S^0. The principal symbol is $s + \alpha_0\chi(s/\alpha_0)$ and the full symbol P of this operator for any system of local coordinates satisfies $P - 1 \in h^N S^{-N}$ near $\xi = \infty$. Clearly $P - 1$ (on the operator level) is of trace class. Further, we have $p \geq 1/C$. Let $P_t = tP + (1-t)$ for $0 \leq t \leq 1$, so that $P_1 = P$, $P_0 = 1$. The principal symbol $p_t = tp + (1-t)$ fulfills $p_t \geq 1/C$ and P_t is uniformly invertible for $0 \leq t \leq 1$ and $h > 0$ small enough. Now

$$\partial_t \ln \det P_t = \operatorname{tr}\left(P_t^{-1} \partial_t P_t \right),$$

and by pseudodifferential operator calculus, we get

$$\begin{aligned} \partial_t \ln \det P_t &= \frac{1}{(2\pi h)^n} \left(\mathcal{O}(h) + \iint p_t^{-1} \partial_t p_t dx d\xi \right) \\ &= \frac{1}{(2\pi h)^n} \left(\mathcal{O}(h) + \iint \partial_t \ln(p_t) dx d\xi \right). \end{aligned}$$

Integration from $t = 0$ to $t = 1$ yields

$$\ln \det P = \frac{1}{(2\pi h)^n} \left(\mathcal{O}(h) + \iint \ln(p) dx d\xi \right),$$

which amounts to (16.3.47). □

Extend χ to be an element of $C_0^\infty(\mathbf{R}; \mathbf{C})$ in such a way that $t + \chi(t) \neq 0$ for all $t \in \mathbf{R}$. Then (cf. [56]), we use that

$$\frac{d}{dt} \ln \left(E + t\chi(\frac{E}{t}) \right) = \frac{1}{t} \psi(\frac{E}{t}), \tag{16.3.48}$$

where

$$\psi(E) = \frac{\chi(E) - E\chi'(E)}{E + \chi(E)}, \tag{16.3.49}$$

so that $\psi \in C_0^\infty(\mathbf{R})$. By standard functional calculus for self-adjoint operators and the classical identity

$$\frac{d}{dt}\ln(\det A_t) = \operatorname{tr} A_t^{-1}\frac{d}{dt}A_t$$

for differentiable trace-class perturbations of the identity,[1] we have

$$\frac{d}{dt}\ln\det(S + t\chi(\frac{S}{t})) = \operatorname{tr}\frac{1}{t}\psi(\frac{S}{t}). \tag{16.3.50}$$

Using (16.3.44) (now with S as in that equation), we get for $t \geq \alpha \geq h > 0$:

$$\frac{d}{dt}\ln\det\left(S + t\chi(\frac{1}{t}S)\right) = \frac{1}{(2\pi h)^n}\left(\iint \frac{1}{t}\psi(\frac{s}{t})dxd\xi + \mathcal{O}(1)\frac{h}{t^2}V_N(t) + \mathcal{O}(h^{\widetilde{N}})\right).$$

Integrating this from $t = \alpha_0$ to $t = \alpha$ and using (16.3.47), (16.3.48), we get

$$\begin{aligned}&\ln\det\left(S + \alpha\chi(\frac{1}{\alpha}S)\right) = \\ &\frac{1}{(2\pi h)^n}\left(\iint \ln\left(s + \alpha\chi(\frac{s}{\alpha})\right)dxd\xi + \mathcal{O}(h + \int_\alpha^{\alpha_0}\frac{h}{t^2}V(t)dt)\right).\end{aligned} \tag{16.3.51}$$

In fact, this follows from an estimation of

$$\begin{aligned}\int_\alpha^{\alpha_0}\frac{h}{t^2}V_N(t)dt &= \int_\alpha^{\alpha_0}\frac{h}{t^2}\int_0^1\frac{1}{(1+\frac{s}{t})^N}dV(s)dt \\ &= \int_0^1 J(s)dV(s),\end{aligned} \tag{16.3.52}$$

where

$$J(s) = \int_\alpha^{\alpha_0}\frac{h}{t^2}\frac{t^N}{(t+s)^N}dt = \frac{h}{s}\int_{\alpha/s}^{\alpha_0/s}\frac{\widetilde{t}^N}{(\widetilde{t}+1)^N}\widetilde{t}^{-2}d\widetilde{t}.$$

[1] Equation (8.4.7) extends to the case when A_0 and A_1 are trace-class operators as in Sect. 8.4 and the identity is valid for finite-rank perturbations of the identity. By Taylor expansion and partitions of unity we can approximate a C^1 family A_t of trace-class perturbations of the identity by a sequence of such perturbations $A_t^{(\nu)}$ so that $A_t^{(\nu)} \to A_t$ uniformly in C_1 and similarly for the derivatives (on any given compact interval), and so that $\mathcal{N}(A_t^{(\nu)})^\perp \cup \mathcal{R}(A_t^{(\nu)}) \subset \mathcal{H}^{(\nu)}$, where $\mathcal{H}^{(\nu)}$ is independent of t and of finite dimension. It then suffices to pass to the limit.

Considering separately the three cases, $\alpha/s \gg 1, \alpha/s \asymp 1, \alpha/s \ll 1$, we see that

$$J(s) \asymp \frac{h}{\max(\alpha, s)}.$$

Thus the term (16.3.52) is

$$\asymp \int_0^1 \frac{h}{\max(\alpha, s)} dV(s) = \frac{h}{1} V(1) + \int_\alpha^1 \frac{h}{s^2} V(s) ds,$$

and (16.3.51) follows.

Write

$$\begin{aligned}
&\iint \ln\left(s(x,\xi) + \alpha\chi(\frac{s(x,\xi)}{\alpha})\right) dx d\xi - \iint \ln s(x,\xi) dx d\xi \\
&= \int_0^\alpha \iint \frac{1}{t}\psi(\frac{s}{t}) dx d\xi dt = \int_0^\alpha \int_0^1 \frac{1}{t}\psi(\frac{\sigma}{t}) dV(\sigma) dt \\
&\le \mathcal{O}(1) \int_0^\alpha \int_0^1 \left(1 + \frac{\sigma}{t}\right)^{-N} dV(\sigma) \frac{dt}{t} = \mathcal{O}(1) \int_0^\alpha \frac{1}{t} V_N(t) dt.
\end{aligned}$$

Here we may notice that for $t \le 1/2$:

$$V_N(t) = \int_0^1 \left(1 + \frac{\sigma}{t}\right)^{-N} dV(\sigma) \ge \int_0^t \left(1 + \frac{\sigma}{t}\right)^{-N} dV(\sigma) \ge 2^{-N} V(t).$$

Combining the above computation with (16.3.51), we get

Proposition 16.3.10 *If* $0 \le \chi \in C_0^\infty([0,\infty[)$, $\chi(0) > 0$, *we have uniformly for* $0 < h \le \alpha \ll 1$

$$\ln\det\left(S + \alpha\chi(\frac{1}{\alpha}S)\right) =$$
$$\frac{1}{(2\pi h)^n} \left(\iint \ln s(x,\xi) dx d\xi + \mathcal{O}(1)\left(h + \int_0^\alpha V_N(t)\frac{dt}{t} + \int_\alpha^{\alpha_0} \frac{h}{t} V(t) \frac{dt}{t}\right)\right). \tag{16.3.53}$$

16.4 Subharmonicity and Symplectic Volume

Recall that P_z is defined by (16.3.2) and has the semi-classical principal symbol

$$p_z(\rho) = (\widetilde{p}(\rho) - z)^{-1}(p - z),\ z \in \Omega.$$

In Sect. 16.3 we encountered the function

$$\phi(z) = \iint \ln |p_z(x,\xi)| dx d\xi = \frac{1}{2} \iint \ln s_z(x,\xi) dx d\xi. \tag{16.4.1}$$

We also considered the function

$$V(t) = V_z(t) = \iint_{|p(x,\xi)-z|^2 \le t} dx d\xi, \tag{16.4.2}$$

and the closely related one

$$\widetilde{V}_z(t) = \iint_{|p_z(x,\xi)|^2 \le t} dx d\xi. \tag{16.4.3}$$

These two functions will be used only for small values of t; they are equivalent, in the sense that

$$\frac{1}{C} V_z(\frac{t}{C}) \le \widetilde{V}_z(t) \le C V_z(Ct),\ 0 \le t \le 1/C,\ z \in \Omega, \tag{16.4.4}$$

for some $C \gg 1$, as we saw after (16.3.46) where the notation was slightly different.

For $\kappa \in]0,1]$ consider the property that

$$V_z(t) = \mathcal{O}(t^\kappa), \quad 0 \le t \ll 1. \tag{16.4.5}$$

Recall that m is the order of the elliptic differential operator P.

Proposition 16.4.1 Equation (16.4.5) *holds uniformly for* $z \in \Omega$ *when* $\kappa = \frac{1}{2m}$.

Proof Because of the ellipticity, the set of points where $|p(x,\xi) - z| \le 1$ is contained in a fixed compact set in T^*X, independent of z, that we can cover by finitely many sets of the form T^*U_j, $j = 1, 2, \dots, N_0$, where U_j are local coordinates charts that we can identify with open bounded subsets of $\mathbf{R}^n$. On each such T^*U_j we know that $\xi_n \mapsto p(x,\xi) - z$ is a polynomial of degree m with leading term $a_j(x)\xi_n^m$, $a_j(x) \neq 0$. By the standard factorization of this polynomial;

$$p(x,\xi) - z = a_j(x)(\xi_n - \lambda_1(x,\xi',z)) \cdots (\xi_n - \lambda_m(x,\xi',z)),$$

we see that $|p(x,\xi) - z| \ge t^{1/2}$ when $\mathbf{C} \ni \xi_n \notin \bigcup_j D(\lambda_j(x,\xi',z), Ct^{1/(2m)})$ where $C > 0$ is independent of x, ξ', z after shrinking U_j slightly. It follows that

$$L(\{\xi_n \in \mathbf{R};\ |p(x,\xi) - z| \le t^{1/2}\}) = \mathcal{O}(1) t^{\frac{1}{2m}}$$

uniformly for $x \in U_j$, $\xi' \in \mathbf{R}^{n-1}$, where now L denotes the Lebesgue measure on the real line. The result then follows from Fubini's theorem. □

Proposition 16.4.2 *The function $\phi(z)$ defined in* (16.4.1) *is continuous and subharmonic on Ω. Moreover,*

$$\frac{\Delta\phi(z)}{2\pi}L(dz) = p_*(dxd\xi), \tag{16.4.6}$$

where the right-hand side is defined to be the direct image under the symbol map: $T^*X \ni (x,\xi) \mapsto p(x,\xi) \in \mathbf{C}$ of the symplectic volume element $dxd\xi$ (which we also denote $d\rho$, $\rho = (x,\xi)$).

Proof The property (16.4.5) for some $\kappa > 0$ (here with $\kappa = 1/(2m)$) implies that the integral (16.4.1) converges. Moreover, with $0 \le \chi \in C_0^\infty(\mathbf{R})$, $\chi(0) > 0$,

$$\phi(z) = \lim_{\alpha\to 0}\frac{1}{2}\iint \ln\left(s_z(x,\xi) + \alpha\chi(\frac{s_z(x,\xi)}{\alpha})\right)dxd\xi,$$

where the convergence is locally uniform, so that $\phi(z)$ is the locally uniform limit of continuous functions and hence continuous.

Next, $p_z(\rho) = (\widetilde{p}(\rho) - z)^{-1}(p(\rho) - z)$, where the first factor is holomorphic and non-vanishing for $z \in \Omega$, so

$$\Delta_z \ln|p_z(\rho)| = \Delta_z \ln|p(\rho) - z| = 2\pi\delta(z - p(\rho)).$$

If $\psi \in C_0^\infty(\Omega)$, we get

$$\begin{aligned}
&\int (\Delta_z\phi(z))\psi(z)L(dz) = \int \phi(z)\Delta_z\psi(z)L(dz)\\
&= \int_{T^*X}\int_{\mathbf{C}} \ln|p_z(\rho)|\Delta_z\psi(z)L(dz)d\rho = \int_{T^*X}\int_{\mathbf{C}} \Delta_z(\ln|p_z(\rho)|)\psi(z)L(dz)d\rho\\
&\qquad = 2\pi\int_{T^*X}\int_{\mathbf{C}} \delta(z - p(\rho))\psi(z)L(dz)d\rho = 2\pi\int_{T^*X}\psi(p(\rho))d\rho,
\end{aligned}$$

which establishes (16.4.6). □

16.5 Extension of the Bounds to the Perturbed Case

For $s > n/2$, we consider the perturbed operator

$$P_\delta = P + \delta(h^{\frac{n}{2}} q_1 + q_2) = P + \delta(Q_1 + Q_2) = P + \delta Q, \tag{16.5.1}$$

where $q_j \in H^s(X)$,

$$\|q_1\|_{H_h^s} \leq 1, \ \|q_2\|_{H^s} \leq 1, \ 0 \leq \delta \ll 1. \tag{16.5.2}$$

According to Propositions 16.2.1 and 16.2.4, $Q = \mathcal{O}(1) : H_h^s \to H_h^s$, and hence by duality and interpolation,

$$Q = \mathcal{O}(1) : H_h^\sigma \to H_h^\sigma, \ -s \leq \sigma \leq s. \tag{16.5.3}$$

Let $\widetilde{P}_\delta := P_\delta + \widetilde{P} - P$. Here

$$(\widetilde{P} - z)^{-1} = \mathcal{O}(1) : \begin{cases} H_h^\sigma \to H_h^{\sigma+m}, \\ H_h^{\sigma-m} \to H_h^\sigma, \end{cases}$$

for $-s \leq \sigma \leq s$ and using (16.5.1), (16.5.3), the resolvent identity, and the fact that $0 \leq \delta \ll 1$, we get the same conclusion for $(\widetilde{P}_\delta - z)^{-1}$. The spectrum of P_δ in Ω is discrete and from Remark 14.1.2 it follows that it coincides with the set of zeros of

$$\det((\widetilde{P}_\delta - z)^{-1}(P_\delta - z)) = \det(1 - (\widetilde{P}_\delta - z)^{-1}(\widetilde{P} - P)). \tag{16.5.4}$$

Put

$$P_{\delta,z} := (\widetilde{P}_\delta - z)^{-1}(P_\delta - z) = 1 - (\widetilde{P}_\delta - z)^{-1}(\widetilde{P} - P) =: 1 - K_{\delta,z}, \tag{16.5.5}$$

$$S_\delta := (P_\delta - z)^*(P_\delta - z), \tag{16.5.6}$$

$$S_{\delta,z} := P_{\delta,z}^* P_{\delta,z} = 1 - (K_{\delta,z} + K_{\delta,z}^* - K_{\delta,z}^* K_{\delta,z}) =: 1 - L_{\delta,z}. \tag{16.5.7}$$

Then

$$K_{\delta,z}, L_{\delta,z} = \mathcal{O}(1) : H_h^{-s-m} \to H_h^{s+m}. \tag{16.5.8}$$

Observe that

$$\|K_{\delta,z}\|_{\mathrm{tr}} \leq \|(\widetilde{P}_\delta - z)^{-1}\| \|\widetilde{P} - P\|_{\mathrm{tr}} \leq \mathcal{O}(h^{-n}), \ \|L_{\delta,z}\|_{\mathrm{tr}} \leq \mathcal{O}(h^{-n}). \tag{16.5.9}$$

We shall extend Proposition 16.3.8, Remark 16.3.9 and Proposition 16.3.10 to the perturbed case for $\delta \geq 0$ small enough.

As in Proposition 16.3.8, let $\chi \in C_0^\infty([0,+\infty[)$ and let $\widetilde{\chi} \in C_0^\infty(\mathbf{C})$ be an almost holomorphic extension of χ. Then, working in the parameter range $0 < h \leq \alpha \ll 1$, we have

$$\chi(\alpha^{-1}S_\delta) = -\frac{1}{\pi}\int (w - S_\delta)^{-1}\frac{1}{\alpha}(\partial_{\overline{z}}\widetilde{\chi})(\frac{w}{\alpha})L(dw). \tag{16.5.10}$$

This is not obviously of trace-class for all values of s, n, m, so we make a modification: We can find $0 \leq \psi \in C_0^\infty(T^*X)$ such that

$$|p(x,\xi) - z|^2 + \psi(x,\xi) \geq \frac{1}{C}\langle\xi\rangle^{2m}, \text{ on } T^*X,$$

for all $z \in \Omega$. Let $\psi = \psi(x, hD_x)$ denote also a corresponding self-adjoint quantization. Then for $h > 0$ small enough,

$$(P - z)^*(P - z) + \psi \geq \frac{1}{C}\langle hD\rangle^{2m}.$$

Now,

$$S_\delta = S_0 + \delta Q^*(P - z) + (P - z)^*\delta Q + \delta^2 Q^* Q, \tag{16.5.11}$$

so for δ, h small enough we get

$$S_\delta + \psi \geq \frac{1}{2C}\langle hD\rangle^{2m}.$$

In particular, $(S_\delta + \psi - w)^{-1}$ exists and is uniformly bounded for w in a small neighborhood of $0 \in \mathbf{C}$. On the other hand,

$$(w - S_\delta)^{-1} = (w - (S_\delta + \psi))^{-1} - (w - S_\delta)^{-1}\psi(w - (S_\delta + \psi))^{-1}. \tag{16.5.12}$$

When inserting this in (16.5.10) the first term gives the contribution zero, since it is holomorphic in a neighborhood of 0, and we get

$$\chi(\alpha^{-1}S_\delta) = \frac{1}{\pi}\int (w - S_\delta)^{-1}\psi(w - (S_\delta + \psi))^{-1}\frac{1}{\alpha}(\partial_{\overline{z}}\widetilde{\chi})(\frac{w}{\alpha})L(dw), \tag{16.5.13}$$

which is of trace class, since ψ is.

Differentiating this relation with respect to δ, we get

$$\begin{aligned}
&\frac{\partial}{\partial\delta}\chi(\alpha^{-1}S_\delta) = \\
&\frac{1}{\pi}\int (w-S_\delta)^{-1}\dot{S}_\delta(w-S_\delta)^{-1}\psi(w-(S_\delta+\psi))^{-1}\frac{1}{\alpha}(\partial_{\overline{z}}\widetilde{\chi})(\frac{w}{\alpha})L(dw) \\
&+\frac{1}{\pi}\int (w-S_\delta)^{-1}\psi(w-(S_\delta+\psi))^{-1}\dot{S}_\delta(w-(S_\delta+\psi))^{-1}\frac{1}{\alpha}(\partial_{\overline{z}}\widetilde{\chi})(\frac{w}{\alpha})L(dw),
\end{aligned} \tag{16.5.14}$$

where

$$\dot{S}_\delta = \frac{\partial}{\partial\delta}S_\delta = Q^*(P-z)+(P-z)^*Q+2\delta Q^*Q = \mathcal{O}(1): H_h^{\sigma+m}\to H_h^{\sigma-m}.$$

From (16.5.11) it follows that $S_\delta = S_0 + R$, where $R = \mathcal{O}(\delta) : H_h^{\sigma+m} \to H_h^{\sigma-m}$, $-s \le \sigma \le s$. If w_0 belongs to some compact set in $\mathbf{C}$ that is disjoint from $\mathbf{R}$, it follows that for δ small enough, $w_0 - S_\delta : H_h^{\sigma+m} \to H_h^{\sigma-m}$ is bijective for $-s \le \sigma \le s$. In fact,

$$w_0 - S_\delta = w_0 - S_0 - R = (w_0 - S_0)(1-(w_0-S_0)^{-1}R),$$

and we see that $(w_0 - S_0)^{-1} = \mathcal{O}(1) : H_h^{\sigma-m} \to H_h^{\sigma+m}$, so $(w_0 - S_0)^{-1}R = \mathcal{O}(\delta) : H_h^{\sigma+m} \to H_h^{\sigma+m}$, and hence

$$(w_0-S_\delta)^{-1} = (1-(w_0-S_0)^{-1}R)^{-1}(w_0-S_0)^{-1} = \mathcal{O}(1): H_h^{\sigma-m}\to H_h^{\sigma+m},$$

as claimed.

Let w belong to a bounded set in $\mathbf{C}$ disjoint from $\mathbf{R}$. Iterating the resolvent identity, we have

$$\begin{aligned}
&(w-S_\delta)^{-1} = (w_0-S_\delta)^{-1} + (w_0-w)(w_0-S_\delta)^{-2}+\cdots \\
&+(w_0-w)^{2N-1}(w_0-S_\delta)^{-2N} + (w_0-w)^{2N}(w-S_\delta)^{-1}(w_0-S_\delta)^{-2N}.
\end{aligned}$$

The last term can also be written in a sandwiched form as

$$(w_0-w)^{2N}(w_0-S_\delta)^{-N}(w-S_\delta)^{-1}(w_0-S_\delta)^{-N}.$$

For N large enough, we have $(w_0 - S_\delta)^{-N} = \mathcal{O}(1) : H_h^{-s-m} \to H_h^{s+m}$. In particular,

$$(w_0-S_\delta)^{-N} = \mathcal{O}(1): \begin{cases} H_h^{\sigma-m}\to H^0, \\ H^0 \to H_h^{\sigma+m}. \end{cases}$$

Since $(w - S_\delta)^{-1} = \mathcal{O}(|\Im w|^{-1}) : H^0 \to H^0$, we conclude that

$$(w - S_\delta)^{-1} = \mathcal{O}(|\Im w|^{-1}) : \ H_h^{\sigma-m} \to H_h^{\sigma+m}. \tag{16.5.15}$$

Similarly, with ψ as in (16.5.12), positive on a sufficiently large set depending on the bounded set where we let w vary,

$$(w - (S_\delta + \psi))^{-1} = \mathcal{O}(1) : \ H_h^{\sigma-m} \to H_h^{\sigma+m}. \tag{16.5.16}$$

Using also that $\|\psi\|_{\mathrm{tr}} = \mathcal{O}(h^{-n})$, $(\partial_{\overline{z}}\widetilde{\chi})(w/\alpha) = \mathcal{O}_N((|\Im w|/\alpha)^N)$ for all $N \geq 0$ and choosing $\sigma = 0$, we see that the trace norm of the first integral in (16.5.14) is

$$\mathcal{O}(1) \int_{|w| \leq \mathcal{O}(\alpha)} |\Im w|^{-2} h^{-n} \frac{1}{\alpha} \left(\frac{|\Im w|}{\alpha}\right)^N L(dw),$$

while the second integral satisfies the better bound with the exponent -2 replaced by -1. Choosing $N = 2$, we get

$$\|\frac{\partial}{\partial \delta} \chi(\alpha^{-1} S_\delta)\|_{\mathrm{tr}} = \mathcal{O}(1)\alpha^{-1} h^{-n}. \tag{16.5.17}$$

Integrating this "from 0 to δ" and applying Proposition 16.3.8, we get

Proposition 16.5.1 *Let $\chi \in C_0^\infty(\mathbf{R})$. For $0 < h \leq \alpha \ll 1$, we have*

$$\|\chi(\alpha^{-1} S_\delta)\|_{\mathrm{tr}} \leq \mathcal{O}(1) h^{-n} \left(V_N(\alpha) + h^{\widetilde{N}} + \frac{\delta}{\alpha}\right), \tag{16.5.18}$$

$$\operatorname{tr} \chi(\alpha^{-1} S_\delta) = \frac{1}{(2\pi h)^n} \iint \chi(\frac{s(x,\xi)}{\alpha}) dx d\xi + \mathcal{O}(1) h^{-n} \left(V_N(\alpha) + h^{\widetilde{N}} + \frac{\delta}{\alpha}\right), \tag{16.5.19}$$

where $s(x, \xi) = |p(x, \xi) - z|^2$.

We next extend Remark 16.3.9 to the perturbed case. Recall the expressions for $K_{\delta,z}$ and $L_{\delta,z}$ in (16.5.5), (16.5.7) and also (16.5.9). We have

$$\dot{K}_{\delta,z} = -(z - \widetilde{P}_\delta)^{-1} Q (z - \widetilde{P}_\delta)^{-1} (\widetilde{P} - P), \tag{16.5.20}$$

so

$$\|\dot{K}_{\delta,z}\| \leq \mathcal{O}(1), \ \|\dot{K}_{\delta,z}\|_{\mathrm{tr}} \leq \mathcal{O}(h^{-n}), \tag{16.5.21}$$

since $\|Q\| = \mathcal{O}(1)$. Here the dot indicates differentiation with respect to δ. It follows that

$$\|\dot{S}_{\delta,z}\|_{\mathrm{tr}} \le \mathcal{O}(h^{-n}). \tag{16.5.22}$$

With $\chi \in C_0^\infty(\mathbf{R})$, we get

$$\frac{\partial}{\partial\delta}\chi(\alpha^{-1}S_{\delta,z}) = -\frac{1}{\pi}\int (w - S_{\delta,z})^{-1}\dot{S}_{\delta,z}(w - S_{\delta,z})^{-1}\frac{1}{\alpha}(\partial_{\overline{z}}\widetilde{\chi})(\frac{w}{\alpha})L(dw), \tag{16.5.23}$$

which again leads to

$$\|\frac{\partial}{\partial\delta}\chi(\alpha^{-1}S_{\delta,z})\|_{\mathrm{tr}} = \mathcal{O}(1)\alpha^{-1}h^{-n} \tag{16.5.24}$$

and combining this with Remark 16.3.9, we get

Proposition 16.5.2 *Proposition* 16.5.1 *remains valid if we replace* (S_δ, s) *with* $(S_{\delta,z}, s_z)$, *where*

$$s_z(x,\xi) = \frac{|p(x,\xi) - z|^2}{|\widetilde{p}(x,\xi) - z|^2}.$$

Finally, we shall extend Proposition 16.3.10 to the perturbed case. Let $0 \le \chi \in C_0^\infty([0,\infty[), \chi(0) > 0$. Put

$$f(t) = t + \alpha\chi(\frac{t}{\alpha}) =: t + g(t),\ g = g_\alpha \in C_0^\infty.$$

For $0 < h \le \alpha \ll 1$, we have

$$f(S_{\delta,z}) = S_{\delta,z} - \frac{1}{\pi}\int (w - S_{\delta,z})^{-1}\overline{\partial}\widetilde{g}(w)L(dw).$$

From this we see that

$$\frac{\partial}{\partial\delta}f(S_{\delta,z}) = \dot{S}_{\delta,z} - \frac{1}{\pi}\int (w - S_{\delta,z})^{-1}\dot{S}_{\delta,z}(w - S_{\delta,z})^{-1}\overline{\partial}\widetilde{g}(w)L(dw).$$

Now,

$$\frac{\partial}{\partial\delta}\ln\det f(S_{\delta,z}) = \mathrm{tr}\, f(S_{\delta,z})^{-1}\frac{\partial}{\partial\delta}f(S_{\delta,z}) =$$
$$\mathrm{tr}\,(f(S_\delta)^{-1}\dot{S}_\delta) - \frac{1}{\pi}\int \mathrm{tr}\,(f(S_{\delta,z})^{-1}(w - S_{\delta,z})^{-1}\dot{S}_{\delta,z}(w - S_{\delta,z})^{-1})\overline{\partial}\widetilde{g}(w)L(dw).$$

Here $f(S_{\delta,z})^{-1}$ and $(w - S_{\delta,z})^{-1}$ commute, so by the cyclicity of the trace and an integration by parts, we see that the last term is equal to

$$\begin{aligned}
&\operatorname{tr}\left(f(S_{\delta,z})^{-1}\frac{(-1)}{\pi}\int (w-S_{\delta,z})^{-2}\overline{\partial}_w\widetilde{g}(w)L(dw)\dot{S}_{\delta,z}\right)\\
&=\operatorname{tr}\left(f(S_{\delta,z})^{-1}\frac{(-1)}{\pi}\int (w-S_{\delta,z})^{-1}\overline{\partial}_w\partial_w\widetilde{g}(w)L(dw)\dot{S}_{\delta,z}\right)\\
&=\operatorname{tr}\left(f(S_{\delta,z})^{-1}g'(S_{\delta,z})\dot{S}_{\delta,z}\right),
\end{aligned}$$

leading to the general identity

$$\frac{\partial}{\partial\delta}\ln\det f(S_{\delta,z})=\operatorname{tr}\left(f(S_{\delta,z})^{-1}f'(S_{\delta,z})\dot{S}_{\delta,z}\right).$$

With the above choice of f we have $f|_{[0,+\infty[}\geq \alpha/C$, so $\|f(S_{\delta,z})^{-1}\|\leq C/\alpha$. Moreover, $f'=\mathcal{O}(1)$, so $\|f'(S_{\delta,z})\|=\mathcal{O}(1)$. Using also (16.5.22), we conclude that

$$\frac{\partial}{\partial\delta}\ln\det f(S_{\delta,z})=\mathcal{O}(\alpha^{-1}h^{-n}).$$

Integrating from 0 to δ and using Proposition 16.3.10, we get

Proposition 16.5.3 *If* $0\leq\chi\in C_0^\infty([0,+\infty[)$, $\chi(0)>0$, *we have, uniformly for* $0<h\leq\alpha\ll 1$,

$$\begin{aligned}
\ln\det(S_{\delta,z}+\alpha\chi(\alpha^{-1}S_{\delta,z}))&=\frac{1}{(2\pi h)^n}\left(\iint \ln s_z(x,\xi)dxd\xi\right.\\
&\left.+\mathcal{O}(1)\left(\int_0^\alpha V_N(t)\frac{dt}{t}+\int_\alpha^{\alpha_0}\frac{h}{t}V(t)\frac{dt}{t}+\frac{\delta}{\alpha}+h\right)\right).
\end{aligned}$$

We end the section with some remarks about H^s-properties of eigenfunctions for low lying eigenvalues of S_δ and $S_{\delta,z}$, here under the only assumption on δ that $0\leq\delta\ll 1$.

We start with S_δ. Let $0\leq t_1^2\leq t_2^2\leq\cdots$ denote the increasing sequence of eigenvalues of S_δ, repeated according to their multiplicity and with $t_j\geq 0$. Let $e_1, e_2,\dots$ be a corresponding orthonormal family of eigenfunctions. Proposition 16.5.1 shows that

$$N_0:=\#\{k;\ t_k^2\leq 1/2\}\leq\mathcal{O}(h^{-n}) \tag{16.5.25}$$

Proposition 16.5.4 *Under the assumptions above we have*

$$\|\sum_{j=1}^{N_0}\lambda_j e_j\|_{H_h^{s+m}} \le \mathcal{O}_s(1)\|\vec{\lambda}\|_{\ell^2}, \tag{16.5.26}$$

where $\vec{\lambda} = \begin{pmatrix}\lambda_1 \ \lambda_2 \ \dots \ \lambda_{N_0}\end{pmatrix}^{\rm t}$.

Proof Using (16.5.15), (16.5.16) in (16.5.13) with $\alpha = 1$, we see that for every $\chi \in C_0^\infty(\mathbf{R})$,

$$\chi(S_\delta) = \mathcal{O}(1) : H_h^{-s-m} \to H_h^{s+m}. \tag{16.5.27}$$

(We choose the set where $\psi > 0$ is sufficiently large depending on the support of χ and of its almost holomorphic extension, and use that $\psi = \mathcal{O}(1) : H_h^{-s+m} \to H_h^{s-m}$.)

Let $\chi \ge 0$ be equal to 1 on $[0, 1/2]$, so that $\chi(S_\delta)(\sum_1^{N_0}\lambda_j e_j) = \sum_1^{N_0}\lambda_j e_j$. Using only that $\chi(S_\delta) = \mathcal{O}(1) : H^0 \to H_h^{s+m}$, we get (16.5.26). □

Next consider $S_{\delta,z}$ in (16.5.7). From (16.5.20), the fact that $\widetilde{P} - P = \mathcal{O}(1) : H_h^{s_1} \to H_h^{s_2}$, for all $s_1,\ s_2$ and the observation after (16.5.3), we see that $\dot{K}_{\delta,z} = \mathcal{O}(1) : H_h^{-s-m} \to H_h^{s+m}$ and by duality and the explicit formula for $L_{\delta,z}$, we get the same facts for $\dot{K}_{\delta,z}^*$ and $\dot{L}_{\delta,z}$. Thus, $\dot{S}_{\delta,z} = \mathcal{O}(1) : H_h^{-s-m} \to H_h^{s+m}$, and we conclude that

$$S_{\delta,z} = S_{0,z} + R_{\delta,z},\ R_{\delta,z} = \mathcal{O}(1) : H_h^{-s-m} \to H_h^{s+m}.$$

We then apply the sandwiched resolvent identity

$$\begin{aligned}(w - S_{\delta,z})^{-1} =&(w - S_{0,z})^{-1} + (w - S_{0,z})^{-1}R_{\delta,z}(w - S_{0,z})^{-1}\\ &+ (w - S_{0,z})^{-1}R_{\delta,z}(w - S_{\delta,z})^{-1}R_{\delta,z}(w - S_{0,z})^{-1},\end{aligned} \tag{16.5.28}$$

to get for $\chi \in C_0^\infty([0, 3/4[\,)$ that

$$\begin{aligned}\chi(S_{\delta,z}) =&\chi(S_{0,z}) - \frac{1}{\pi}\int (w - S_{0,z})^{-1}R_{\delta,z}(w - S_{0,z})^{-1}\overline{\partial}\widetilde{\chi}(w)\,L(dw)\\ &- \frac{1}{\pi}\int (w - S_{0,z})^{-1}R_{\delta,z}(w - S_{\delta,z})^{-1}R_{\delta,z}(w - S_{0,z})^{-1}\overline{\partial}\widetilde{\chi}(w)L(dw).\end{aligned}$$

Here $\chi(S_{0,z})$ is an h-pseudodifferential operator of order 0 in h and of order $-\infty$ in ξ, so $\chi(S_{0,\delta}) = \mathcal{O}(1) : H_h^{-s-m} \to H_h^{s+m}$. Each of the two integrals define operators that are $\mathcal{O}(1) : H_h^{-s-m} \to H_h^{s+m}$. In fact, $(w - S_{0,z})^{-1} = \mathcal{O}(|\Im w|^{-1}) : H_h^\sigma \to H_h^\sigma$ for every real σ.

This is the analogue of (16.5.27), and completing the discussion as at the end of the proof of Proposition 16.5.4, we get

Proposition 16.5.5 *Proposition* 16.5.4 *remains valid if we let* $e_1, \dots, e_{N_0}$ *(still with* $N_0 = \mathcal{O}(h^{-n})$*) be an orthonormal family of eigenfunctions corresponding to the spectrum of* $S_{\delta,z}$ *in* $[0, 1/2]$.

Notes This chapter gives the first part of the proof of Theorem 15.3.1, following closely [137,139]. See Proposition 16.5.3 and the corresponding upper bound on $\ln \det S_{\delta,z}$, which is bounded from above by $\ln \det(S_{\delta,z} + \alpha\chi(S_{\delta,z}/\alpha))$. In Chap. 17 we give corresponding lower bounds, valid with probability close to 1—while the upper bounds are valid for all sufficiently small perturbations.

17 Proof II: Lower Bounds

In this chapter we give a lower bound on ln det $S_{\delta,z}$ which is valid with high probability, and then using also the upper bounds of Chap. 16, we conclude the proof of Theorem 15.3.1 with the help of Theorem 12.1.2.

17.1 Singular Values and Determinants of Certain Matrices Associated to δ Potentials

As in Chap. 16, let X be a compact smooth manifold of dimension n, equipped with a positive smooth density of integration dx. Let $e_1(x), \dots, e_N(x)$ be continuous functions on X. The discussion will remain valid if we replace X by a compact subset with non-empty interior and smooth boundary. We would like to find a continuous function $q : X \to \mathbf{C}$ such that the matrix $M_q = (M_{q;j,k})_{1\le j,k\le N}$, given by

$$M_{q;j,k} = \int_X q(x)e_j(x)e_k(x)dx,$$

has nice lower bounds for its singular values. In the present section, we shall achieve such a goal when q is replaced by a sum of Dirac masses of the form $\delta_a = \sum_1^N \delta(x - a_j)$ for a suitable $a = (a_1, a_2, \dots, a_N) \in X^N$. Notice that $M_{\delta_a} = (M_{j,k})$, where

$$M_{j,k} = \sum_{\nu=1}^{N} e_j(a_\nu)e_k(a_\nu) \tag{17.1.1}$$

J. Sjöstrand, *Non-Self-Adjoint Differential Operators, Spectral Asymptotics and Random Perturbations*, Pseudo-Differential Operators 14,
https://doi.org/10.1007/978-3-030-10819-9_17

and that

$$M = E \circ E^{\mathrm{t}}, \tag{17.1.2}$$

where

$$E = (e_j(a_k))_{1 \le j,k \le N}.$$

If we introduce the column vectors

$$\vec{e}\,(x) = \begin{pmatrix} e_1(x) \\ e_2(x) \\ \vdots \\ \vdots \\ e_N(x) \end{pmatrix} \in \mathbf{C}^N,$$

then $E = (\vec{e}\,(a_1),\ \vec{e}\,(a_2),\ \dots,\ \vec{e}\,(a_N))$.

Lemma 17.1.1 *Let $M \in [1, N]$ be an integer and let $L \subset \mathbf{C}^N$ be a linear subspace of dimension $M-1$, for some $1 \le M \le N$. Then there exists $x \in X$ such that*

$$\operatorname{dist}(\vec{e}\,(x), L)^2 \ge \frac{1}{\operatorname{vol}(X)} \operatorname{tr}((1-\pi_L)\mathcal{E}_X). \tag{17.1.3}$$

Here $\mathcal{E}_X = ((e_j|e_k)_{L^2(X)})_{1 \le j,k \le N}$ is the Gramian of $(e_1, \dots, e_N)$ and π_L denotes the orthogonal projection from $\mathbf{C}^N$ onto L.

Proof Let $\nu_1, \dots, \nu_N$ be an orthonormal basis in $\mathbf{C}^N$ such that L is spanned by $\nu_1, \dots, \nu_{M-1}$ (and equal to 0 when $M = 1$). Let $(\cdot|\cdot)$ denote the usual scalar product on $\mathbf{C}^N$ and let $(\cdot|\cdot)_X$ be the scalar product on $L^2(X)$. Write

$$\nu_\ell = \begin{pmatrix} \nu_{1,\ell} \\ \vdots \\ \vdots \\ \nu_{N,\ell} \end{pmatrix}.$$

We have

$$\begin{aligned}\operatorname{dist}(\overrightarrow{e}(x), L)^2 &= \sum_{\ell=M}^{N} |(\overrightarrow{e}(x)|\nu_\ell)|^2 \\ &= \sum_{\ell=M}^{N} |\sum_j e_j(x)\overline{\nu}_{j,\ell}|^2 \\ &= \sum_{\ell=M}^{N} \sum_{j,k} \overline{\nu}_{j,\ell} e_j(x) \overline{e}_k(x) \nu_{k,\ell}.\end{aligned}$$

It follows that

$$\int_X \operatorname{dist}(\overrightarrow{e}(x), L)^2 dx = \sum_{\ell=M}^{N} (\mathcal{E}_X \nu_\ell|\nu_\ell) = \operatorname{tr}((1-\pi_L)\mathcal{E}_X).$$

It then suffices to estimate the integral from above by

$$\operatorname{vol}(X) \max_{x\in X} \operatorname{dist}(\overrightarrow{e}(x), L)^2.$$

□

If we make the assumption that

$$e_1, \dots, e_N \text{ is an orthonormal family in } L^2(X), \tag{17.1.4}$$

then $\mathcal{E}_X = 1$ and (17.1.3) simplifies to

$$\max_{x\in X} \operatorname{dist}(\overrightarrow{e}(x), L)^2 \geq \frac{N-M+1}{\operatorname{vol}(X)}. \tag{17.1.5}$$

In the general case, let $0 \leq \varepsilon_1 \leq \varepsilon_2 \leq \cdots \leq \varepsilon_N$ denote the eigenvalues of $\mathcal{E}_X$. Then we have

$$\inf_{\dim L = M-1} \operatorname{tr}((1-\pi_L)\mathcal{E}_X) = \varepsilon_1 + \varepsilon_2 + \cdots + \varepsilon_{N-M+1} =: E_M. \tag{17.1.6}$$

Indeed, the min-max principle shows that

$$\varepsilon_k = \inf_{\dim L' = k} \sup_{\substack{\nu \in L' \\ \|\nu\|=1}} (\mathcal{E}_X \nu|\nu),$$

and so for a general subspace L of dimension $M-1$, the eigenvalues of $(1-\pi_L)\mathcal{E}_X(1-\pi_L)$ are $\varepsilon_1' \leq \cdots \leq \varepsilon_{N-M+1}'$, with $\varepsilon_j' \geq \varepsilon_j$.

Now, we can use the lemma to choose successively $a_1, \ldots, a_N \in X$ such that

$$\|\vec{e}(a_1)\|^2 \geq \frac{E_1}{\mathrm{vol}(X)},$$

$$\mathrm{dist}(\vec{e}(a_2), \mathbf{C}\vec{e}(a_1))^2 \geq \frac{E_2}{\mathrm{vol}(X)},$$

$$\ldots$$

$$\mathrm{dist}(\vec{e}(a_M), \mathbf{C}\vec{e}(a_1) \oplus \cdots \oplus \mathbf{C}\vec{e}(a_{M-1}))^2 \geq \frac{E_M}{\mathrm{vol}(X)},$$

$$\ldots$$

Let $\nu_1, \nu_2, \ldots, \nu_N$ be the Gram–Schmidt orthonormalization of the basis $\vec{e}(a_1)$, $\vec{e}(a_2), \ldots, \vec{e}(a_N)$, so that

$$\vec{e}(a_M) \equiv c_M \nu_M \bmod (\nu_1, \ldots, \nu_{M-1}), \text{ where } |c_M| \geq \left(\frac{E_M}{\mathrm{vol}(X)}\right)^{\frac{1}{2}}. \tag{17.1.7}$$

Recall that E is the $N \times N$ matrix with the columns $\vec{e}(a_j)$. Expressing these vectors in the basis $\nu_1, \ldots, \nu_N$ will not change the absolute value of the determinant and E now becomes an upper triangular matrix with diagonal entries $c_1, \ldots, c_N$. Hence

$$|\det E| = |c_1 \cdots c_N|, \tag{17.1.8}$$

and (17.1.7) implies that

$$|\det E| \geq \frac{(E_1 E_2 \cdots E_N)^{1/2}}{(\mathrm{vol}(X))^{N/2}}. \tag{17.1.9}$$

We now return to M in (17.1.1), (17.1.2). Since $\det M = (\det E)^2$, we have

$$|\det M| \geq \frac{E_1 E_2 \cdots E_N}{\mathrm{vol}(X)^N}. \tag{17.1.10}$$

Under the assumption (17.1.4), this simplifies to

$$|\det M| \geq \frac{N!}{\mathrm{vol}(X)^N}. \tag{17.1.11}$$

It will also be useful to estimate the singular values $s_1(M) \geq s_2(M) \geq \cdots \geq s_N(M)$ of the matrix M (by definition, the decreasing sequence of eigenvalues of the matrix $(M^*M)^{\frac{1}{2}}$). Clearly,

$$s_1^N \geq s_1^{k-1} s_k^{N-k+1} \geq \prod_{j=1}^{N} s_j = |\det M|, \quad 1 \leq k \leq N, \tag{17.1.12}$$

and we recall that

$$s_1 = \|M\|. \tag{17.1.13}$$

Combining (17.1.10) and (17.1.12), we get

Proposition 17.1.2 *Under the above assumptions,*

$$s_1 \geq \frac{(E_1 \cdots E_N)^{\frac{1}{N}}}{\mathrm{vol}\,(X)}, \tag{17.1.14}$$

$$s_k \geq s_1 \left(\prod_{j=1}^{N} \left(\frac{E_j}{s_1 \mathrm{vol}\,(X)} \right) \right)^{\frac{1}{N-k+1}}. \tag{17.1.15}$$

17.2 Singular Values of Matrices Associated to Suitable Admissible Potentials

We shall now carry over the results of the preceding section to potentials that are linear combinations of eigenfunctions of the auxiliary operator $h^2\widetilde{R}$. Recall the definition of ϵ_k, μ_k, in Sect. 16.2: $\widetilde{R}\epsilon_k = (\mu_k^0)^2\epsilon_k$, $\mu_k = h\mu_k^0$. Also recall that $D := \#\{k;\ \mu_k \leq L\} = \mathcal{O}(\langle L\rangle^{n/2}/h^n)$ by Weyl asymptotics for elliptic self-adjoint operators on a compact manifold.

Definition 17.2.1 An admissible potential is a potential of the form

$$q(x) = \sum_{0<\mu_k\leq L} \alpha_k \epsilon_k(x),\ \alpha \in \mathbf{C}^D. \tag{17.2.1}$$

We shall construct an admissible potential q for which the singular values of $P + \delta q$ satisfy nice lower bounds. For that purpose, we shall approximate δ-potentials in H^{-s} with admissible ones and then apply the results of the preceding section. Let us start with the approximation. Let $s > n/2 + \epsilon$, $\epsilon > 0$, as in (15.2.5).

Proposition 17.2.2 *Let $a \in X$. Then there exist $\alpha_k \in \mathbf{C}$, $1 \le k < \infty$, such that for $L \ge 1$, there exists $r \in H^{-s}$ such that*

$$\delta_a(x) = \sum_{\mu_k \le L} \alpha_k \epsilon_k + r(x), \tag{17.2.2}$$

where

$$\|r\|_{H_h^{-s}} \le C_{s,\epsilon} L^{-(s-\frac{n}{2}-\epsilon)} h^{-\frac{n}{2}}, \tag{17.2.3}$$

$$(\sum |\alpha_k|^2)^{\frac{1}{2}} \le \langle L \rangle^{\frac{n}{2}+\epsilon} \left(\sum_{\mu_k \le L} \langle \mu_k \rangle^{-2(\frac{n}{2}+\epsilon)} |\alpha_k|^2 \right)^{\frac{1}{2}} \le C L^{\frac{n}{2}+\epsilon} h^{-\frac{n}{2}}. \tag{17.2.4}$$

Proof Observe first that

$$\|\delta_a\|_{H_h^{-s}} = \mathcal{O}(1) \|\langle h\xi \rangle^{-s}\|_{L^2} = \mathcal{O}_s(1) h^{-\frac{n}{2}}. \tag{17.2.5}$$

In general, if $u \in H^{-s_1}(\widetilde{X})$, then by Proposition 16.2.2 and the subsequent observation (where s was arbitrary) we have

$$u = \sum_1^\infty \alpha_k \epsilon_k, \quad \sum \langle \mu_k \rangle^{-2s_1} |\alpha_k|^2 \asymp \|u\|^2_{H_h^{-s_1}}.$$

Thus, if $s > s_1 > n/2$:

$$u = \sum_{\mu_k \le L} \alpha_k \epsilon_k + r, \tag{17.2.6}$$

where

$$\|r\|^2_{H_h^{-s}} \asymp \sum_{\mu_k > L} \langle \mu_k \rangle^{-2s} |\alpha_k|^2 \le C L^{-2(s-s_1)} \|u\|^2_{H_h^{-s_1}}, \tag{17.2.7}$$

$$\left(\sum_{\mu_k \le L} |\alpha_k|^2 \right)^{\frac{1}{2}} \le \langle L \rangle^{s_1} \left(\sum_{\mu_k \le L} \langle \mu_k \rangle^{-2s_1} |\alpha_k|^2 \right)^{\frac{1}{2}} \le C L^{s_1} \|u\|^2_{H_h^{-s_1}}. \tag{17.2.8}$$

In particular, when $u = \delta_a$ get the proposition by taking $s_1 = \epsilon + n/2$. □

Let P_δ be as in (16.5.1), (16.5.2). Let $0 < h \le \alpha \le 1/2$, and let $t_1^2 \le t_2^2 \le \cdots \le t_{N_\alpha}^2$ be the eigenvalues of $P_\delta^* P_\delta$ in $[0, \alpha[$ with $t_j > 0$. We call t_j the singular values of P_δ, arranged in increasing order. Let $e_1, \dots, e_{N_\alpha}$ be a corresponding system of eigenfunctions. From (16.5.18), we know that

$$N_\alpha \le \mathcal{O}(1)h^{-n}\left(V_N(\alpha) + h^{\widetilde{N}} + \frac{\delta}{\alpha}\right),$$

but for the moment we will only use that $N_\alpha \le N_1 \le \mathcal{O}(h^{-n})$ (when $\delta \le \alpha$).

Let $a = (a_1, \dots, a_{N_\alpha}) \in X^{N_\alpha}$ and put

$$q_a(x) = \sum_{j=1}^{N_\alpha} \delta_{a_j}(x), \tag{17.2.9}$$

$$M_{q_a;j,k} = \int q_a(x)e_k(x)e_j(x)dx,\ 1 \le j,k \le N_\alpha. \tag{17.2.10}$$

We lighten the notation by writing N instead of N_α, when possible. Then using (16.2.2), (16.5.26), and the fact that $\|q_a\|_{H_h^{-s}} = \mathcal{O}(1)Nh^{-n/2}$, we get for all $\lambda, \mu \in \mathbf{C}^N$,

$$\langle M_{q_a}\lambda|\mu\rangle = \int q_a(x)\left(\sum \lambda_k e_k\right)\left(\sum \mu_j e_j\right)dx = \mathcal{O}(1)Nh^{-n}\|\lambda\|\|\mu\|$$

and hence

$$s_1(M_{q_a}) = \|M_{q_a}\|_{\mathcal{L}(\mathbf{C}^N,\mathbf{C}^N)} = \mathcal{O}(1)Nh^{-n}. \tag{17.2.11}$$

We now choose a so that (17.1.14) and (17.1.15) hold, where we recall that $s_1 \ge s_2 \ge \dots \ge s_N$ are the singular values of M_{q_a} and E_j is defined in (17.1.6), where $0 \le \varepsilon_1 \le \varepsilon_2 \le \dots \le \varepsilon_N$ are the eigenvalues of the Gramian $\mathcal{E}_X = ((e_j|e_k)_{L^2(X)})_{1\le j,k\le N}$.

Since $\{e_j\}$ is an orthonormal system,

$$E_j = N - j + 1. \tag{17.2.12}$$

Then (17.1.14) gives the lower bound

$$s_1 \ge \frac{(N!)^{\frac{1}{N}}}{\mathrm{vol}\,(X)} = (1 + \mathcal{O}(\frac{\ln N}{N}))\frac{N}{e\,\mathrm{vol}\,(X)}, \tag{17.2.13}$$

where the last identity follows from Stirling's formula.

Rewriting (17.1.15) as

$$s_k \ge s_1^{-\frac{k-1}{N-k+1}}\left(\prod_1^N \frac{E_j}{\mathrm{vol}\,(X)}\right)^{\frac{1}{N-k+1}},$$

and using (17.2.11), we get

$$s_k \geq \frac{1}{C^{\frac{k-1}{N-k+1}} (\mathrm{vol}\,(X))^{\frac{N}{N-k+1}}} \left(\frac{h^n}{N}\right)^{\frac{k-1}{N-k+1}} (N!)^{\frac{1}{N-k+1}}. \tag{17.2.14}$$

Summing up, we get

Proposition 17.2.3 *There exist $a_1, \dots, a_N \in X$, with $N = N_\alpha$ defined above, such that if $q_a = \sum_1^N \delta(x - a_j)$ and $M_{q_a;j,k} = \int q_a(x) e_k(x) e_j(x) dx$, then the singular values $s_1 \geq s_2 \geq \dots \geq s_N$ of M_{q_a}, satisfy* (17.2.11), (17.2.13) *and* (17.2.14).

We shall next approximate q_a by an admissible potential. Apply Proposition 17.2.2 to each δ-function in q_a, to see that

$$q_a = q + r,\ q = \sum_{\mu_k \leq L} \alpha_k \epsilon_k, \tag{17.2.15}$$

where

$$\|q\|_{H_h^{-s}} \leq C h^{-\frac{n}{2}} N, \tag{17.2.16}$$

$$\|r\|_{H_h^{-s}} \leq C_\epsilon L^{-(s-\frac{n}{2}-\epsilon)} h^{-\frac{n}{2}} N, \tag{17.2.17}$$

$$\left(\sum |\alpha_k|^2\right)^{\frac{1}{2}} \leq C L^{\frac{n}{2}+\epsilon} h^{-\frac{n}{2}} N. \tag{17.2.18}$$

In order to estimate M_r, we write

$$\langle M_r \beta | \gamma \rangle = \int r(x) \left(\sum \beta_k e_k\right) \left(\sum \gamma_j e_j\right) dx,$$

so that

$$|\langle M_r \beta | \gamma \rangle| \leq C \|r\|_{H_h^{-s}} h^{-\frac{n}{2}} \left\| \sum \beta_k e_k \right\|_{H_h^s} \left\| \sum \gamma_j e_j \right\|_{H_h^s}.$$

Applying (16.5.26) to the last two factors, we get with a new constant $C > 0$:

$$|\langle M_r \beta | \gamma \rangle| \leq C \|r\|_{H_h^{-s}} h^{-\frac{n}{2}} \|\beta\| \|\gamma\|,$$

so

$$\|M_r\| \leq C h^{-\frac{n}{2}} \|r\|_{H_h^{-s}}. \tag{17.2.19}$$

Using (17.2.17), we get for every $\epsilon > 0$

$$\|M_r\| \le C_\epsilon L^{-(s-\frac{n}{2}-\epsilon)} h^{-n} N. \tag{17.2.20}$$

For the admissible potential q in (17.2.15), we thus obtain from (17.2.14), (17.2.20) and the fact that $s_k(M_q) \ge s_k(M_{\delta_a}) - \|M_r\|$:

$$s_k(M_q) \ge \frac{1}{C^{\frac{k-1}{N-k+1}} (\mathrm{vol}\,(X))^{\frac{N}{N-k+1}}} \left(\frac{h^n}{N}\right)^{\frac{k-1}{N-k+1}} (N!)^{\frac{1}{N-k+1}} - C_\epsilon L^{-(s-\frac{n}{2}-\epsilon)} h^{-n} N. \tag{17.2.21}$$

Similarly, from (17.2.11), (17.2.20) we get for $L \ge 1$:

$$\|M_q\| \le C N h^{-n}. \tag{17.2.22}$$

Using Proposition 16.2.2 and the subsequent observations, we get for all $s_1 > n/2$,

$$\begin{aligned}
\|q\|_{H_h^s} &\le \mathcal{O}(1) \left(\sum_{\mu_k \le L} \langle \mu_k \rangle^{2s} |\alpha_k|^2 \right)^{\frac{1}{2}} \\
&\le \mathcal{O}(1) \left(\sum_{\mu_k \le L} \langle \mu_k \rangle^{-2s_1} |\alpha_k|^2 \right)^{\frac{1}{2}} L^{s+s_1} \\
&\le \mathcal{O}(1) h^{-\frac{n}{2}} N L^{s+s_1},
\end{aligned}$$

where we used (17.2.16) with s replaced by s_1 in the last step. Thus for every $\epsilon > 0$,

$$\|q\|_{H_h^s} \le \mathcal{O}(1) N L^{s+\frac{n}{2}+\epsilon} h^{-\frac{n}{2}}, \ \forall \epsilon > 0. \tag{17.2.23}$$

Summing up, we have obtained

Proposition 17.2.4 *We can find an admissible potential q as in* (17.2.15), (17.2.18) *such that the matrix M_q, defined by*

$$M_{q;j,k} = \int q e_k e_j dx,$$

satisfies (17.2.21), (17.2.22). *Moreover the H_h^s-norm of q satisfies* (17.2.23). *Here $N \le \mathcal{O}(h^{-n})$ is the number of singular values of P_δ in $[0, \alpha^{1/2}[$ and we assume that $\delta \le \alpha$.*

Notice also that if we choose $\widetilde{R}$ with real coefficients, then we can choose q real-valued.

17.3 Lower Bounds on the Small Singular Values for Suitable Perturbations

Let P, p be as in (15.1.1)–(15.1.8). Fix $z \in \Omega$. Our unperturbed operator will be as in (16.5.1)–(16.5.3), where we change the notations slightly:

$$\begin{aligned} &P_0 = P + \delta_0 Q_0, \quad 0 \leq \delta_0 \ll 1, \\ &Q_0 = \mathcal{O}(1) : H_h^\sigma \to H_h^\sigma \text{ uniformly for } -s \leq \sigma \leq s. \end{aligned} \tag{17.3.1}$$

We also assume that Q_0 is the operator of multiplication by a potential, so that P_0 also satisfies the symmetry assumption (15.1.7). Proposition 16.5.1 gives:

Proposition 17.3.1 *For $0 < h \leq \alpha \ll 1$, the number $N_0(\alpha^{1/2})$ of eigenvalues of $(P_0 - z)^*(P_0 - z)$ in $[0, \alpha]$ satisfies*

$$N_0(\alpha^{1/2}) = \mathcal{O}_{N,\widetilde{N}}(1) h^{-n} \left(V_N(\alpha) + h^{\widetilde{N}} + \frac{\delta_0}{\alpha} \right). \tag{17.3.2}$$

We can get a slightly different and more interesting bound when δ_0/α is not very small: Recall the max-min formula (cf. (17.A.6))

$$t_j(P_0 - z) = \sup_{\operatorname{codim} L = j-1} \inf_{\substack{u \in L \\ \|u\|=1}} \|(P_0 - z)u\|,$$

and the similar one for $t_j(P - z)$. Now, $\|(P_0 - z)u\| \geq \|(P - z)u\| - \delta_0 \|Q_0\|$ when $\|u\| = 1$. Thus (which can also be viewed as a special case of the Ky Fan inequalities),

$$t_j(P_0 - z) \geq t_j(P - z) - \delta_0 \|Q_0\|.$$

In particular, if $t_j(P_0 - z) \leq \alpha^{1/2}$, then $t_j(P - z) \leq \alpha^{1/2} + \delta_0 \|Q_0\|$, leading to

$$N_0(\alpha^{1/2}) \leq N(\alpha^{1/2} + \delta_0 \|Q_0\|),$$

where the right-hand side denotes the number of singular values of $P - z$ in $[0, \alpha^{1/2} + \delta_0 \|Q_0\|]$, or equivalently, the number of eigenvalues of $(P - z)^*(P - z)$ in $[0, (\alpha^{1/2} + \delta_0 \|Q_0\|)^2]$. Combining this with Proposition 17.3.1 for δ_0 there equal to 0 and assuming that $\|Q\| \leq 1$ for simplicity, we get

Proposition 17.3.2 *For all $M, \widetilde{N} > 0$,*

$$N_0(\alpha^{1/2}) \leq \mathcal{O}_{M,\widetilde{N}}(1) h^{-n} (V_M((\alpha^{1/2} + \delta_0)^2) + h^{\widetilde{N}}).$$

This is of interest when $\delta_0 \leq \alpha^{1/2}$.

Assume from now on that

$$0 \le \delta_0 \le h^{\frac{1}{2}}. \tag{17.3.3}$$

Choose $\tau_0 \in]0, h^{1/2}]$, $\alpha = h$, and let $N = N_0(\tau_0)$,

$$N \le \mathcal{O}(1)(h^{M-n} + h^{-n} V_M(4h)) =: M_0 \tag{17.3.4}$$

be the number of singular values of $P_0 - z$ in $[0, \tau_0[$, labeled in increasing order so that $0 \le t_1(P_0 - z) \le \dots \le t_N(P_0 - z) < \tau_0 \le t_{N+1}(P_0 - z)$, and let $e_1, e_2, \dots, e_N$ be an associated family of eigenfunctions of $(P_0 - z)^*(P - z)$. Fix $\theta \in]0, 1/4[$.

Proposition 17.3.3 *If q is an admissible potential as in* (17.2.1) *with $L \ge 1$, then*

$$\|q\|_\infty \le Ch^{-\frac{n}{2}} \|q\|_{H_h^s} \le \widetilde{C} h^{-\frac{n}{2}} L^s |\alpha|. \tag{17.3.5}$$

If N is sufficiently large, there exists an admissible potential q as in (17.2.1) *such that*

a)

$$\begin{aligned} |\alpha| &\le L^{\frac{n}{2}+\epsilon} h^{-\frac{n}{2}} N, \\ \|q\|_{H_h^s} &\le \mathcal{O}(1) h^{-\frac{n}{2}} N L^{s+\frac{n}{2}+\epsilon}. \end{aligned} \tag{17.3.6}$$

$$L \asymp M_0^{\frac{1}{s-\frac{n}{2}-\epsilon}} h^{-\frac{2n}{s-\frac{n}{2}-\epsilon}}. \tag{17.3.7}$$

b) *The conclusion of Proposition* 17.2.4 *applies to q, so* (17.2.21) *and* (17.2.22) *hold. Put $P_\delta = P_0 + \delta q$, $0 \le \delta \ll 1$. Then*

c)

$$t_\nu(P_\delta - z) \ge t_\nu(P_0 - z) - C\delta h^{-n} L^{\frac{n}{2}+\epsilon+s} N, \ \ \nu \ge 1.$$

d) *There exists $N_2 > 0$ depending only on n, s, ϵ, that we can choose arbitrarily large, such that if $\delta = C\tau_0 h^{N_2 - n}$, $C \gg 1$, then either for q as above or for $q = 0$, we have,*

$$t_\nu(P_\delta) \ge \tau_0 h^{N_2}, \quad 1 + [(1-\theta)N] \le \nu \le N. \tag{17.3.8}$$

Here $[x] = \sup(]-\infty, x] \cap \mathbf{Z})$ denotes the integer part of $x \in \mathbf{R}$.

When N belongs to a bounded interval, a)–d) *are still valid, if we restrict ν in* (17.3.8) *to the value $\nu = N$.*

Since $N \le M_0 \lesssim h^{-n}$, we get from (17.3.6), (17.3.7) that

$$L \lesssim h^{-M_{\min}}, \quad M_{\min} := \frac{3n}{s - \frac{n}{2} - \epsilon},$$
$$|\alpha| \le R := h^{-\frac{3n}{2}} h^{-(\frac{n}{2}+\epsilon)M_{\min}},$$

corresponding to the minimal orders of magnitude in (15.2.5).

Proof We choose q as in Proposition 17.2.4 so that (17.3.5), (17.3.6) follow from (17.2.18) and the estimates leading to (17.2.23). Then we also have (17.2.21) and we choose L large enough to guarantee that the first term to the right is dominant for $1 \le k \le N/2$: We have for such k:

$$s_k(M_q) \ge \frac{1}{C}\frac{h^n}{N}(N!)^{\frac{1}{N}} - C_\epsilon L^{-(s-\frac{n}{2}-\epsilon)}\frac{N}{h^n}.$$

By Stirling's formula, $(N!)^{1/N} \ge N/\mathrm{Const}$, so we get with a new constant $C > 0$:

$$s_k(M_q) \ge \frac{h^n}{C} - C_\epsilon L^{-(s-\frac{n}{2}-\epsilon)}\frac{N}{h^n}, \quad 1 \le k \le \frac{N}{2}. \tag{17.3.9}$$

Thus,

$$s_k(M_q) \ge \frac{h^n}{2C}, \tag{17.3.10}$$

provided that $L^{s-\frac{n}{2}-\epsilon} \gg Nh^{-2n}$. In view of (17.3.4), this is achieved with a sufficiently large L satisfying (17.3.7).

We have then verified all the statements up to and including b). Statement c) follows from the max-min description of $t_\nu(P_\delta)$ and the fact that

$$\|q\|_\infty \le Ch^{-n}L^{\frac{n}{2}+\epsilon+s}N. \tag{17.3.11}$$

Let $e_1, \dots, e_N$ be an orthonormal family of eigenfunctions corresponding to $t_\nu(P_0 - z)$, so that

$$(P_0 - z)^*(P_0 - z)e_j = (t_j(P_0 - z))^2 e_j. \tag{17.3.12}$$

Using the symmetry assumption (15.1.7) which extends to P_0, we see that a corresponding orthonormal family of eigenfunctions of $(P_0 - z)(P_0 - z)^*$ is given by

$$\widetilde{f}_j = \Gamma e_j. \tag{17.3.13}$$

If the non-vanishing t_j are not all distinct it is not immediately clear that we can arrange so that $\widetilde{f}_j = f_j$ as in (17.A.10) (with P replaced by P_0), but we know that $\widetilde{f}_1, \dots, \widetilde{f}_N$ and $f_1, \dots, f_N$ are orthonormal families that span the same space F_N. Let E_N be the span of $e_1, \dots, e_N$. Then

$$(P_0 - z) : E_N \to F_N \text{ and } (P_0 - z)^* : F_N \to E_N \tag{17.3.14}$$

have the same singular values $0 \le t_1 \le t_2 \le \dots \le t_N$.

Define $R_+ : L^2 \to \mathbf{C}^N$, $R_- : \mathbf{C}^N \to L^2$ by

$$R_+ u(j) = (u|e_j), \quad R_- u_- = \sum_1^N u_-(j)\widetilde{f}_j. \tag{17.3.15}$$

Then

$$\mathcal{P} = \begin{pmatrix} P_0 - z & R_- \\ R_+ & 0 \end{pmatrix} : H_h^m \times \mathbf{C}^N \to L^2 \times \mathbf{C}^N \tag{17.3.16}$$

has a bounded inverse

$$\mathcal{E} = \begin{pmatrix} E^0 & E_+^0 \\ E_-^0 & E_{-+}^0 \end{pmatrix}.$$

Since we do not necessarily have (17.A.10), we cannot assert that $E_{-+}^0 = \operatorname{diag}(t_j)$, only that E_{-+}^0 is unitarily equivalent to $\operatorname{diag}(t_j)$, but the singular values of E_{-+}^0 are given by $t_j(E_{-+}^0) = t_j(P_0 - z)$, $1 \le j \le N$, or equivalently by $s_j(E_{-+}^0) = t_{N+1-j}(P_0 - z)$, for $1 \le j \le N$. We here follow the convention of using the letter s for labelling the singular values in decreasing order and t for the increasing order.

We will apply Sect. 17.A, and recall that N is assumed to be sufficiently large and that θ has been fixed in $]0, 1/4[$. Since z is fixed, it will also be notationally convenient to assume that $z = 0$. We consider two cases.

Case 1. $s_j(E_{-+}^0) \ge 8\tau_0 h^{N_2}$, for $1 \le j \le N - [(1-\theta)N]$. Then from (17.A.37) we get the conclusion in d) with $q = 0$, $P_\delta = P_0$.

Case 2.

$$s_j(E_{-+}^0) < 8\tau_0 h^{N_2} \text{ for some } j \text{ such that } 1 \le j \le N - [(1-\theta)N]. \tag{17.3.17}$$

From (17.2.22) and the fact that $N \le M_0$ in (17.3.4) we get

$$s_1(M_q) \le \|M_q\| \le CM_0 h^{-n}. \tag{17.3.18}$$

Put

$$\begin{aligned} P_\delta &= P_0 + \delta q = P_0 + \widetilde{\delta} Q, \\ \widetilde{\delta} &= \delta C h^{-n} L^{\frac{n}{2}+\epsilon+s} M_0, \ Q = \delta q/\widetilde{\delta}, \ \|Q\| \le 1, \end{aligned} \tag{17.3.19}$$

where the last estimate comes from (17.3.11). Then, if $\widetilde{\delta} \le \tau_0/2$, we can replace P_0 by P_δ in (17.3.16) and we still have a well-posed problem with inverse as in (17.A.25)–(17.A.31), satisfying (17.A.35)–(17.A.37) with $Q_\omega = Q$ as above and with δ replaced by $\widetilde{\delta}$. Here $E_-^0 Q E_+^0 = M_q/(Ch^{-n}L^{\frac{n}{2}+\epsilon+s}M_0)$ and according to (17.3.10), we have with a new constant C

$$s_k(\delta E_-^0 q E_+^0) = s_k(\widetilde{\delta} E_-^0 Q E_+^0) \ge \frac{\delta h^n}{C}, \quad 1 \le k \le \frac{N}{2}. \tag{17.3.20}$$

Playing with the general estimate (17.A.15), we get

$$s_\nu(A+B) \ge s_{\nu+k-1}(A) - s_k(B),$$

and for a sum of three operators

$$s_\nu(A+B+C) \ge s_{\nu+k+\ell-2}(A) - s_k(B) - s_\ell(C).$$

We apply this to $E_{-+}^{\widetilde{\delta}}$ in (17.A.36) and get

$$s_\nu(E_{-+}^{\widetilde{\delta}}) \ge s_{\nu+k-1}(\widetilde{\delta} E_-^0 Q E_+^0) - s_k(E_{-+}^0) - 2\frac{\widetilde{\delta}^2}{\tau_0}. \tag{17.3.21}$$

Here we use (17.3.17), (17.3.20) with $j = k = N - [(1-\theta)N]$, to get for $\nu \le N - [(1-\theta)N]$

$$s_\nu(E_{-+}^{\widetilde{\delta}}) \ge \frac{\delta h^n}{C} - 8\tau_0 h^{N_2} - 2\frac{\widetilde{\delta}^2}{\tau_0}. \tag{17.3.22}$$

Recall that $\theta < \frac{1}{4}$ and that $\widetilde{\delta}$ is given in (17.3.19). The first term to the right dominates when

$$\delta h^n \gg \tau_0 h^{N_2}, \ \delta h^n \gg \frac{\widetilde{\delta}^2}{\tau_0}, \tag{17.3.23}$$

i.e., when

$$\tau_0 h^{N_2-n} \ll \delta \ll \tau_0 \frac{h^{3n}}{L^{n+2\epsilon+2s} M_0^2}, \tag{17.3.24}$$

where the last term is

$$\asymp \tau_0 \frac{h^{3n+2n\frac{n+2\epsilon+2s}{s-\frac{n}{2}-\epsilon}}}{M_0^{2+\frac{n+2\epsilon+2s}{s-\frac{n}{2}-\epsilon}}}$$

by (17.3.7) and $M_0 \le \mathcal{O}(h^{-n})$ by (17.3.4), so the last term in (17.3.24) is $\ge \tau_0 h^{N_3}/\mathcal{O}(1)$ for some $N_3 = N_3(n, s, \epsilon) > 0$ that can be explicitly computed. It suffices to choose $N_2 > N_3 + n$ and we have a non-empty range of δ satisfying (17.3.24). In particular we can take $\delta = C\tau_0 h^{N_2-n}$ for some large enough C.

Then we get from (17.3.22)

$$s_\nu(E_{-+}^{\widetilde{\delta}}) \ge 8\tau_0 h^{N_2}, \ 1 \le \nu \le N - [(1-\theta)N].$$

Equivalently, for $t_\nu(E_{-+}^{\widetilde{\delta}}) = s_{N+1-\nu}(E_{-+}^{\widetilde{\delta}})$ we get

$$t_\nu(E_{-+}^{\widetilde{\delta}}) \ge 8\tau_0 h^{N_2}, \ 1 + [(1-\theta)N] \le \nu \le N. \tag{17.3.25}$$

From (17.3.25) and (17.A.37), we get (17.3.8).

When $N = \mathcal{O}(1)$, we still get (17.3.22) with $\nu = k = 1$ and this gives the last statement. □

The conclusion of c) in the proposition (which of course is valid also when we take $q = 0$ in d)) implies that with the choice of δ as in d), we have

$$t_\nu(P_\delta) \ge t_\nu(P_0) - \tau_0 h^{N_2-K},$$

for some K which depends only on n, s, ϵ. Choosing N_2 sufficiently large, we get $t_\nu(P_\delta) \ge t_\nu(P_0) - \tau_0 h^{N_1}$, where $N_1 > 0$ can be taken arbitrarily large (if we choose N_2 sufficiently large).

The construction can now be iterated. Assume that $N \gg 1$ and replace (P_0, N, τ_0) by $(P_\delta, N^{(1)}, \tau_0 h^{N_2}) =: (P^{(1)}, N^{(1)}, \tau_0^{(1)})$, where $N^{(1)} \le [(1-\theta)N]$ is the number of singular values of $P^{(1)}$ in $[0, \tau_0^{(1)}]$ and keep on, using the same values for the exponents N_1, N_2. Then we get a sequence $(P^{(k)}, N^{(k)}, \tau_0^{(k)})$, $k = 0, 1, \dots, k(N)$, where the last value $k(N)$ is determined by the fact that $N^{(k(N))}$ is of the order of magnitude of a large constant. Moreover,

$$t_\nu(P^{(k)}) \ge \tau_0^{(k)}, \ N^{(k)} < \nu \le N^{(k-1)}, \tag{17.3.26}$$

$$t_\nu(P^{(k+1)}) \ge t_\nu(P^{(k)}) - \tau_0^{(k)} h^{N_1}, \ \nu > N^{(k)}, \tag{17.3.27}$$

$$\tau_0^{(k+1)} = \tau_0^{(k)} h^{N_2}, \tag{17.3.28}$$

$$\begin{aligned} &N^{(k+1)} \le [(1-\theta)N^{(k)}] \text{ is the number of} \\ &\text{singular values of } P^{(k+1)} \text{ in } [0, \tau_0^{(k+1)}], \end{aligned} \tag{17.3.29}$$

$$P^{(0)} = P_0, \; N^{(0)} = N, \; \tau_0^{(0)} = \tau_0.$$

Here,

$$P^{(k+1)} = P^{(k)} + \delta^{(k+1)} q^{(k+1)}, \quad \delta^{(k+1)} = C\tau_0^{(k)} h^{N_2 - n}, \quad \|q^{(k+1)}\|_{H_h^s} \le Ch^{-K}.$$

Notice that $N^{(k)}$ decays exponentially fast with k:

$$N^{(k)} \le (1-\theta)^k N, \tag{17.3.30}$$

so we get the condition on k that $(1-\theta)^k N \ge C \gg 1$ which gives,

$$k \le \frac{\ln \frac{N}{C}}{\ln \frac{1}{1-\theta}}. \tag{17.3.31}$$

We also have

$$\tau_0^{(k)} = \tau_0 \left(h^{N_2}\right)^k. \tag{17.3.32}$$

For $\nu > N$, we iterate (17.3.27) to get

$$\begin{aligned} t_\nu(P^{(k)}) &\ge t_\nu(P) - \tau_0 h^{N_1}\left(1 + h^{N_2} + h^{2N_2} + \cdots\right) \\ &\ge t_\nu(P) - 2\tau_0 h^{N_1}. \end{aligned} \tag{17.3.33}$$

For $1 \ll \nu \le N$, let $\ell = \ell(N)$ be the unique value for which $N^{(\ell)} < \nu \le N^{(\ell-1)}$, so that

$$t_\nu(P^{(\ell)}) \ge \tau_0^{(\ell)}, \tag{17.3.34}$$

by (17.3.26). If $k > \ell$, we get

$$t_\nu(P^{(k)}) \ge t_\nu(P^{(\ell)}) - 2\tau_0^{(\ell)} h^{N_1}. \tag{17.3.35}$$

The iteration above works until we reach a value $k = k_0 = \mathcal{O}(\frac{\ln \frac{N}{C}}{\ln \frac{1}{1-\theta}})$ for which $N^{(k_0)} = \mathcal{O}(1)$. After that, we continue the iteration further by decreasing $N^{(k)}$ by one unit at each step.

Let $k_1 = \mathcal{O}\left(\ln(\frac{N}{C})/\ln(\frac{1}{1-\theta})\right) + \mathcal{O}(1)$, be the last k-value we get in this way. Then, by construction,

$$P^{(k_1+1)} = P_0 + \left(\delta^{(1)}q^{(1)} + \ldots + \delta^{(k_1+1)}q^{(k_1+1)}\right)$$

$$P_0 + \delta\left(q^{(1)} + h^{N_2}q^{(2)} + \ldots + h^{k_1N_2}q^{(k_1+1)}\right),$$

where $\delta = C\tau_0^{(0)}h^{N_2-n}$ and $q^{(j)}$ satisfy (17.3.6), (17.3.7) uniformly. In particular, $\|q^{(k+1)}\|_{H_h^s} \le Ch^{-K}$ and this also holds for $q^{(1)} + h^{N_2}q^{(2)} + \cdots + h^{k_1N_2}q^{(k_1+1)}$.

Summing up the discussion so far, we have obtained

Proposition 17.3.4 *Let P_0, z, $s > \epsilon + n/2$, $\epsilon > 0$ be as in Proposition* 17.3.3. *Choose $\tau_0 \in]0, h^{1/2}]$, so that the number N of singular values of $P_0 - z$ in $[0, \tau_0[$ satisfies* (17.3.4). *Let $N_2 > 0$ be large enough. Let $0 < \theta < \frac{1}{4}$ and let $N(\theta) \le \mathcal{O}(1)$ be sufficiently large.*

We can find $N = N^{(0)} > N^{(1)} > \cdots > N^{(k_1)} = 1$ and an admissible potential $q = q_h(x)$ as in (17.2.1), *satisfying* (17.3.5)–(17.3.7), *such that if $P_\delta = P_0 + \delta q$, $\delta = C\tau_0 h^{N_2-n}$, the following holds:*

As long as $N^{(k)} \ge N(\theta)$, we have $N^{(k+1)} \le [(1-\theta)N^{(k)}]$, $N^{(0)} = N$. Let $k_0 \ge 0$ be the last k value we get in this way. For $k > k_0$ we have $N^{(k+1)} = N^{(k)} - 1$, until we reach the value k_1 for which $N^{(k_1)} = 1$.

Put $\tau_0^{(k)} = \tau_0 h^{kN_2}$, $1 \le k \le k_1 + 1$. We have the following estimates on the singular values of $P_\delta - z$:

- *If $\nu > N^{(0)}$, then $t_\nu(P_\delta - z) \ge (1 - \frac{h^{N_2-K}}{C})t_\nu(P_0 - z)$.*
- *If $N^{(k)} < \nu \le N^{(k-1)}$, $1 \le k \le k_1$, then $t_\nu(P_\delta - z) \ge (1 - \mathcal{O}(h^{N_2-K}))\tau_0^{(k)}$.*
- *Finally, if $\nu = N^{(k_1)} = 1$, then $t_1(P_\delta - z) \ge (1 - \mathcal{O}(h^{N_2-K}))\tau_0^{(k_1+1)}$.*

Here $K = K(n, s, \epsilon)$ is independent of the other parameters.

We shall now obtain the corresponding estimates for the singular values of $P_{\delta,z} = (\widetilde{P}_\delta - z)^{-1}(P_\delta - z)$, where $\widetilde{P}$ is defined as in the beginning of Sect. 16.3 and $\widetilde{P}_\delta = P_\delta + \widetilde{P} - P$. Let $e_1, \ldots, e_N$ be an orthonormal family corresponding to the singular values $t_j(P_\delta)$ in $[0, \sqrt{h}[$, put $f_j = \overline{e}_j$, and let

$$(P_\delta - z)u + R_-u_- = v, \quad R_+u = v_+$$

be the corresponding Grushin problem so that the solution operators fulfill

$$\|E\| \le \frac{1}{\sqrt{h}}, \ \|E_\pm\| \le 1, \quad t_j(E_{-+}) = t_j(P_\delta - z) \le \sqrt{h}, \ 1 \le j \le N. \tag{17.3.36}$$

In order to shorten the notation, we shall assume that $z = 0$. Put $\widetilde{R}_- = \widetilde{P}_\delta^{-1} R_-$. Then the problem

$$P_{\delta,z}u + \widetilde{R}_- u_- = v, \quad R_+ u = v_+,$$

is well posed, with the solution

$$u = \widetilde{E}v + \widetilde{E}_+ v_+, \quad u_- = \widetilde{E}_- v + \widetilde{E}_{-+} v_+,$$

where

$$\begin{aligned} \widetilde{E} &= E\widetilde{P}_\delta, \ \widetilde{E}_+ = E_+, \\ \widetilde{E}_- &= E_-\widetilde{P}_\delta, \ \widetilde{E}_{-+} = E_{-+}. \end{aligned}$$

Adapting the estimate (17.A.18) to our situation, we get

$$t_k(P_{\delta,z}) \ge \frac{t_k(P_\delta)}{\|E\widetilde{P}_\delta\| t_k(P_\delta) + \|E_+\| \|E_-\widetilde{P}_\delta\|}, \ 1 \le k \le N, \tag{17.3.37}$$

where we also recall that $t_k(P_\delta) \le \sqrt{h}$.

Write

$$\begin{aligned} E\widetilde{P}_\delta &= EP_\delta + E(\widetilde{P} - P), \\ E_-\widetilde{P}_\delta &= E_-P_\delta + E_-(\widetilde{P} - P), \end{aligned}$$

and use that

$$\begin{aligned} EP_\delta &= 1 - E_+R_+ = \mathcal{O}(1) \text{ in } \mathcal{L}(L^2, L^2), \\ E_-P_\delta &= -E_{-+}R_+ = \mathcal{O}(\sqrt{h}) \text{ in } \mathcal{L}(L^2, \ell^2), \end{aligned}$$

together with (17.3.36) and the fact that $\|\widetilde{P} - P\| = \mathcal{O}(1)$. It follows that

$$\|E\widetilde{P}_\delta\| = \mathcal{O}(\frac{1}{\sqrt{h}}), \ \|E_-\widetilde{P}_\delta\| = \mathcal{O}(1).$$

Using this in (17.3.37), we get

$$t_k(P_{\delta,z}) \geq \frac{t_k(P_\delta)}{C\frac{t_k(P_\delta)}{\sqrt{h}}+C} \geq \frac{t_k(P_\delta)}{2C}, \tag{17.3.38}$$

where used that $t_k(P_\delta) \leq \sqrt{h}$ when $1 \leq k \leq N$. From Proposition 17.3.4 we get:

Proposition 17.3.5 *Let P_0, z, $s > \epsilon + n/2$, $\epsilon > 0$ be as in Proposition 17.3.3. Choose $\tau_0 \in]0, h^{1/2}]$, so that the number N of singular values of $P_0 - z$ in $[0, \tau_0[$ satisfies (17.3.4). Let $N_2 > 0$ be large enough. Let $0 < \theta < \frac{1}{4}$ and let $N(\theta) \leq \mathcal{O}(1)$ be sufficiently large.*

Let $N = N^{(0)} > N^{(1)} > \cdots > N^{(k_1)} = 1$, $q = q_h(x)$, $\tau_0^{(k)} = \tau_0 h^{kN_2}$, $1 \leq k \leq k_1 + 1$ be as in Proposition 17.3.4. We have the following estimates on the singular values of $P_{\delta,z}$ (defined after Proposition 17.3.4)

- *If $\nu > N^{(0)}$, then $t_\nu(P_{\delta,z}) \geq (1 - \frac{h^{N_2-K}}{C})t_\nu(P_{0,z})$.*
- *If $N^{(k)} < \nu \leq N^{(k-1)}$, $1 \leq k \leq k_1$, then $t_\nu(P_{\delta,z}) \geq \frac{(1-\mathcal{O}(h^{N_2-K}))\tau_0^{(k)}}{(2C)}$.*
- *Finally, if $\nu = N^{(k_1)} = 1$, then $t_1(P_\delta - z) \geq \frac{(1-\mathcal{O}(h^{N_2-K}))\tau_0^{(k_1+1)}}{(2C)}$.*

Here $K = K(n, s, \epsilon)$ is independent of the other parameters.

We recall that, by Proposition 16.5.3,

$$\begin{aligned} &\ln\det(S_{\delta,z} + \alpha\chi(\alpha^{-1}S_{\delta,z})) = \\ &\frac{1}{(2\pi h)^n}\left(\iint \ln s_z(x,\xi)dxd\xi\right. \\ &\left. + \mathcal{O}(1)\left(\int_0^\alpha V_N(t)\frac{dt}{t} + \int_\alpha^{\alpha_0}\frac{h}{t}V(t)\frac{dt}{t} + \frac{\delta}{\alpha} + h\right)\right), \end{aligned}$$

when $P_\delta = P + \delta(h^{\frac{n}{2}}q_1 + q_2) = P + \delta Q$ and $\|q_1\|_{H_h^s}$, $\|q_2\|_{H_1^s} \leq 1$, $0 < h \leq \alpha \ll 1$. Here $0 \leq \chi \in C_0^\infty([0, +\infty[)$, $\chi(0) > 0$ and $S_{\delta,z} = (\widetilde{P_\delta} - z)^{-1}(P_\delta - z)$.

We apply this with $P_0 = P$ in (17.3.1) and P_δ as in Proposition 17.3.4 and with δ above replaced by $\widetilde{\delta}$ in (17.3.19), so that $P_\delta = P + \widetilde{\delta}Q$, $Q = \mathcal{O}(1) : H_h^\sigma \to H_h^\sigma$ for $-s \leq \sigma \leq s$. Thus we get

$$\begin{aligned} &\ln\det(S_{\delta,z} + \alpha\chi(\alpha^{-1}S_{\delta,z})) = \\ &\frac{1}{(2\pi h)^n}\left(\iint \ln s_z(x,\xi)dxd\xi\right. \\ &\left. + \mathcal{O}(1)\left(\int_0^\alpha V_N(t)\frac{dt}{t} + \int_\alpha^{\alpha_0}\frac{h}{t}V(t)\frac{dt}{t} + \frac{\widetilde{\delta}+\delta_0}{\alpha} + h\right)\right), \end{aligned} \tag{17.3.39}$$

where $\widetilde{\delta} = \delta C h^{-n} L^{\frac{n}{2}+\epsilon+s} M_0$ with M_0 given in (17.3.4), L is given in (17.3.7), and δ is given in d) in Proposition 17.3.3. Notice that

$$\frac{\widetilde{\delta}}{\alpha} = \mathcal{O}(h^{N_2-K}\tau_0), \tag{17.3.40}$$

where K is independent of N_2.

Now choose $\alpha = h$ and set out to estimate $\ln\det S_{\delta,z}$. First we have the upper bound

$$\ln\det S_{\delta,z} \leq \ln\det(S_{\delta,z} + \alpha\chi(\alpha^{-1}S_{\delta,z})),$$

which can be bounded from above by the right hand side of (17.3.39).

As for the lower bound, we use that

$$\ln\det S_{\delta,z} = \ln\det(S_{\delta,z} + \alpha\chi(\alpha^{-1}S_{\delta,z})) - \sum_1^\infty \ln\left(\frac{t_k^2 + \alpha\chi(t_k^2/\alpha)}{t_k^2}\right), \tag{17.3.41}$$

where $0 < t_1 \leq t_2 \leq \cdots$ are the singular values of $P_{\delta,z}$, treated in Proposition 17.3.5. The last sum in (17.3.41) has at most finitely many non-vanishing terms and is equal to $\sum_1^\infty \ln(1 + \alpha\chi(t_k^2/\alpha)t_k^{-2})$. Assuming that $0 \leq \chi \leq 1$, it can be bounded from above by

$$\sum_{t_k^2 \in \alpha \operatorname{supp}\chi} \ln(1 + \alpha t_k^{-2}) \leq (\#\{k;\ t_k^2 \leq \alpha \sup\operatorname{supp}\chi\}) \ln(1 + \alpha t_1^{-2}). \tag{17.3.42}$$

Combining (17.3.38) with Proposition 17.3.2 (cf. (17.3.4)), we have for every $N > 0$,

$$\#\{k;\ t_k^2 \leq \alpha \sup\operatorname{supp}\chi\} \leq \mathcal{O}_N(1)(h^{N-n} + h^n V_N(Ch)). \tag{17.3.43}$$

From Proposition 17.3.5 and the fact that $k_1 = \mathcal{O}(\ln(1/h))$ we see that

$$t_1 \geq \tau_0 h^{N_2 \mathcal{O}(1)\ln\frac{1}{h}} \geq \tau_0 h^{\mathcal{O}(1)\ln\frac{1}{h}},$$

so

$$t_1^{-2} \leq \tau_0^{-2}\left(\frac{1}{h}\right)^{\mathcal{O}(1)\ln\frac{1}{h}},$$

and hence

$$\ln(1 + \alpha t_1^{-2}) \leq \mathcal{O}(1)\left(\ln\frac{1}{h}\right)^2 \ln\frac{1}{\tau_0},$$

which in conjunction with (17.3.43) and (17.3.42) gives

$$\sum_{t_k^2 \le \alpha \sup \operatorname{supp} \chi} \ln(1+\alpha t_k^{-2}) \le \mathcal{O}(1)\left(\ln \frac{1}{\tau_0}\right)\left(\ln \frac{1}{h}\right)^2 (h^{N-n}+h^{-n}V_N(Ch)). \tag{17.3.44}$$

Finally, (17.3.41), (17.3.39) and (17.3.40) provide the lower bound for every $N > 0$,

$$\begin{aligned}\ln \det S_{\delta,z} \ge (2\pi h)^{-n}\Bigg[&\iint \ln s_z(x,\xi)dx d\xi \\ &+ \mathcal{O}(1)\left(\int_0^h V_N(t)\frac{dt}{t} + \int_h^{\alpha_0} \frac{h}{t}V(t)\frac{dt}{t}\right) + \mathcal{O}(\tau_0 h^{N_2-K} + \frac{\delta_0}{h} + h)\Bigg] \\ &- \mathcal{O}(1)\left(\ln \frac{1}{\tau_0}\right)\left(\ln \frac{1}{h}\right)^2 \left(h^{N-n} + h^{-n}V_N(Ch)\right).\end{aligned} \tag{17.3.45}$$

Here we have chosen $\alpha = h$ in (17.3.39), motivated by the fact that

$$\alpha \longmapsto \int_0^{\alpha} V(t)\frac{dt}{t} + \int_{\alpha}^{\alpha_0} \frac{h}{t}V(t)\frac{dt}{t}$$

is minimal at $\alpha = h$.

In the following, we assume that $\delta_0 \le \mathcal{O}(h^2)$ and that N_2 is large enough so that

$$\tau_0 h^{N_2-K} + \frac{\delta_0}{h} \le \mathcal{O}(h). \tag{17.3.46}$$

17.4 Estimating the Probability that det $P_{\delta,z}$ Is Small

In this section we keep the assumptions on (P, z) made in the beginning of Sect. 17.3 and choose P_0 as in (17.3.1) with $0 \le \delta_0 \le h^2$ as in (15.3.4). As in the preceding section, we consider perturbations of P_0 of the form

$$P_\delta = P_0 + \delta q, \quad \delta = C\tau_0 h^{N_2-n},\ C \gg 1, \tag{17.4.1}$$

where $N_2 \ge N_2(n, s, \epsilon, M, \widetilde{M})$ and q is an admissible potential of the form (17.2.1)

$$q = \sum_{0<\mu_k\le L} \alpha_k \epsilon_k, \quad \alpha \in \mathbf{C}^D; \tag{17.4.2}$$

here $\epsilon_1, \epsilon_2, \ldots$ is an orthonormal basis of eigenfunctions of the auxiliary operator $h^2\widetilde{R} > 0$, which is an elliptic semi-classical differential operator of the form (15.2.1) with smooth coefficients, and $h^2\widetilde{R}\epsilon_k = \mu_k^2\epsilon_k$, $\mu_k \nearrow +\infty$. We will assume that $L \geq 1$ and recall that

$$D := \sharp\{k;\ \mu_k \leq L\} = \mathcal{O}((L/h)^n) = \mathcal{O}(h^{-N_3}),\ N_3 = (M+1)n.$$

Here we impose the following bounds on the sum in (17.4.2) (cf. (15.2.5)):

$$L_{\min} \leq L \leq h^{-M},\quad L_{\min} \asymp M_0^{\frac{1}{s-\frac{n}{2}-\epsilon}} h^{-\frac{2n}{s-\frac{n}{2}-\epsilon}},\quad M \geq \frac{3n}{s-\frac{n}{2}-\epsilon}, \tag{17.4.3}$$

$$|\alpha| \leq R,\ R_{\min} \leq R \leq h^{-\widetilde{M}},\quad R_{\min} := L_{\min}^{\frac{n}{2}+\epsilon} M_0 h^{-\frac{n}{2}},\quad \widetilde{M} \geq \frac{3n}{2} + (\frac{n}{2}+\epsilon)M. \tag{17.4.4}$$

Recall from (17.3.4) that

$$M_0 = h^{M-n} + h^{-n}V_M(4h) \leq \mathcal{O}(h^{-n}), \tag{17.4.5}$$

and that the number N of singular values of $P_\delta - z$ in $[0, \tau_0[$ fulfills (17.3.4):

$$N \leq \mathcal{O}(1)M_0, \tag{17.4.6}$$

for any $M > 0$.

The main conclusion of the preceding section is that there exists q as in (17.2.1) with $L = L_{\min}$, $|\alpha| \leq R_{\min}$, such that (17.3.45) holds (and q is the random potential in P_δ in the Propositions 17.3.4 and 17.3.5).

We next review the upper bounds of Chap. 16 culminating in Proposition 16.5.3 that we apply with $\alpha = h$, recalling that "δ" there refers to

$$\sup_{-s\leq\sigma\leq s} \|P_\delta - P\|_{\mathcal{L}(H_h^\sigma, H_h^\sigma)}.$$

With the notations of the present chapter, we get for general q as in (17.4.2)–(17.4.4):

$$\ln\det(S_{\delta,z}) \leq$$

$$\frac{1}{(2\pi h)^n}\Bigg[\iint \ln s_z(x,\xi)dxd\xi + \mathcal{O}(1)\left(\int_0^h V_N(t)\frac{dt}{t} + \int_h^{\alpha_0}\frac{h}{t}V(t)\frac{dt}{t}\right) + \mathcal{O}(1)\left(h + \frac{\delta}{h}h^{-\frac{n}{2}}\|q\|_{H_h^s}\right)\Bigg]. \tag{17.4.7}$$

Recall that δ is given in (17.4.1) and that

$$\|q\|_{H_h^s} \leq \mathcal{O}(1)RL^s \leq \mathcal{O}(1)h^{-\widetilde{M}-sM},$$

so (17.4.7) gives

$$\begin{aligned}&\ln\det(S_{\delta,z}) \leq \\ &\frac{1}{(2\pi h)^n}\left[\iint \ln s_z(x,\xi)dxd\xi + \mathcal{O}(1)\left(\int_0^h V_N(t)\frac{dt}{t} + \int_h^{\alpha_0}\frac{h}{t}V(t)\frac{dt}{t}\right)\right. \\ &\qquad\left. + \mathcal{O}(1)\left(h + \tau_0 h^{N_2-n-1-\frac{n}{2}-\widetilde{M}-sM}\right)\right],\end{aligned} \tag{17.4.8}$$

and this still holds if we allow α to vary in the larger ball

$$|\alpha|_{\mathbf{C}^D} \leq 2R = \mathcal{O}(h^{-\widetilde{M}}). \tag{17.4.9}$$

(Our probability measure will be supported in $B_{\mathbf{C}^D}(0,R)$, but we will need to work in a larger ball.)

We consider the holomorphic function

$$F(\alpha) = (\det P_{\delta,z})\exp\left(-\frac{1}{(2\pi h)^n}\iint \ln|p_z|dxd\xi\right), \tag{17.4.10}$$

where we recall that $\det S_{\delta,z} = |\det P_{\delta,z}|^2$, $s_z(x,\xi) = |p_z(x,\xi)|^2$. Then by (17.4.8), we have

$$\ln|F(\alpha)| \leq \epsilon_0(h)h^{-n},\ |\alpha| < 2R, \tag{17.4.11}$$

and for one particular value $\alpha = \alpha^0$ with $|\alpha^0| \leq \frac{1}{2}R$, corresponding to the special potential in (17.3.45):

$$\ln|F(\alpha^0)| \geq -\epsilon_0(h)h^{-n}, \tag{17.4.12}$$

where we put with $C > 0$ sufficiently large,

$$\begin{aligned}\epsilon_0(h)/C = &\int_0^h V_N(t)\frac{dt}{t} + \int_h^{\alpha_0}\frac{h}{t}V(t)\frac{dt}{t} \\ &+ h + \ln\frac{1}{\tau_0}\left(\ln\frac{1}{h}\right)^2 (h^{N-n} + V_N(Ch)) + \tau_0 h^{N_2-K},\end{aligned} \tag{17.4.13}$$

for an arbitrary fixed $N > 0$. Here K is independent of the choice of $N_2 \gg 1$.

Let $\alpha^1 \in \mathbf{C}^D$ with $|\alpha^1| = R$ and consider the holomorphic function of one complex variable

$$f(w) = F(\alpha^0 + w\alpha^1). \tag{17.4.14}$$

We will mainly consider this function for w in the disc $D_{\alpha^0,\alpha^1} = D(w_0, r_0)$ determined by the condition $|\alpha^0 + w\alpha^1| < R$:

$$D_{\alpha^0,\alpha^1} : \left| w + \left(\frac{\alpha^0}{R} | \frac{\alpha^1}{R}\right)\right|^2 < 1 - \left|\frac{\alpha^0}{R}\right|^2 + \left|\left(\frac{\alpha^0}{R} | \frac{\alpha^1}{R}\right)\right|^2 =: r_0^2, \tag{17.4.15}$$

with center $w_0 = -(\alpha^0/R|\alpha^1/R)$ and radius $r_0 \in [\sqrt{3}/2, \sqrt{5}/2]$.

From (17.4.11), (17.4.12) we get

$$\ln |f(0)| \geq -\epsilon_0(h)h^{-n}, \ \ln |f(w)| \leq \epsilon_0(h)h^{-n}. \tag{17.4.16}$$

By (17.4.11), we may assume that the last estimate holds in a larger disc, say $D(-(\frac{\alpha^0}{R}|\frac{\alpha^1}{R}), 2r_0)$. Let $w_1, \dots, w_M$ be the zeros of f in $D(-(\frac{\alpha^0}{R}|\frac{\alpha^1}{R}), 3r_0/2)$. Then it is standard to get the factorization

$$f(w) = e^{g(w)} \prod_{j=1}^{M} (w - w_j), \quad w \in D(-(\frac{\alpha^0}{R}|\frac{\alpha^1}{R}), 4r_0/3), \tag{17.4.17}$$

together with the bounds

$$|\Re g(w)| \leq \mathcal{O}(\epsilon_0(h)h^{-n}), \quad M = \mathcal{O}(\epsilon_0(h)h^{-n}). \tag{17.4.18}$$

See for instance Section 5 in [135] where further references are given.

For $0 < \epsilon \ll 1$, put

$$\Omega(\epsilon) = \{r \in [0, 2r_0[;\ \exists w \in D_{\alpha^0,\alpha^1} \text{ such that } |w| = r \text{ and } |f(w)| < \epsilon\}. \tag{17.4.19}$$

If $r \in \Omega(\epsilon)$ and w is a corresponding point in D_{α^0,α^1}, we have with $r_j = |w_j|$,

$$\prod_{j=1}^{M} |r - r_j| \leq \prod_{j=1}^{M} |w - w_j| \leq \epsilon \exp(\mathcal{O}(\epsilon_0(h)h^{-n})). \tag{17.4.20}$$

Then at least one of the factors $|r - r_j|$ is bounded by $(\epsilon e^{\mathcal{O}(\epsilon_0(h)h^{-n})})^{1/M}$. In particular, the Lebesgue measure $\lambda(\Omega(\epsilon))$ of the set $\Omega(\epsilon)$ is bounded by $2M(\epsilon e^{\mathcal{O}(\epsilon_0(h)h^{-n})})^{1/M}$.

Noticing that the last bound increases with M when the last member of (17.4.20) is ≤ 1 and that $\{r \in [0, +\infty[\cap D_{\alpha^0,\alpha^1};\ f(r) < \epsilon\} \subset \Omega(\epsilon)$, we get

Proposition 17.4.1 *Let* $\alpha^1 \in \mathbf{C}^D$ *with* $|\alpha^1| = R$ *and assume that* $\epsilon > 0$ *is small enough so that the last member of* (17.4.20) *is* ≤ 1. *Then*

$$\lambda(\{r \in [0, +\infty[;\ |\alpha^0 + r\alpha^1| < R,\ |F(\alpha^0 + r\alpha^1)| < \epsilon\}) \leq \frac{\epsilon_0(h)}{h^n} \exp\left(\mathcal{O}(1) + \frac{h^n}{\mathcal{O}(1)\epsilon_0(h)} \ln \epsilon\right). \tag{17.4.21}$$

Here and in the following, the symbol $\mathcal{O}(1)$ *in a denominator indicates a bounded positive quantity.*

Typically, we can choose $\epsilon = \exp\left(-\frac{\epsilon_0(h)}{h^{n+\alpha}}\right)$ for some small $\alpha > 0$ and then the upper bound in (17.4.21) becomes

$$\frac{\epsilon_0(h)}{h^n} \exp\left(\mathcal{O}(1) - \frac{1}{\mathcal{O}(1)h^\alpha}\right).$$

Now we equip $B_{\mathbf{C}^D}(0, R)$ with a probability measure of the form

$$\mathbf{P}(d\alpha) = C(h)e^{\Phi(\alpha)}L(d\alpha), \tag{17.4.22}$$

where $L(d\alpha)$ is the Lebesgue measure, Φ is a C^1 function which depends on h and satisfies

$$|\nabla\Phi| = \mathcal{O}(h^{-N_4}), \tag{17.4.23}$$

and $C(h)$ is the appropriate normalization constant.

Writing $\alpha = \alpha^0 + Rr\alpha^2$, $0 \leq r < r_0(\alpha^2)$, $\alpha^2 \in S^{2D-1}$, $\frac{1}{2} \leq r_0 \leq \frac{3}{2}$, we get

$$\mathbf{P}(d\alpha) = \widetilde{C}(h)e^{\phi(r)}r^{2D-1}dr\,S(d\alpha^2), \tag{17.4.24}$$

where $\phi(r) = \phi_{\alpha^0,\alpha^2}(r) = \Phi(\alpha^0 + rR\alpha^2)$ so that $\phi'(r) = \mathcal{O}(h^{-N_5})$, $N_5 = N_4 + \widetilde{M}$. Here $S(d\alpha^2)$ denotes the Lebesgue measure on S^{2D-1}.

For a fixed α^2, we consider the normalized measure

$$\mu(dr) = \widehat{C}(h)e^{\phi(r)}r^{2D-1}dr \tag{17.4.25}$$

on $[0, r_0(\alpha^2)]$ and we want to show an estimate similar to (17.4.21) for μ instead of λ. Write $e^{\phi(r)}r^{2D-1} = \exp(\phi(r) + (2D-1)\ln r)$ and consider the derivative of the exponent:

$$\phi'(r) + \frac{2D-1}{r}.$$

This derivative is ≥ 0 for $r \leq 2\widetilde{r}_0$, where $\widetilde{r}_0 = C^{-1}\min(1, Dh^{N_5})$ for some large constant C, and we may assume that $2\widetilde{r}_0 \leq 1/2$. Introduce the measure $\widetilde{\mu} \geq \mu$ by

$$\widetilde{\mu}(dr) = \widehat{C}(h)e^{\phi(r_{\max})}r_{\max}^{2D-1}dr, \quad r_{\max} := \max(r, \widetilde{r}_0). \tag{17.4.26}$$

Since $\widetilde{\mu}([0, \widetilde{r}_0]) \leq \mu([\widetilde{r}_0, 2\widetilde{r}_0])$, we get

$$\widetilde{\mu}([0, r(\alpha^2)]) \leq \mathcal{O}(1). \tag{17.4.27}$$

We can write

$$\widetilde{\mu}(dr) = \widehat{C}(h)e^{\psi(r)}dr, \tag{17.4.28}$$

where

$$\begin{aligned} &\psi'(r) = \mathcal{O}(\max(D, h^{-N_5})) = \mathcal{O}(h^{-N_6}), \\ &N_6 = \max(N_3, N_5), \end{aligned} \tag{17.4.29}$$

where $D = \mathcal{O}(h^{-N_3})$ by the estimate on D prior to (17.4.3).

Decompose $[0, r_0(\alpha^2)]$ into $\asymp h^{-N_6}$ intervals of length $\asymp h^{N_6}$. If I is such an interval, we see that

$$\frac{\lambda(dr)}{C\lambda(I)} \leq \frac{\widetilde{\mu}(dr)}{\widetilde{\mu}(I)} \leq C\frac{\lambda(dr)}{\lambda(I)} \quad \text{on } I. \tag{17.4.30}$$

From (17.4.21), (17.4.30) we get when the right-hand side of (17.4.20) is ≤ 1,

$$\begin{aligned} &\widetilde{\mu}(\{r \in I;\ |F(\alpha^0 + rR\alpha^2)| < \epsilon\})/\widetilde{\mu}(I) \leq \frac{\mathcal{O}(1)}{\lambda(I)}\frac{\epsilon_0(h)}{h^n}\exp\left(\frac{h^n}{\mathcal{O}(1)\epsilon_0(h)}\ln\epsilon\right) \\ &= \mathcal{O}(1)h^{-N_6}\frac{\epsilon_0(h)}{h^n}\exp\left(\frac{h^n}{\mathcal{O}(1)\epsilon_0(h)}\ln\epsilon\right). \end{aligned}$$

Use that $\widetilde{\mu}([0, r_0(\alpha^2)]) = \mathcal{O}(1)$, multiply with $\widetilde{\mu}(I)$ and sum the estimates over I, to get

$$\begin{aligned} &\widetilde{\mu}(\{r \in [0, r(\alpha^2)];\ |F(\alpha^0 + rR\alpha^1)| < \epsilon\}) \\ &\leq \mathcal{O}(1)h^{-N_6}\frac{\epsilon_0(h)}{h^n}\exp\left(\frac{h^n}{\mathcal{O}(1)\epsilon_0(h)}\ln\epsilon\right). \end{aligned} \tag{17.4.31}$$

Since $\mu \leq \widetilde{\mu}$, we get the same estimate with $\widetilde{\mu}$ replaced by μ. Then from (17.4.24) we get

Proposition 17.4.2 *Let $\epsilon > 0$ be small enough for the right-hand side of* (17.4.20) *to be* ≤ 1. *Then*

$$\mathbf{P}(|F(\alpha)| < \epsilon) \leq \mathcal{O}(1)h^{-N_6}\frac{\epsilon_0(h)}{h^n}\exp\left(\frac{h^n}{\mathcal{O}(1)\epsilon_0(h)}\ln\epsilon\right). \tag{17.4.32}$$

Remark 17.4.3 In the case when $\widetilde{R}$ has real coefficients, we may assume that the eigenfunctions ϵ_j are real, and from the observation after Proposition 17.2.4 we see that we can choose α_0 above to be real. The foregoing discussion can then be restricted to the case of real α^2 and hence to real α. We can then introduce the probability measure P as in (17.4.22) on the real ball $B_{\mathbf{R}^D}(0, R)$. The subsequent discussion goes through without any changes, and we still have the conclusion of Proposition 17.4.2.

Remark 17.4.4 Choosing N_2 sufficiently large in the definition of δ, we see that $\epsilon_0(h)$ in (17.4.13) satisfies

$$\epsilon_0(h) \leq C\left(\int_0^h V_M(t)\frac{dt}{t} + \int_h^{\alpha_0}\frac{h}{t}V(t)\frac{dt}{t} + \ln\frac{1}{\tau_0}\left(\ln\frac{1}{h}\right)^2 V_M(Ch) + h\right), \tag{17.4.33}$$

$C = C(M)$, $M > n$.

Remark 17.4.5 Let us write $\epsilon = \exp(-\widetilde{\epsilon}/h^n)$. Proposition 17.4.2 shows that if $\widetilde{\epsilon} \geq C\epsilon_0(h) > 0$ for $C > 0$ sufficiently large, then

$$\mathbf{P}(|F(\alpha)| < \exp(-\widetilde{\epsilon}/h^n)) \leq \mathcal{O}(1)h^{-N_6-n}\epsilon_0(h)\exp\left(-\frac{\widetilde{\epsilon}}{\mathcal{O}(1)\epsilon_0(h)}\right). \tag{17.4.34}$$

17.5 End of the Proof

We first have a closer look at the two integrals appearing in the definition of $\epsilon_0(h)$ in (17.4.13). By (16.3.37),

$$\int_0^h V_N(t)\frac{dt}{t} = \int_0^h\int_0^1\left(1+\frac{\sigma^2}{t}\right)^{-N} dV(\sigma^2)\frac{dt}{t}. \tag{17.5.1}$$

Recall from Proposition 16.4.2 (using also Proposition 16.4.1, see also [56, 104]) that the function

$$\phi(z) := \frac{1}{2}\iint \ln s_z(x,\xi)dxd\xi = \iint \ln|p_z(x,\xi)|dxd\xi$$

is continuous and subharmonic with

$$\frac{\Delta\phi}{2\pi}(w)L(dw) = p_*(dxd\xi). \tag{17.5.2}$$

Hence,

$$V(\sigma^2) = \int_{|w-z|\leq\sigma} \frac{\Delta\phi}{2\pi}(w)L(dw),$$

so

$$V_N(t) = \int_{|w-z|\leq 1} \left(1+\frac{|w-z|^2}{t}\right)^{-N} \frac{\Delta\phi}{2\pi}(w)L(dw). \tag{17.5.3}$$

This yields

$$\begin{aligned}\int_0^h V_N(t)\frac{dt}{t} &= \int_0^h \int_{|w-z|\leq 1} \left(1+\frac{|w-z|^2}{t}\right)^{-N} \frac{\Delta\phi}{2\pi}(w)L(dw)\frac{dt}{t}\\ &= \int_{|w-z|\leq 1} f_N\left(\frac{|w-z|^2}{h}\right)\frac{\Delta\phi}{2\pi}(w)L(dw),\end{aligned} \tag{17.5.4}$$

where

$$f_N(\frac{|w-z|^2}{h}) = \int_0^h \left(1+\frac{|w-z|^2}{t}\right)^{-N} \frac{dt}{t} = \int_0^{\frac{h}{|w-z|^2}} \left(1+\frac{1}{\tau}\right)^{-N}\frac{d\tau}{\tau},$$

$$\begin{aligned}f_N(\lambda) &= \int_0^{1/\lambda} \left(1+\frac{1}{\tau}\right)^{-N}\frac{d\tau}{\tau}\\ &= \int_0^{1/\lambda}\left(\frac{\tau}{1+\tau}\right)^N \frac{d\tau}{\tau} \asymp \begin{cases}(1+\ln\frac{1}{\lambda}), & \text{if } 0<\lambda\leq 1,\\ \lambda^{-N}, & \text{if } \lambda\geq 1.\end{cases}\end{aligned} \tag{17.5.5}$$

Next, we have a look at

$$\begin{aligned}\int_h^1 \frac{h}{t}V(t)\frac{dt}{t} &= \int_{\sqrt{h}}^1 \frac{h}{\tau^2}V(\tau^2)2\frac{d\tau}{\tau} = \int_{\sqrt{h}}^1 \int_{|w-z|\leq\tau} \frac{\Delta\phi}{2\pi}(w)L(dw)\frac{2h}{\tau^2}\frac{d\tau}{\tau}\\ &= \int_{|w-z|\leq 1}\int_{\max(|w-z|,\sqrt{h})}^1 \frac{2h}{\tau^2}\frac{d\tau}{\tau}\frac{\Delta\phi}{2\pi}(w)L(dw)\\ &= \int_{|w-z|\leq 1}\left(\min\left\{\frac{h}{|w-z|^2},1\right\}-h\right)\frac{\Delta\phi}{2\pi}(w)L(dw).\end{aligned}$$

Thus,

$$\int_h^1 \frac{h}{t} V(t) \frac{dt}{t} = \int g_0 \left(\frac{|w-z|^2}{h} \right) \frac{\Delta \phi}{2\pi}(w) L(dw), \tag{17.5.6}$$

$$g_0(\lambda) = \left\{ \min(\frac{1}{\lambda}, 1) - h \right\}_+ = \begin{cases} 1-h, & \text{if } \lambda \le 1, \\ (\frac{1}{\lambda} - h)_+, & \text{if } \lambda > 1. \end{cases} \tag{17.5.7}$$

Combining (17.5.4), (17.5.5), (17.5.3), (17.5.6) and (17.5.7) we get (cf (17.4.13), (17.4.33)):

$$\begin{aligned} J_N(z;h) :&= \int_0^h V_N(t) \frac{dt}{t} + \int_h^{\alpha_0} \frac{h}{t} V(t) \frac{dt}{t} + V_N(Ch) \\ &= \int_{|z-w|\le 1} k_N \left(\frac{|z-w|^2}{h} \right) \frac{\Delta \phi}{2\pi}(w) L(dw), \end{aligned} \tag{17.5.8}$$

$$k_N(\lambda) \lesssim k(\lambda) := \frac{1}{1+\lambda} + \ln_+ \frac{1}{\lambda} \asymp \begin{cases} (1 + \ln \frac{1}{\lambda}), & \text{if } 0 < \lambda \le 1, \\ \frac{1}{\lambda}, & \text{if } \lambda \ge 1. \end{cases} \tag{17.5.9}$$

Applying (17.4.33),we get

$$\epsilon_0(h) \lesssim \ln \frac{1}{\tau_0} \left(\ln \frac{1}{h} \right)^2 \int_{|z-w|\le 1} k \left(\frac{|z-w|^2}{h} \right) \frac{\Delta \phi}{2\pi}(w) L(dw) + \mathcal{O}(h). \tag{17.5.10}$$

Let $\Gamma \Subset \Omega$ be a domain with Lipschitz boundary as in the beginning of Chap. 12, with constant scale $r = \sqrt{h}$. Let $z_1^0, z_2^0, \dots, z_{\widetilde{N}}^0 \in \partial\Gamma$ be distributed on the boundary as in that chapter, so that $\partial\Gamma \subset \bigcup_1^{\widetilde{N}} D(z_j^0, \sqrt{h})$, $\operatorname{dist}(z_j^0, z_{j+1}^0) \asymp \sqrt{h}$, with the cyclic convention that $\widetilde{N} + 1 = 1$. Let $J = J_N(z;h)$ be defined in (17.5.8). We are interested in

$$\sum_{j=1}^{\widetilde{N}} J(z_j^0; h) \lesssim \int G^0(w) \frac{\Delta \phi}{2\pi}(w) L(dw), \tag{17.5.11}$$

where

$$G^0(w) = \sum_{j=1}^{\widetilde{N}} k \left(\frac{|z_j^0 - w|^2}{h} \right). \tag{17.5.12}$$

We apply the averaging argument at the end of Sect. 12.3 with r equal to $\sqrt{h}$ and $\mu_+(dw) = (2\pi)^{-1}\Delta\phi(w)L(dw)$ and conclude that there exist $\widetilde{z}_j \in D(z_j^0, \epsilon\sqrt{h})$ for any fixed $\epsilon > 0$, such that

$$\int_{D(\widetilde{z}_j,\sqrt{h})} k\left(\frac{|\widetilde{z}_j - w|^2}{h}\right) \frac{\Delta\phi}{2\pi}(w)L(dw) = \mathcal{O}(1)\int_{D(\widetilde{z}_j,\sqrt{h})} \frac{\Delta\phi}{2\pi}(w)L(dw). \tag{17.5.13}$$

We now replace the sum in (17.5.11) by

$$\sum_1^{\widetilde{N}} J(\widetilde{z}_j; h) = \int G^1(w)\frac{\Delta\phi}{2\pi}(w)L(dw), \tag{17.5.14}$$

where

$$G^1(w) = \sum_1^{\widetilde{N}} k\left(\frac{|\widetilde{z}_j - w|^2}{h}\right), \tag{17.5.15}$$

so that by the preceding observation

$$\int G^1(w)\frac{\Delta\phi}{2\pi}(w)L(dw) \le \mathcal{O}(1)\int G^2(w)\frac{\Delta\phi}{2\pi}(w)L(dw),$$

$$G^2(w) = \sum_1^{\widetilde{N}} \widehat{k}\left(\frac{|\widetilde{z}_j - w|^2}{h}\right), \tag{17.5.16}$$

$$\widehat{k}\left(\frac{|w - z|^2}{h}\right) = \frac{h}{h + |w - z|^2}. \tag{17.5.17}$$

From the form of $\widehat{k}$, we see that the order of magnitude of the right-hand side in (17.5.14) will not change if we replace $\widetilde{z}_j$ in (17.5.15) by z_j^0. Thus,

$$\sum_{j=1}^{\widetilde{N}} J(\widetilde{z}_j; h) \le \mathcal{O}(1)\int G(w)\frac{\Delta\phi}{2\pi}(w)L(dw), \tag{17.5.18}$$

where

$$G(w)=\sum_{j=1}^{\widetilde{N}}\widehat{k}\left(\frac{|z_j^0-w|^2}{h}\right). \tag{17.5.19}$$

From the geometric assumptions on Γ, we see that the order of magnitude of G and hence the validity of (17.5.18) will not change if we replace G in (17.5.19) by

$$G(w)=G(w,\Gamma;h)=\int_{\partial\Gamma}\widehat{k}\left(\frac{|z-w|^2}{h}\right)\frac{|dz|}{\sqrt{h}}. \tag{17.5.20}$$

By a change of variables,

$$G(w,\Gamma;h)=G(\frac{w}{\sqrt{h}},\frac{1}{\sqrt{h}}\Gamma;1). \tag{17.5.21}$$

Here $\widetilde{\Gamma}=h^{-1/2}\Gamma$ is a Lipschitz domain as in Chap. 12 with constant scale $r=1$, so the problem of studying G is reduced to that of studying

$$G(\widetilde{w},\widetilde{\Gamma};1)=\int_{\partial\widetilde{\Gamma}}\frac{1}{1+|z-\widetilde{w}|^2}|dz|. \tag{17.5.22}$$

Consider the following regularity property for $\partial\widetilde{\Gamma}$: For some fixed $\kappa\in[0,1[$, we have

$$|\partial\widetilde{\Gamma}\cap D(\widetilde{w},\widetilde{R})|\le\mathcal{O}(1)\widetilde{R}^{1+\kappa},\quad \widetilde{R}\ge 1, \tag{17.5.23}$$

uniformly with respect to $\widetilde{w}$. Then by writing

$$G(\widetilde{w},\widetilde{\Gamma};1)=\int_{d(\widetilde{w},\partial\widetilde{\Gamma})}^{+\infty}\frac{1}{1+\widetilde{R}^2}d|\partial\widetilde{\Gamma}\cap D(\widetilde{w},\widetilde{R})|,$$

we see that

$$G(\widetilde{w},\widetilde{\Gamma},1)=\mathcal{O}(1)(1+d(\widetilde{w},\partial\widetilde{\Gamma}))^{\kappa-1}. \tag{17.5.24}$$

The most regular case is that with $\kappa=0$.

Returning to Γ, we may assume that for some $\kappa\in[0,1[$:

$$|\partial\Gamma\cap D(w,R)|\le\mathcal{O}(1)\sqrt{h}\left(\frac{R}{\sqrt{h}}\right)^{1+\kappa},\quad R\ge\sqrt{h}. \tag{17.5.25}$$

Then relations (17.5.21)–(17.5.24) lead to

$$G(w, \Gamma; h) = \mathcal{O}(1) \left(1 + \frac{d(x, \partial\Gamma)}{\sqrt{h}}\right)^{\kappa-1}. \tag{17.5.26}$$

Remark 17.5.1 The estimate (17.5.26) can be improved when $d(w, \Gamma) \geq \mathrm{diam}\,(\Gamma)$. Let us first observe that (17.5.25) with $w \in \partial\Gamma$ and $R = \mathrm{diam}\,(\Gamma)$ gives the estimate

$$|\partial\Gamma| \leq \mathcal{O}(1)\sqrt{h} \left(\frac{\mathrm{diam}\,(\Gamma)}{\sqrt{h}}\right)^{1+\kappa}. \tag{17.5.27}$$

When $d(w, \partial\Gamma) \geq \mathrm{diam}\,(\Gamma)$ we get from (17.5.20)

$$G(w, \Gamma; h) \asymp \frac{|\partial\Gamma|}{\sqrt{h}} \frac{1}{1 + \frac{d(w,\partial\Gamma)^2}{h}}. \tag{17.5.28}$$

From (17.5.27), we see that this estimate is sharper than (17.5.26) when $d(w, \Gamma) \geq \mathrm{diam}\,(\Gamma)$.

Let us sum up the results obtained so far, in order to apply Theorem 12.1.2. We choose z_j^0 as in that theorem with r there equal to $\sqrt{h}$. Apply Remark 17.4.5 with z there replaced by $\widetilde{z}_j$ and $\epsilon_0(h)$ replaced by $\epsilon(\widetilde{z}_j; h)$, where

$$\epsilon(z; h) = C \ln \frac{1}{\tau_0} \left(\ln \frac{1}{h}\right)^2 \int_{|z-w|\leq 1} k\left(\frac{|z-w|^2}{h}\right) \frac{\Delta\phi}{2\pi}(w) L(dw) + \mathcal{O}(h), \tag{17.5.29}$$

cf (17.5.10) and where k can be replaced by $\widehat{k}$, thanks to the choice of $\widetilde{z}_j$.[1] By Remark 17.4.5 we know that if $\widetilde{\epsilon}_j \geq C\epsilon(\widetilde{z}_j; h) \gg 0$, then

$$P\left(|\det P_{\delta,\widetilde{z}_j}| < e^{\frac{1}{(2\pi h)^n}(\phi(\widetilde{z}_j) - \widetilde{\epsilon}_j)}\right) \leq \mathcal{O}(1) h^{-N_6-n} \epsilon(\widetilde{z}_j; h) e^{-\frac{\widetilde{\epsilon}_j}{\mathcal{O}(1)\epsilon(\widetilde{z}_j; h)}}. \tag{17.5.30}$$

On the other hand, by Proposition 16.5.3, we know that for every admissible perturbation

$$|\det P_{\delta,z}| \leq \exp\left(\frac{1}{(2\pi h)^n}(\phi(z) + \epsilon(z; h))\right). \tag{17.5.31}$$

[1] We could also replace $1_{D(0,1)}(z - w)$ with a smooth cutoff $\chi(z - w)$, where $1_{D(0,1)} \leq \chi \in C_0^\infty(D(0, 1+\theta))$ for some $0 < \theta \ll 1$, in order to make ϵ continuous in z.

In order to fix the ideas, we choose $\widetilde{\epsilon}_j = h^{-\delta}\epsilon(\widetilde{z}_j; h)$ for some fixed small $\delta > 0$. We can then apply Theorem 12.1.2 with the following substitutions:

- "h" in the theorem should be replaced by $(2\pi h)^n$,
- "ϕ" in Theorem 12.1.2 should be replaced by the function $\phi(z) + \epsilon(z; h)$,
- ϵ_j in the theorem should be replaced with $\widetilde{\epsilon}_j = h^{-\delta}\epsilon(\widetilde{z}_j; h)$.

We also recall the passage from Theorem 12.1.1 to Theorem 12.1.2 by an averaging procedure, and that we have applied the same procedure to see that $\widetilde{z}_j$ can be chosen so that

$$\epsilon(\widetilde{z}_j; h) = \mathcal{O}(1) \ln \frac{1}{\tau_0} \left(\ln \frac{1}{h} \right)^2 \int_{|\widetilde{z}_j - w| \leq 1} \widehat{k}\left(\frac{|\widetilde{z}_j - w|^2}{h} \right) \frac{\Delta \phi}{2\pi}(w) L(dw) + \mathcal{O}(h). \tag{17.5.32}$$

In view of Remark 12.3.2, we can choose the same $\widetilde{z}_j$ as in the proof of Theorem 12.1.2 (in our situation).

From Theorems 12.1.2 and 17.5.30 we now conclude that with probability

$$\geq 1 - \mathcal{O}(1) h^{-N_6 - n} \left(\sum \epsilon(\widetilde{z}_j; h) \right) e^{-\frac{h^{-\delta}}{\mathcal{O}(1)}}, \tag{17.5.33}$$

we have

$$\begin{aligned} &\left| \#(\sigma(P_\delta) \cap \Gamma) - \frac{1}{(2\pi h)^n} \int_\Gamma \frac{\Delta(\phi + \epsilon(z; h))}{2\pi} L(dz) \right| \\ &\leq \frac{C_2}{h^n} \left(\int_{\widetilde{\gamma}_r} (\Delta\phi + |\Delta\epsilon|) L(dw) + h^{-\delta} \sum_{j=1}^{\widetilde{N}} \epsilon(\widetilde{z}_j; h) \right), \end{aligned} \tag{17.5.34}$$

where (as defined in Theorem 12.1.1)

$$\widetilde{\gamma}_r = \bigcup_{z \in \gamma} D(z, r(z)), \; \gamma = \partial \Gamma,$$

here with $r = \sqrt{h}$. It is also clear that the number of points $\widetilde{z}_j$ satisfies $\widetilde{N} = \mathcal{O}(h^{-1})$.

Let us next review the remainder terms in (17.5.34):

Combining (17.5.32) and (17.5.18), we see that

$$\sum \epsilon(\widetilde{z}_j; h) \leq \mathcal{O}(1) \ln \frac{1}{\tau_0} \left(\ln \frac{1}{h} \right)^2 \int_\Omega G(w) \frac{\Delta\phi}{2\pi}(w) L(dw) + \mathcal{O}(\widetilde{N} h), \tag{17.5.35}$$

where G is given by (17.5.20). Here, $\widetilde{N} h = \mathcal{O}(|\partial\Gamma| h^{1/2})$.

Noticing that the $\mathcal{O}(h)$ term in (17.5.32) can be assumed to be constant, we next look at

$$\begin{aligned}&\int_{\Gamma} \Delta\epsilon(z;h)L(dz)\\ &= \mathcal{O}(1)\ln\frac{1}{\tau_0}\left(\ln\frac{1}{h}\right)^2 \int_{z\in\Gamma}\int_{w\in\Omega} \Delta_z\left(k\left(\frac{|z-w|^2}{h}\right)\right)\frac{\Delta\phi(w)}{2\pi}L(dw)L(dz)\\ &= \mathcal{O}(1)\ln\frac{1}{\tau_0}\left(\ln\frac{1}{h}\right)^2 \int_{\Omega} H(w;h)\frac{\Delta\phi(w)}{2\pi}L(dw),\end{aligned}$$

where

$$H(w;h) = \int_{z\in\Gamma} \Delta_z\left(k\left(\frac{|z-w|^2}{h}\right)\right)L(dz),$$

and Ω is a fixed neighborhood of the closure of Γ.

By Green's formula,

$$H(w;h) = \int_{\partial\Gamma} \frac{\partial}{\partial n_z}k\left(\frac{|z-w|^2}{h}\right)|dz|,$$

where n_z is the exterior unit normal. An easy calculation shows that

$$\frac{\partial}{\partial n_z}k\left(\frac{|z-w|^2}{h}\right) = \mathcal{O}(1)\left(\frac{h}{(h+|z-w|^2)^{3/2}} + 1_{\{|z-w|\le\sqrt{h}\}}\frac{1}{|z-w|}\right),$$

so

$$H(w;h) = \mathcal{O}(1)\int_{\partial\Gamma}\left(\frac{h}{(h+|z-w|^2)^{3/2}} + 1_{\{|z-w|\le\sqrt{h}\}}\frac{1}{|z-w|}\right)|dz|.$$

Comparing with (17.5.20) and (17.5.17),

$$G(w;h) = \int_{\partial\Gamma}\frac{h}{h+|z-w|^2}\frac{|dz|}{\sqrt{h}},$$

we see that

$$H \le \mathcal{O}(1)G + \ln_+\left(\frac{h}{\mathrm{dist}\,(w,\partial\Gamma)^2}\right)$$

and conclude that

$$\left|\int_{\Gamma} \Delta\epsilon(z; h) L(dz)\right| \lesssim \ln \frac{1}{\tau_0} \left(\ln \frac{1}{h}\right)^2 \left(\int_{\Omega} G(w) \Delta\phi(w) L(dw) \right.$$
$$\left. + \int_{\Omega} \ln_+ \left(\frac{h}{\operatorname{dist}(w, \partial\Gamma)^2}\right) \Delta\phi(w) L(dw)\right). \quad (17.5.36)$$

The first term in the right-hand side is the same as in (17.5.35), while the second one will require an averaging argument.

Finally, we look at

$$\int_{\widetilde{\gamma}_r} |\Delta\epsilon| L(dz),$$

still with $r = \sqrt{h}$ and again with the $\mathcal{O}(h)$ term in (17.5.29) constant. This time, we use (17.5.29) with k in (17.5.9) regularized near $\lambda = 1$. Apply Δ_z to get

$$\Delta_z \epsilon = \mathcal{O}(1) \ln \frac{1}{\tau_0} \left(\ln \frac{1}{h}\right)^2 \left(\int_{\Omega} \frac{h}{(h + |z - w|^2)^2} \Delta\phi(w) L(dw) + \Delta\phi(z)\right).$$

Consequently,

$$\int_{\widetilde{\gamma}_r} |\Delta\epsilon| L(dz) = \mathcal{O}(1) \ln \frac{1}{\tau_0} \left(\ln \frac{1}{h}\right)^2$$
$$\left(\int_{\Omega} K(w) \Delta\phi(w) L(dw) + \int_{\widetilde{\gamma}_r} \Delta\phi(z) L(dz)\right), \quad (17.5.37)$$

where

$$K(w) = \int_{\widetilde{\gamma}_r} \frac{h}{(h + |z - w|^2)^2} L(dz) = \mathcal{O}(1) \int_{\partial\Gamma} \frac{h^{3/2}}{(h + |z - w|^2)^2} |dz| \le \mathcal{O}(1) G(w).$$

Since $G \gtrsim 1$ on $\widetilde{\gamma}_r$, we have

$$\int_{\widetilde{\gamma}_r} |\Delta\epsilon(z)| L(dz) \lesssim \ln \frac{1}{\tau_0} \left(\ln \frac{1}{h}\right)^2 \int_{\Omega} G(z) \Delta\phi(z) L(dz). \quad (17.5.38)$$

Combining (17.5.34)–(17.5.38), we get

$$\left|\#(\sigma(P_\delta)\cap\Gamma) - \frac{1}{(2\pi h)^n}\int_\Gamma \frac{\Delta\phi}{2\pi}L(dz)\right| \le$$
$$\frac{\mathcal{O}(1)}{h^n}\left(h^{-\delta}\ln\frac{1}{\tau_0}\left(\ln\frac{1}{h}\right)^2\int_\Omega G(w,\Gamma)\Delta\phi(w)L(dw)+\mathcal{O}(|\partial\Gamma|h^{\frac{1}{2}})\right)$$
$$+\frac{\mathcal{O}(1)}{h^n}\int_\Omega \ln_+\left(\frac{h}{\operatorname{dist}(w,\partial\Gamma)^2}\right)\Delta\phi(w)L(dw), \qquad (17.5.39)$$

valid with a probability as in (17.5.33).

For $t\in]-\sqrt{h},\sqrt{h}[$, let

$$\Gamma_t = \begin{cases}\{z\in\Gamma;\ \operatorname{dist}(z,\partial\Gamma)>|t|\}, & \text{if } t\le 0,\\ \{z\in\Omega;\ \operatorname{dist}(z,\Gamma)<t\}, & \text{if } t>0.\end{cases}$$

Then $\partial\Gamma_t$ is a uniformly Lipschitz curve of scale $\sqrt{h}$. Locally after rotation near any given point $z_0\in\partial\Gamma$, we have in $D(z_0,2\sqrt{h})$

$$\Gamma_t = \{z\in \operatorname{neigh}(z_0,\mathbf{C});\ \Im z < f(t,\Re z)\}$$

where $f(t,\cdot)$ is Lipschitz, uniformly with respect to t, and

$$f(t,\Re z) - f(s,\Re z) \asymp t-s.$$

Let $\gamma_t=\partial\Gamma_t$. By construction, the curves γ_t are mutually disjoint and fill up $\widetilde{\gamma}_r$ ($r=\sqrt{h}$). Clearly,

$$\left|\int_{\Gamma_t}\frac{\Delta\phi}{2\pi}L(dz)-\int_\Gamma\frac{\Delta\phi}{2\pi}L(dz)\right| \le \int_{\widetilde{\gamma}_r}\frac{\Delta\phi}{2\pi}L(dz). \qquad (17.5.40)$$

Applying the proof of (17.5.39) with Γ replaced by $\widetilde{\gamma}_r$, we can omit the troublesome last term there, coming from (17.5.36), and we conclude that with a probability as in (17.5.33),

$$|\#(\sigma(P_\delta)\cap\widetilde{\gamma}_r| \lesssim$$
$$\frac{\mathcal{O}(1)}{h^n}\left(h^{-\delta}\ln\frac{1}{\tau_0}\left(\ln\frac{1}{h}\right)^2\int_\Omega G(w,\Gamma)\Delta\phi(w)L(dw)+\mathcal{O}(|\partial\Gamma|h^{\frac{1}{2}})\right). \qquad (17.5.41)$$

Consequently for each $t \in]-\sqrt{h}, \sqrt{h}[$, after first replacing Γ by Γ_t in (17.5.39), we have with a probability as in (17.5.33):

$$\left|\#(\sigma(P_\delta)\cap\Gamma) - \frac{1}{(2\pi h)^n}\int_\Gamma \frac{\Delta\phi}{2\pi}L(dz)\right| \leq$$
$$\frac{\mathcal{O}(1)}{h^n}\left(h^{-\delta}\ln\frac{1}{\tau_0}\left(\ln\frac{1}{h}\right)^2\int_\Omega G(w,\Gamma)\Delta\phi(w)L(dw) + \mathcal{O}(|\partial\Gamma|h^{\frac{1}{2}})\right)$$
$$+\frac{\mathcal{O}(1)}{h^n}\int_\Omega \ln_+\left(\frac{h}{\mathrm{dist}\,(w,\gamma_t)^2}\right)\Delta\phi(w)L(dw). \qquad (17.5.42)$$

Now,

$$\frac{1}{2\sqrt{h}}\int_{-\sqrt{h}}^{\sqrt{h}}\ln_+\left(\frac{h}{\mathrm{dist}\,(w,\gamma_t)^2}\right)dt \lesssim 1_{\widetilde{\gamma}_{2\sqrt{h}}},$$

so

$$\frac{1}{2\sqrt{h}}\int_{-\sqrt{h}}^{\sqrt{h}}dt\int_\Omega \ln_+\left(\frac{h}{\mathrm{dist}\,(w,\gamma_t)^2}\right)\Delta\phi(w)L(dw) \lesssim \int_{\widetilde{\gamma}_{2\sqrt{h}}}\Delta\phi(w)L(dw),$$

so there exists $t \in]-\sqrt{h}, \sqrt{h}[$ such that

$$\int_\Omega \ln_+\left(\frac{h}{\mathrm{dist}\,(w,\gamma_t)^2}\right)\Delta\phi(w)L(dw) \lesssim \int_{\widetilde{\gamma}_{2\sqrt{h}}}\Delta\phi(w)L(dw).$$

As in (17.5.38), the last integral is

$$\mathcal{O}(1)\int_\Omega G(w,\Gamma)\Delta\phi(w)L(dw),$$

so choosing this value of t in (17.5.42), we can get rid of the last term there, and this completes the proof of Theorem 15.3.1. □

We finally explain the extensions in Remark 15.3.2. The case when (15.3.4) holds is already included in the proof above. When we merely assume (15.3.5) for some $\vartheta \in]0, 1/2[$ we get the corresponding weakening of the theorem after some modifications, basically just by multiplying some powers of h in some remainders with $h^{-\vartheta}$:

- Before (17.3.46): Weaken the assumption to $\delta_0 \leq h^{2-\vartheta}$.
- In (17.3.46), replace the last h with $h^{1-\vartheta}$.
- First sentence in Sect. 17.4: $0 \leq \delta_0 \leq h^{2-\vartheta}$.
- Equation (17.4.7): Replace the first h with $h^{1-\vartheta}$ in the last $\mathcal{O}(\ldots)$.

- Equation (17.4.8): Idem.
- Equation (17.4.13): Replace the third term h to the right with $h^{1-\vartheta}$.
- Equation (17.4.33): Replace the last term h to the right with $h^{1-\vartheta}$.
- Equation (17.5.10): Replace the last term $\mathcal{O}(h)$ to the right with $\mathcal{O}(h^{1-\vartheta})$.
- Equation (17.5.29): Idem.
- Equation (17.5.32): Idem.
- In (17.5.35) and on the following line: Read $\widetilde{N}h^{1-\vartheta} = \mathcal{O}(|\partial\Gamma|h^{\frac{1}{2}-\vartheta})$.
- On the following line: "Noticing that the $\mathcal{O}(h^{1-\vartheta})$ term ..."
- Equation (17.5.39): Replace $\mathcal{O}(|\partial\Gamma|h^{\frac{1}{2}})$ with $\mathcal{O}(|\partial\Gamma|h^{\frac{1}{2}-\vartheta})$
- Equation (17.5.41): Idem.
- Equation (17.5.42): Idem.

17.A Appendix: Grushin Problems and Singular Values

We mainly consider the case of an unbounded operator

$$P = P_0 + V,$$

where P_0 is an elliptic differential operator on X and $V \in L^\infty(X)$. The underlying Hilbert space is $\mathcal{H} = L^2(X) = H^0$ and we will view P as an operator from H^m to H^0. The dual of H^m is H^{-m} and we shall keep in mind the variational point of view with the triple

$$H^m \subset H^0 \subset H^{-m}.$$

Consider $\Box = P^*P : H^m \to H^{-m}$. For $u \in H^m$, we have

$$(\Box u|u) = \|Pu\|^2 \geq 0. \tag{17.A.1}$$

Proposition 17.A.1 *If $w \in]0, +\infty, [$, then $\Box + w : H^m \to H^{-m}$ is bijective with bounded inverse.*

Proof The injectivity is clear since $((\Box + w)u|u) = w\|u\|^2 + \|Pu\|^2$, $u \in H^m$, and combining this with the standard elliptic apriori estimate, $\|u\|_{H^m} \leq C(\|Pu\| + \|u\|)$, we get

$$\|u\|_{H^m}^2 \leq C(w)((\Box + w)u|u),$$

implying

$$\|u\|_{H^m} \leq C(w)\|(\Box + w)u\|_{H^{-m}}.$$

From this estimate and the fact that $\Box^* = \Box$, when we take the adjoint in the sense of bounded operators $H^m \to H^{-m}$, it is standard to get the desired conclusion. $\square$

Notice that when P is injective, then by ellipticity and compactness, we have $(\Box u|u) \geq \frac{1}{C}\|u\|^2$ for some $C > 0$ and we get the conclusion of the proposition also when $w = 0$.

The operator $(\Box + w)^{-1} : H^{-m} \to H^m$ induces a compact self-adjoint operator $H^0 \to H^0$. The range consists of all $u \in H^m$ such that $Pu \in H^m$. The spectral theorem for compact self-adjoint operators tells us that there is an orthonormal basis of eigenfunctions $e_1, e_2, \dots$ in H^0 such that

$$(\Box + w)^{-1} e_j = \mu_j^2(w) e_j, \tag{17.A.2}$$

where $0 < \mu_j \searrow 0$ when $j \to +\infty$. Clearly $e_j \in H^m$, $Pe_j \in H^m$, so we can apply $\Box + w$ to (17.A.2) and get

$$(\Box + w) e_j = \mu_j(w)^{-2} e_j, \tag{17.A.3}$$

which we write as

$$\Box e_j = (\mu_j(w)^{-2} - w) e_j. \tag{17.A.4}$$

Here $\mu_j(w)^{-2} - w = (\Box e_j | e_j) \geq 0$, so we have found an orthonormal basis $e_1, e_2, \dots \in H^0$ with e_j, $Pe_j \in H^m$ such that

$$\Box e_j = t_j^2 e_j, \quad 0 \leq t_j \nearrow +\infty. \tag{17.A.5}$$

It is easy to check that t_j^2 are independent of w, that e_j can be chosen independent of w, and we have

$$\mu_j(w)^2 = \frac{1}{t_j^2 + w}.$$

From Proposition 17.A.1 and its proof we know that $\Box + w$ is self-adjoint as a bounded operator: $H^m \to H^{-m}$. Consider $\Box$ as a closed unbounded operator $\Box_{\mathrm{sa}} : L^2 \to L^2$ with domain $\mathcal{D}(\Box_{\mathrm{sa}}) = \{u \in H^m;\ \Box u \in L^2\}$. Then $\mathcal{D}(\Box_{\mathrm{sa}}) = (\Box + w)^{-1} L^2$, which is dense in $(\Box + w)^{-1} H^{-m} = H^m$ and hence dense in L^2. $\Box_{\mathrm{sa}}$ (or equivalently $\Box_{\mathrm{sa}} + w$) is closed: If $(\Box_{\mathrm{sa}} + w) u_j = v_j$, $u_j \to u$, $v_j \to v$ in L^2, then $v_j \to v$ in H^{-m}, $u_j \to u$ in H^m, hence $(\Box + w) u = v$ and since $v \in L^2$, we get $u \in \mathcal{D}_{\mathrm{sa}}$. Similar arguments show that $\Box_{\mathrm{sa}}$ is self-adjoint. We also know that $\Box_{\mathrm{sa}}$ has a purely discrete spectrum and that $\{e_j\}_{j=1}^{\infty}$ is an orthonormal basis of eigenfunctions.

We have the max-min principle

$$t_j^2 = \sup_{\operatorname{codim} L = j-1} \inf_{\substack{u \in L, \\ \|u\|=1}} (\Box u|u), \tag{17.A.6}$$

where L varies in the set of closed subspaces of H^0 that are also contained in H^m. Similarly, from (17.A.3), we have the mini-max principle

$$\mu_j^2 = \inf_{\operatorname{codim} L = j-1} \sup_{\substack{v \in L \\ \|v\|=1}} ((\Box + w)^{-1} v|v), \tag{17.A.7}$$

where L varies in the set of closed subspaces of H^0. When $0 \notin \sigma(P)$, so that $P : H^m \to H^0$ is bijective, we can extend (17.A.7) to the case $w = 0$ and then, as we have seen, $\mu_j(0)^2 = t_j^{-2}$.

Now assume

$$P : H^m \to H^0 \text{ is a Fredholm operator of index } 0. \tag{17.A.8}$$

The discussion above applies also to PP^* when P is viewed as an operator $H^0 \to H^{-m}$ so that $P^* : H^m \to H^0$. Put

$$\Box^\dagger = PP^* : H^m \longrightarrow H^{-m}.$$

Then as in (17.A.5) we have an orthonormal basis $f_1, f_2, \dots$ in H^0 with $f_j,\ P^* f_j \in H^m$ such that

$$\Box^\dagger f_j = \widetilde{t}_j^2 f_j, \quad 0 \le \widetilde{t}_j \nearrow +\infty. \tag{17.A.9}$$

Proposition 17.A.2 *We have* $\widetilde{t}_j = t_j$ *and we can choose* f_j *so that*

$$Pe_j = t_j f_j, \quad P^* f_j = t_j e_j. \tag{17.A.10}$$

Proof We have $\mathcal{N}(\Box) = \mathcal{N}(P)$, $\mathcal{N}(\Box^\dagger) = \mathcal{N}(P^*)$ when P and P^* are viewed as operators $H^m \to H^0$. Notice however that by elliptic regularity, the kernel of $P^* : H^0 = H^{-m}$ is the same as the one of $P^* : H^m \to H^0$. Since P is Fredholm of index 0, the kernels of P and P^* have the same dimension, and consequently

$$\dim \mathcal{N}(\Box) = \dim \mathcal{N}(\Box^\dagger).$$

Let $t_0^2 = t_{j_0}^2$ be a non-vanishing eigenvalue of $\Box$ of multiplicity k_0, so that

$$t_{j_0 - 1} < t_{j_0} = \dots = t_{j_0 + k_0 - 1} < t_{j_0 + k_0}$$

for some $j_0, k_0 \in \mathbf{N}^*$ and with the convention that the first inequality is absent when $j_0 = 1$. If $0 \neq u \in \mathcal{N}(\Box - t_0^2)$, we know that u, $Pu \in H^m$, $Pu \neq 0$ and we notice that

$$\Box^\dagger Pu = P\Box u = t_j^2 Pu \quad \text{in } H^{-m}.$$

Thus $v := Pu \in H^m$ is non-zero and satisfies

$$\Box^\dagger v = t_j^2 v,$$

so P gives an injective map from $\mathcal{N}(\Box - t_0^2)$ into $\mathcal{N}(\Box^\dagger - t_0^2)$.

By the same argument P^* is injective from $\mathcal{N}(\Box^\dagger - t_0^2)$ to $\mathcal{N}(\Box - t_0^2)$ so the two spaces have the same dimension. It follows that $\widetilde{t}_j = t_j$ for all j.

Let e_j, $j_0 \leq j \leq j_0 + k_0 - 1$ be an orthonormal basis for $\mathcal{N}(\Box - t_0^2)$ and put $f_j = t_0^{-1} Pe_j \in \mathcal{N}(\Box^\dagger - t_0^2)$. Then

$$(f_j | f_k) = t_0^{-2}(Pe_j | Pe_k) = t_0^{-2}(\Box e_j | e_k) = \delta_{j,k},$$

so f_j, $j_0 \leq j \leq j_0 + k_0 - 1$ form an orthonormal basis for $\mathcal{N}(\Box^\dagger - t_0^2)$. Also notice that

$$P^* f_j = t_0^{-1} P^* Pe_j = t_0 e_j,$$

and we get (17.A.10) in the non-trivial case when $t_j \neq 0$. □

Write $t_j(P) = t_j$, so that $t_j(P^*) = t_j$ by the proposition. When P has a bounded inverse let $s_1(P^{-1}) \geq s_2(P^{-1}) \geq \cdots$ be the singular values of the inverse (as a compact operator in L^2). We have

$$s_j(P^{-1}) = \frac{1}{t_j(P)}. \tag{17.A.11}$$

Let $1 \leq N < \infty$ and let $R_+ : H^m \to \mathbf{C}^N$, $R_- : \mathbf{C}^N \to H^0$ be bounded operators. Assume that

$$\mathcal{P} = \begin{pmatrix} P & R_- \\ R_+ & 0 \end{pmatrix} : H^m \times \mathbf{C}^N \to H^0 \times \mathbf{C}^N \tag{17.A.12}$$

is bijective with a bounded inverse

$$\mathcal{E} = \begin{pmatrix} E & E_+ \\ E_- & E_{-+} \end{pmatrix}. \tag{17.A.13}$$

Recall that P has a bounded inverse precisely when E_{-+} does, and when this happens we have the relations,

$$P^{-1} = E - E_+ E_{-+}^{-1} E_-, \quad E_{-+}^{-1} = -R_+ P^{-1} R_-. \tag{17.A.14}$$

Cf. Sects. 3.2, 5.3, Chap. 6, Sect. 8.1 and Chap. 13. Recall ([48] and Proposition 8.2.2) that if $A, B : \mathcal{H}_1 \to \mathcal{H}_2$ and $C : \mathcal{H}_2 \to \mathcal{H}_3$ are bounded operators, where $\mathcal{H}_j$ are complex Hilbert spaces then we have the general estimates

$$s_{n+k-1}(A+B) \le s_n(A) + s_k(B), \tag{17.A.15}$$

$$s_{n+k-1}(CA) \le s_n(C) s_k(A), \tag{17.A.16}$$

In particular, for $k = 1$, we get

$$s_n(CA) \le \|C\| s_n(A), \ s_n(CA) \le s_n(C)\|A\|, \ s_n(A+B) \le s_n(A) + \|B\|.$$

Applying this to the second part of (17.A.14), we get

$$s_k(E_{-+}^{-1}) \le \|R_-\| \|R_+\| s_k(P^{-1}), \quad 1 \le k \le N$$

whence

$$t_k(P) \le \|R_-\| \|R_+\| t_k(E_{-+}), \quad 1 \le k \le N. \tag{17.A.17}$$

By a perturbation argument, we see that this holds also in the case when P, E_{-+} are non-invertible.

Similarly, from the first part of (17.A.14) we get

$$s_k(P^{-1}) \le \|E\| + \|E_+\| \|E_-\| s_k(E_{-+}^{-1}),$$

leading to

$$t_k(P) \ge \frac{t_k(E_{-+})}{\|E\| t_k(E_{-+}) + \|E_+\| \|E_-\|}. \tag{17.A.18}$$

Again this can be extended to the non-necessarily invertible case by means of small perturbations.

Generalizing Sect. 3.2 (as in [56]), we get a natural construction of a Grushin problem associated to a given operator. Let $P = P^0 : H^m \to H^0$ be a Fredholm operator of index 0 as above. Choose N so that $t_{N+1}(P^0)$ is strictly positive. In the following we sometimes write t_j instead of $t_j(P^0)$ for short.

Recall that t_j^2 are the first eigenvalues both for $P^{0*}P^0$ and P^0P^{0*}. Let $e_1, \ldots, e_N$ and $f_1, \ldots, f_N$ be corresponding orthonormal systems of eigenvectors of $P^{0*}P^0$ and P^0P^{0*}, respectively. They can be chosen so that

$$P^0 e_j = t_j f_j, \quad P^{0*} f_j = t_j e_j. \tag{17.A.19}$$

Define $R_+ : L^2 \to \mathbf{C}^N$ and $R_- : \mathbf{C}^N \to L^2$ by

$$R_+ u(j) = (u|e_j), \quad R_- u_- = \sum_1^N u_-(j) f_j. \tag{17.A.20}$$

It is easy to see that the Grushin problem

$$\begin{cases} P^0 u + R_- u_- = v, \\ R_+ u = v_+, \end{cases} \tag{17.A.21}$$

has a unique solution $(u, u_-) \in L^2 \times \mathbf{C}^N$ for every $(v, v_+) \in L^2 \times \mathbf{C}^N$, given by

$$\begin{cases} u = E^0 v + E_+^0 v_+, \\ u_- = E_-^0 v + E_{-+}^0 v_+, \end{cases} \tag{17.A.22}$$

where

$$\begin{cases} E_+^0 v_+ = \sum_1^N v_+(j) e_j, & E_-^0 v(j) = (v|f_j), \\ E_{-+}^0 = -\mathrm{diag}\,(t_j), & \|E^0\| \le \frac{1}{t_{N+1}}. \end{cases} \tag{17.A.23}$$

E^0 can be viewed as the inverse of P^0 as an operator from the orthogonal space $(e_1, e_2, \ldots, e_N)^\perp$ to $(f_1, f_2, \ldots, f_N)^\perp$.

We notice that in this case the norms of R_+ and R_- are equal to 1, so (17.A.17) tells us that $t_k(P^0) \le t_k(E_{-+}^0)$ for $1 \le k \le N$, but of course the expression for E_{-+}^0 in (17.A.23) implies equality.

Let $Q \in \mathcal{L}(H^0, H^0)$ and put $P^\delta = P^0 - \delta Q$ (where we sometimes put a minus sign in front of the perturbation for notational convenience). We are particularly interested in the case when $Q = Q_\omega u = qu$ is the operator of multiplication by a function q. Here $\delta > 0$ is a small parameter. Choose $R_\pm$ as in (17.A.20). Then if $\delta < t_{N+1}$ and $\|Q\| \le 1$, the perturbed Grushin problem

$$\begin{cases} P^\delta u + R_- u_- = v, \\ R_+ u = v_+, \end{cases} \tag{17.A.24}$$

is well posed and has the solution

$$\begin{cases} u = E^{\delta} v + E^{\delta}_{+} v_{+}, \\ u_{-} = E^{\delta}_{-} + E^{\delta}_{-+} v_{+}, \end{cases} \tag{17.A.25}$$

where

$$\mathcal{E}^{\delta} = \begin{pmatrix} E^{\delta} & E^{\delta}_{+} \\ E^{\delta}_{-} & E^{\delta}_{-+} \end{pmatrix} \tag{17.A.26}$$

is obtained from $\mathcal{E}^{0}$ by the formula

$$\mathcal{E}^{\delta} = \mathcal{E}^{0} \left(1 - \delta \begin{pmatrix} QE^{0} & QE^{0}_{+} \\ 0 & 0 \end{pmatrix} \right)^{-1}. \tag{17.A.27}$$

Using the Neumann series, we get

$$\begin{aligned} E^{\delta}_{-+} &= E^{0}_{-+} + E^{0}_{-} \delta Q (1 - E^{0} \delta Q)^{-1} E^{0}_{+} \\ &= E^{0}_{-+} + \delta E^{0}_{-} Q E^{0}_{+} + \delta^{2} E^{0}_{-} Q E^{0} Q E^{0}_{+} + \delta^{3} E^{0}_{-} Q (E^{0} Q)^{2} E^{0}_{+} + \cdots \end{aligned} \tag{17.A.28}$$

We also get

$$E^{\delta} = E^{0}(1 - \delta Q E^{0})^{-1} = E^{0} + \sum_{k=1}^{\infty} \delta^{k} E^{0} (Q E^{0})^{k}, \tag{17.A.29}$$

$$E^{\delta}_{+} = (1 - E^{0} \delta Q)^{-1} E^{0}_{+} = E^{0}_{+} + \sum_{k=1}^{\infty} \delta^{k} (E^{0} Q)^{k} E^{0}_{+}, \tag{17.A.30}$$

$$E^{\delta}_{-} = E^{0}_{-}(1 - \delta Q E^{0})^{-1} = E^{0}_{-} + \sum_{k=1}^{\infty} \delta^{k} E^{0}_{-} (Q E^{0})^{k}. \tag{17.A.31}$$

The leading perturbation in E^{δ}_{-+} is δM, where $M = E^{0}_{-} Q E^{0}_{+} : \mathbf{C}^{N} \to \mathbf{C}^{N}$ has the matrix

$$M(\omega)_{j,k} = (Q e_{k} | f_{j}), \tag{17.A.32}$$

which in the multiplicative case reduces to

$$M(\omega)_{j,k} = \int q(x) e_{k}(x) \overline{f_{j}(x)} dx. \tag{17.A.33}$$

Put $\tau_0 = t_{N+1}(P^0)$ and recall the assumption

$$\|Q\| \leq 1. \tag{17.A.34}$$

Then, if $\delta \leq \tau_0/2$, the new Grushin problem is well posed with an inverse $\mathcal{E}^\delta$ given in (17.A.26)–(17.A.31). We get

$$\|E^\delta\| \leq \frac{1}{1-\frac{\delta}{\tau_0}}\|E^0\| \leq \frac{2}{\tau_0}, \quad \|E_\pm^\delta\| \leq \frac{1}{1-\frac{\delta}{\tau_0}} \leq 2, \tag{17.A.35}$$

$$\|E_{-+}^\delta - (E_{-+}^0 + \delta E_-^0 Q E_+^0)\| \leq \frac{\delta^2}{\tau_0}\frac{1}{1-\frac{\delta}{\tau_0}} \leq 2\frac{\delta^2}{\tau_0}. \tag{17.A.36}$$

Using this in (17.A.17), (17.A.18) together with the fact that $t_k(E_{-+}^\delta) \leq 2\tau_0$,[2] we get

$$\frac{t_k(E_{-+}^\delta)}{8} \leq t_k(P^\delta) \leq t_k(E_{-+}^\delta). \tag{17.A.37}$$

Notes This chapter concludes the proof of Theorem 15.3.1 and we have very much followed [137, 139]. Notice that the probabilistic lower bounds on the singular values of certain matrices in the Sects. 17.1 and 17.2 play a crucial role. It would be very interesting and probably very important to enrich this passage with different, perhaps more efficient methods. That would make it possible to generalize Theorem 15.3.1 and treat larger classes of operators and perturbations.

[2]Indeed,

$$\begin{aligned} t_k(E_{-+}^\delta) &\leq t_k(E_{-+}^\delta) + \|E_{-+}^\delta - E_{-+}^0\| \leq t_k(E_{-+}^0) + \delta + 2\frac{\delta^2}{\tau_0} \\ &\leq t_k(E_{-+}^0) + 2\delta \leq t_k(E_{-+}^0) + \tau_0 \leq 2\tau_0. \end{aligned}$$

Distribution of Large Eigenvalues for Elliptic Operators

18

18.1 Introduction

In this chapter we consider elliptic differential operators on a compact manifold and rather than taking the semi-classical limit ($h \to$), we let $h = 1$ and study the distribution of large eigenvalues. Bordeaux Montrieux [16, 17] studied elliptic systems of differential operators on S^1 with random perturbations of the coefficients, and under some additional assumptions, he showed that the large eigenvalues obey the Weyl law *almost surely*. His analysis was based on a reduction to the semi-classical case, where he could use and extend the methods of Hager [55].

In [19] Bordeaux Montrieux and the author considered scalar elliptic operators on a general smooth compact manifold, using the semi-classical results of [139]. The present chapter follows closely, [19], but we replace [139] by the results of Sect. 15.3, which leads to some modifications.

Let X be a smooth compact manifold of dimension n. Let P^0 be an elliptic differential operator on X of order $m \geq 2$ with smooth coefficients and with classical principal symbol $p(x, \xi)$. In local coordinates we have,

$$P^0 = \sum_{|\alpha| \leq m} a_\alpha^0(x) D^\alpha, \quad p(x, \xi) = \sum_{|\alpha| = m} a_\alpha^0(x) \xi^\alpha. \tag{18.1.1}$$

The ellipticity of P^0 means that $p(x, \xi) \neq 0$ for real $\xi \neq 0$. We assume that

$$p(T^*X) \neq \mathbf{C}. \tag{18.1.2}$$

Let dx be a strictly positive smooth density of integration dx on X and use it to define the L^2 norm $\| \cdot \|$ and the inner product $(\cdot|\cdot)$. Let $\Gamma : L^2(X) \to L^2(X)$ be the antilinear

J. Sjöstrand, *Non-Self-Adjoint Differential Operators, Spectral Asymptotics and Random Perturbations*, Pseudo-Differential Operators 14,
https://doi.org/10.1007/978-3-030-10819-9_18

operator of complex conjugation, given by $\Gamma u = \overline{u}$. We need the symmetry assumption (cf. (15.1.7)),

$$(P^0)^* = \Gamma P^0 \Gamma, \tag{18.1.3}$$

where $(P^0)^*$ is the formal complex adjoint of P^0. As in Sect. 15.1 we observe that the property (18.1.3) implies that

$$p(x, -\xi) = p(x, \xi), \tag{18.1.4}$$

and conversely, if (18.1.4) holds, then the operator $\frac{1}{2}(P^0 + \Gamma(P^0)^*\Gamma)$ has the same principal symbol p and satisfies (18.1.3).

Let $\widetilde{R}$ be an elliptic second-order differential operator on X with smooth coefficients, which is self-adjoint and strictly positive. Let $\epsilon_0, \epsilon_1, \dots$ be an orthonormal basis of eigenfunctions of $\widetilde{R}$ so that

$$\widetilde{R}\epsilon_j = (\mu_j^0)^2 \epsilon_j, \quad 0 < \mu_0^0 < \mu_1^0 \le \mu_2^0 \le \cdots \tag{18.1.5}$$

Our randomly perturbed operator is

$$P_\omega^0 = P^0 + q_\omega^0(x), \tag{18.1.6}$$

where ω is the random parameter and

$$q_\omega^0(x) = \sum_{j=0}^{\infty} \alpha_j^0(\omega)\epsilon_j. \tag{18.1.7}$$

Here we assume that $\alpha_j^0(\omega)$ are independent complex Gaussian random variables of variance σ_j^2 and mean value 0:

$$\alpha_j^0 \sim \mathcal{N}_{\mathbf{C}}(0, \sigma_j^2), \tag{18.1.8}$$

where

$$\frac{1}{\mathcal{O}(1)}(\mu_j^0)^{-\rho} e^{-(\mu_j^0)^{\frac{\beta}{M+1}}} \le \sigma_j \le \mathcal{O}(1)(\mu_j^0)^{-\rho}, \tag{18.1.9}$$

$$M = \frac{3n - \frac{1}{2}}{s - \frac{n}{2} - \epsilon}, \quad 0 \le \beta < \frac{1}{2}, \quad \rho > n, \tag{18.1.10}$$

with s, ρ, ϵ fixed constants such that

$$\frac{n}{2} < s < \rho - \frac{n}{2},\ 0 < \epsilon < s - \frac{n}{2}.$$

The existence of such a sequence of random variables is stated in [87, Theorem 2.19].

It would be very interesting to explore more general "invariantly defined" random perturbations, e.g., of Gaussian type. cf. [90].

We will see below that $q_\omega^0 \in H^s(X)$ almost surely since $s < \rho - \frac{n}{2}$. Hence, $q_\omega^0 \in L^\infty$ almost surely, and it follows that P_ω^0 has purely discrete spectrum.

Consider the function $F(w) = \arg p(w)$ on the cosphere bundle S^*X. For given $\theta_0 \in S^1 \simeq \mathbf{R}/(2\pi\mathbf{Z})$, $N_0 \in \dot{\mathbf{N}} := \mathbf{N} \setminus \{0\}$, we introduce the property $\mathrm{P}(\theta_0, N_0)$:

$$\sum_{k=1}^{N_0} |\nabla^k F(w)| \neq 0 \text{ on } \{w \in S^*X;\ F(w) = \theta_0\}. \tag{18.1.11}$$

Notice that if $\mathrm{P}(\theta_0, N_0)$ holds, then $\mathrm{P}(\theta, N_0)$ holds for all θ in some neighborhood of θ_0. Also, if X is connected and X, p are analytic and the analytic function F is not constant, then there exists $N_0 \in \dot{\mathbf{N}}$ such that $\mathrm{P}(\theta_0, N_0)$ holds for all θ_0.

The main result of this chapter is basically the one of [19].

Theorem 18.1.1 *Assume that $m \geq 2$. Let $0 \leq \theta_1 \leq \theta_2 \leq 2\pi$ and assume that* $\mathrm{P}(\theta_1, N_0)$ *and* $\mathrm{P}(\theta_2, N_0)$ *hold for some $N_0 \in \dot{\mathbf{N}}$. Also assume that $\beta \in [0, 1/(2N_0)[$* (cf. (18.1.9).) *Let $g \in C^\infty([\theta_1, \theta_2];]0, \infty[)$ and put*

$$\Gamma^g_{\theta_1,\theta_2;0,\lambda} = \{re^{i\theta};\, \theta_1 \leq \theta \leq \theta_2,\ 0 \leq r \leq \lambda g(\theta)\}.$$

Then for every $\delta \in]0, 1/(2N_0) - \beta[$ there exists $C > 0$ such that almost surely: there exists $C(\omega) < \infty$ such that for all $\lambda \in [1, \infty[$:

$$\begin{aligned}\left|\#(\sigma(P_\omega^0) \cap \Gamma^g_{\theta_1,\theta_2;0,\lambda}) - \frac{1}{(2\pi)^n} \operatorname{vol} p^{-1}(\Gamma^g_{\theta_1,\theta_2;0,\lambda})\right| \\ \leq C(\omega) + C\lambda^{\frac{n}{m} - \frac{1}{m}(\frac{1}{2N_0} - \beta - \delta)}.\end{aligned} \tag{18.1.12}$$

Here $\sigma(P_\omega^0)$ denotes the spectrum and $\#(A)$ denotes the number of elements in the set A. In (18.1.12) *the eigenvalues are counted with their algebraic multiplicities.*

By the same proof we have an almost certain conclusion for a whole family of θ_1, θ_2, g:

Theorem 18.1.2 *Assume that $m \geq 2$. Let Θ be a compact subset of* $[0, 2\pi]$. *Let $N_0 \in \mathbf{N}$ and assume that* $\mathrm{P}(\theta, N_0)$ *holds uniformly for $\theta \in \Theta$. Also assume that $\beta \in [0, \frac{1}{2N_0}[$.*

Let $\mathcal{G}$ be a subset of $\{(g, \theta_1, \theta_2);\ \theta_j \in \Theta, \theta_1 \le \theta_2,\ g \in C^\infty([\theta_1, \theta_2];\ ]0, \infty[)\}$ *with the property that* g *and* $1/g$ *are uniformly bounded in* $C^\infty([\theta_1, \theta_2];\ ]0, \infty[)$ *when* (g, θ_1, θ_2) *varies in* $\mathcal{G}$. *Then for every* $\delta \in]0, \frac{1}{2N_0} - \beta[$ *there exists* $C > 0$ *such that almost surely: there exists* $C(\omega) < \infty$ *such that for all* $\lambda \in [1, \infty[$ *and all* $(g, \theta_1, \theta_2) \in \mathcal{G}$, *we have the estimate* (18.1.12).

In (18.1.9) we can choose σ_j decaying faster than any negative power of μ_j^0. The discussion below will imply that $q_\omega(x)$ is almost surely a smooth function. Very roughly, Theorem 18.1.2 then implies that for almost every elliptic operator of order ≥ 2 with smooth coefficients on a compact manifold which satisfies the conditions (18.1.2), (18.1.3), the large eigenvalues distribute according to Weyl's law in sectors with limiting directions that satisfy a weak non-degeneracy condition.

18.2 Some Examples

Let $f \in C^\infty(S^1)$ be non-vanishing and take its values in a closed sector $\Gamma \subset \mathbf{C}$ of angle $< \pi$. In other words, there exist $\theta_0 \in \mathbf{R}$, $\alpha \in [0, \pi/2[$ such that

$$\arg f(S^1) = [\theta_0 - \alpha, \theta_0 + \alpha]. \tag{18.2.1}$$

Assume for simplicity that $\theta_0 = 0$. Then (see [16, 17, 124]) the spectrum of $f(x)D$ can be computed directly and it consists of the simple eigenvalues

$$\lambda_k = \frac{k}{\langle 1/f \rangle},\ k \in \mathbf{Z}, \tag{18.2.2}$$

where $\langle 1/f \rangle$ denotes the mean-value of the function $1/f$. Since $1/f$ is non-vanishing with values in the sector $\overline{\Gamma}$, the same holds for $\langle 1/f \rangle$.

The antisymmetric operator $f^{1/2} D f^{1/2} = f^{-1/2}(fD)f^{1/2}$ has the same spectrum, and the elliptic symmetric operator

$$P^0 = (f^{1/2} D f^{1/2})^2 = D f^2 D - \frac{1}{4}(f')^2 - \frac{1}{2} f f'' \tag{18.2.3}$$

therefore has the spectrum

$$\{\mu_0, \mu_1, \mu_2, \dots\}, \quad \mu_k = \lambda_k^2 = \frac{k^2}{\langle 1/f \rangle^2}, \tag{18.2.4}$$

where μ_0 is a simple eigenvalue and $\mu_1, \mu_2, \dots$ are double. The principal symbol of P^0 is given by

$$p(x,\xi) = f(x)^2\xi^2, \tag{18.2.5}$$

and its range is the sector

$$[0,\infty[\, e^{i[-2\alpha,2\alpha]} \tag{18.2.6}$$

(having chosen $\theta_0 = 0$) which does not intersect the open negative half-line. The eigenvalues μ_k are situated on a half-line inside the range (18.2.6), and unless $\arg f =$ const., we see that the Weyl asymptotics does not hold for P^0. On the other hand, if we add the non-degeneracy assumption

$$\sum_{k=1}^{N_0} \left| \left(\frac{d}{dx}\right)^k \arg f(x) \right| \neq 0,\ x \in S^1, \quad \text{for some } N_0 \in \mathbf{N}\setminus\{0\}, \tag{18.2.7}$$

then the property P(θ, N_0) holds for all θ and we know from the Theorems 18.1.1 and 18.1.2 that the Weyl asymptotics holds almost surely for the random perturbations $P^0_\omega = P^0 + q^0_\omega$ if q^0_ω is given in (18.1.7)–(18.1.10).

Despite the fact that (in some sense and with the additional conditions in our main theorems) almost all symmetric elliptic differential operators obey Weyl asymptotics, it is probably a difficult task to find explicit operators with this property outside the class of normal operators and operators with principal symbol having constant argument. To find such examples one would probably like to assume the coefficients to be analytic, but in that case the Weyl asymptotics is unlikely to hold. Indeed, in the analytic case there is the possibility of making an analytic distorsion (for instance by replacing the underlying compact analytic manifold by a small deformation) which will not change the spectrum (by ellipticity and analyticity), but which will replace the given real phase space by a deformation, likely to change the Weyl law. In one and two dimensions analytic distorsions have been used to determine the spectrum (by making the operator more normal): in the two-dimensional semi-classical case this was done in [71–73, 104, 105], and in [142] it was shown that the resulting law is in general different from the Weyl law (naively because a complex Bohr–Sommerfeld law relies on going out in the complex domain while the Weyl law only uses the real cotangent space).

To illustrate this, let us consider the second-order differential operator on S^1,

$$P^0 = a(x)D^2 + b(x)D + c(x), \tag{18.2.8}$$

where the coefficients a, b, c are smooth (and 2π-periodic when considered as functions on $\mathbf{R}$). We assume that P^0 is elliptic, so that $a(x) \neq 0$, and that the range of $\arg a$ is the interval $[-2\alpha, 2\alpha]$ for some $\alpha \in [0, \frac{\pi}{2}[$. Then $a(x) = f(x)^2$, where f is smooth, non-vanishing and the range of $\arg f$ is $[-\alpha, \alpha]$. The Bohr–Sommerfeld quantization condition, which correctly describes the large eigenvalues when P^0 is self-adjoint and more generally when $a > 0$, would predict that the large eigenvalues μ_k^2 are determined by the condition

$$I(\mu_k) = 2\pi k + \mathcal{O}(1), \quad k \in \mathbf{Z}, \ |k| \gg 1, \tag{18.2.9}$$

where $I(\mu)$ is the action, defined by $I(\mu) = \int_0^{2\pi} \xi(x, \mu) dx$, $\xi(x, \mu) = \mu/f(x)$, so that $p(x, \xi(x, \mu)) = \mu^2$, where $p(x, \xi) = f(x)^2\xi^2$ is the principal symbol of P^0. Notice that this simplifies to

$$\mu_k = \frac{k}{\langle 1/f \rangle} + \mathcal{O}(1). \tag{18.2.10}$$

We also recall that the remainder has a complete asymptotic expansion in negative powers of k. As we have seen, this rule is correct in the special case of the operator (18.2.3) and as we have noticed it becomes almost surely false if we add a random smooth zero-order term (at least in the symmetric case in the sense of (18.1.3)).

However, the Bohr–Sommerfeld rule is correct under analyticity assumptions, as we shall now review (cf. Chap. 7): Look for a complex change of variables $x = x(t)$ with $0 = x(0)$ such that $f(x)D_x = \kappa D_t$, for a suitable $\kappa \in \mathbf{C} \setminus \{0\}$. We get

$$\frac{dt}{dx} = \frac{\kappa}{f(x)},$$

so the inverse $t(x)$ is given by

$$t = \kappa \int_0^x \frac{dy}{f(y)}. \tag{18.2.11}$$

If f is merely smooth, we can still define a complex curve $t(x)$ by (18.2.11) for real x. We now determine κ by the condition that $t(2\pi) = 2\pi$, i.e.,

$$\kappa = \frac{1}{\langle 1/f \rangle}. \tag{18.2.12}$$

Assume that f extends to a holomorphic non-vanishing function in a 2π-periodic simply connected neighborhood Ω of $\mathbf{R}$ in $\mathbf{C}$. Then $t(x)$ extends to a holomorphic function on Ω, and we assume that the set $\{x \in \Omega;\ t(x) \in \mathbf{R}\}$ contains (the image of) a smooth curve $\gamma : \mathbf{R} \to \Omega$ such that γ' is 2π-periodic, $\gamma(0) = 0$, $\gamma(2\pi) = 2\pi$. Also assume that b, c extend to holomorphic functions on Ω. Notice that if $f_0 > 0$ is an analytic 2π-periodic function and if f is a small perturbation of f_0 in a fixed neighborhood of $\mathbf{R}$, then f fulfills the assumptions above. In a small neighborhood of γ we can replace the variable x by t and thus get the operator

$$\widetilde{P} = \kappa^2 D_t^2 + \widetilde{b}(t)D_t + \widetilde{c}(t), \tag{18.2.13}$$

which is well defined in a small neighborhood of $\mathbf{R}_t$. For this operator it is quite easy to justify the Bohr–Sommerfeld rule by some version of the complex WKB-method (cf. [50] and Chap. 7). Now the Bohr–Sommerfeld rule is clearly invariant under the change of variables above. Moreover, eigenfunctions of $\widetilde{P}$ defined near $\mathbf{R}_t$ are also eigenfunctions of P^0 with respect to the x-variables in a neighborhood of γ, and since P^0 is elliptic in Ω, they extend to holomorphic functions in Ω and by restriction become eigenfunctions on $\mathbf{R}_x$. The same remark holds for generalized eigenfunctions. Hence the eigenvalues of $\widetilde{P}$ are also eigenvalues of P^0. This argument works equally well in the other direction so we can identify completely the spectra of $\widetilde{P}$ and of P^0, and this completes the (review of the) justification of the Bohr–Sommerfeld rule (and hence of the non-validity of Weyl asymptotics when $\arg f$ is non-constant) for the operator (18.2.8) in the analytic case.

18.3 Volume Considerations

The remainder of the chapter is mainly devoted to the proof of Theorem 18.1.1. In the next section we shall perform a reduction to a semi-classical situation and work with $h^m P^0$, which has the semi-classical principal symbol p in (18.1.1). An important quantity is

$$\mathrm{vol}\, p^{-1}(\gamma + D(0,t)), \tag{18.3.1}$$

where $D(0,t) = \{z \in \mathbf{C};\ |z| < t\}$, $\gamma = \partial\Gamma$ and $\Gamma \Subset \dot{\mathbf{C}} := \mathbf{C} \setminus \{0\}$ is assumed to have piecewise smooth boundary.

Proposition 18.3.1 *Let γ be the curve $\{\tau e^{i\theta} \in \mathbf{C};\ \tau = g(\theta),\ \theta \in S^1\}$, where $0 < g \in C^1(S^1)$. Then*

$$\mathrm{vol}\,(p^{-1}(\gamma + D(0,t))) = \mathcal{O}(t),\ t \longrightarrow 0.$$

Proof This follows from the fact that the radial derivative of p is $\neq 0$. More precisely, write $T^*X \setminus 0 \ni \rho = rw,\ w \in S^*X,\ r > 0$, so that $p(\rho) = r^m p(w),\ p(w) \neq 0$. If $\rho \in p^{-1}(\gamma + D(0,t))$, for $0 \le t \ll 1$, we have $|p(rw) - g(\theta)e^{i\theta}| < t$ for some θ, so

$$||p(rw)| - g(\theta)| \le \mathcal{O}(t), \quad |\arg p(w) - \theta| \le \mathcal{O}(t),$$

which implies that $|p(rw)| - g(\arg p(w)| \leq \mathcal{O}(t)$. Hence, we have for some $C \geq 1$, independent of t,

$$g(\arg p(w)) - Ct \leq r^m |p(w)| \leq g(\arg p(w)) + Ct,$$

$$\left(\frac{g(\arg p(w)) - Ct}{|p(w)|}\right)^{\frac{1}{m}} \leq r \leq \left(\frac{g(\arg p(w)) + Ct}{|p(w)|}\right)^{\frac{1}{m}},$$

when t is small. So for every $w \in S^*X$, r has to belong to an interval of length $\mathcal{O}(t)$. □

We next study the volume in (18.3.1) when γ is a radial segment of the form $[r_1, r_2]e^{i\theta_0}$, where $0 < r_1 < r_2$ and $\theta_0 \in S^1$.

Proposition 18.3.2 *Let $\theta_0 \in S^1$, $N_0 \in \dot{\mathbf{N}}$ and assume that $P(\theta_0, N_0)$ holds. Then if $0 < r_1 < r_2$ and γ is the radial segment $[r_1, r_2]e^{i\theta_0}$, we have*

$$\mathrm{vol}\,(p^{-1}(\gamma + D(0,t))) = \mathcal{O}(t^{1/N_0}),\ t \longrightarrow 0.$$

Proof We first observe that it suffices to show that

$$\mathrm{vol}_{S^*X} F^{-1}([\theta_0 - t, \theta_0 + t]) = \mathcal{O}(t^{1/N_0}).$$

At every point $w_0 \in F^{-1}(\theta_0)$ we can choose coordinates $w_1, \dots, w_{2n-1}$, centered at w_0, such that for some $k \in \{1, \dots, N_0\}$, we have $\partial_{w_1}^j (F - \theta_0)(w_0) = 0$ when $0 \leq j \leq k-1$, and $\neq 0$ when $j = k$. Put $J(s, \rho) = |F(w) - \theta_0|^2$, $s = w_1$, $\rho = (w_2, \dots, w_{2n-1})$ and assume for simplicity that w_0 corresponds to $(s, \rho) = (0, \rho_0)$. Then $J \geq 0$,

$$\partial_s^j J(0, \rho_0) \begin{cases} = 0, & \text{if } j \leq 2k-1, \\ > 0, & \text{if } j = 2k. \end{cases}$$

Applying Proposition 10.2.2 to the two functions $J(\pm s, \rho)$, we see that

$$J(s, \rho) \geq s^{2k}/C,\quad (s, \rho) \in \mathrm{neigh}\,((0, \rho_0), \mathbf{R}^{2n-1}).$$

It follows that

$$\mathrm{vol}\,(F^{-1}([\theta_0 - t, \theta_0 + t]) \cap \mathrm{neigh}\,(w_0)) = \mathrm{vol}\,(J^{-1}([0, t^2]) \cap \mathrm{neigh}\,(w_0)) = \mathcal{O}(t^{1/k}).$$

It then suffices to use a simple covering argument. □

Now, let $0 \le \theta_1 < \theta_2 \le 2\pi$, $g \in C^\infty([\theta_1, \theta_2];]0, \infty[\,)$ and put

$$\Gamma^g_{\theta_1,\theta_2;r_1,r_2} = \{re^{i\theta};\ \theta_1 \le \theta \le \theta_2,\ r_1 g(\theta) \le r \le r_2 g(\theta)\}, \tag{18.3.2}$$

for $0 \le r_1 \le r_2 < \infty$. If $0 < r_1 < r_2 < +\infty$ and $\mathrm{P}(\theta_j, N_0)$ hold for $j = 1, 2$, then the last two propositions imply that

$$\mathrm{vol}\, p^{-1}(\partial \Gamma^g_{\theta_1,\theta_2;r_1,r_2} + D(0,t)) = \mathcal{O}(t^{1/N_0}),\ t \longrightarrow 0. \tag{18.3.3}$$

18.4 Semiclassical Reduction

We are interested in the distribution of large eigenvalues ζ of P^0_ω, and write

$$\zeta = \frac{z}{h^m},\ |z| \asymp 1,\ h \asymp |\zeta|^{-1/m},\ 0 < h \ll 1. \tag{18.4.1}$$

Then

$$h^m(P^0_\omega - \zeta) = h^m P^0_\omega - z = P + h^m q^0_\omega - z, \tag{18.4.2}$$

where

$$P := h^m P^0 = \sum_{|\alpha| \le m} a_\alpha(x; h)(hD)^\alpha. \tag{18.4.3}$$

Here

$$\begin{aligned} a_\alpha(x; h) &= \mathcal{O}(h^{m-|\alpha|}) \text{ in } C^\infty, \\ a_\alpha(x; h) &= a^0_\alpha(x) \text{ when } |\alpha| = m. \end{aligned} \tag{18.4.4}$$

P is a semi-classical differential operator with semi-classical principal symbol $p(x, \xi)$ in (18.1.1).

Our strategy will be to decompose the random perturbation as

$$h^m q^0_\omega = \delta q_\omega + k_\omega(x),$$

where the two terms are independent, and with probability very close to 1, δq_ω will be a semi-classical random perturbation as in Sect. 15.2, while

$$\|k_\omega\|_{H^s} \le h, \tag{18.4.5}$$

and

$$s \in]\frac{n}{2}, \rho - \frac{n}{2}[\tag{18.4.6}$$

is fixed. Then $h^m P_\omega^0$ will be viewed as a random perturbation of $h^m P^0 + k_\omega$ and we will apply Theorem 15.3.1 with the extension indicated right thereafter. To achieve this without extra assumptions on the order m, we will also have to represent some of our random variables $\alpha_j^0(\omega)$ as sums of two independent Gaussian random variables.

We start by examining when

$$\|h^m q_\omega^0\|_{H^s} \le h^{2-\vartheta}, \tag{18.4.7}$$

for a fixed $\vartheta \in [0, 1/2[$.

Proposition 18.4.1 *There is a constant $C > 0$ such that* (18.4.7) *holds with probability*

$$\ge 1 - \exp\left(C - \frac{1}{2Ch^{2(m-2+\vartheta)}}\right).$$

Proof We have

$$h^m q_\omega^0 = \sum_{j=0}^{\infty} \alpha_j(\omega)\epsilon_j, \quad \alpha_j = h^m \alpha_j^0 \sim \mathcal{N}_{\mathbf{C}}(0, (h^m \sigma_j)^2), \tag{18.4.8}$$

and the α_j are independent. Now, from Proposition 16.2.2, applied with "h" there equal to 1, we see that

$$\|h^m q_\omega^0\|_{H^s}^2 \asymp \sum_{j=0}^{\infty} |(\mu_j^0)^s \alpha_j(\omega)|^2, \tag{18.4.9}$$

where $(\mu_j^0)^s \alpha_j \sim \mathcal{N}_{\mathbf{C}}(0, (\widetilde{\sigma}_j)^2)$ are independent random variables and $\widetilde{\sigma}_j = (\mu_j^0)^s h^m \sigma_j$.

Now recall Proposition 3.4.1, cf. [17], which has a straightforward extension to the case of countably many random variables: Let $d_0, d_1, \ldots$ be a finite or infinite family of independent complex Gaussian random variables, $d_j \sim \mathcal{N}_{\mathbf{C}}(0, (\widehat{\sigma}_j)^2)$, $0 < \widehat{\sigma}_j < \infty$, and assume that $\sum \widehat{\sigma}_j^2 < \infty$. Then for every $t > 0$,

$$\mathbf{P}\left(\sum_{j=0}^{\infty} |d_j|^2 \ge t\right) \le \exp\left(\frac{1}{2 \max \widehat{\sigma}_j^2}\left(C_0 \sum_{j=0}^{\infty} \widehat{\sigma}_j^2 - t\right)\right). \tag{18.4.10}$$

Here $\mathbf{P}(A)$ denotes the probability of the event A and $C_0 > 0$ is a universal constant. The estimate is interesting only when $t > C_0 \sum \widehat{\sigma}_j^2$ and for such values of t it improves if we replace $\{d_0, d_1, \dots\}$ by a subfamily. Indeed, $\sum \widehat{\sigma}_j^2$ will then decrease, and so will $\max \widehat{\sigma}_j^2$.

Apply this to (18.4.9) with $d_j = (\mu_j^0)^s \alpha_j$, $t = h^{4-2\vartheta}$. Here, we recall that $\widetilde{\sigma}_j = (\mu_j^0)^s h^m \sigma_j$, and get from (18.1.9), (18.4.6) that

$$\max \widetilde{\sigma}_j^2 \asymp h^{2m}, \tag{18.4.11}$$

while

$$\sum_{j=0}^{\infty} \widetilde{\sigma}_j^2 \lesssim h^{2m} \sum_{j=0}^{\infty} (\mu_j^0)^{2(s-\rho)}. \tag{18.4.12}$$

Let $N(\mu) = \#\left(\sigma\left(\sqrt{\widetilde{R}}\right) \cap]0, \mu]\right)$ be the number of eigenvalues of $\sqrt{\widetilde{R}}$ in $]0, \mu]$, so that $N(\mu) \asymp \mu^n$, $\mu \to \infty$, by the standard Weyl asymptotics for positive elliptic operators on compact manifolds. The last sum in (18.4.12) is equal to

$$\int_0^\infty \mu^{2(s-\rho)} dN(\mu) = \int_0^\infty 2(\rho - s)\mu^{2(s-\rho)-1} N(\mu) d\mu,$$

which is finite since $2(s-\rho) + n < 0$ by (18.4.6). Thus

$$\sum_{j=0}^{\infty} \widetilde{\sigma}_j^2 \lesssim h^{2m}, \tag{18.4.13}$$

and the proposition follows by applying (18.4.9), (18.4.11) and (18.4.13) to (18.4.10) with $t = h^{4-2\vartheta}$. □

We next review the choice of parameters for the random perturbation in Theorem 15.3.1. This perturbation is of the form δq_ω,

$$\delta = \tau_0 h^{N_2 - n}, \quad 0 < \tau_0 \le h^2, \tag{18.4.14}$$

where $N_2 \gg 1$ is constant,

$$q_\omega(x) = \sum_{0 < h\mu_k^0 \le L} \alpha_k(\omega)\epsilon_k(x), \ |\alpha|_{\mathbf{C}^D} \le R, \tag{18.4.15}$$

and a possible choice of L, R is

$$L = Ch^{-M}, \quad R = Ch^{-\widetilde{M}}, \tag{18.4.16}$$

with $C \gg 1$ as in (15.2.5) and

$$M = \frac{3n}{s - \frac{n}{2} - \epsilon}, \quad \widetilde{M} = \frac{3n}{2} + \left(\frac{n}{2} + \epsilon\right) M. \tag{18.4.17}$$

Here $\epsilon > 0$ is any fixed parameter in $]0, s - \frac{n}{2}[$ and q_ω should be subject to a probability density on $B_{\mathbf{C}^D}(0, R)$ of the form $C(h)e^{\Phi(\alpha;h)}L(d\alpha)$, where

$$|\nabla_\alpha \Phi| = \mathcal{O}(h^{-N_4}), \tag{18.4.18}$$

for some constant $N_4 \geq 0$.

Write

$$q_\omega^0 = q_\omega^1 + q_\omega^2, \tag{18.4.19}$$

$$q_\omega^1 = \sum_{0 < h\mu_j^0 \leq L} \alpha_j^0(\omega)\epsilon_j, \quad q_\omega^2 = \sum_{h\mu_j^0 > L} \alpha_j^0(\omega)\epsilon_j. \tag{18.4.20}$$

From Proposition 18.4.1 and its proof, especially the observation after (18.4.10), we know that

$$\|h^m q_\omega^2\|_{H^s} \leq h^{2-\vartheta} \text{ with probability } \geq 1 - \exp\left(C_0 - \frac{1}{2Ch^{2(m-2+\vartheta)}}\right). \tag{18.4.21}$$

When $m = 2$ we will take $\vartheta \in]0, 1/2[$ and for $m > 2$ (i.e. $m \geq 4$), we will take $\vartheta = 0$. Write

$$P + h^m q_\omega^0 = (P + h^m q_\omega^2) + h^m q_\omega^1$$

and recall Theorem 15.3.1 and Remark 15.3.2.

The next question is then wether $h^m q_\omega^1$ can be written as $\tau_0 h^{N_2-n} q_\omega$ where $q_\omega = \sum_{0<h\mu_j^0 \leq L} \alpha_j \epsilon_j$ and $|\alpha|_{\mathbf{C}^D} \leq R$ with probability close to 1. We get

$$\alpha_j = \frac{1}{\tau_0} h^{m-N_2+n} \alpha_j^0(\omega) \sim \mathcal{N}(0, \widehat{\sigma}_j^2),$$

$$\frac{1}{\tau_0} h^{m-N_2+n} (\mu_j^0)^{-\rho} e^{-(\mu_j^0)^{\frac{\beta}{M+1}}} \lesssim \widehat{\sigma}_j \lesssim \frac{1}{\tau_0} h^{m-N_2+n} (\mu_j^0)^{-\rho}.$$

Applying (18.4.10), we get

$$\mathbf{P}(|\alpha|_{\mathbf{C}^D}^2 \geq R^2) \leq \exp\left(C - \frac{R^2 \tau_0^2}{Ch^{2(m-N_2-n)}}\right), \tag{18.4.22}$$

which is $\mathcal{O}(1)\exp(-h^{-\delta})$ provided that

$$-2\widetilde{M}+2\frac{\ln(1/\tau_0)}{\ln(1/h)}+2(N_2-n-m)<-\delta. \tag{18.4.23}$$

Here $\tau_0 \le h^2$ and if we choose $\tau_0 = h^2$ or more generally bounded from below by some power of h, we see that (18.4.23) holds for any fixed δ, provided that m is sufficiently large.

In order to avoid such an extra assumption, we represent α_j^0 for $h\mu_j^0 \le L$ as the sum of two independent Gaussian random variables. Let $j_0 = j_0(h)$ be the largest j for which $h\mu_j^0 \le L$. Put

$$\sigma' = \frac{1}{C}h^K e^{-Ch^{-\beta}}, \text{ where } K \ge \rho(M+1),\ C \gg 1 \tag{18.4.24}$$

so that $\sigma' \le \frac{1}{2}\sigma_j$ for $1 \le j \le j_0(h)$. The factor h^K is needed only when $\beta = 0$.

For $j \le j_0$, we may assume that $\alpha_j^0(\omega) = \alpha'_j(\omega) + \alpha''_j(\omega)$, where $\alpha'_j \sim \mathcal{N}(0,(\sigma')^2)$, $\alpha''_j \sim \mathcal{N}_{\mathbf{C}}(0,(\sigma''_j)^2)$ are independent random variables and

$$\sigma_j^2 = (\sigma')^2 + (\sigma''_j)^2,$$

so that

$$\sigma''_j = \sqrt{\sigma_j^2 - (\sigma')^2} \asymp \sigma_j.$$

Put $q_\omega^1 = q'_\omega + q''_\omega$, where

$$q'_\omega = \sum_{h\mu_j^0 \le L} \alpha'_j(\omega)\epsilon_j, \quad q''_\omega = \sum_{h\mu_j^0 \le L} \alpha''_j(\omega)\epsilon_j.$$

Now (cf. (18.4.19)) we write

$$P + h^m q_\omega^0 = (P + h^m(q''_\omega + q_\omega^2)) + h^m q'_\omega.$$

Theorem 15.3.1 is valid for random perturbations of

$$P_0 := P + h^m(q''_\omega + q_\omega^2),$$

provided that $\|h^m(q''_\omega + q_\omega^2)\|_{H^s} \le h^2$, which again holds with a probability as in (18.4.21) and when $\|h^m(q''_\omega + q_\omega^2)\|_{H^s} \le h^{2-\vartheta}$, we have the weakened variant in Remark 15.3.2. The new random perturbation is $h^m q'_\omega$, which we write as $\tau_0 h^{N_2-n}\widetilde{q}_\omega$, where $\widetilde{q}_\omega$ takes the form

$$\widetilde{q}_\omega(x) = \sum_{0<h\mu_j^0\leq L} \vartheta_j(\omega)\epsilon_j, \tag{18.4.25}$$

with new independent random variables

$$\vartheta_j = \frac{1}{\tau_0}h^{m-N_2+n}\alpha_j'(\omega) \sim \mathcal{N}\left(0, \left(\frac{1}{\tau_0}h^{m-N_2+n}\sigma'(h)\right)^2\right). \tag{18.4.26}$$

Now, by (18.4.10),

$$\mathbf{P}(|\vartheta|^2_{\mathbf{C}^D} > R^2) \leq \exp\left(\mathcal{O}(1)D - \frac{R^2\tau_0^2}{\mathcal{O}(1)(h^{m-N_2+n}\sigma'(h))^2}\right).$$

Here by Weyl's law for the distribution of eigenvalues of elliptic self-adjoint differential operators, we have $D \asymp (L/h)^n$. Moreover, L and R behave like certain powers of h.

- In the case when $\beta = 0$, we choose $\tau_0 = h^2$. Then for any $a > 0$ we get

$$\mathbf{P}(|\vartheta|_{\mathbf{C}^D} > R) \leq C\exp\left(-\frac{1}{Ch^a}\right)$$

  for any given fixed a, provided we choose K large enough in (18.4.24).
- In the case $\beta > 0$ we get the same conclusion with $\tau_0 = h^{-\widetilde{K}}\sigma'$ provided $\widetilde{K}$ is large enough.

In both cases, we see that the independent random variables ϑ_j in (18.4.25), (18.4.26) have a joint probability density $C(h)e^{\Phi(\alpha;h)}L(d\alpha)$, satisfying (18.4.18) for some N_4 depending on K.

Recall the choice of τ_0 above, depending on whether $\beta = 0$ or $\beta > 0$. Also recall that $\sigma' = \frac{1}{C}h^K e^{-Ch^{-\beta}}$, $K \geq \rho(M+1)$. Let $\Gamma \Subset \dot{\mathbf{C}}$ have piecewise smooth boundary, so that G satisfies (15.3.8) with $\kappa = 0$:

$$G = \mathcal{O}(1)\left(1 + \frac{d(w, \partial\Gamma)}{\sqrt{h}}\right)^{-1}.$$

When $m > 2$ (i.e., ≥ 4), we have $\|h^m(q''_\omega + q^2_\omega)\|_{H^s} \leq h^2$, with a probability as in (18.4.21), where we take $\vartheta = 0$ and apply Theorem 15.3.1 to see that with probability

$$\geq 1 - \frac{\mathcal{O}(1)e^{-\frac{h^{-\widetilde{\delta}}}{\mathcal{O}(1)}}}{h^{N_6+n}}\left(\left(h^{-\beta} + \ln\frac{1}{h}\right)\left(\ln\frac{1}{h}\right)^2 \int_\Omega G(w,\Gamma)\Delta\phi(w)L(dw) + |\partial\Gamma|h^{\frac{1}{2}}\right), \tag{18.4.27}$$

we have

$$\left|\#(\sigma(P_\delta)\cap\Gamma) - \frac{1}{(2\pi h)^n}\mathrm{vol}\,(p^{-1}(\Gamma))\right| \leq$$
$$\frac{\mathcal{O}(1)}{h^n}\left(\left(h^{-\beta} + \ln\frac{1}{h}\right)h^{-\widetilde{\delta}}\left(\ln\frac{1}{h}\right)^2 \int_\Omega G(w,\Gamma)\Delta\phi(w)L(dw) + |\partial\Gamma|h^{\frac{1}{2}}\right). \tag{18.4.28}$$

Here $0 < \widetilde{\delta} \ll 1$ is a free parameter and q''_ω, q^2_ω are fixed and satisfy $\|h^m(q''_\omega + q^2_\omega)\|_{H^s} \leq h^2$ which holds with a probability as in (18.4.27).

When $m = 2$, we settle for $\|h^m(q''_\omega + q^2_\omega)\|_{H^s} \leq h^{2-\vartheta}$ and get the weaker statement with $|\partial\Gamma|h^{\frac{1}{2}}$ replaced by $|\partial\Gamma|h^{\frac{1}{2}-\vartheta}$ in (18.4.27), (18.4.28). (Cf. Remark 15.3.2.)

Now let $\Gamma = \Gamma^g_{\theta_1,\theta_2;r_1,r_2}$ be as in (18.3.2) and assume $\mathrm{P}(\theta_j, N_0)$ for $j = 1, 2$ so that (18.3.3) holds. Then, using also (15.3.8) with $\kappa = 0$;

$$G(w,\Gamma) = \mathcal{O}(1)\left(1 + \frac{d(w,\partial\Gamma)}{\sqrt{h}}\right)^{-1},$$

we see that

$$\int_\Gamma G(w,\Gamma)\Delta\phi(w)L(dw) = \mathcal{O}\left(h^{\frac{1}{2N_0}}\right).$$

In conclusion,

Proposition 18.4.2 *Let $\Gamma = \Gamma_{\theta_1,\theta_2;r_1,r_2}$ be as in* (18.3.2) *and assume property* $\mathrm{P}(\theta_j, N_0)$ *holds for $j = 1, 2$. Choose $\delta = \tau_0 h^{N_2-n}$ with τ_0 as above. Let q''_ω, q^2_ω satisfy $\|h^m(q''_\omega + q^2_\omega)\|_{H^s} \leq h^{2-\vartheta}$, which is fulfilled with a probability as in* (18.4.21). *Here $\vartheta \in [0, 1/2[$ and we take $\vartheta = 0$ when $m \geq 4$ and $\vartheta > 0$ when $m = 2$.*

Then with a probability for q^1_ω that is

$$\geq 1 - \mathcal{O}(h^{-\mathcal{O}(1)})e^{-h^{-\widetilde{\delta}}/\mathcal{O}(1)} \tag{18.4.29}$$

we have when $m \geq 4$ that

$$
\begin{aligned}
&\left|\left(\#\sigma(P_\delta) \cap \Gamma - \frac{1}{(2\pi h)^n} \mathrm{vol}\,(p^{-1}(\Gamma))\right)\right| \leq \\
&\frac{\mathcal{O}(1)}{h^n}\left(\left(h^{-\beta} + \ln\frac{1}{h}\right) h^{-\widetilde{\delta}} \left(\ln\frac{1}{h}\right)^2 h^{\frac{1}{2N_0}} + h^{\frac{1}{2}}\right).
\end{aligned}
\tag{18.4.30}
$$

When $m = 2$, the same conclusion is valid, provided that $h^{\frac{1}{2}}$ is replaced with $h^{(\frac{1}{2}-\vartheta)}$.

The estimate (18.4.30) is of interest when $\beta + \widetilde{\delta} < 1/(2N_0)$.

As noticed after Theorem 15.3.1, with probability as in (18.4.29) (with $\mathcal{O}(h^{-\mathcal{O}(1)})$ replaced with $\mathcal{O}(h^{-\mathcal{O}(1)-1/2})$), the conclusion (18.4.30) holds simultaneously for all $\Gamma = \Gamma_{\theta_1,\theta_2;r_1,r_2}$ with θ_1, θ_2 fixed as in the proposition and with $r_1 < r_2$ varying in any fixed compact interval in $]0, +\infty[$.

18.5 End of the Proof

Let θ_1, θ_2, N_0 be as in Theorem 18.1.1, so that $\mathrm{P}(\theta_1, N_0)$ and $\mathrm{P}(\theta_2, N_0)$ hold. Then, choosing $\vartheta = 0$ for $m \geq 4$, $\vartheta = \widetilde{\delta}/2$ for $m = 2$, we get from Proposition 18.4.2 that

$$
\begin{aligned}
&\left|\#(\sigma(h^m P_\omega) \cap \Gamma^g_{\theta_1,\theta_2;1,\lambda}) - \frac{1}{(2\pi h)^n} \mathrm{vol}\,(p^{-1}(\Gamma^g_{\theta_1,\theta_2;1,\lambda}))\right| \\
&\qquad\qquad \leq \frac{\mathcal{O}(1)}{h^{n+\beta+\widetilde{\delta}}} h^{\frac{1}{2N_0}} \left(\ln\frac{1}{h}\right)^3,
\end{aligned}
\tag{18.5.1}
$$

simultaneously for $1 \leq \lambda \leq 2$ and all (θ_1, θ_2) in a set where $\mathrm{P}(\theta_1, N_0)$ and $\mathrm{P}(\theta_2, N_0)$ hold uniformly, with probability as in (18.4.29).

Assuming $\mathrm{P}(\theta_1, N_0)$ and $\mathrm{P}(\theta_2, N_0)$, we want to count the number of eigenvalues of P^0_ω in

$$
\Gamma_{1,\lambda} = \Gamma^g_{\theta_1,\theta_2;1,\lambda}
$$

when $\lambda \to \infty$. Let $k(\lambda)$ be the largest integer k for which $2^k \leq \lambda$ and decompose

$$
\Gamma_{1,\lambda} = \left(\bigcup_0^{k(\lambda)-1} \Gamma_{2^k,2^{k+1}}\right) \cup \Gamma_{2^{k(\lambda)},\lambda}.
$$

In order to count the eigenvalues of P_ω^0 in $\Gamma_{2^k,2^{k+1}}$ we define h by $h^m 2^k = 1$, $h = 2^{-k/m}$, so that

$$\#(\sigma(P_\omega^0) \cap \Gamma_{2^k,2^{k+1}}) = \#(\sigma(h^m P_\omega^0) \cap \Gamma_{1,2}),$$
$$\frac{1}{(2\pi)^n}\mathrm{vol}\,(p^{-1}(\Gamma_{2^k,2^{k+1}})) = \frac{1}{(2\pi h)^n}\mathrm{vol}\,(p^{-1}(\Gamma_{1,2})).$$

Thus, with probability $\geq 1 - Ce^{-2^{\frac{k\widetilde{\delta}}{m}}/C}$ we have

$$\begin{aligned}\Big|\#(\sigma(P_\omega^0) \cap \Gamma_{2^k,2^{k+1}}) - \frac{1}{(2\pi)^n}\mathrm{vol}\,p^{-1}(\Gamma_{2^k,2^{k+1}})\Big| \\ \leq C_\delta 2^{(n+\beta+\widetilde{\delta})\frac{k}{m}} 2^{-\frac{k}{2N_0 m}} \left(\frac{k}{m}\right)^3.\end{aligned} \tag{18.5.2}$$

Similarly, with probability $\geq 1 - Ce^{-2^{k(\lambda)\widetilde{\delta}/m}/C}$, we have

$$\begin{aligned}\Big|\#(\sigma(P_\omega^0) \cap \Gamma_{2^{k(\lambda)},\widetilde{\lambda}}) - \frac{1}{(2\pi)^n}\mathrm{vol}\,p^{-1}(\Gamma_{2^{k(\lambda)},\widetilde{\lambda}})\Big| \\ \leq C_\delta \lambda^{\frac{n}{m}} \lambda^{\frac{\beta+\widetilde{\delta}}{m}} \lambda^{-\frac{1}{2N_0}} (\ln \lambda)^3,\end{aligned} \tag{18.5.3}$$

simultaneously for all $\widetilde{\lambda} \in [\lambda, 2\lambda[$.

Now, we proceed as in [17], using essentially the Borel–Cantelli lemma. Use that

$$\sum_{k=\ell}^{\infty} e^{-2^{k\widetilde{\delta}/m}/C} = \mathcal{O}(1)e^{-2^{\ell\widetilde{\delta}/m}/C},$$

$$\sum_{2^k \leq \lambda} 2^{(n+\beta+\widetilde{\delta})\frac{k}{m}} 2^{-\frac{1}{2N_0}\frac{k}{m}} \left(\frac{k}{m}\right)^3 = \mathcal{O}(1)\lambda^{\frac{1}{m}(n+\beta+\widetilde{\delta})}\lambda^{-\frac{1}{2N_0}\frac{1}{m}} (\ln \lambda)^3,$$

to conclude that with probability $\geq 1 - Ce^{-2^{\widetilde{\delta}\frac{\ell}{m}}/C}$, we have

$$|\#(\sigma(P_\omega^0) \cap \Gamma_{2^\ell,\lambda}) - (2\pi)^{-n}\mathrm{vol}\,p^{-1}(\Gamma_{2^\ell,\lambda})| \leq C\lambda^{\frac{n}{m}}\lambda^{-\frac{1}{m}(\frac{1}{2N_0}-\beta-\widetilde{\delta})}(\ln \lambda)^3$$

for all $\lambda \geq 2^\ell$. Hence with the same probability, there exists a constant $C(\omega)$ such that

$$|\#(\sigma(P_\omega^0) \cap \Gamma_{0,\lambda}) - (2\pi)^{-n}\mathrm{vol}\,p^{-1}(\Gamma_{0,\lambda})| \leq C\lambda^{\frac{n}{m}}\lambda^{-\frac{1}{m}(\frac{1}{2N_0}-\beta-\widetilde{\delta})}(\ln \lambda)^3 + C(\omega).$$

This statement implies Theorem 18.1.1. □

Proof of Theorem 18.1.2 This is just a minor modification of the proof of Theorem 18.1.1. Indeed, we already used the second part of Proposition 18.4.2 to get (18.5.3) with the probability indicated there. In that estimate we are free to vary (g, θ_1, θ_2) in $\mathcal{G}$ and the same holds for the estimate (18.5.2). With these modifications, the same proof gives Theorem 18.1.2. □

Notes The present chapter follows closely [19] with only minor modifications, such as referring to Sect. 15.3 instead of [139] in the proof. The results are basically the same as in [19].

Spectral Asymptotics for $\mathcal{PT}$ Symmetric Operators

19

19.1 Introduction

$\mathcal{PT}$-symmetry has been proposed as an alternative to self-adjointness in quantum physics, see Bender et al. [13], Bender and Mannheim [14]. Thus for instance, if we consider a Schrödinger operator on $\mathbf{R}^n$,

$$P = -h^2\Delta + V(x), \tag{19.1.1}$$

the usual assumption of self-adjointness (implying that the potential V is real valued) can be replaced by that of $\mathcal{PT}$-symmetry:

$$V \circ \iota = \overline{V}, \tag{19.1.2}$$

where $\iota : \mathbf{R}^n \to \mathbf{R}^n$ is an isometry with $\iota^2 = 1 \neq \iota$. If we introduce the **p**arity operator $\mathcal{P}_\iota u(x) = u(\iota(x))$ and the **t**ime reversal operator $\mathcal{T}u = \overline{u}$, then this can be written

$$[P, \mathcal{P}_\iota \mathcal{T}] = 0. \tag{19.1.3}$$

Under very weak assumptions it is easy to see that the spectrum of a $\mathcal{PT}$-symmetric operator is invariant under reflexion in the real line. (We only need P to be closed and to commute with $\mathcal{PT}$ in the natural sense, including that the domain of P is invariant under $\mathcal{PT}$.) From the point of view of physics it seems important that the spectrum is real, and a natural mathematical question is then to determine when this is the case. Results on reality and non-reality of the spectrum of $\mathcal{PT}$-symmetric operators can be found in [14, 26, 27, 126].

J. Sjöstrand, *Non-Self-Adjoint Differential Operators, Spectral Asymptotics and Random Perturbations*, Pseudo-Differential Operators 14,
https://doi.org/10.1007/978-3-030-10819-9_19

In Sects. 19.2 and 19.3 we consider operators with random perturbations and show, by adapting the results of Chaps. 15 and 18, that most such operators have non-real eigenvalues, in the semi-classical case and in the case of large eigenvalues. These two sections are an adaptation of the note [143]. In Sect. 19.4 we describe without proofs some recent results for the semi-classical Schrödinger operator with one or two potential wells: In the case of one well the spectrum is real provided that the potential is analytic,[1] while in the double-well case, we get non-real spectrum when the coupling constant is larger than some exponentially small quantity. A very interesting question is to give criteria for $\mathcal{PT}$ symmetric operators with analytic coefficients in any dimension to have real spectrum or not.

19.2 The Semi-Classical Case with Random Perturbations

Let X be a compact smooth manifold of dimension n. Let $\iota : X \to X$ be a smooth involution: $\iota^2 = \mathrm{id}$, with $\iota \neq \mathrm{id}$. Fix a smooth positive density dx on X which is invariant under ι and let us take L^2 norms with respect to dx. Let P be a differential operator on X of order $m \geq 2$ with smooth coefficients as in (15.1.1)–(15.1.8). The operator Γ in (15.1.7) is now denoted $\mathcal{T}$, to stick to the $\mathcal{PT}$-terminology. Also assume that

$$P\mathcal{P} = \mathcal{P}P^*, \text{ where } \mathcal{P}u(x) = \mathcal{P}_\iota u(x) := u(\iota(x)). \tag{19.2.1}$$

It follows from (15.1.7) and (19.2.1) that P is $\mathcal{PT}$ symmetric:

$$[\mathcal{PT}, P] = 0. \tag{19.2.2}$$

Example 19.2.1 $P = -h^2\Delta + V(x)$ on $\mathbf{T}^n$ where $\Re V$ is even and $\Im V$ is odd, $V(-x) = \overline{V}(x)$. Then P is symmetric in the sense of (15.1.7) and $\mathcal{PT}$-symmetric with $\iota(x) = -x$.

Let $\widetilde{R}$ be an auxiliary h-independent positive second-order elliptic differential operator on X which commutes with $\mathcal{P}$. We also assume that $\widetilde{R}$ is real, or equivalently that

$$[\mathcal{T}, \widetilde{R}] = 0. \tag{19.2.3}$$

Then $\widetilde{R}$ has an orthonormal basis of real eigenfunctions e_j such that $\mathcal{P}e_j = (-1)^{k(j)}e_j$ where $k(j) = 1$ or $k(j) = -1$. We say that e_j is even in the first case and odd in second case. Put $\epsilon_j = e_j$ when e_j is even, and $\epsilon_j = ie_j$ when e_j is odd. Then $\{\epsilon_j\}$ is also

[1] This is not in contradiction with the result in Sects. 19.2, 19.3, since the random perturbation typically destroys uniform analyticity.

an orthonormal basis and a linear combination $V = \sum \alpha_j \epsilon_j$ is $\mathcal{PT}$ symmetric iff the coefficients α_j are real: $\mathcal{P}(V) = \overline{V}$.

Let $\Omega \Subset \mathbf{C}$ be a fixed open, simply connected set and define ϕ up to a harmonic function as in (15.3.1).

By $B_{\mathbf{R}^d}(0, r)$ we denote the open ball in $\mathbf{R}^d$ with center 0 and radius r. Let q_ω be a random potential of the form,

$$q_\omega(x) = \sum_{0<\mu_k\le L} \alpha_k(\omega)\epsilon_k(x), \quad \alpha(\omega) = (\alpha_k(\omega))_{0<\mu_k\le L} \in B_{\mathbf{R}^D}(0, R), \tag{19.2.4}$$

where $\mu_k > 0$ are the square roots of the eigenvalues of $h^2\widetilde{R}$. We choose $L = L(h)$, $R = R(h)$ in the intervals (15.2.5) for some $\epsilon \in]0, s - \frac{n}{2}[$, $s > \frac{n}{2}$, and recall that the dimension D in (15.2.4) is of the order of magnitude $(L/h)^n$. We introduce the small parameter $\delta = \tau_0 h^{N_2 - n}$, $0 < \tau_0 \le h^2$, where $N_2 \ge N_2(n, s, \epsilon, M, \widetilde{M})$ is sufficiently large.

The randomly perturbed $\mathcal{PT}$ symmetric operator is

$$P_\delta = P + \delta q_\omega. \tag{19.2.5}$$

N_2 is chosen large enough so that

$$\|h^{N_2-n} q_\omega\|_{L^\infty} \le \mathcal{O}(1) h^{-n/2} \|h^{N_2-n} q_\omega\|_{H_h^s} \le \mathcal{O}(1).$$

The random variables $\alpha_j(\omega)$ will have a joint probability distribution

$$\mathbf{P}(d\alpha) = C(h) e^{\Phi(\alpha;h)} L(d\alpha), \tag{19.2.6}$$

where for some $N_4 > 0$,

$$|\nabla_\alpha \Phi| = \mathcal{O}(h^{-N_4}), \tag{19.2.7}$$

and $L(d\alpha)$ is the Lebesgue measure. ($C(h)$ is the normalizing constant ensuring that the probability of $B_{\mathbf{R}^D}(0, R)$ is equal to 1.)

Let $\Gamma \Subset \Omega$ be a Lipschitz domain of constant scale $\sqrt{h}$ and define G as in (15.3.15). With these modifications, the main result of this section reads as Theorem 15.3.1. We repeat the formulation for convenience.

Theorem 19.2.2 *Let $\widetilde{\delta} > 0$. Then with probability*

$$\ge 1 - \mathcal{O}(1) h^{-N_6 - n} \left(\ln \frac{1}{\tau_0} \left(\ln \frac{1}{h} \right)^2 \int_\Omega G(w, \Gamma) \Delta\phi(w) L(dw) + |\partial\Gamma| h^{\frac{1}{2}} \right) e^{-\frac{h^{-\widetilde{\delta}}}{\mathcal{O}(1)}},$$

the number of eigenvalues of P_δ in Γ (counted with their algebraic multiplicities) satisfies

$$\left|\#(\sigma(P_\delta)\cap\Gamma)-\frac{1}{(2\pi h)^n}\mathrm{vol}\,(p^{-1}(\Gamma))\right|\leq$$
$$\frac{\mathcal{O}(1)}{h^n}\left(h^{-\widetilde{\delta}}\ln\frac{1}{\tau_0}\left(\ln\frac{1}{h}\right)^2\int_\Omega G(w,\Gamma)\Delta\phi(w)L(dw)+|\partial\Gamma|h^{\frac{1}{2}}\right). \tag{19.2.8}$$

Remark 15.3.2 remains valid also and we do not repeat it.

From Example 15.3.5 we see that it is easy to find $\mathcal{PT}$-symmetric operators in any given dimension which have plenty of non-real eigenvalues.

Proof of Theorem 19.2.2 We just have to make some small modifications in the proof of Theorem 15.3.1 and only mention the points where a difference appears. The proof uses three ingredients:

(1) The construction of a special perturbation of the form δq_ω with q_ω as in (19.2.4) but with α in the *complex* ball $B_{\mathbf{C}^D}(0, R)$ for which we have nice lower bounds on the small singular values of P_δ in (19.2.5), see Proposition 17.3.4.
(2) A complex variable argument in the α variables using the existence of the special perturbation in step 1, which permits to conclude that we have nice lower bounds on a relative determinant for $P_\delta - z$, with probability close to 1.
(3) Application of Theorem 12.1.2 about the number of zeros of holomorphic functions with exponential growth.

In the present situation we want our special perturbation $\delta q_\omega(x)$ to be $\mathcal{PT}$-symmetric, that is, we want the coefficients α in (19.2.4) to be real. All the parts of the proofs in step 1 carry over to the case of real α except the following result, which is the basic ingredient in the iterative process leading to the propositions mentioned above:

Let $e_1, \dots, e_N$ be an orthonormal family in $L^2(X)$ such that

$$\|\sum_{j=1}^N \lambda_j e_j\|_{H_h^s}\leq \mathcal{O}(1)\|\lambda\|_{\mathbf{C}^N}$$

where the constant $\mathcal{O}(1)$ is independent of the family and especially of N. Then there exists

$$q=\sum_{0<\mu_j\leq L}\alpha_j\epsilon_j,\ \alpha_j\in\mathbf{C}, \tag{19.2.9}$$

with $\|\alpha\|_{\mathbf{C}^D} \leq R$ with the parameters as in (15.2.5), such that (17.2.23) holds,

$$\|q\|_{H_h^s} \leq \mathcal{O}(1)h^{-\frac{n}{2}}NL^{s+\frac{n}{2}+\epsilon},$$

and such that the matrix

$$M_q = \left(\int q(x)e_j(x)e_k(x))dx\right)_{1\leq j,k\leq N}$$

and its singular values

$$\|M_q\| = s_1(M_q) \geq \cdots \geq s_N(M_q)$$

satisfy (17.2.22), (17.3.10)

$$\|M_q\| \leq \mathcal{O}(1)Nh^{-n},$$

$$s_k(M_q) \geq h^n/\mathcal{O}(1), \text{ for } 1 \leq k \leq N/2. \tag{19.2.10}$$

Write $q = q_1 + iq_2$, where $q_1 = \sum(\Re\alpha_j)\epsilon_j$, $q_2 = \sum(\Im\alpha_j)\epsilon_j$, so that q_1 and q_2 are $\mathcal{PT}$-symmetric. The upper bounds on $\|q\|_{H_h^s}$ and on $\|M_q\|$ follow from the bound $\|\alpha\| \leq R$ and therefore carry over to q_j. Since $M_q = M_{q_1} + iM_{q_2}$ we can apply the Ky Fan inequalities (Corollary 8.2.2 [48]) and get

$$\frac{h^n}{\mathcal{O}(1)} \leq s_{2k-1}(M_q) \leq s_k(M_{q_1}) + s_k(M_{q_2}), \quad 1 \leq k \leq \frac{N}{4}.$$

Since the singular values are enumerated in decreasing order, it follows that for j equal to 1 or 2, we have

$$s_k(M_{q_j}) \geq \frac{h^n}{2\mathcal{O}(1)}, \quad 1 \leq k \leq \frac{N}{4}. \tag{19.2.11}$$

This means that step 1 can be carried out and we get a $\mathcal{PT}$ symmetric operator P_δ as in Proposition 17.3.4, the only slight difference is that rather than taking θ in $]0, 1/4[$ we have to confine this parameter to the smaller interval $]0, 1/8[$.

Step 2 now works because of Remark 15.3.3 and its proof, where the main point is the reality of the coefficients α_j while the assumption of reality of the basis elements is not necessary, and was made there only because we mainly have in mind a real perturbation for resonance theory (not treated in this book).

Step 3 can be carried out without any modifications. □

19.3 Weyl Asymptotics for Large Eigenvalues

Let P^0 be an elliptic differential operator on X of order $m \geq 2$ with smooth coefficients and with principal symbol $p_m(x, \xi)$. In local coordinates we get, using standard multi-index notation,

$$P^0 = \sum_{|\alpha|\leq m} a_\alpha^0(x) D^\alpha, \quad p_m(x, \xi) = \sum_{|\alpha|=m} a_\alpha^0(x)\xi^\alpha. \tag{19.3.1}$$

Recall that the ellipticity of P^0 means that $p_m(x, \xi) \neq 0$ for $\xi \neq 0$. We assume that

$$p_m(T^*X) \neq \mathbf{C}. \tag{19.3.2}$$

As before we assume symmetry,

$$(P^0)^* = \mathcal{T} P^0 \mathcal{T}, \tag{19.3.3}$$

and that

$$P^0 \mathcal{P} = \mathcal{P}(P^0)^*, \tag{19.3.4}$$

with $\mathcal{P} = \mathcal{P}_\iota$ as in Sect. 19.2. Then P^0 is $\mathcal{PT}$ symmetric.

Let $\widetilde{R}$ be a reference operator as in and around (19.2.3) and define ϵ_j as there. Write

$$\widetilde{R}\epsilon_j = (\mu_j^0)^2 \epsilon_j, \quad 0 < \mu_0^0 < \mu_1^0 \leq \mu_2^0 \leq \dots \tag{19.3.5}$$

so that $\mu_k = h\mu_k^0$ where μ_k are given after (19.2.4). Our randomly perturbed operator is

$$P_\omega^0 = P^0 + q_\omega^0(x), \tag{19.3.6}$$

where ω is the random parameter and

$$q_\omega^0(x) = \sum_0^\infty \alpha_j^0(\omega)\epsilon_j. \tag{19.3.7}$$

Here we assume that $\alpha_j^0(\omega)$ are independent real Gaussian random variables of variance σ_j^2 and mean value 0:

$$\alpha_j^0 \sim \mathcal{N}(0, \sigma_j^2), \tag{19.3.8}$$

where (as in (18.1.9), (18.1.10)),

$$\frac{1}{\mathcal{O}(1)}(\mu_j^0)^{-\rho}e^{-(\mu_j^0)^{\frac{\beta}{M+1}}} \leq \sigma_j \leq \mathcal{O}(1)(\mu_j^0)^{-\rho}, \tag{19.3.9}$$

$$M = \frac{3n-\frac{1}{2}}{s-\frac{n}{2}-\epsilon},\ 0 \leq \beta < \frac{1}{2},\ \rho > n, \tag{19.3.10}$$

where s, ρ, ϵ are fixed constants such that

$$\frac{n}{2} < s < \rho - \frac{n}{2},\ 0 < \epsilon < s - \frac{n}{2}.$$

Let $H^s(X)$ be the standard Sobolev space of order s. As we saw in Sect. 18.1 (where the random variables α_j^0 were complex valued), $q_\omega^0 \in H^s(X)$ almost surely since $s < \rho - \frac{n}{2}$. Hence $q_\omega^0 \in L^\infty$ almost surely, implying that P_ω^0 has purely discrete spectrum.

Consider the function $F(w) = \arg p_m(w)$ on S^*X. For given $\theta_0 \in S^1 \simeq \mathbf{R}/(2\pi\mathbf{Z})$, $N_0 \in \dot{\mathbf{N}} := \mathbf{N} \setminus \{0\}$, we recall the property $\mathrm{P}(\theta_0, N_0)$:

$$\sum_1^{N_0} |\nabla^k F(w)| \neq 0 \text{ on } \{w \in S^*X;\ F(w) = \theta_0\}. \tag{19.3.11}$$

We can now state the main result of this section, which is an adaptation of Theorem 18.1.1 (cf. [19]).

Theorem 19.3.1 *Assume that $m \geq 2$. Let $0 \leq \theta_1 \leq \theta_2 \leq 2\pi$ and assume that $\mathrm{P}(\theta_1, N_0)$ and $\mathrm{P}(\theta_2, N_0)$ hold for some $N_0 \in \dot{\mathbf{N}}$. Also assume that $\beta \in [0, 1/(2N_0)[$. Let $g \in C^\infty([\theta_1, \theta_2];\ ]0, \infty[)$ and put*

$$\Gamma^g_{\theta_1,\theta_2;0,\lambda} = \{re^{i\theta};\ \theta_1 \leq \theta \leq \theta_2,\ 0 \leq r \leq \lambda g(\theta)\}.$$

Then for every $\delta \in]0, \frac{1}{2N_0} - \beta[$ there exists $C > 0$ such that almost surely there exists $C(\omega) < \infty$ such that for all $\lambda \in [1, \infty[$:

$$\begin{aligned} &\left|\#(\sigma(P_\omega^0) \cap \Gamma^g_{\theta_1,\theta_2;0,\lambda}) - \frac{1}{(2\pi)^n}\mathrm{vol}\, p_m^{-1}(\Gamma^g_{\theta_1,\theta_2;0,\lambda})\right| \\ &\leq C(\omega) + C\lambda^{\frac{n}{m} - \frac{1}{m}(\frac{1}{2N_0}-\beta-\delta)}. \end{aligned} \tag{19.3.12}$$

We also have an extension to families $\Gamma^g_{\theta_1,\theta_2;0,\lambda}$ that satisfy the assumptions uniformly. Cf. Theorem 18.1.2.

Theorem 19.3.2 *Assume that $m \geq 2$. Let Θ be a compact subset of $[0, 2\pi]$. Let $N_0 \in \mathbf{N}$ and assume that* $\mathrm{P}(\theta, N_0)$ *holds uniformly for $\theta \in \Theta$. Let $\mathcal{G}$ be a subset of $\{(g, \theta_1, \theta_2);\ \theta_j \in \Theta,\ \theta_1 \leq \theta_2,\ g \in C^\infty([\theta_1, \theta_2];]0, \infty[)\}$ with the property that g and $1/g$ are uniformly bounded in $C^\infty([\theta_1, \theta_2];]0, \infty[)$ when (g, θ_1, θ_2) varies in $\mathcal{G}$. Then for every $\delta \in]0, \frac{1}{2N_0} - \beta[$ there exists $C > 0$ such that almost surely: there exists $C(\omega) < \infty$ such that for all $\lambda \in [1, \infty[$ and all $(g, \theta_1, \theta_2) \in \mathcal{G}$, we have the estimate* (19.3.12).

The Condition (19.3.9) allows us to choose σ_j decaying faster than any negative power of μ_j^0. Then as in Chap. 18 it follows that $q_\omega(x)$ is almost surely a smooth function. Theorem 19.3.2 says roughly that for almost every $\mathcal{PT}$ symmetric elliptic operator of order ≥ 2 with smooth coefficients on a compact manifold which satisfies the conditions (19.3.2)–(19.3.4), the large eigenvalues distribute according to Weyl's law in sectors with limiting directions that satisfy a weak non-degeneracy condition.

Proof of Theorem 19.3.1 As already mentioned, the theorem is a variant of Theorem 18.1.1 (cf. Theorem 1.1 in [19]). The difference is just that we now use real random variables in the perturbation q_ω^0 in order to guarantee the $\mathcal{PT}$-symmetry and use Theorem 19.2.2 instead of Theorem 18.1.1. □

The proof of Theorem 19.3.2 is a modification as in the proof of Theorem 18.1.2.

19.4 $\mathcal{PT}$-Symmetric Potential Wells

In this section we discuss some recent results about real and non-real eigenvalues of $\mathcal{PT}$-symmetric deterministic perturbations of a $\mathcal{P}$-symmetric self-adjoint Schrödinger operator in the semi-classical limit. Let X be either $\mathbf{R}^n$ or a compact smooth Riemannian manifold of dimension n. As the unperturbed operator, we take

$$P_0 = -h^2\Delta + V_0(x) \tag{19.4.1}$$

on X, where Δ is the Laplace-Beltrami operator and $V_0 \in C^\infty(X; \mathbf{R})$. Assume that

$$V_0 \circ \iota = V_0, \tag{19.4.2}$$

where $\iota : X \to X$ is an isometry with

$$\iota^2 = \mathrm{id} \neq \iota. \tag{19.4.3}$$

We are interested in the spectrum of $\mathcal{PT}$-symmetric perturbations of P_0 near some fixed real energy E_0. To ensure that the spectrum is discrete near E_0 when $X = \mathbf{R}^n$, we assume in that case that

$$\alpha := \liminf_{x\to\infty} V_0(x) > E_0. \tag{19.4.4}$$

The perturbed operator is

$$P_\delta = -h^2\Delta + V_\delta(x), \tag{19.4.5}$$

where

$$V_\delta(x) = V_0(x) + i\delta W(x), \tag{19.4.6}$$

$\delta \in \mathbf{R}$, $|\delta| \ll 1$, and $W \in C^\infty(X; \mathbf{R})$ is bounded and odd with respect to ι,

$$W \circ \iota = -W. \tag{19.4.7}$$

(We will also allow W to be unbounded in one of the results below.) We define P_0 as a self-adjoint operator by taking the Friedrichs extension of (19.4.1) from $C_0^\infty(X)$. Then $\sigma_{\mathrm{ess}}(P_0) \subset [\alpha, +\infty[$ and $\sigma(P_\delta)$ is contained in $\mathbf{R} + \overline{D(0, \delta\|W\|_{L^\infty})}$ and is purely discrete in a fixed neighborhood of E_0 when $|\delta|$, $h \ll 1$.

In the self-adjoint case ($\delta = 0$) very detailed informations about the eigenvalues can be obtained from "tunneling analysis": Assume that

$$V_0^{-1}(]-\infty, 0]) = \bigcup_{j\in J} U_j, \quad \#J < \infty, \tag{19.4.8}$$

where the potential wells U_j are closed (and hence compact by (19.4.4)) and pairwise disjoint:

$$U_j \cap U_k = \emptyset, \quad j \neq k. \tag{19.4.9}$$

We refer to [59] where further references can be found. An important ingredient here is the Lithner–Agmon metric $V_0(x)_+ dx^2$ [3, 95], where dx^2 denotes the Riemannian metric on X and as usual $a_+ = \max(a, 0)$ for $a \in \mathbf{R}$. The corresponding distance $d(x, y) \geq 0$ is symmetric and satisfies the triangle inequality, but may be degenerate in the sense that

$$d(x, y) = 0 \not\Rightarrow x = y.$$

We assume that U_j are "connected" in the sense that

$$\mathrm{diam}_d(U_j) = 0, \ j \in J. \tag{19.4.10}$$

19.4.1 Simple Well in One Dimension

We describe a recent result by Boussekkine and Mecherout [23] which very roughly says that when $X = \mathbf{R}$, $\iota(x) = -x$, $J = \{0\}$ and V_0 and W are real analytic near U_0, then for $|\delta|$, h small enough, the spectrum of P_δ in a fixed complex neighborhood of E_0 is purely real. See also Rouby [120] for a more general result. We give a more detailed formulation.

Let $V_0 \in C^\infty(\mathbf{R}; \mathbf{R})$ be smooth and satisfy (19.4.2)–(19.4.4), with $\iota(x) = -x$. Assume

(H1) there exists $m_0 \geq 0$ such that for any $\alpha \in \mathbf{N}$, there exists $C_\alpha > 0$ such that

$$|\partial_x^\alpha V_0| \leq C_\alpha \langle x \rangle^{m_0 - \alpha}, \qquad \text{on } \mathbf{R}.$$

When $m_0 > 0$ we strengthen (19.4.4) by assuming that

$$V_0(x) \geq \frac{1}{C_0} |x|^{m_0} \quad \text{for } |x| \geq C_0,$$

form some positive constant C_0,

(H2) $V_0 - E_0$ has exactly one potential well, more precisely,

$$V_0^{-1}(]-\infty, E_0]) = [\alpha_0^0, \beta_0^0] =: U_0,$$

$$V_0^{-1}(]-\infty, E_0[) =]\alpha_0^0, \beta_0^0[,$$

where $-\infty < \alpha_0^0 < \beta_0^0 < \infty$. Moreover,

$$V_0'(\alpha_0^0) < 0, \ V_0'(\beta_0^0) > 0.$$

Notice that $\alpha_0^0 = -\beta_0^0$.

(H3) V_0 is analytic near U_0.

As before, $V_\delta(x) = V_0(x) + \delta W(x)$, where we assume

(H4) $W \in C^\infty(\mathbf{R}; \mathbf{R})$ is odd, $W(-x) = -W(x)$,
(H5) W satisfies the first part of (H1) with the same m_0,
(H6) W is analytic near U_0.

Let V_0, W also denote holomorphic extensions to a complex neighborhood of U_0. Then for $E, \delta \in \mathbf{C}$ with $|E - E_0|$ and $|\delta|$ small enough, we have unique solutions $\alpha_0(E, \delta)$, $\beta_0(E, \delta)$ in complex neighborhoods of α_0^0, β_0^0, respectively, of the equations

$$V_\delta(\alpha_0(E, \delta)) = E, \quad V_\delta(\beta_0(E, \delta)) = E,$$

and α_0, β_0 are holomorphic functions of (E, δ).

For $(E, \delta) \in \text{neigh}\,((E_0, 0), \mathbf{C}^2)$, we put

$$I_0(E, \delta) = 2 \int_{\alpha_0(E,\delta)}^{\beta_0(E,\delta)} (E - V_\delta(x))^{\frac{1}{2}} dx, \tag{19.4.11}$$

where we integrate along the straight line segment from $\alpha_0(E, \delta)$ to $\beta_0(E, \delta)$ and choose the branch of the square root which has argument close to 0. (When E is real and $\delta = 0$, this is a real integral with a positive integrand.) It is easy to check that $I_0(E, \delta)$ is a holomorphic function of (E, δ).

We define $P_\delta = -h^2 \Delta + V_\delta$ as the unbounded closed operator $L^2(\mathbf{R}) \to L^2(\mathbf{R})$ with domain,

$$\mathcal{D}(P_\delta) = \{u \in L^2;\ u', u'', \langle x \rangle^{m_0} u \in L^2\},$$

and again the spectrum of P_δ is purely discrete in a fixed complex neighborhood of E_0, when $|\delta|$, h are small enough.

Theorem 19.4.1 (Boussekkine and Mecherout [23]) *Let V_0, $W \in C^\infty(\mathbf{R}; \mathbf{R})$ be even and odd, respectively, and satisfy* (19.4.2)–(19.4.4), (H1)–(H6). *Define $I_0(E, \delta)$ as above, where $V_\delta(x) = V_0(x) + i\delta W(x)$ and $(E, \delta) \in$* neigh $((E_0, 0), \mathbf{C}^2)$. *There exists a function $I(E, \delta; h)$ on* neigh $((E_0, 0, 0), \mathbf{C} \times \mathbf{C} \times \mathbf{R})$, *holomorphic in (E, δ) with*

$$I(E, \delta; h) \sim I_0(E, \delta) + h I_1(E, \delta) + \cdots \tag{19.4.12}$$

in the space of holomorphic functions defined near $(E_0, 0)$, such that $I(E, \delta; h)$ is real when E, δ are real and such that for $\delta \in]-\delta_0, \delta_0[$ with $0 < \delta_0 \ll 1$, the eigenvalues E in neigh $(E_0, \mathbf{C})$ *are given by the Bohr–Sommerfeld condition*

$$I(E, \delta; h) = 2\pi(k + 1/2)h, \ k \in \mathbf{Z}. \tag{19.4.13}$$

These eigenvalues are real.

Outline of the Proof This is an application of the complex WKB method, explained in Chap. 7. Let $A > \sup U_0$, so that $-A < \inf U_0$ by parity and assume that A is small enough so that V_0, W are analytic near $[-A, A]$. We can then consider the Dirichlet realization

$P_\delta^{\rm D}$ of P_δ on $L^2([-A,A])$. Then, as explained at the end of Sect. 7.2, the eigenvalues of $P_\delta^{\rm D}$ near E_0 are given by a Bohr–Sommerfeld condition

$$I^{\rm D}(E,\delta;h)=2\pi(k+1/2)h,\ k\in\mathbf{Z},$$

where $I^{\rm D}$, denoted by $\widetilde{I}(E;h)$ in (7.2.50), has the general properties of I in the theorem. Keeping track of the $\mathcal{PT}$-symmetry, it is easy to show that $I^{\rm D}(E,\delta;h)$ is real when E,δ are real, and we get the conclusion of the theorem for $P_\delta^{\rm D}$.

To get the theorem, we need to complete the complex WKB method with a study of the exponential asymptotics of null solutions of $P_\delta-E$ that are of class L^2 near $+\infty$ or near $-\infty$. This can be done by applying the fundamental idea of the complex WKB-method, as explained in Sect. 7.1 and it is here that we use (H1), (H5). This extension is indicated in [23] and carried out in detail in the work [101] and allows to make an asymptotic study of the Wronskian of two null solutions of $\mathrm{P}_\delta-E$ that are L^2 near $-\infty$ and $+\infty$ respectively. □

The analyticity assumptions on V_0,W are essential, as can be seen by applying Theorem 19.2.2[2] to $-h^2\Delta+V_0+i\delta_0(W+\delta q_\omega)$, where $\delta_0>0$ is small and fixed and δq_ω is a random perturbation. (When extending Theorem 19.2.2 to the case of $\mathbf{R}^n$, a suitable cutoff function has to be inserted in the random perturbation and we refer to [137].)

On the other hand, if V,W are merely smooth but satisfy the other assumptions in Theorem 19.4.1, then it seems clear that the conclusion there remains valid for $|\delta|\le\mathcal{O}(1)h\ln(1/h)$.

19.4.2 Double Wells in Arbitrary Dimension

We consider the general situation described in the beginning of this section, in particular in (19.4.1)–(19.4.10). Also assume that V_0 has a double-well structure at energy $E_0=0$, and that the two wells are exchanged by ι. More precisely, $J=\{-1,1\}$, $U_{\pm1}\neq\emptyset$, and

$$\iota(U_{-1})=U_1. \tag{19.4.14}$$

To describe the spectrum of P_0, we introduce two self-adjoint reference operators. Let $\chi_{\pm1}\in C_0^\infty(X;[0,1])$ satisfy:

$$\chi_j=1 \text{ near } U_j, \tag{19.4.15}$$

$$\operatorname{supp}\chi_j\subset B(U_j,\rho)=:U_j^\rho \tag{19.4.16}$$

[2]We here neglect the fact that Theorem 19.2.2 was established for operators on compact manifolds and not on $\mathbf{R}^n$.

where $\rho > 0$ is small. Here

$$B(U_j, \rho) = \{x \in X;\ d(U_j, x) < \rho\}.$$

Put

$$\widetilde{P}_j = \widetilde{P}_{0,j} = P_0 + \lambda\chi_{-j},\ \ j = \pm 1, \tag{19.4.17}$$

where $\lambda > 0$ is a constant, large enough so that

$$\{x \in X;\ V_0(x) + \lambda\chi_{-j}(x) \le 0\} = U_j.$$

If we define

$$\mathcal{P}u = u \circ \iota,\ \ u \in L^2(X), \tag{19.4.18}$$

then $\mathcal{P}$ is unitary on $L^2(X)$ with $\mathcal{P} \neq 1 = \mathcal{P}^2$ and we have

$$\mathcal{P} \circ P_0 = P_0 \circ \mathcal{P}, \tag{19.4.19}$$

$$\mathcal{P} \circ \widetilde{P}_j = \widetilde{P}_{-j} \circ \mathcal{P},\ \ j = \pm 1. \tag{19.4.20}$$

The last relation implies that $\widetilde{P}_{-1}$ and $\widetilde{P}_1$ have the same spectrum.

Assume that

$$\widetilde{\mu}(h) = o(h) \tag{19.4.21}$$

is a simple eigenvalue of $\widetilde{P}_1$ (and hence of $\widetilde{P}_{-1}$), and that there exist $C_0,\ N_0 > 0$ such that

$$\sigma(\widetilde{P}_{\pm 1}) \cap \,]\,\widetilde{\mu}(h) - h^{N_0}/C_0, \widetilde{\mu}(h) + h^{N_0}/C_0\,[= \{\widetilde{\mu}(h)\}. \tag{19.4.22}$$

For $h > 0$ small enough, P_0 has exactly two eigenvalues in the interval

$$]\,\widetilde{\mu}(h) - h^{N_0}/(2C_0), \widetilde{\mu}(h) + h^{N_0}/(2C_0)\,[\,,$$

namely the eigenvalues $\mu(h) \pm |t(h)|$ of the interaction matrix,

$$\begin{pmatrix} \mu(h) & t(h) \\ \overline{t(h)} & \mu(h) \end{pmatrix},$$

where $\mu(h) \in \mathbb{R}$ and $t(h) \in \mathbb{C}$, satisfy

$$\mu(h) = \widetilde{\mu}(h) + \mathcal{O}_\rho(e^{(\epsilon(\rho)-2S_0)/h}), \ \ \epsilon(\rho) \to 0, \ \rho \to 0,$$

$$\forall \alpha > 0, \ t(h) = \mathcal{O}_\alpha(e^{(\alpha-S_0)/h}).$$

Here,

$$S_0 = d(U_1, U_{-1}) \tag{19.4.23}$$

is the Lithner–Agmon distance between the two wells $U_{\pm 1}$. Quite often we also have a lower bound on $|t(h)|$:

$$\forall \alpha > 0, \ |t(h)|^{-1} = \mathcal{O}_\alpha(e^{(\alpha+S_0)/h}).$$

See for example [59, 130] or the review paper [133] and the references therein.

Concerning the perturbation W, we assume (19.4.7).

The result of Benbernou et al. [12] is the following:

Theorem 19.4.2 ([12]) *Under the above assumptions, the operator P_δ has exactly two eigenvalues (counted with their algebraic multiplicity) in $D(\widetilde{\mu}, h^{N_0}/C)$ for $C \gg 0$ and for δ real such that $|\delta| \ll h^{N_0}$. These eigenvalues are equal to the eigenvalues of the matrix*

$$X_\delta = \begin{pmatrix} a(\delta) & b(\delta) \\ \overline{b}(\delta) & \overline{a}(\delta) \end{pmatrix},$$

and hence of the form

$$\lambda_\pm = \Re a \pm \sqrt{|b|^2 - (\Im a)^2}.$$

Here $a(\delta) = a(\delta; h)$, $b(\delta) = b(\delta; h)$ satisfy,

$$a(0; h) = \mu(h), \ \ b(0; h) = t(h),$$

$$\partial_\delta a = i \int W(x)|\widetilde{e}_1(x)|^2 dx + \mathcal{O}(\delta h^{-N_0}) + \mathcal{O}_\delta(e^{(\epsilon(\delta)-2S_0)/h}),$$

$$\partial_\delta b = \mathcal{O}_\delta e^{(\epsilon(\delta)-S_0)/h},$$

for all $\delta > 0$, where $\epsilon(\delta) \to 0$, $\delta \to 0$. Further, $\widetilde{e}_1$ is the normalized eigenfunction with $(\widetilde{P}_1 - \widetilde{\mu}(h))\widetilde{e}_1 = 0$.

If $W > 0$ *on* U_1, *then*

$$\int W(x)|\widetilde{e}_1(x)|^2 dx \asymp 1, \tag{19.4.24}$$

and if we assume that (19.4.24) *holds, then there exists* $\delta_+ \geq 0$ *with the asymptotics,*

$$\delta_+ = (1 + \mathcal{O}_\delta(e^{(\epsilon(\delta)-S_0)/h}))\frac{|t(h)|}{\int W(x)|\widetilde{e}_1(x)|^2 dx}, \quad \epsilon(\delta) \to 0, \ \delta \to 0,$$

such that

- *The two eigenvalues are real and distinct for* $|\delta| < \delta_+$.
- *They are double and real when* $|\delta| = \delta_+$.
- *They are non-real and complex conjugate, when* $\delta_+ < |\delta| \ll h^{N_0}$.

The proof is based on the analysis of multi-well Hamiltonians in [59] (cf. [41]) that we extend slightly to non-self-adjoint operators, keeping track of the $\mathcal{PT}$-symmetry. See [12].

Notes In this chapter we have reviewed some recent results on the eigenvalues of $\mathcal{PT}$-symmetric operators. The spectrum of such an operator is invariant under reflection in the real line and from the point of view of Physics it is important to know when it is real or not. Our general results on Weyl asymptotics for differential operators with random perturbations in the lowest order coefficient tend to show that most of the time the spectrum is non-real. In the case of a Schrödinger operator we review a result of [23] generalized in [120] which states that the eigenvalues are real in the case of an analytic single-well potential with not too large imaginary part. We also recall a result of [12] for double-well potentials showing that the eigenvalues bifurcate out into the complex plane already for very small perturbations of a real potential (regardless of the analyticity or lack of it).

Numerical Illustrations 20

by

Agnès E. Sjöstrand[1] and Johannes Sjöstrand

In this chapter we give some numerical illustrations to the results about the asymptotic distribution of eigenvalues. Such calculations have already been carried out by many people, Trefethen [151], Trefethen and Embree [152], Davies [33, 35], Davies and Hager [38], Zworski [161, 162] and many others. Numerical calculations with special attention to Weyl asymptotics have been carried out by Hager [54, 55], Bordeaux Montrieux [16], Vogel [155, 156]. Many of the illustrations below are therefore well-known and even though we wrote our own Matlab programs, we have clearly benefitted from the preceding works.

20.1 Jordan Block

The easiest case is that of a large Jordan block. Let J be the Jordan block of size N (denoted A_0 in (2.4.1)). In Fig. 20.1 we let $N = 20$ and trace 31 level curves for the norm of the resolvent, corresponding to the levels e^k, $k = 0, 1, \dots, 30$. We see that the norm of the resolvent grows very fast towards the center of the unit disc.

Let $J_\delta = J + \delta Q$, where Q is an $N \times N$-matrix whose entries are real independent Gaussian random variables $\sim \mathcal{N}(0, 1)$. In Fig. 20.2 we trace the spectrum of J_δ and the level curves of the norm of the resolvent. There is a dramatic improvement.

[1] agnes.sjo@gmail.com

J. Sjöstrand, *Non-Self-Adjoint Differential Operators, Spectral Asymptotics and Random Perturbations*, Pseudo-Differential Operators 14,
https://doi.org/10.1007/978-3-030-10819-9_20

It is probably a general result, that the presence of a random perturbation improves the resolvent norm in the spectral region. This appears quite clearly in the proofs of Weyl asymptotics, and an explicit result in this direction in the one-dimensional case can be found in [16]. For this reason we can be quite confident that numerical calculations of spectra of perturbed operators are accurate, while the corresponding calculations for the unperturbed operators can give strange results deep inside the pseudospectrum.

Naturally, the reduction of a differential operator to a large matrix can be a source of errors as well.

In Figs. 20.3, 20.4, and 20.5 we give a series of computations for larger Jordan blocks ($N = 100$). The calculation of corresponding pseudospectral level curves becomes time consuming and is presented only in the last figure.

In accordance with the results of Chap. 13, most of the eigenvalues are very close to a circle with radius close to 1. The radius of this circle decreases slightly when we decrease δ. This can be viewed as a result on Weyl asymptotics, with $J = 1_{]0,N]\cap\mathbf{Z}}e^{iD}$, on $\ell^2([1,\dots,N]) \simeq \mathbf{C}^N$, with $h = 2\pi/N$. The semiclassical symbol of J is now $e^{i\xi}$, whose range is equal to S^1. In addition to the eigenvalues close to a circle there is a "uniformly bounded" number of eigenvalues inside. The expected density of those interior eigenvalues has been determined in [147].

The results and corresponding simulations can be generalized to bidiagonal operators of the form $P = 1_{]0,N]\cap\mathbf{Z}}(ae^{iD} + be^{-iD})$. The symbol is now $ae^{i\xi} + be^{-i\xi}$ whose range is an ellipse. See Sjöstrand and Vogel [145, 146].

20.2 Hager's Operator

We consider Hager's operator $hD + g(x)$ on S^1 (cf. Chap. 3). We will study two choices of g.

20.2.1 The Case When $g(x) = i\cos x$

Then,

$$P = hD_x + i\cos x. \tag{20.2.1}$$

As we saw in the beginning of Chap. 3, the spectrum of P is an arithmetic progression of real eigenvalues (since the mean value of $\Im g$ is zero). As a numerical approximation, we restrict the operator to the space $\mathcal{H}_N = \{u = \sum_{-N}^{N} \widehat{u}_k e^{ikx}\}$ and then project to the same space. In other words, we study the spectrum of $\Pi_N P : \mathcal{H}_N \to \mathcal{H}_N$. In Fig. 20.6, we have plotted the eigenvalues of P or rather $\Pi_N P$ in the case when N is small $= 20$. By choosing h with $Nh = 2$ we can arrange so that the real parts of the eigenvalues are situated in $[-2, 2]$. We notice that the real eigenvalues in the middle (roughly the ones with real part between -0.7 and 0.7) try to mimic the true eigenvalues of (20.2.1), while the ones on the diagonal segments are influenced by the fact that we do numerics with a projected "Toeplitz" operator. (The ones to the left are reminiscent of the eigenvalues of $(hD)^2 + i\cos(x)$, which we study in the next section. An even better model, within reach of semi-classical analysis would be $1_{\mathbf{R}_+}(hD)(hD + i\cos x)$.)

The same Fig. 20.6 also shows pseudospectral curves, more precisely the level curves $\|(z-P)^{-1}\| = e^k$, for $0 \le k \le K$, with $K = 20$. Naturally each eigenvalue is encircled by 21 such curves, but in order to save computing time, we have chosen a rather low resolution, so these fine details are not visible. What is important though and clearly visible is the very strong pseudospectral effect in the interior of the rectangle $[-2, 2] + i[-1, 1]$. Near the middle the norm of the resolvent is of the order at least $e^{20} = 4.8517 \times 10^8$. Hence there is a corresponding spectral instability under perturbations. For even stronger effects we expect Matlab to have difficulties and to make numerical errors.

When adding a random perturbation, the resolvent becomes reasonable in the new spectral region, as can be seen in Fig. 20.7. Here we consider

$$P_\delta = P + \delta Q, \tag{20.2.2}$$

where

$$Qu(x) = \sum_{j,k=-N}^{N} (2\pi)^{-1} q_{j,k}(\omega)(u|e^{ik(\cdot)})e^{ijx} \tag{20.2.3}$$

and $q_{j,k}$ are independent complex random variables $\sim \mathcal{N}_{\mathbf{C}}(0, 1)$. (Again the numerical computations concern $\Pi_N P_\delta$.) In Fig. 20.7 we still take $N = 20$, $h = 0.1$. The parameter δ is equal to 0.05. The norm of the particular random perturbation Q is 12.3104. This norm varies very little from one simulation to another. As we see, the eigenvalues now wander around in the range of the symbol and the resolvent norm no longer attains extremely large values (except right near the eigenvalues, though not visible because of the limited resolution).

In Figs. 20.8 and 20.9 we present results of analogous simulations for $N = 50$ and $h = 0.04$ (still with $hN = 2$). The pseudospectral effect is now much stronger and Matlab gives several warning messages about the extreme size of the resolvent norm, when calculating the pseudospectral curves. There are hardly no real numerical eigenvalues in the unperturbed case. One possible explanation is mathematical: This may be the case already for the exact operator $\Pi_N P$. We rather think that it is a pseudospectral effect. The spectral instability is so extreme that we are far from the true eigenvalues in the interior region. In the perturbed case, the pseudospectral effect is moderate and we are confident that we get the true eigenvalues of $\Pi_N P_\delta$, and perhaps even those of P_δ when restricting the real part of the spectral parameter to an interval $]-2 + 1/\mathcal{O}(1), 2 - 1/\mathcal{O}(1)[$.

We also did some simulations with $N = 500$, $h = 0.004$. The pseudospectral effect now becomes very strong even close to the boundary of the range of the symbol, and we did not try to draw pseudospectral curves. In Fig. 20.10 we have plotted the numerical eigenvalues when $\delta = 0$. Apparently, the numerical instability sets in already for dist $((\Im z), \{-1, 1\}) \approx 0.15$, and only the eigenvalues stretching in from the corners are likely to be true eigenvalues of $\Pi_N P_0$. (One might be able to justify them semi-classically by comparing them with the eigenvalues of $1_{\mathbf{R}_+}(hD)(hD + i\cos(x))$.)

In Figs. 20.11, 20.12, 20.13, 20.14, 20.15, and 20.16 we have plotted the eigenvalues of the perturbed operator with δ ranging from 10^{-2} down to 10^{-12}. When $\delta = 10^{-2}$ the eigenvalues fill up the range $[-2, 2] + i[-1, 1]$ quite well and some are even slightly outside. (For larger perturbations the random part becomes dominant and we get a cloud of eigenvalues around this rectangle. We are then in the theory of random matrices which has less to do with Weyl asymptotics for differential operators.)

In accordance with the Weyl asymptotics, we see that the eigenvalues are distributed more densely near the upper and lower boundaries of the rectangle. For smaller values of δ they are no longer able to fill up the whole rectangle, and consequently the Weyl law will no longer hold right near the upper and lower boundaries. One can say that some eigenvalues try to obey the Weyl law up to the boundary, get stuck by the weakness of the perturbation and accumulate quite densely on a curve. Inside the region formed by the two curves we have Weyl asymptotics (apparently). The precise shape of the accumulation was studied by Vogel [155] in terms of the expected eigenvalue density. He also verified the Weyl law by letting Matlab count the eigenvalues in different rectangles.

In Figs. 20.17, 20.18, 20.19, 20.20, 20.21, and 20.22 we plotted the perturbed eigenvalues together with the unperturbed ones (dd stands for δ).

20.2.2 The Case $g(x) = e^{ix}$

The unperturbed operator is now

$$P = hD_x + e^{ix}. \tag{20.2.4}$$

We define the corresponding perturbed operators $P_\delta = P + \delta Q$ as before. In Fig. 20.23 we have plotted the eigenvalues of $\Pi_N P_\delta$ when $N = 500$ and $\delta = 10^{-2}$. In the middle the picture is similar to the case of the potential $i\cos x$, but there are two obvious differences: The spectrum now fills up a stadium shaped domain with two discs at the ends, where the eigenvalue density is lower. This is not a numerical artefact, but rather a new case of the Weyl law, and there is a difference between $\Pi_N P_\delta$ and P_δ in the two discs.

The reason becomes clear if we plot the range of the symbol $\xi + e^{ix}$ restricted to $x \in S^1$, $-2 \le \xi \le 2$. In Fig. 20.24 we plotted the values of the symbol $\xi + ie^{ix}$, by letting first x take values in a grid in S^1 of step "Big" and letting ξ go through a denser grid in $[-2, 2]$, then reversing the procedure, letting ξ go through the points $-2 + k\,\mathrm{Big} \in [-2, 2]$, and for each such ξ, letting x go through a denser grid in S^1. Each point in the middle region has two preimages, and in the discs near the ends there is only one preimage per point.

We then plotted the eigenvalues (in blue) against the range for $\delta = 10^{-k}$, $k = 2, 4, 6, 8, 10, 12$, see Figs. 20.25, 20.26, 20.27, 20.28, 20.29, and 20.30.

20.3 $(hD)^2 + i\cos x$

The operator

$$P = (hD)^2 + i\cos(x),\ x \in S^1, \tag{20.3.5}$$

appears naturally as a model when applying the secular method to certain non-self-adjoint perturbations of completely integrable quantum Hamiltonians in two dimensions (Hitrik–Sjöstrand [73]). The range of the symbol $p = \xi^2 + i\cos(x)$ is equal to $[0; +\infty[+ i[-1, 1]$. The lower corner of this image corresponds to $(x, \xi) = (\pi, 0)$ and the harmonic approximation of P near that point is equal to $-i$ plus the non-self-adjoint oscillator $(hD)^2 + ix^2/2$. Near the point $(0, 0)$ the harmonic approximation is equal to i plus $(hD)^2 - ix^2/2$.

The random perturbation is now $P_\delta = P + \delta Q$, where

$$Qu(x) = q(x)u(x),\ \ q(x) = \sum_{k=-N}^{N} \widehat{q}_k e^{ikx}, \tag{20.3.6}$$

and $\widehat{q}_k$ are independent complex random variables $\sim \mathcal{N}_{\mathbb{C}}(0, 1)$ The numerical calculations concern the matrix $\Pi_N P_\delta : \mathcal{H}_N \to \mathcal{H}_N$, with Π_N, $\mathcal{H}_N$ as defined in the preceding section.

In Fig. 20.31 we plot the spectrum in the unperturbed case, when $N = 50$, $h = 0.04$. Except for the branching to the right (which would require more analysis), the eigenvalues behave as one could expect. To the left we have approximations of the eigenvalues of the two harmonic oscillators, and it has been proved by Hitrik [67] that the exact eigenvalues of P do indeed have that behaviour here. Further inside this is only an expectation, which should be possible to verify, using the complex WKB-method. We have also traced 21 pseudospectral curves in a subregion, where the first one happened to fall outside the figure. Already for $N = 50$, there is a strong spectral instability.

In Fig. 20.32, we consider the unperturbed case with $N = 100$, $h = 0.02$ and trace the same number of pseudospectral curves. It seems to us that the "eigenvalues" in the middle are non-real because of very strong numerical instability.

In Fig. 20.33 we see that the pseudospectral behaviour improves and we can have confidence in the computed eigenvalues, at least as eigenvalues of $\Pi_N P_\delta$. In the remaining Figs. 20.34, 20.35, 20.36 20.37 20.38 20.39, and 20.40, we take $N = 500$, $h = 0.004$. In Fig. 20.34 we consider the unperturbed case. Then in

Figs. 20.35, 20.36 20.37 20.38 20.39, and 20.40 we consider the perturbed cases with δ decreasing from 0.01 to 10^{-12}. When $\delta = 0.01$ the perturbation is so strong that many eigenvalues fall outside the range of the symbol and we can hardly say that Weyl asymptotics holds in that case. When $\delta = 10^{-4}$ there is a tendency towards the Weyl law with a higher density of eigenvalues near $\Re z = 0$ and near $\Im z = \pm 1$ mimicking the growth of the density of the direct image of the symplectic volume $dxd\xi$ under the symbol map. In the other figures this is still true to some extent, but the weakness of the perturbation keeps the eigenvalues away from $\Im z = \pm 1$.

20.4 Second-Order Differential Operators on S^1

In the case of Hager's operator and of $(hD)^2 + i\cos$ the critical values of the symbol belong to the boundary of the range. They are folds, except for the two corner points in the range of $\xi^2 + i\cos(x)$, where the differential of the symbol vanishes at the preimages.

In this section and the next one we give examples where the symbol has critical points inside its range, so that the Weyl law will predict an accumulation of eigenvalues near such points in the interior of the spectral region.

In this section we consider elliptic operators on S^1 of the form

$$P = hD \circ (1 + ae^{ix} + a_{\mathrm{m}}e^{-ix} + be^{2ix} + b_{\mathrm{m}}e^{-2ix}) \circ hD + ce^{ix} + c_{\mathrm{m}}e^{-ix} \tag{20.4.7}$$

We then looked for suitable values of the coefficient, leading to nice critical values in the interior of the range of the symbol

$$p = (1 + ae^{ix} + a_{\mathrm{m}}e^{-ix} + be^{2ix} + b_{\mathrm{m}}e^{-2ix})y^2 + ce^{ix} + c_{\mathrm{m}}e^{-ix}, \tag{20.4.8}$$

where for programming reasons the dual variable is called "y" rather than "ξ". Possibly one could do without the terms be^{2ix} and $b_m e^{-2ix}$, but having found some interesting choices with b and $b_m \neq 0$, we decided to keep them.

We computed the range as before, letting x belong to a grid in $]0, 2\pi]$ of step "Big", then for each such x, plotting the image curve by letting $y \in [0, y_{\mathrm{M}}]$ go through a finer grid. Then we reversed the order of the operations, for each y in a grid of step Big in $[0, y_{\mathrm{M}}]$, and we draw the image curve by letting x go through a finer grid. We did choose y_{M} suitably in order to capture the critical values.

When the coefficients in (20.4.7), (20.4.8) are real, the operator P is $\mathcal{PT}$-symmetric respect to with the parity $x \to -x$. Correspondingly, the range of p is symmetric with respect to the real line.

As in Sect. 20.3, we consider random perturbations of the form $P_\delta = P + \delta Q$, where $Qu = q(x)u(x)$, $q(x) = \sum_{-N}^{N} q_k(\omega)e^{ikx}$ and q_k are independent random variables $\sim \mathcal{N}(0, 1)$. In the $\mathcal{PT}$-symmetric case we take q_k real-valued and otherwise complex-valued, so that Q is $\mathcal{PT}$-symmetric in the first case. Then as in the preceding sections we approximate P_δ with $\Pi_N P_\delta : \mathcal{H}_N \to \mathcal{H}_N$, where $\Pi_N, \mathcal{H}_N$ are introduced in Sect. 20.2.

20.4.1 A $\mathcal{PT}$-Symmetric Case

A suitable choice of parameters turned out to be

$$(a,\ a_{\rm m},\ b,\ b_{\rm m},\ c,\ c_{\rm m},\ \mathrm{Big},\ y_{\rm M}) = (-0.4,\ 0.1,\ 0.1,\ 0.01,\ 1,\ 0.5,\ 0.08976,\ 2.3562).$$

$\mathrm{Big} = \pi/35$, $y_{\rm M} = 3\pi/4$. Figure 20.41 shows the range.

We distinguish clearly some folds inside the range and two stronger singular values.

Figures 20.42, 20.43, 20.44, 20.45, and 20.46 show the (numerical) spectrum of P_δ for $N = 1500$ with the range of p in the background, for $\delta =$"dd" ranging from 2.5×10^{-3} down to 2.5×10^{-9}. Weyl asymptotics seems to hold in the sense that the density of eigenvalues is stronger where the range is denser. Apparently, there is more spectral stability near the folds and caustic points, and many eigenvalues can't make it up to these points, when δ becomes smaller. The choice of $y_{\rm M}$ was made to have some approximate fitting with the spectrum. Of course, near the right end of the picture, the plotted eigenvalues seem to represent only a fraction of the true eigenvalues, since the truncation Π_N here cuts away some eigenvalues (that would be visible with a larger value of N).

20.4.2 A Non-$\mathcal{PT}$-Symmetric Case

Here we chose the parameters

$$(a,\ a_{\rm m},\ b,\ b_{\rm m},\ c,\ c_{\rm m},\ \mathrm{Big},\ y_{\rm M}) = (-0.5,\ 0.08i,\ 0,\ 0,\ 1,\ -0.5,\ 0.08976,\ 2.7826),$$

then did the same calculations as in $\mathcal{PT}$-symmetric case with $N = 1500$. See Figs. 20.47, 20.48, 20.49, 20.50, 20.51, and 20.52

20.5 Finite Difference Operators on S^1

Here, we replace hD, y with $\sin(hD)$, $\sin(y)$ in (20.4.7), (20.4.8):

$$P = \sin(hD) \circ (1 + ae^{ix} + a_{\mathrm{m}}e^{-ix} + be^{2ix} + b_{\mathrm{m}}e^{-2ix}) \circ \sin(hD) + ce^{ix} + c_{\mathrm{m}}e^{-ix}, \tag{20.5.9}$$

$$p = (1 + ae^{ix} + a_{\mathrm{m}}e^{-ix} + be^{2ix} + b_{\mathrm{m}}e^{-2ix})(\sin y)^2 + ce^{ix} + c_{\mathrm{m}}e^{-ix}. \tag{20.5.10}$$

This operator acts on $\ell^2(h\mathbf{Z})$. For a given $N \in \mathbf{N} \setminus \{0\}$, we choose $h = 2\pi/N$, so that $h > 0$ is the smallest positive number with $Nh = 2\pi$. Then P acts on the N-dimensional space $\ell^2(S^1_N)$, $S^1_N = h\mathbf{Z}/2\pi\mathbf{Z}$. It is well known that functions u on S^1_N have Fourier sum expansions of the form

$$u = \sum_{k \in \mathbf{Z}/N\mathbf{Z}} \widehat{u}_k e^{ikx} \tag{20.5.11}$$

and

$$\sin(hD)u = \sum_{k \in \mathbf{Z}/N\mathbf{Z}} \sin(hk)\widehat{u}_k e^{ikx}.$$

It is therefore simple to write down the matrix of P in terms of the Fourier coefficients in (20.5.11). Contrary to the case of differential operators, this is an exact reduction to a finite matrix, and no projection operator is necessary. It is also well known that the phase space associated to our operators above is $S^1_x \times S^1_y$ and p in (20.5.10) is naturally a function on that space.

The random perturbation is again multiplicative of the form

$$q = \sum_{k \in \mathbf{Z}/N\mathbf{Z}} q_k(\omega)e^{ikx}, \tag{20.5.12}$$

where q_k are independent $\mathcal{N}(0, 1)$ random variables.

As before, we can represent graphically the range by discretizing in x and y, and then study the eigenvalues of the $N \times N$-matrix of P. Again we choose values of the coefficients that give rise to interesting folds and focal points in the range. Below we present the results in two cases, one with $\mathcal{PT}$-symmetry and one without. In both cases we took the random perturbation real (contrary to the non-$\mathcal{PT}$ symmetric case for differential operators). Recall however that the main results on Weyl asymptotics are the same for real and complex random perturbations (cf. Remark 15.3.3).

20.5.1 A $\mathcal{PT}$-Symmetric Case

To find suitable parameters took much less effort, maybe because the compactness of the phase space. We made the following choice:

$$(a,\ a_{\mathrm{m}},\ b,\ b_{\mathrm{m}},\ c,\ c_{\mathrm{m}}) = (-0.4,\ 0.1,\ 0.1,\ 0.02,\ 0.5,\ 0.25)$$

The range is given in Fig. 20.53. In Figs. 20.54, 20.55, 20.56, and 20.57, we give the spectrum for δ("dd") ranging from 2.5×10^{-4} down to 2.5×10^{-9} with $N = 3000$.

20.5.2 A Non-$\mathcal{PT}$-Symmetric Case

We did the same computations with $N = 3000$ and

$$(a,\ a_{\mathrm{m}},\ b,\ b_{\mathrm{m}},\ c,\ c_{\mathrm{m}}) = (-0.4 + 0.2i,\ 0.1,\ 0,\ 0,\ 0.5,\ 0.25).$$

See Fig. 20.58 for the range and Figs. 20.59, 20.60, 20.61, 20.62, 20.63, and 20.64 for the spectrum.

20.6 $\mathcal{PT}$-Symmetric Double Well

Let $Z = x + h\partial_x$ be the annihilation operator and $Z^* = x - h\partial_x$ the creation operator acting on functions of one real variable. If $e_0, e_1, e_2, \dots$ is the standard orthonormal basis of semi-classical Hermite functions in $L^2(\mathbf{R})$ (obtained by the scaling $\widetilde{x} = h^{1/2}x$ from the usual Hermite functions), then we know that $Ze_k = \sqrt{2kh}e_{k-1}$ ($= 0$ by definition when $k = 0$), $Z^*e_k = \sqrt{2(k+1)}e_{k+1}$. Thus, the infinite matrices of Z and Z^* are easy to write down explicitly. We can then write down the infinite matrices for the multiplication operator $x = (Z + Z^*)/2$ and the semi-classical differentiation, $h\partial_x = (Z - Z^*)/2$ For numerical purposes, we restrict our attention to the N-dimensional space $\mathcal{H}_N$, spanned by $e_0, e_1, \dots, e_{N-1}$ and correspondingly we replace x with $x_N := \Pi_N \circ x : \mathcal{H}_N \to \mathcal{H}_N$ and $h\partial_x$ with $(h\partial)_N := \Pi_N h\partial : \mathcal{H}_N \to \mathcal{H}_N$. In other words, we replace the infinite matrices of x and $h\partial_x$ by the $N \times N$-matrices obtained by restricting the row and column indices to $\{0, \dots, N-1\}$. Thus to treat spectrally the non-self-adjoint harmonic operator, we can replace it by $-(h\partial)_N^2 + ix_N^2$ and this gives the correct spectrum near 0. Here, we give a numerical illustration of Theorems 19.4.1 and 19.4.2 by considering

$$P_\epsilon = -(h\partial_x)^2 + x^4 - x^2 + i\epsilon x, \tag{20.6.13}$$

for $\epsilon > 0$ small. When $\epsilon = 0$ this is a self-adjoint Schrödinger operator with an even double-well potential. The height of the separating barrier is 0 and we know (in the limit $h \to 0$) that the eigenvalues in any region $]-\infty, -1/\mathcal{O}(1)]$ form pairs of values such that the distance between the two elements in the same pair is exponentially small, while the distance between a pair and the next higher one is of the order of h. In any region $[1/\mathcal{O}(1), \mathcal{O}(1)[$ the eigenvalues are simple with a distance $\asymp h$ from any such value to the next higher neighbor. We made a series of computations to find a value of ϵ for which there are two complex conjugate eigenvalues, very close to be a double real eigenvalue. See Fig. 20.65. We see a number of eigenvalues with small real part, that have split off the real line. The remaining eigenvalues are real, except the ones fairly far to the right. The non-reality to the right is certainly due to errors introduced by the spectral truncation. The real eigenvalues up to roughly 0.7 seem to be regularly spaced and to obey the Bohr–Sommerfeld rule. However, a careful look at the picture shows that the 12th real eigenvalue from the left with real part between 0.34 and 0.35 is actually constituted by two eigenvalues very close to each other. On the pdf version we only see that the blue dot is slightly widened horizontally. On the .fig-file this is seen more clearly after several enlargements. Again we think that this is due to the spectral truncation.

We therefore improved the program to get first the matrix of P in a range $1 \le j, k \le \widetilde{N}$ with $\widetilde{N} > N$ and truncating this matrix down to one of size $N \times N$. Then, we get $\Pi_N P_\delta$ with only one spectral truncation. The result obtained with the new program is given in Fig. 20.66. The anomaly near 0.34 has now disappeared, but there is a new close to double real eigenvalue $\approx$0.6. We then plotted both sets of eigenvalues in Fig. 20.67, marking the original ones with a dot and the "improved" ones with an "$\times$". The coincidence in the region $\Re z < 0.34$ is remarkable and forgetting about the anomaly near 0.34 for the originally computed ones, we have a nice coincidence up to $\Re z \approx 0.6$. The second, more refined computation gives apparently a slightly better overall result, but the improvement mainly concerns relatively few eigenvalues situated to the right, where the spectral truncation plays a rule. Since the coincidence is so great up to about 0.6, we conclude (boldly) that Matlab computes the eigenvalues of the differential operator up to that level with quite a good accuracy. In particular, this is the case for the eigenvalues with real part $\lesssim 0.3$, and in particular the splitting of eigenvalues out into the complex is nicely illustrated.

20.7 The Figures

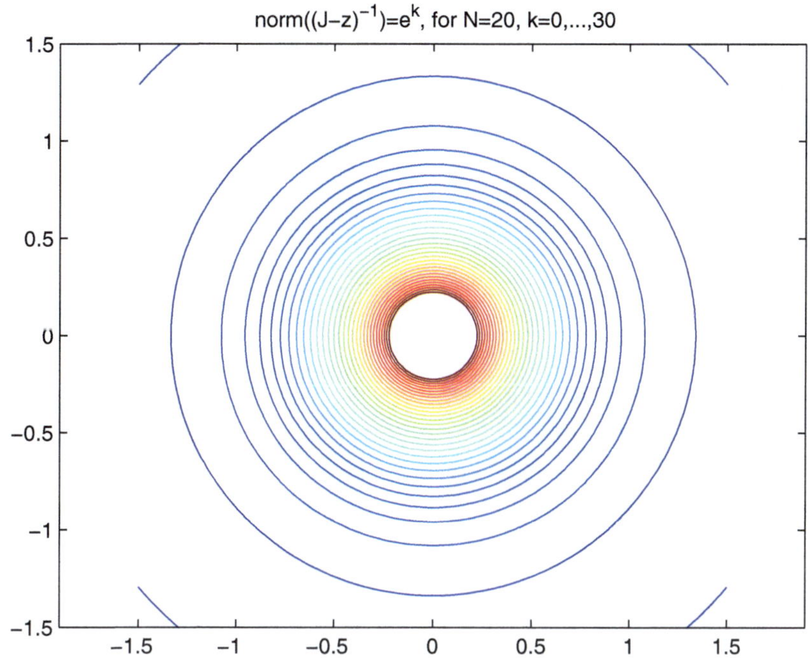

Fig. 20.1 The level curves $\|(J - z)^{-1}\| = e^k$, for $k = 0, \ldots, 30$, $N = 20$

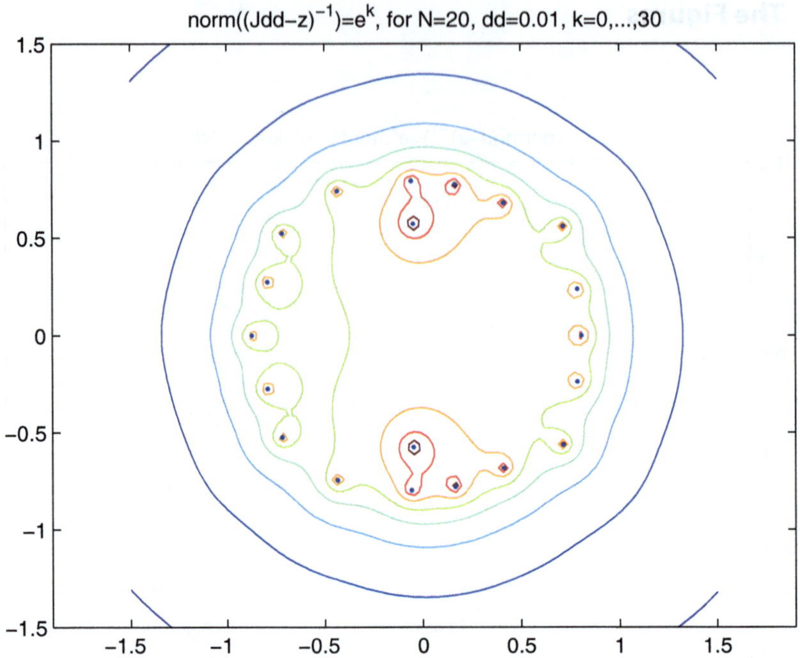

Fig. 20.2 The level curves $\|(J_\delta - z)^{-1}\| = e^k$, for $k = 0, \ldots, 30$, $\delta = 0.01$, $N = 20$

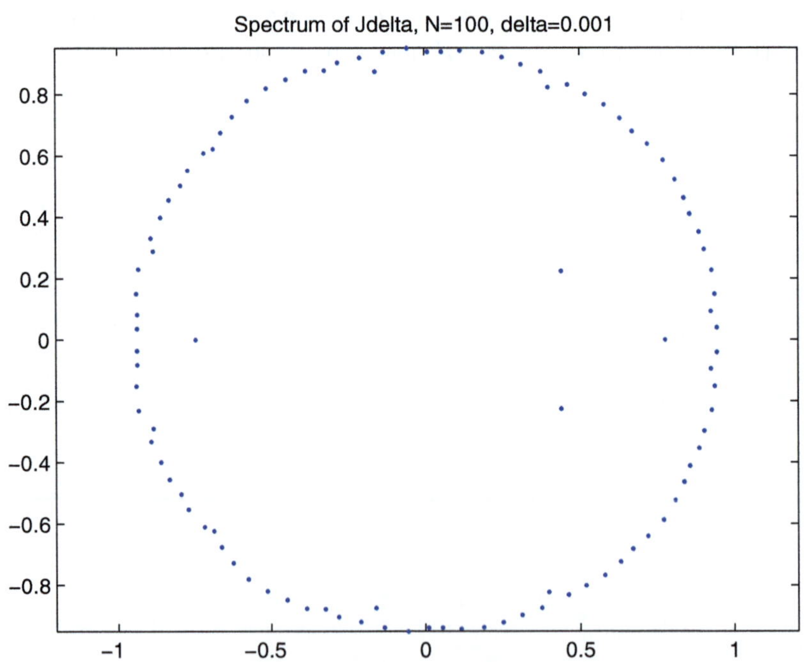

Fig. 20.3 Spectrum of J_δ, $N = 100$, $\delta = 0.001$

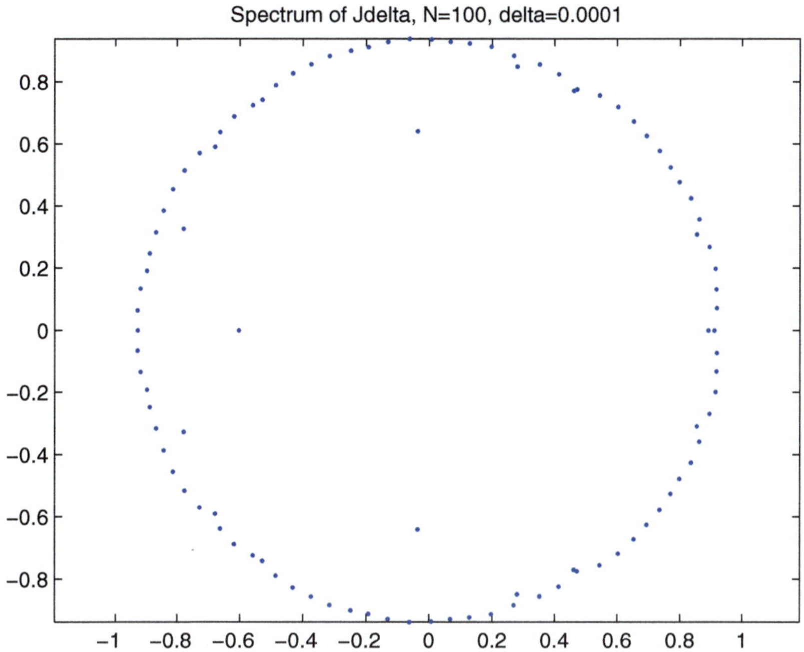

Fig. 20.4 Spectrum of J_δ, $N = 100$, $\delta = 10^{-4}$

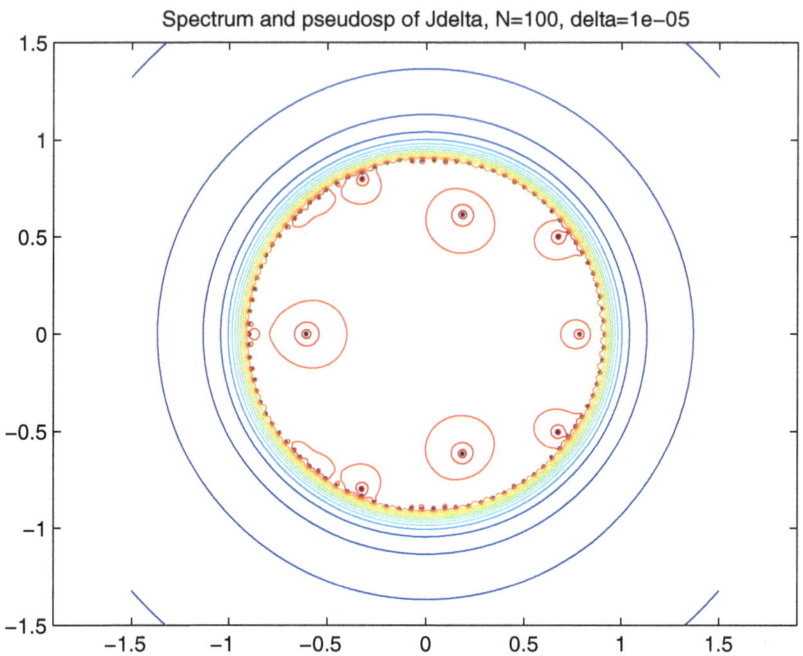

Fig. 20.5 Spectrum of J_δ, $N = 100$, $\delta = 10^{-5}$ and pseudospectral level curves as in Fig. 20.2

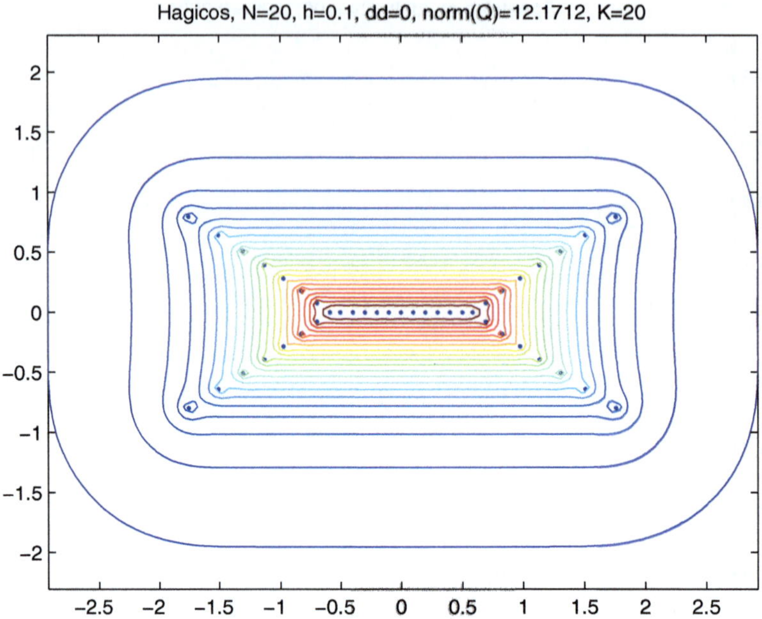

Fig. 20.6 Numerical eigenvalues and pseudospectral curves $\|(P-z)^{-1}\| = e^k$, $k = 0, 1, \dots, 20$, for P in (20.2.1), $h = 10^{-1}$, $N = 20$

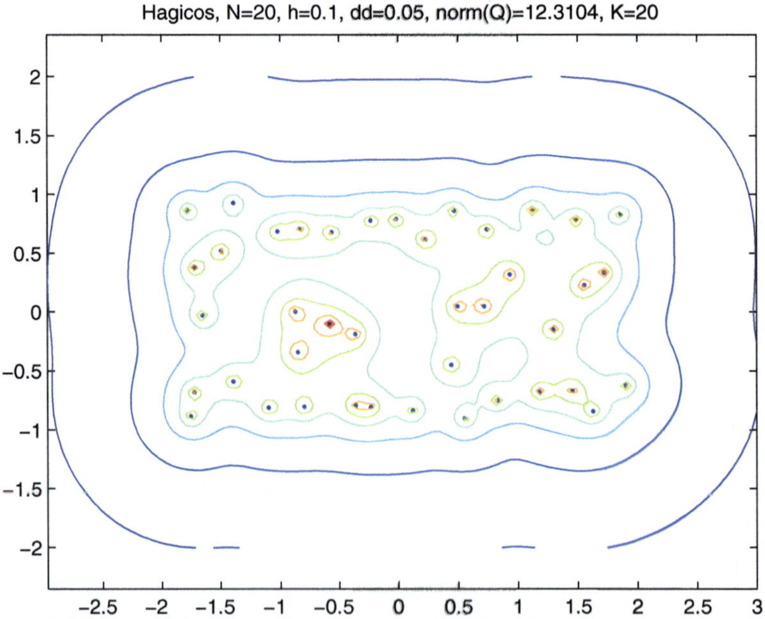

Fig. 20.7 Numerical eigenvalues and $K + 1 = 21$ pseudospectral curves for P_δ in (20.2.2), $h = 10^{-1}$, $N = 20$, $\delta = 0.05$. $\|Q\| = 12.3104$

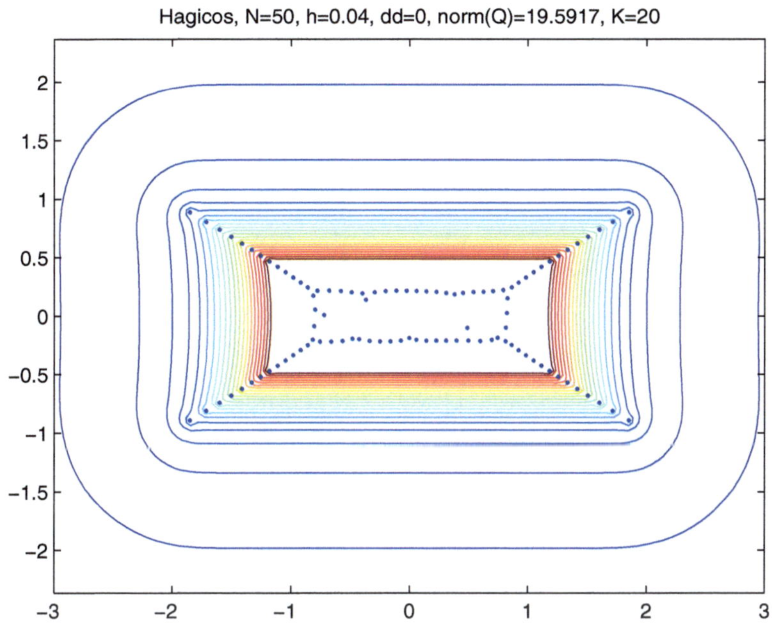

Fig. 20.8 Numerical eigenvalues and pseudospectral curves for P in (20.2.1), $h = 0.04$, $N = 50$, $K = 20$

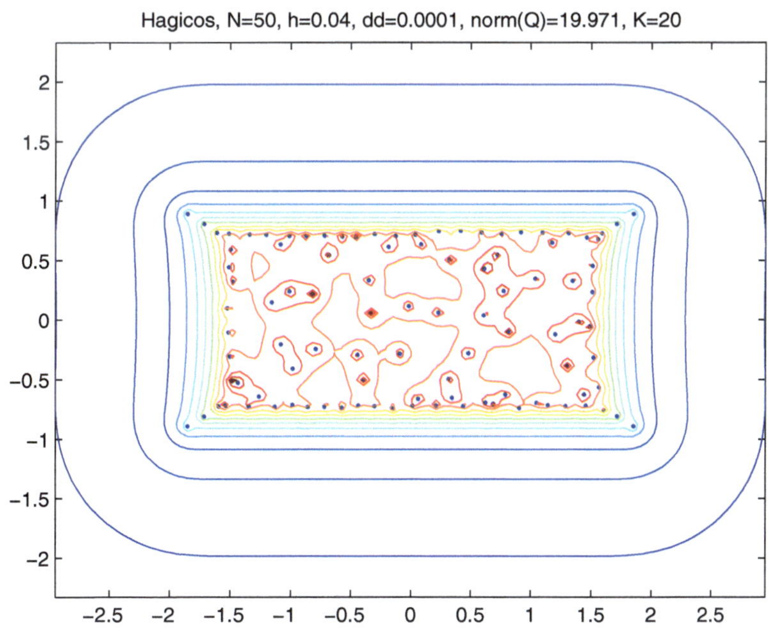

Fig. 20.9 Numerical eigenvalues and pseudospectral curves for P_δ in (20.2.2), $h = 0.04$, $N = 50$, $\delta = 10^{-4}$, $K = 20$. $\|Q\| = 19.971$

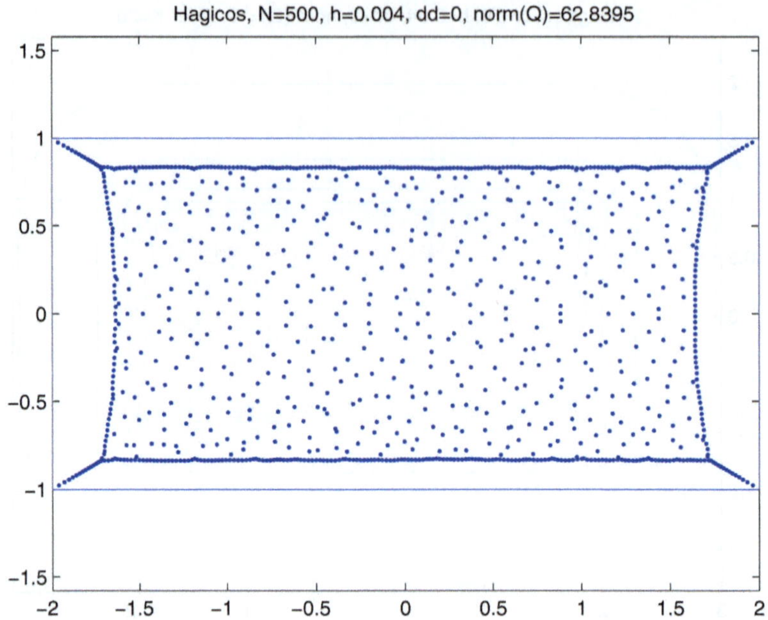

Fig. 20.10 Numerical eigenvalues for P in (20.2.1), $h = 0.004$, $N = 500$

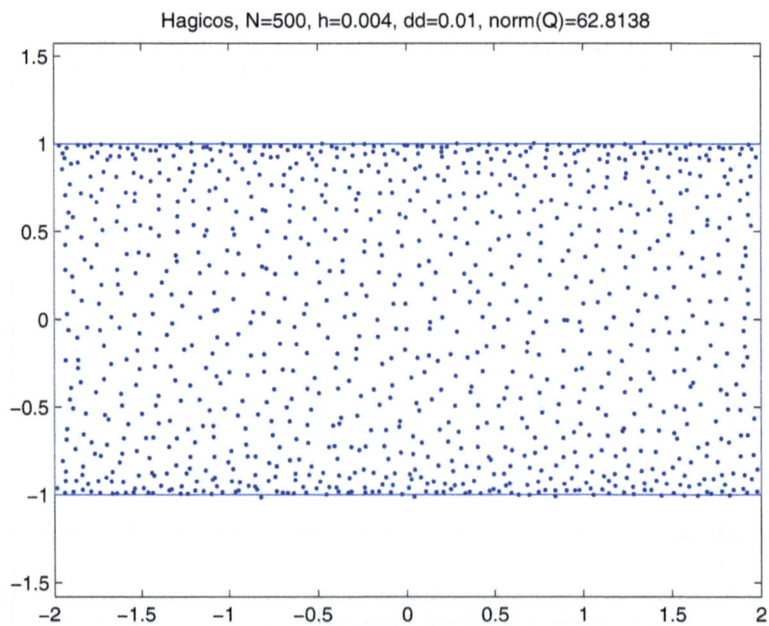

Fig. 20.11 Numerical eigenvalues for P_δ in (20.2.2), $h = 0.004$, $N = 500$, $\delta = 10^{-2}$. $\|Q\| = 62.8138$

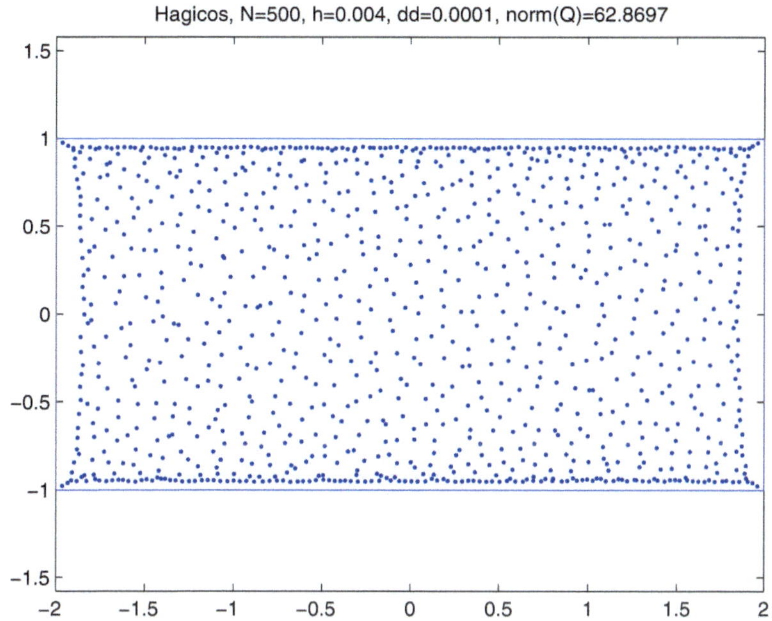

Fig. 20.12 Numerical eigenvalues for P_δ in (20.2.2), $h = 0.004$, $N = 500$, $\delta = 10^{-4}$. $\|Q\| = 62.8697$

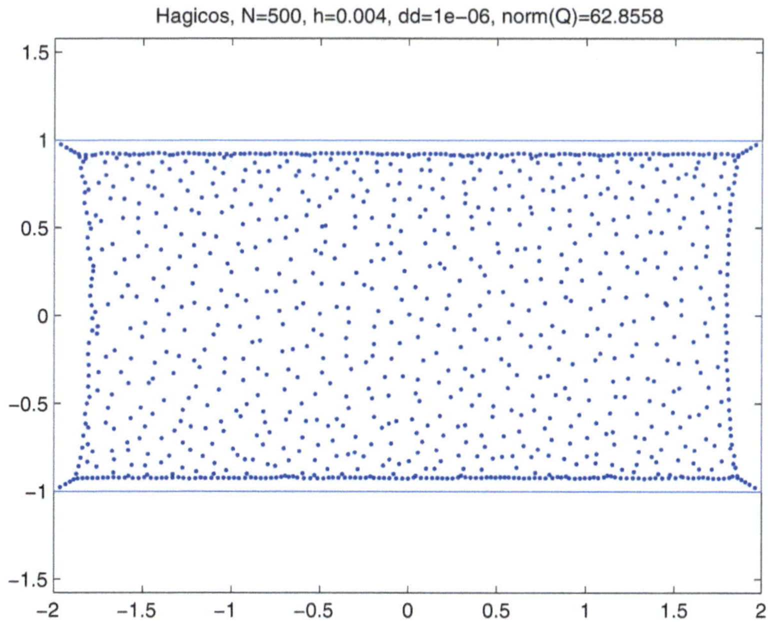

Fig. 20.13 Numerical eigenvalues for P_δ in (20.2.2), $h = 0.004$, $N = 500$, $\delta = 10^{-6}$. $\|Q\| = 62.8558$

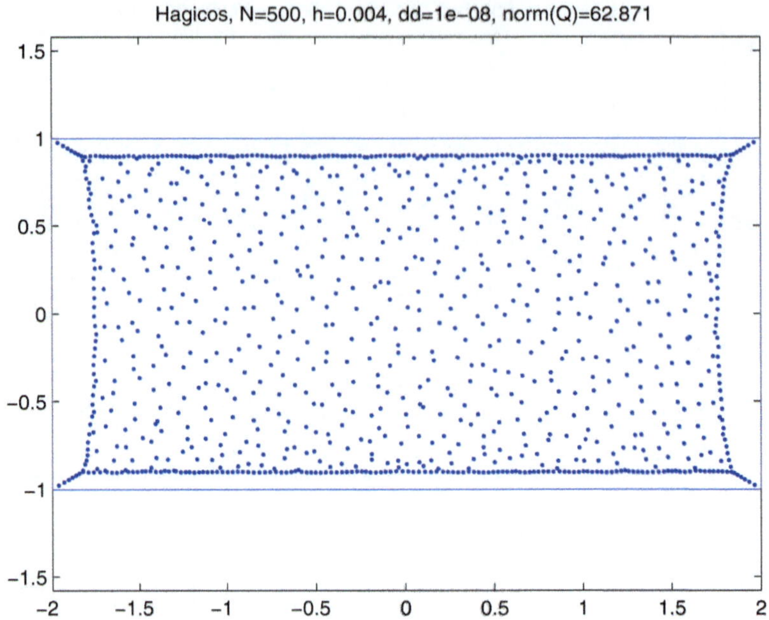

Fig. 20.14 Numerical eigenvalues for Q_δ in (20.2.2), $h = 0.004$, $N = 500$, $\delta = 10^{-8}$. $\|Q\| = 62.871$

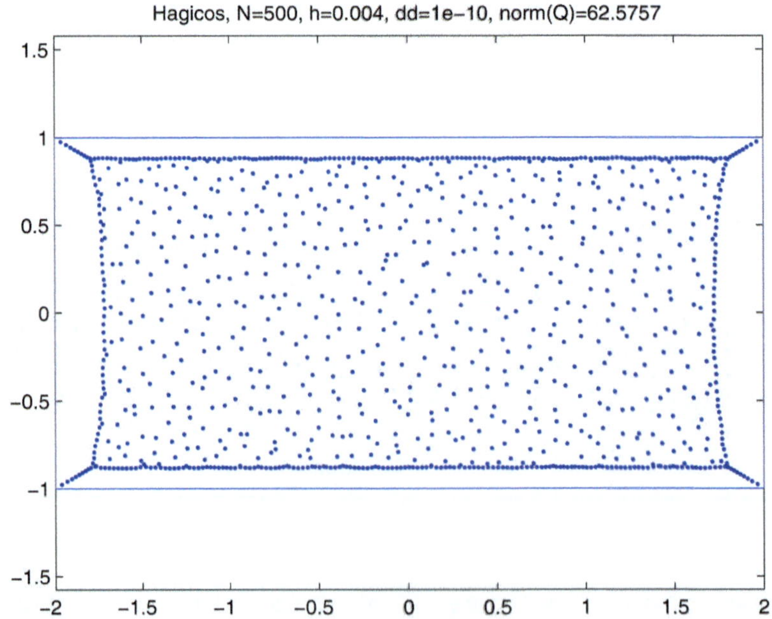

Fig. 20.15 Numerical eigenvalues for Q_δ in (20.2.2), $h = 0.004$, $N = 500$, $\delta = 10^{-10}$. $\|Q\| = 62.5757$

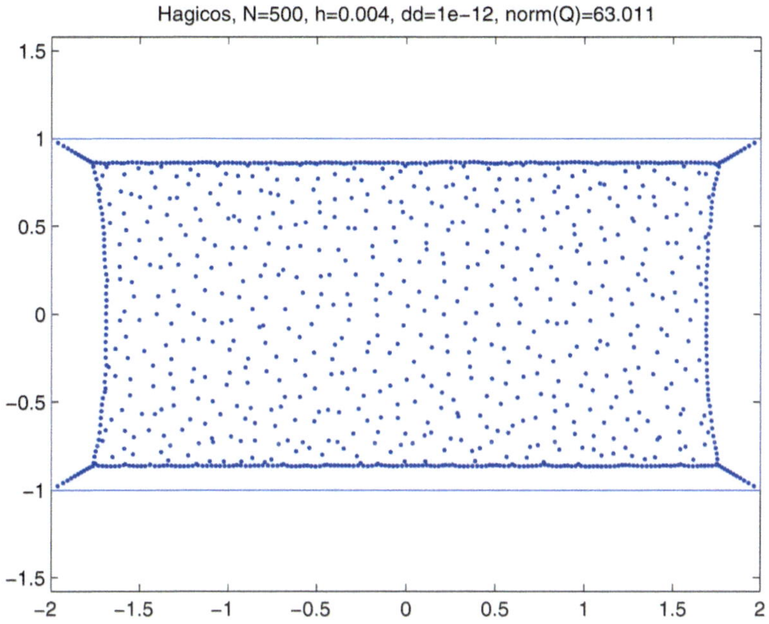

Fig. 20.16 Numerical eigenvalues for P_δ in (20.2.2), $h = 0.004$, $N = 500$, $\delta = 10^{-12}$. $\|Q\| = 63.011$

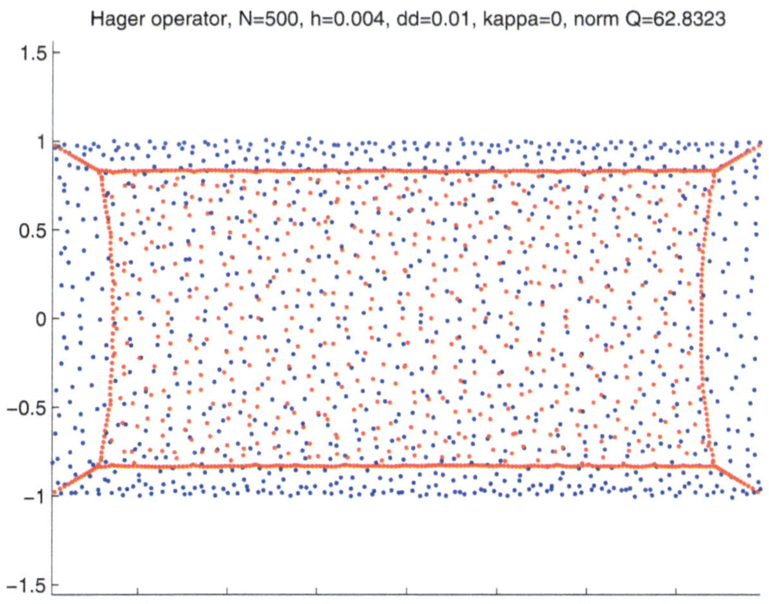

Fig. 20.17 Numerical eigenvalues for P_δ in (20.2.2), $h = 0.004$, $N = 500$, $\delta = 10^{-2}$ (blue) and $\delta = 0$ (red). $\|Q\| = 62.8323$

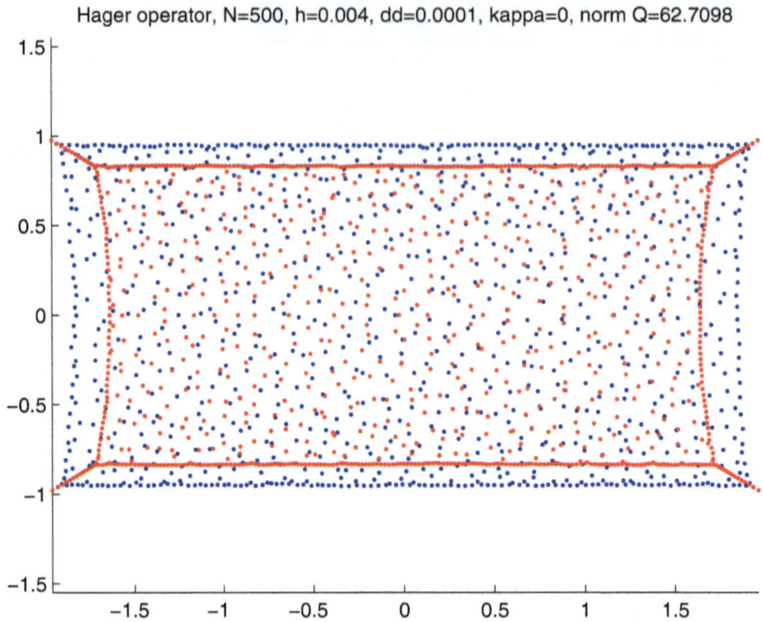

Fig. 20.18 Numerical eigenvalues for P_δ in (20.2.2), $h = 0.004$, $N = 500$, $\delta = 10^{-4}$ (blue) and $\delta = 0$ (red). $\|Q\| = 62.7098$

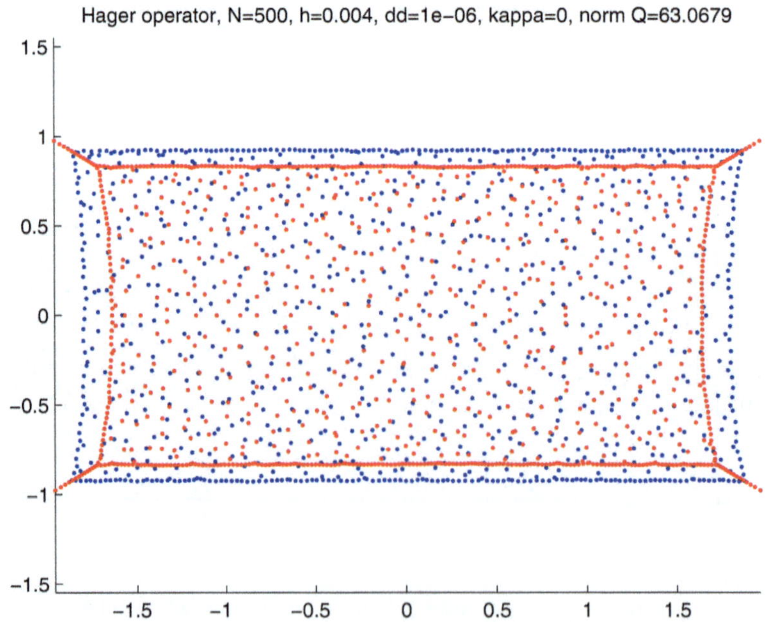

Fig. 20.19 Numerical eigenvalues for P_δ in (20.2.2), $h = 0.004$, $N = 500$, $\delta = 10^{-6}$ (blue) and $\delta = 0$ (red). $\|Q\| = 63.0679$

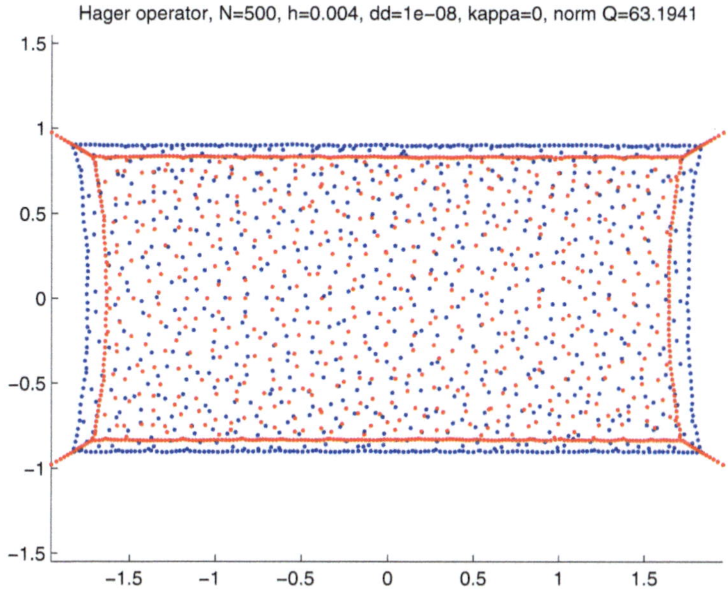

Fig. 20.20 Numerical eigenvalues for P_δ in (20.2.2), $h = 0.004$, $N = 500$, $\delta = 10^{-8}$ (blue) and $\delta = 0$ (red). $\|Q\| = 63.1941$

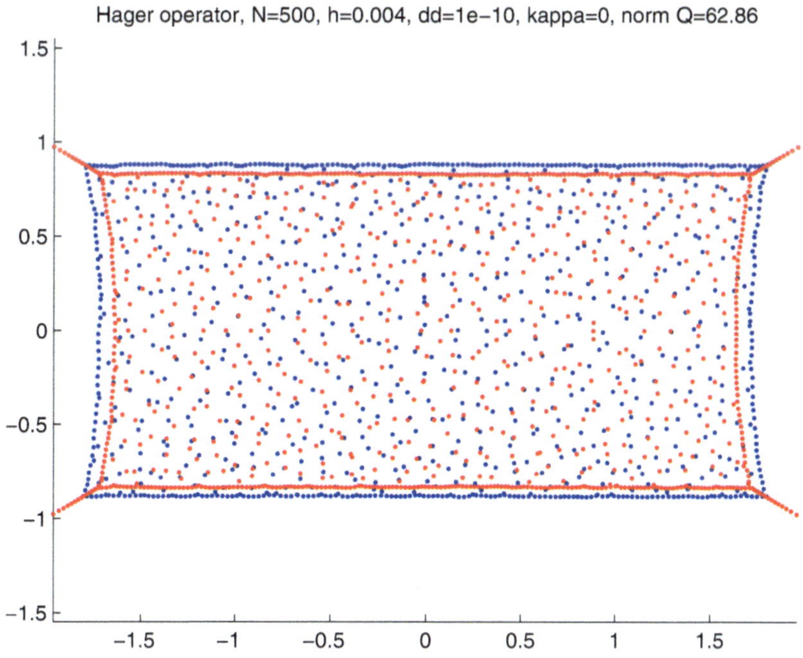

Fig. 20.21 Numerical eigenvalues for P_δ in (20.2.2), $h = 0.004$, $N = 500$, $\delta = 10^{-10}$ (blue) and $\delta = 0$ (red). $\|Q\| = 62.86$

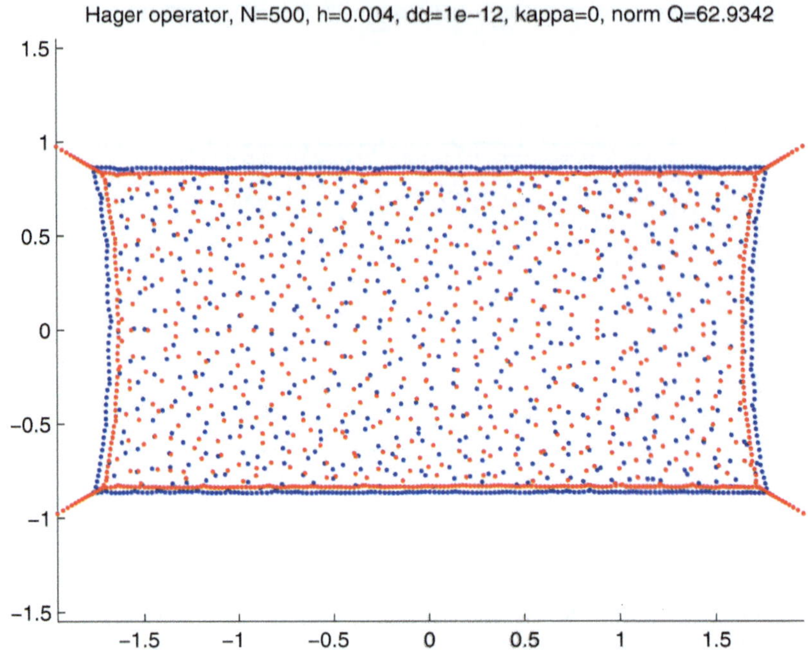

Fig. 20.22 Numerical eigenvalues for P_δ in (20.2.2), $h = 0.004$, $N = 500$, $\delta = 10^{-12}$ (blue) and $\delta = 0$ (red). $\|Q\| = 62.9342$

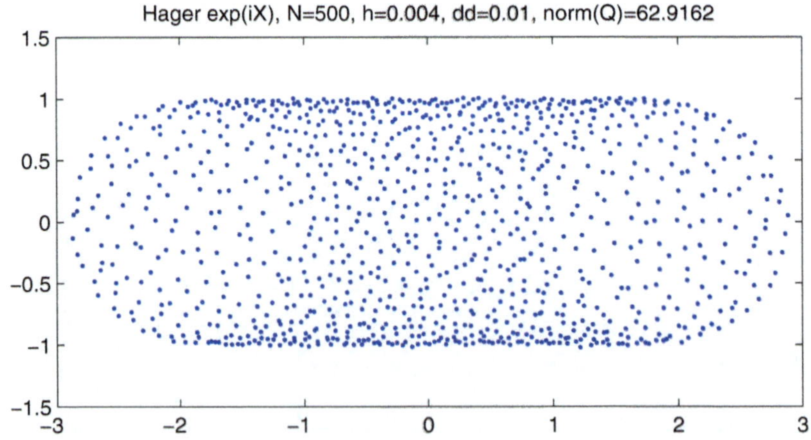

Fig. 20.23 Numerical eigenvalues for $P_\delta = P + \delta Q$ with P as in (20.2.4), $h = 0.004$, $N = 500$, $\delta = 10^{-2}$. $\|Q\| = 62.9162$

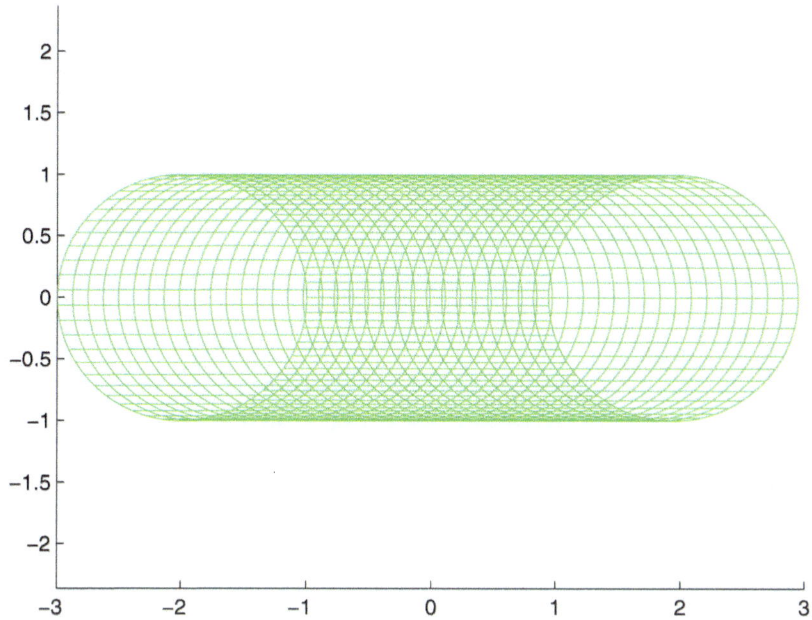

Fig. 20.24 Range of $\xi + ie^x$, restricted to $S^1_x \times [-2, 2]$

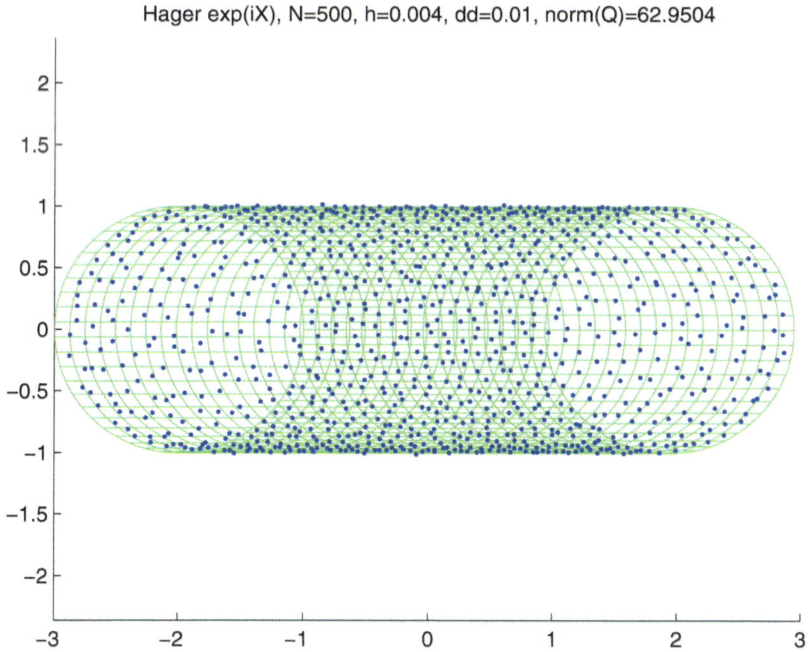

Fig. 20.25 Numerical eigenvalues of $P_\delta = P + \delta Q$ with P as in (20.2.4) for $N = 500$, $h = 0.004$, $\delta = 10^{-2}$. $\|Q\| = 62.9504$

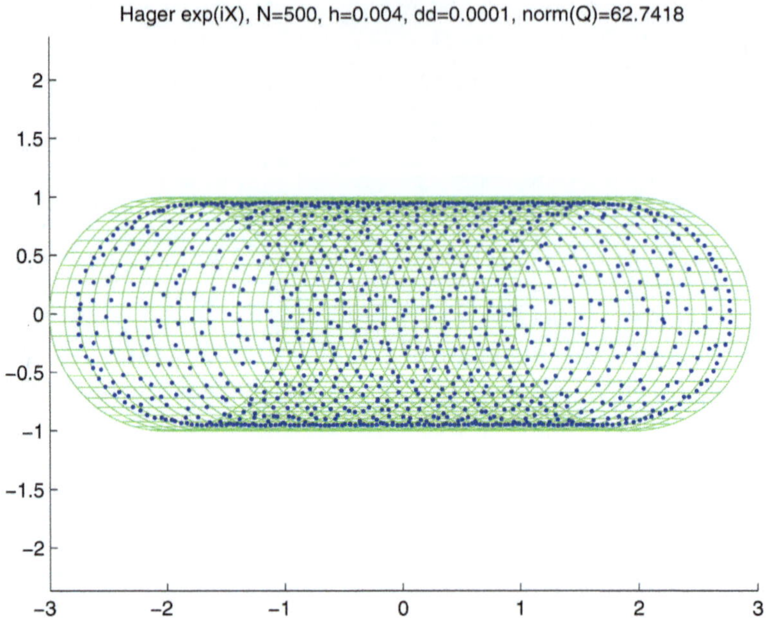

Fig. 20.26 Numerical eigenvalues of $P_\delta = P + \delta Q$ with P as in (20.2.4) for $N = 500$, $h = 0.004$, $\delta = 10^{-4}$. $\|Q\| = 62.7418$

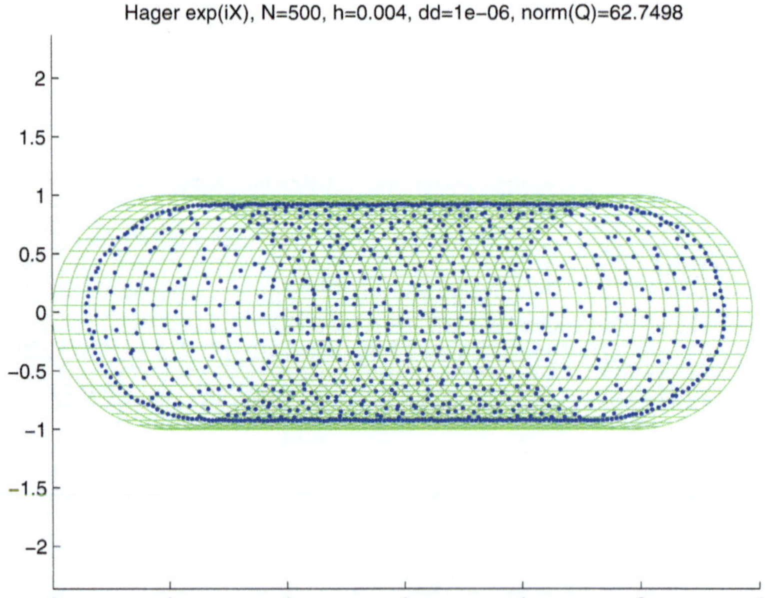

Fig. 20.27 Numerical eigenvalues of $P_\delta = P + \delta Q$ with P as in (20.2.4) for $N = 500$, $h = 0.004$, $\delta = 10^{-6}$. $\|Q\| = 62.7498$

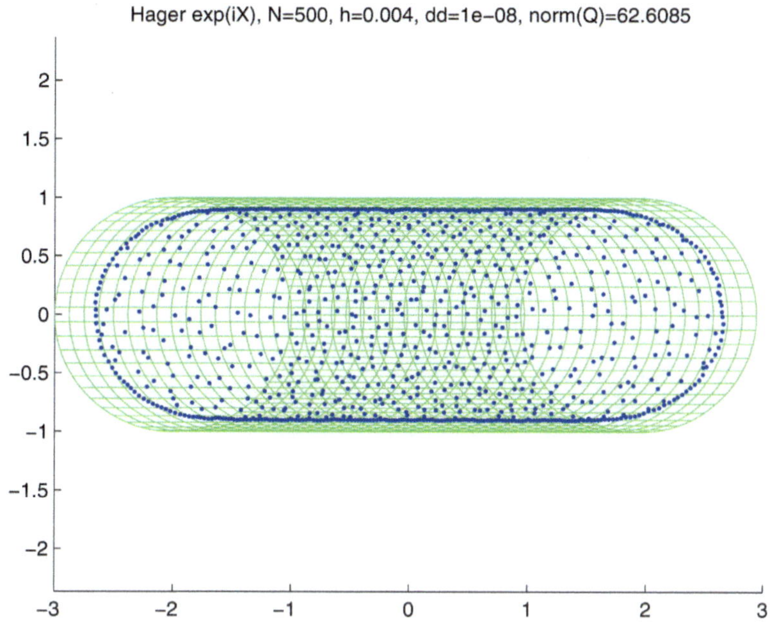

Fig. 20.28 Numerical eigenvalues of $P_\delta = P + \delta Q$ with P as in (20.2.4) for $N = 500$, $h = 0.004$, $\delta = 10^{-8}$. $\|Q\| = 62.6085$

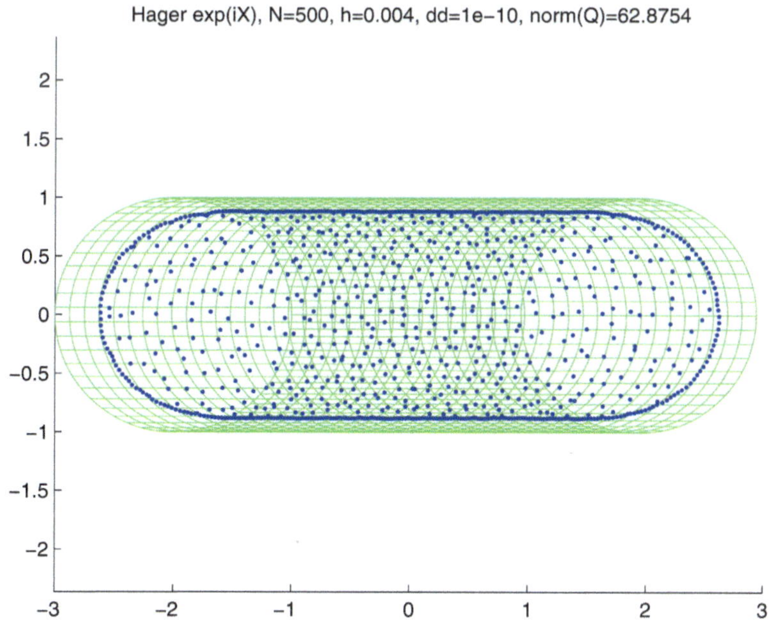

Fig. 20.29 Numerical eigenvalues of $P_\delta = P + \delta Q$ with P as in (20.2.4) for $N = 500$, $h = 0.004$, $\delta = 10^{-10}$. $\|Q\| = 62.8754$

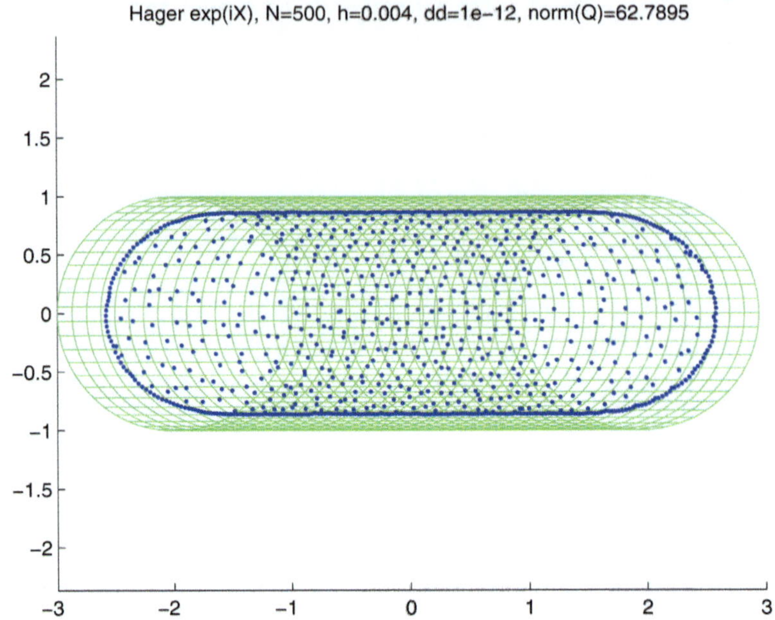

Fig. 20.30 Numerical eigenvalues of $P_\delta = P + \delta Q$ with P as in (20.2.4) for $N = 500$, $h = 0.004$, $\delta = 10^{-12}$. $\|Q\| = 62.7895$

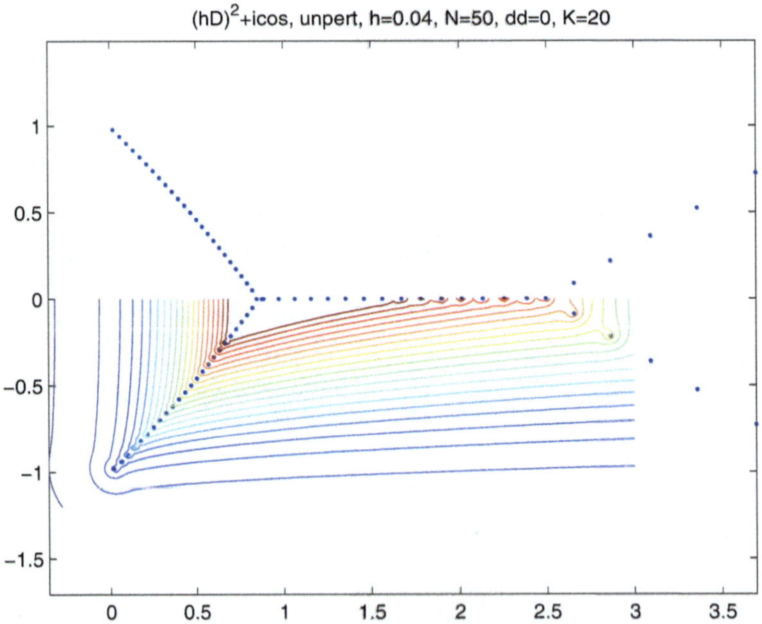

Fig. 20.31 Spectrum and pseudospectral level curves for P in (20.3.5), $N = 50$, $h = 0.04$, $K = 20$

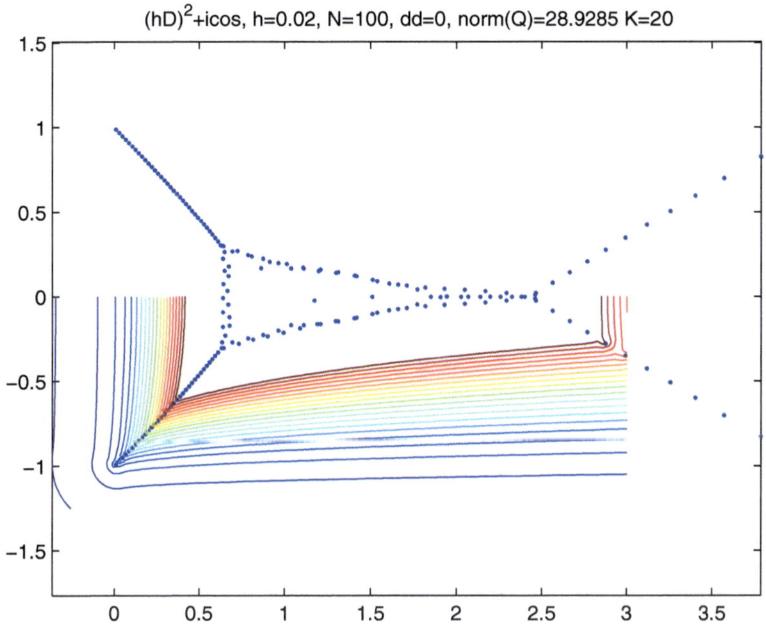

Fig. 20.32 Spectrum and pseudospectral level curves for P in (20.3.5), $N = 100$, $h = 0.02$, $K = 20$

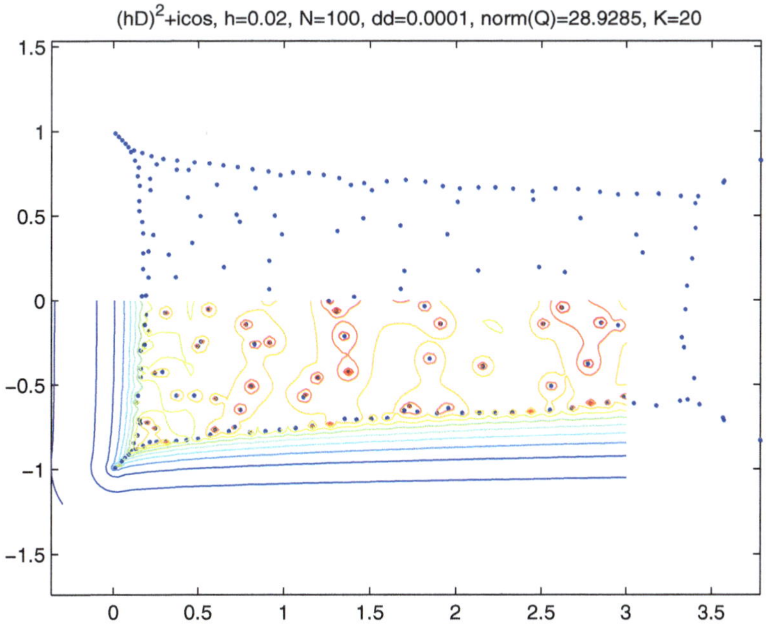

Fig. 20.33 Spectrum and pseudospectral level curves for $P_\delta = P + \delta Q$ with P, Q as in (20.3.5), (20.3.6), $N = 100$, $h = 0.02$, $\delta = 10^{-4}$, $K = 20$. $\|Q\| = 28.9285$

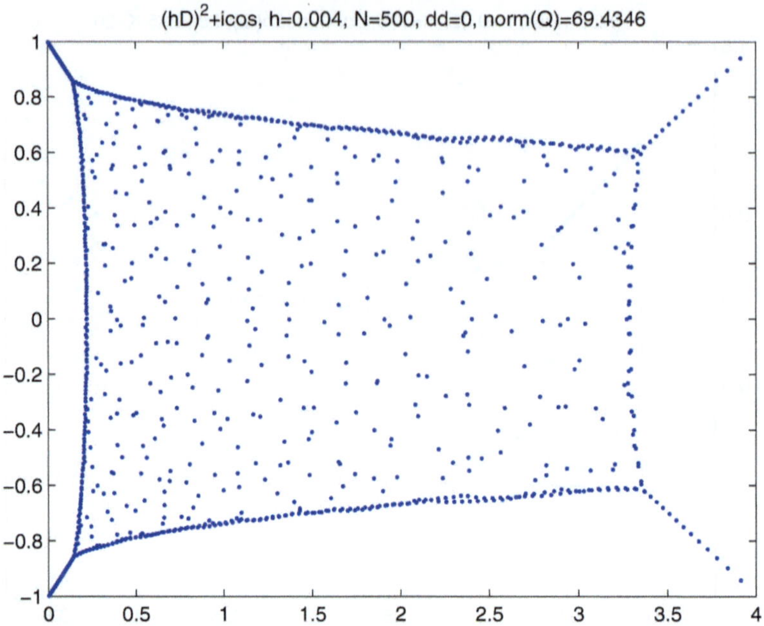

Fig. 20.34 Spectrum for P as in (20.3.5). $N = 500$, $h = 0.004$, $\|Q\| = 69.4346$

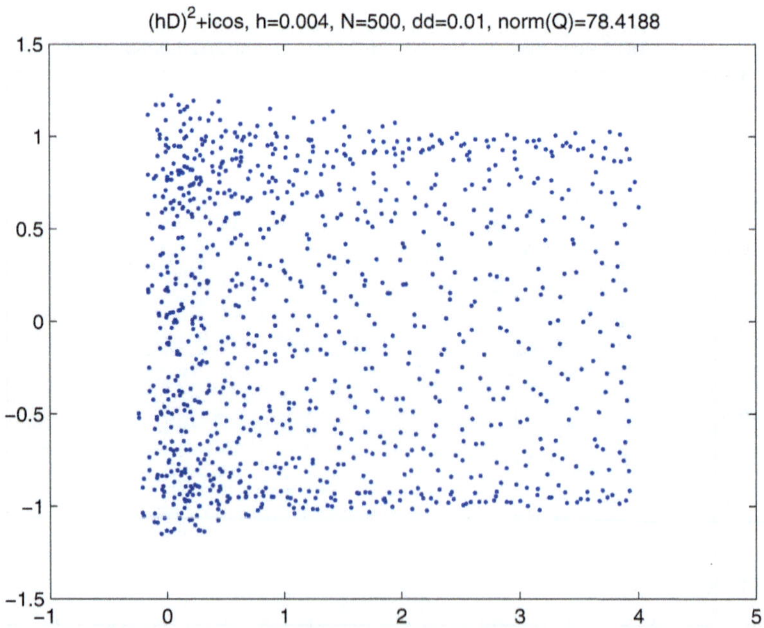

Fig. 20.35 Spectrum for $P_\delta = P + \delta Q$ with P, Q as in (20.3.5), (20.3.6), $N = 500$, $h = 0.004$, $\delta = 0.01$. $\|Q\| = 78.4188$

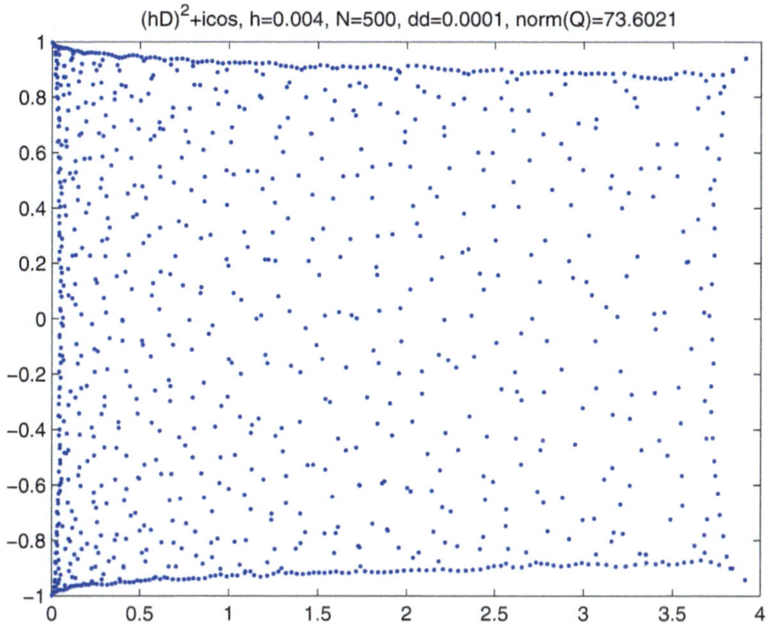

Fig. 20.36 Spectrum for $P_\delta = P + \delta Q$ with P, Q as in (20.3.5), (20.3.6), $N = 500$, $h = 0.004$, $\delta = 10^{-4}$. $\|Q\| = 73.6021$

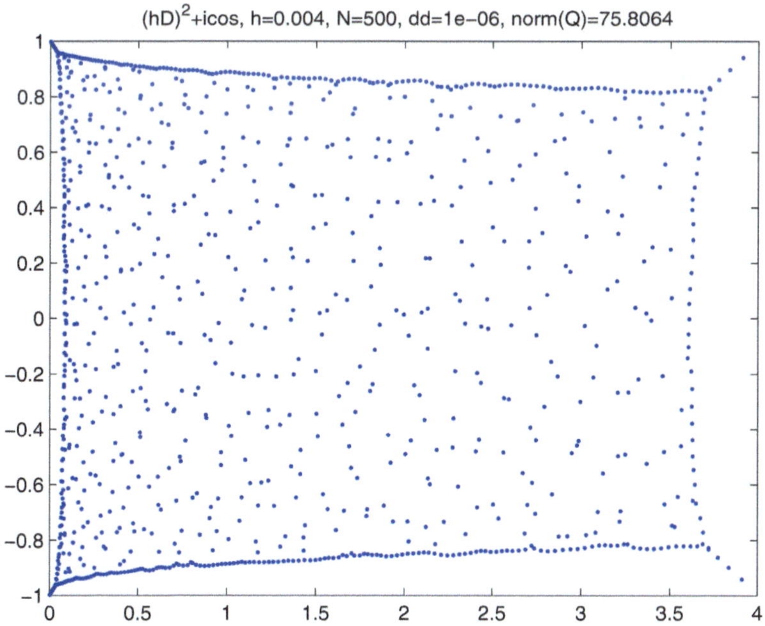

Fig. 20.37 Spectrum for $P_\delta = P + \delta Q$ with P, Q as in (20.3.5), (20.3.6), $N = 500$, $h = 0.004$, $\delta = 10^{-6}$. $\|Q\| = 75.8064$

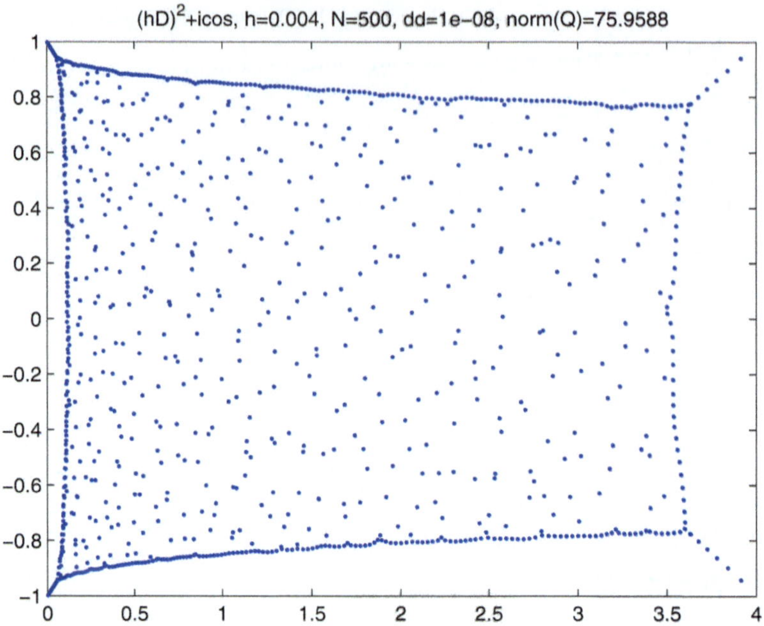

Fig. 20.38 Spectrum for $P_\delta = P + \delta Q$ with P, Q as in (20.3.5), (20.3.6), $N = 500$, $h = 0.004$, $\delta = 10^{-8}$. $\|Q\| = 75.9588$

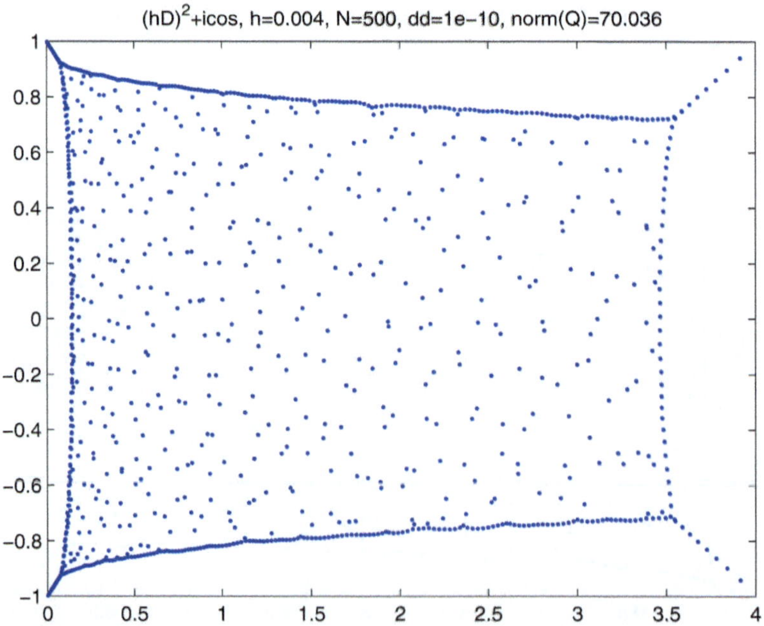

Fig. 20.39 Spectrum for $P_\delta = P + \delta Q$ with P, Q as in (20.3.5), (20.3.6), $N = 500$, $h = 0.004$, $\delta = 10^{-10}$. $\|Q\| = 70.036$

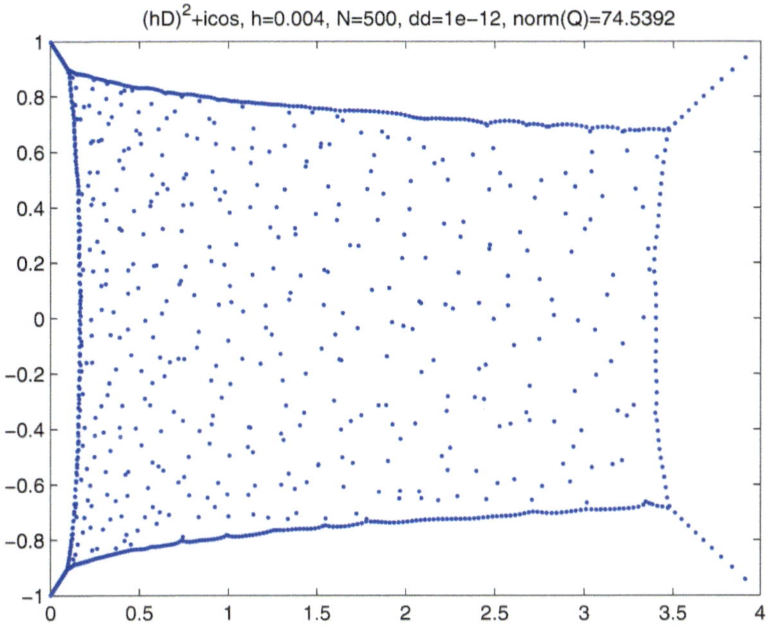

Fig. 20.40 Spectrum for $P_\delta = P + \delta Q$ with P, Q as in (20.3.5), (20.3.6), $N = 500$, $h = 0.004$, $\delta = 10^{-12}$. $\|Q\| = 74.5392$

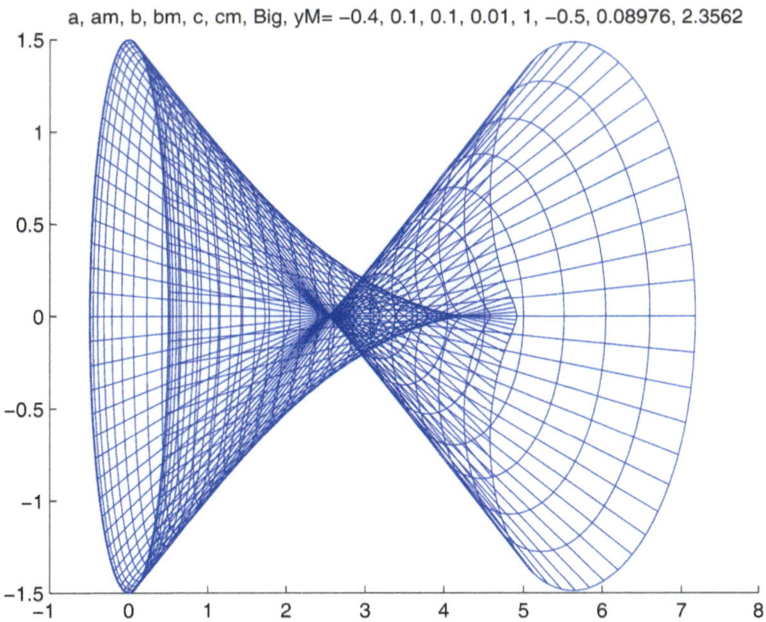

Fig. 20.41 Range of p in (20.4.8), $(a, a_{\mathrm{m}}, b, b_{\mathrm{m}}, c, c_{\mathrm{m}}, \mathrm{Big}, y_{\mathrm{M}}) = (-0.4, 0.1, 0.1, 0.01, 1, -0.5, 0.08976, 2.3562)$

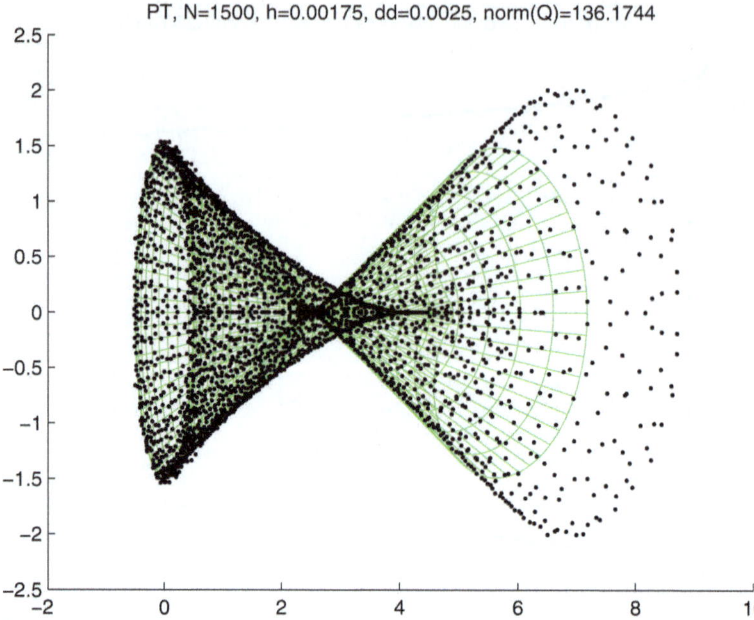

Fig. 20.42 Eigenvalues of $P_\delta = P + \delta Q$ as in (20.4.7), $h = 0.00175$, $N = 1500$, $\delta = 2.5 \times 10^{-3}$. $\|Q\| = 136.1744$

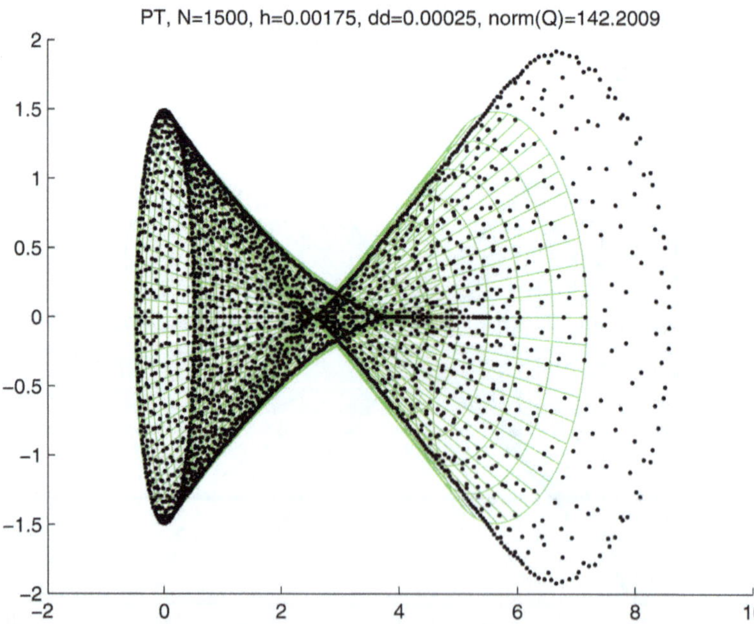

Fig. 20.43 Eigenvalues of $P_\delta = P + \delta Q$ as in (20.4.7), $h = 0.00175$, $N = 1500$, $\delta = 2.5 \times 10^{-4}$. $\|Q\| = 142.2009$

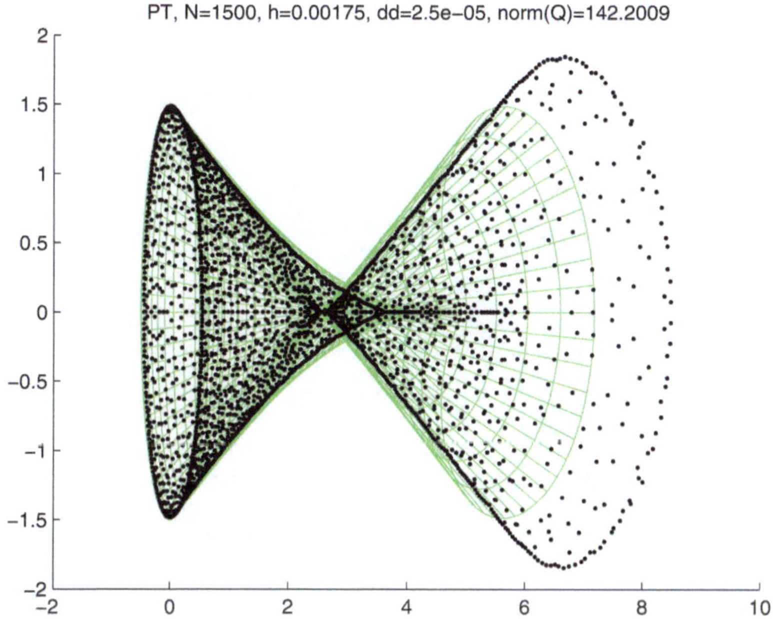

Fig. 20.44 Eigenvalues of $P_\delta = P + \delta Q$ as in (20.4.7), $h = 0.00175$, $N = 1500$, $\delta = 2.5 \times 10^{-5}$. $\|Q\| = 142.2009$

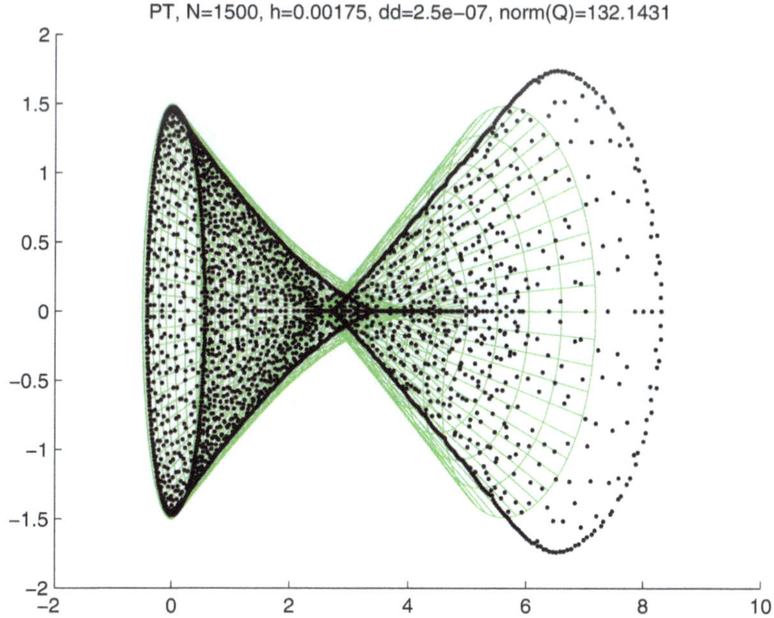

Fig. 20.45 Eigenvalues of $P_\delta = P + \delta Q$ as in (20.4.7), $h = 0.00175$, $N = 1500$, $\delta = 2.5 \times 10^{-7}$. $\|Q\| = 132.1431$

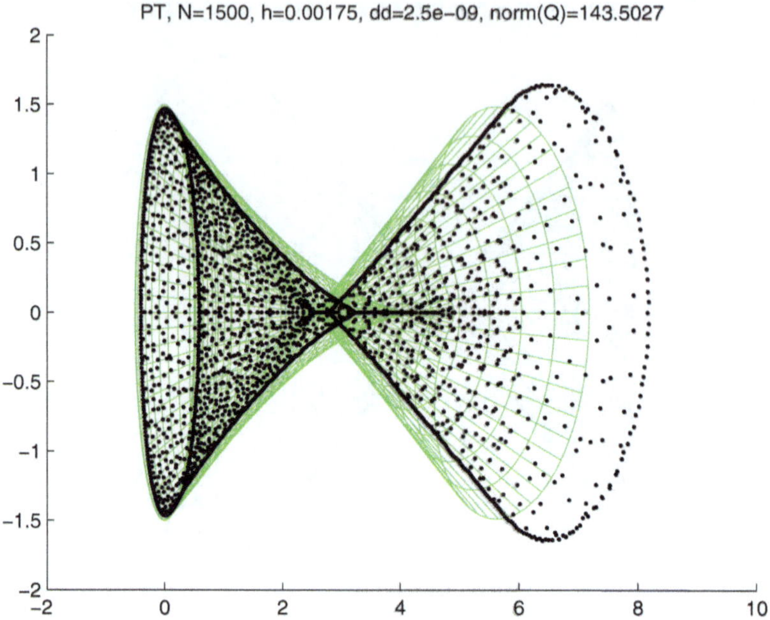

Fig. 20.46 Eigenvalues of $P_\delta = P + \delta Q$ as in (20.4.7), $h = 0.00175$, $N = 1500$, $\delta = 2.5 \times 10^{-9}$. $\|Q\| = 143.5027$

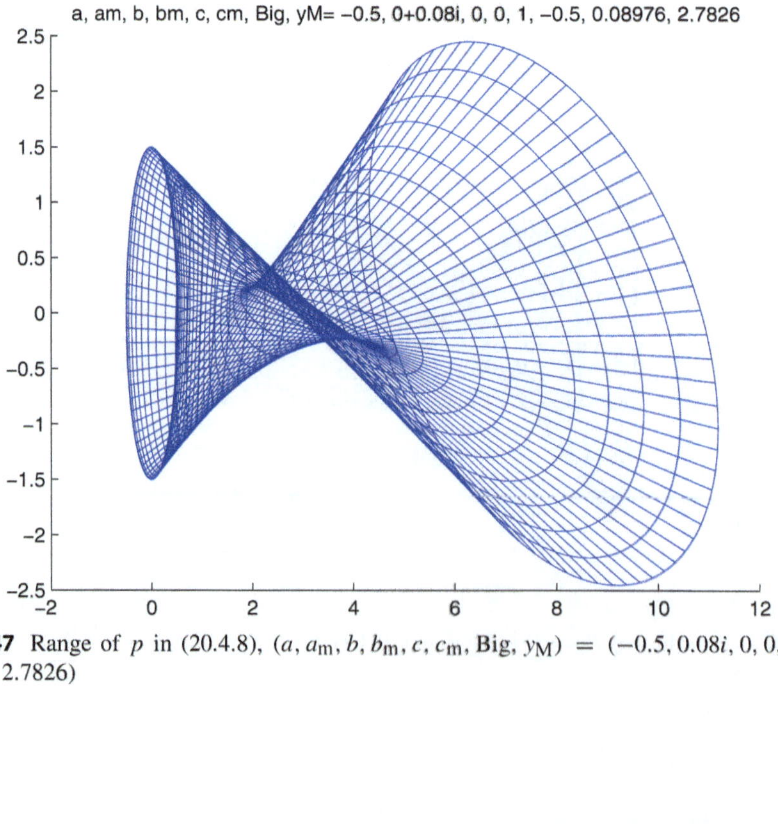

Fig. 20.47 Range of p in (20.4.8), $(a, a_{\rm m}, b, b_{\rm m}, c, c_{\rm m}, {\rm Big}, y_{\rm M}) = (-0.5, 0.08i, 0, 0, 1, -0.5, 0.08976, 2.7826)$

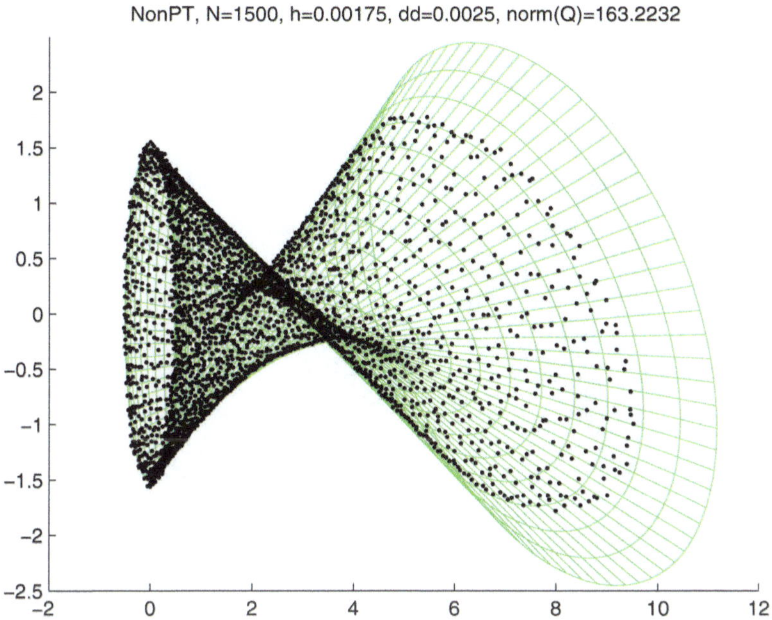

Fig. 20.48 Eigenvalues of $P_\delta = P + \delta Q$ as in (20.4.7), $h = 0.00175$, $N = 1500$, $\delta = 2.5 \times 10^{-3}$. $\|Q\| = 163.2232$

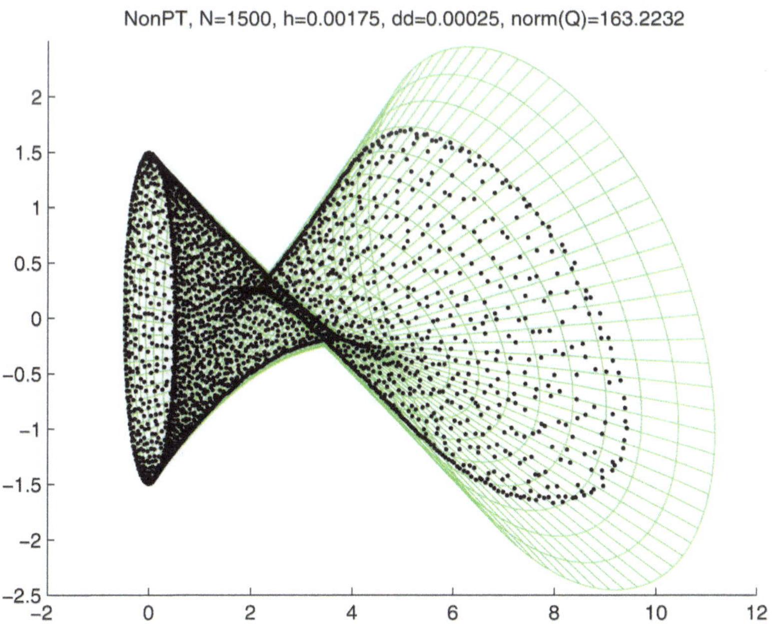

Fig. 20.49 Eigenvalues of $P_\delta = P + \delta Q$ as in (20.4.7), $h = 0.00175$, $N = 1500$, $\delta = 2.5 \times 10^{-4}$. $\|Q\| = 163.2232$

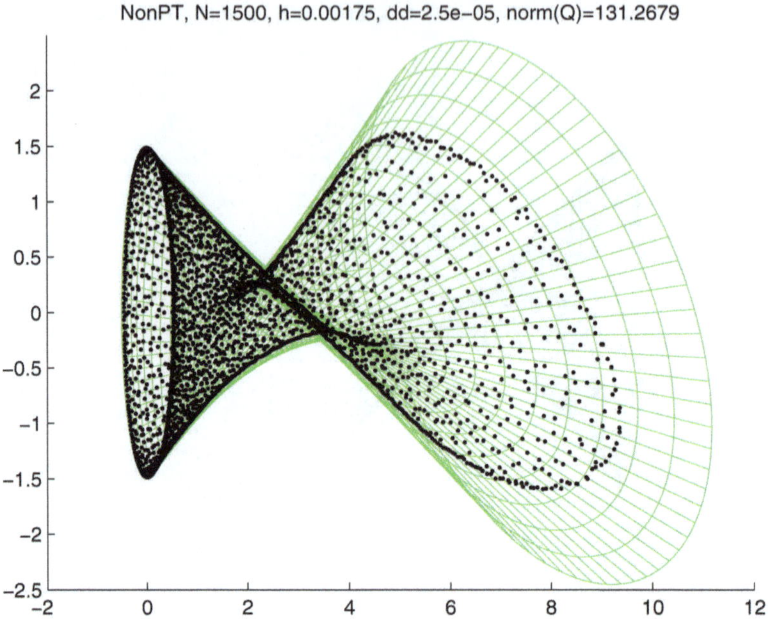

Fig. 20.50 Eigenvalues of $P_\delta = P + \delta Q$ as in (20.4.7), $h = 0.00175$, $N = 1500$, $\delta = 2.5 \times 10^{-5}$. $\|Q\| = 131.2679$

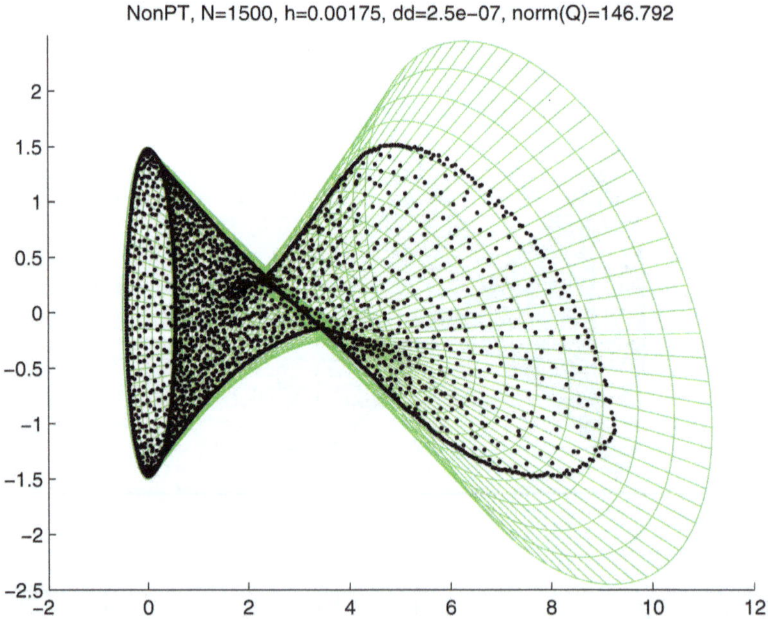

Fig. 20.51 Eigenvalues of $P_\delta = P + \delta Q$ as in (20.4.7), $h = 0.00175$, $N = 1500$, $\delta = 2.5 \times 10^{-7}$. $\|Q\| = 146.792$

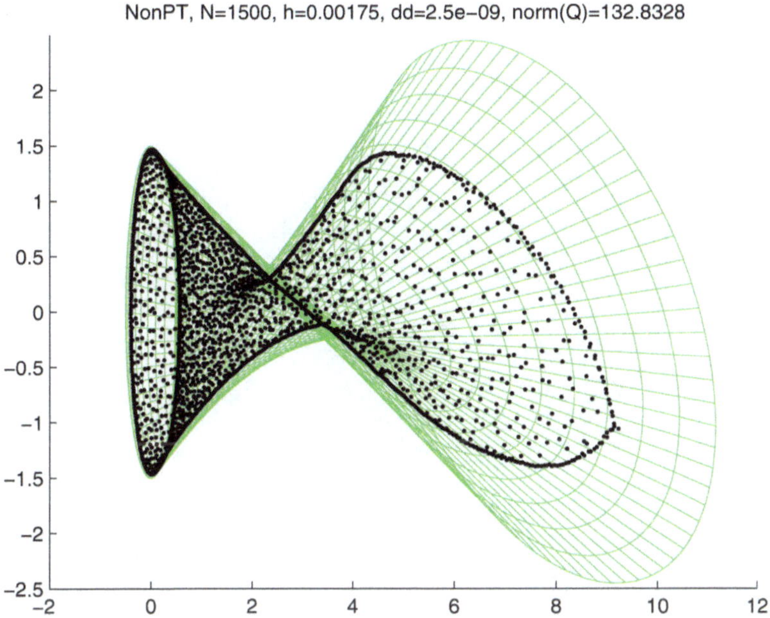

Fig. 20.52 Eigenvalues of $P_\delta = P + \delta Q$ as in (20.4.7), $h = 0.00175$, $N = 1500$, $\delta = 2.5 \times 10^{-9}$. $\|Q\| = 132.8328$

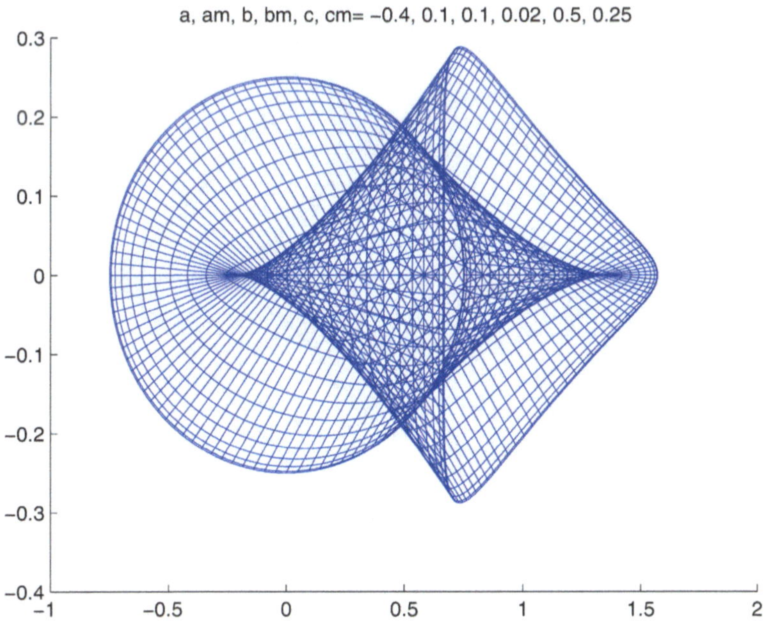

Fig. 20.53 Range of p in (20.5.10), $(a, a_{\mathrm{m}}, b, b_{\mathrm{m}}, c, c_{\mathrm{m}}) = (-0.4, 0.1, 0.1, 0.02, 0.5, 0.25)$

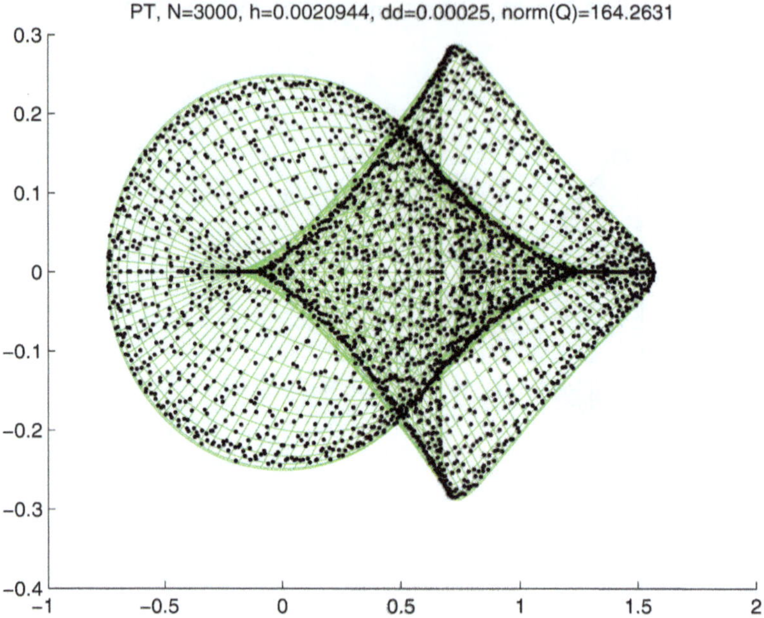

Fig. 20.54 Eigenvalues of $P_\delta = P + \delta Q$ as in (20.5.9) ff., $h = 0.0020944$, $N = 3000$, $\delta = 2.5 \times 10^{-4}$. $\|Q\| = 164.2631$

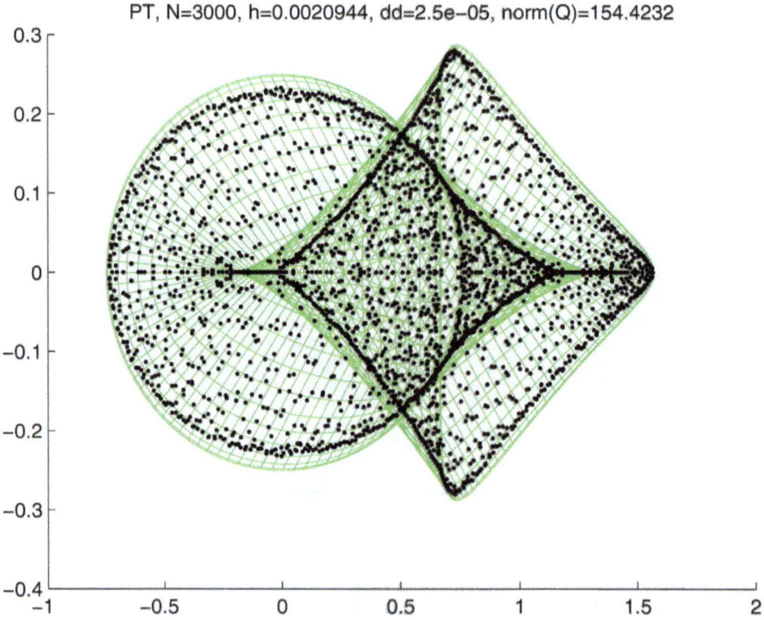

Fig. 20.55 Eigenvalues of $P_\delta = P + \delta Q$ as in (20.5.9) ff., $h = 0.0020944$, $N = 3000$, $\delta = 2.5 \times 10^{-5}$. $\|Q\| = 154.4232$

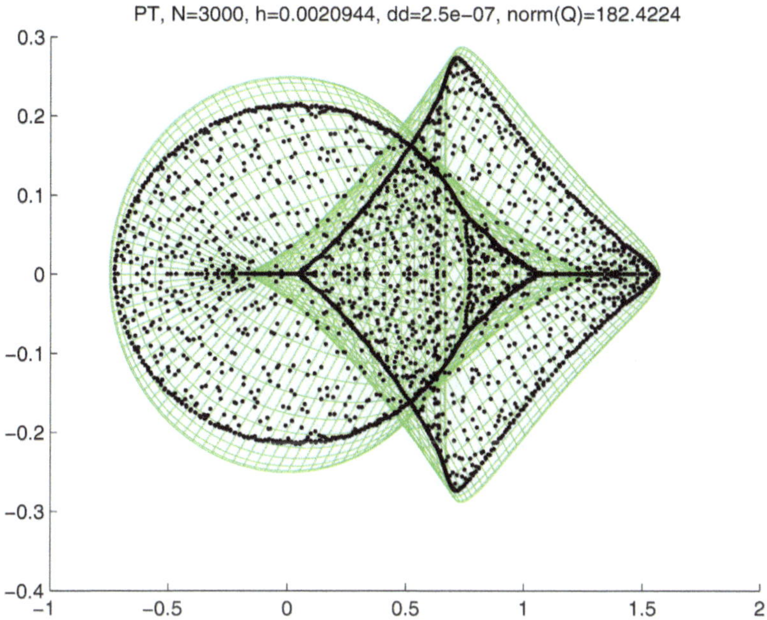

Fig. 20.56 Eigenvalues of $P_\delta = P + \delta Q$ as in (20.5.9) ff., $h = 0.0020944$, $N = 3000$, $\delta = 2.5 \times 10^{-7}$. $\|Q\| = 182.4224$

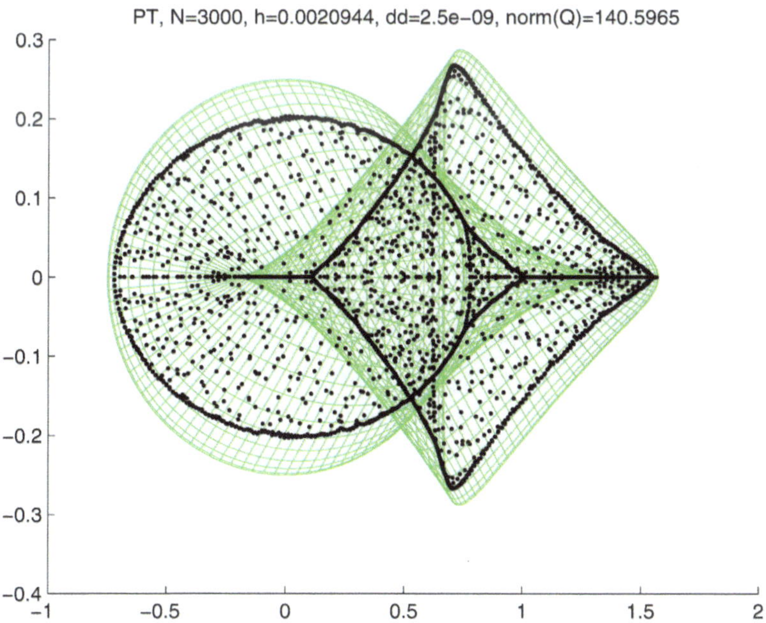

Fig. 20.57 Eigenvalues of $P_\delta = P + \delta Q$ as in (20.5.9) ff., $h = 0.0020944$, $N = 3000$, $\delta = 2.5 \times 10^{-9}$. $\|Q\| = 140.5965$

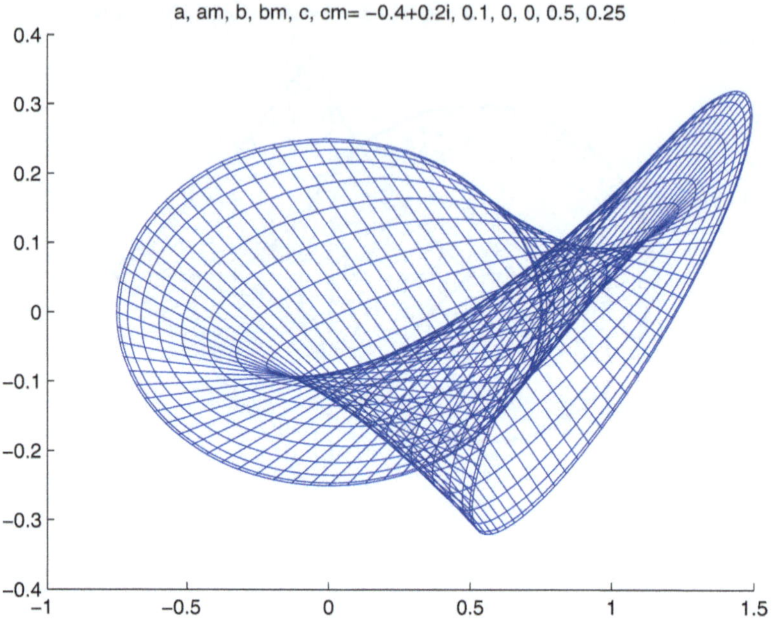

Fig. 20.58 Range of p in (20.5.10), $(a, a_{\mathrm{m}}, b, b_{\mathrm{m}}, c, c_{\mathrm{m}}) = (-0.4 + 0.2i, 0.1, 0, 0, 0.5, 0.25)$

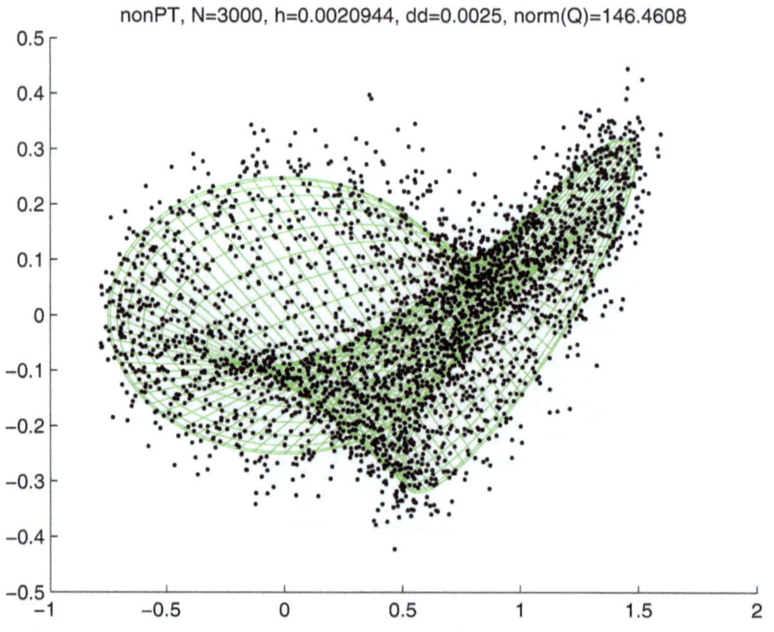

Fig. 20.59 Eigenvalues of $P_\delta = P + \delta Q$ as in (20.5.9) ff., $h = 0.0020944$, $N = 3000$, $\delta = 2.5 \times 10^{-3}$. $\|Q\| = 146.4608$

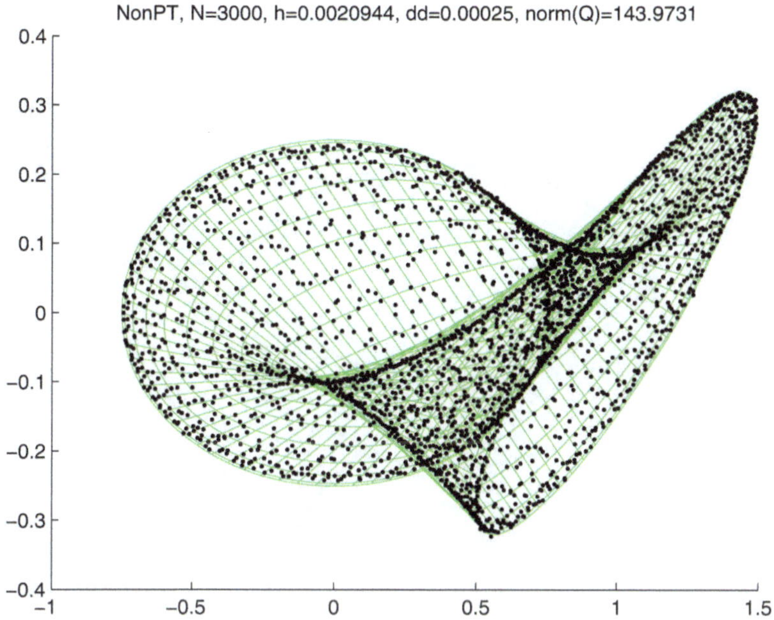

Fig. 20.60 Eigenvalues of $P_\delta = P + \delta Q$ as in (20.5.9) ff., $h = 0.0020944$, $N = 3000$, $\|Q\| = 143.9731$. $\delta = 2.5 \times 10^{-4}$

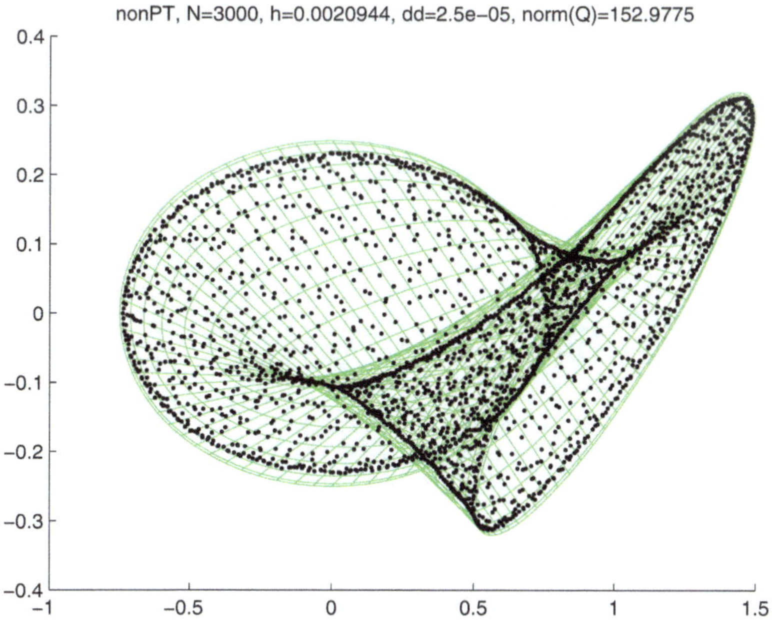

Fig. 20.61 Eigenvalues of $P_\delta = P + \delta Q$ as in (20.5.9) ff., $h = 0.0020944$, $N = 3000$, $\delta = 2.5 \times 10^{-5}$. $\|Q\| = 152.9775$

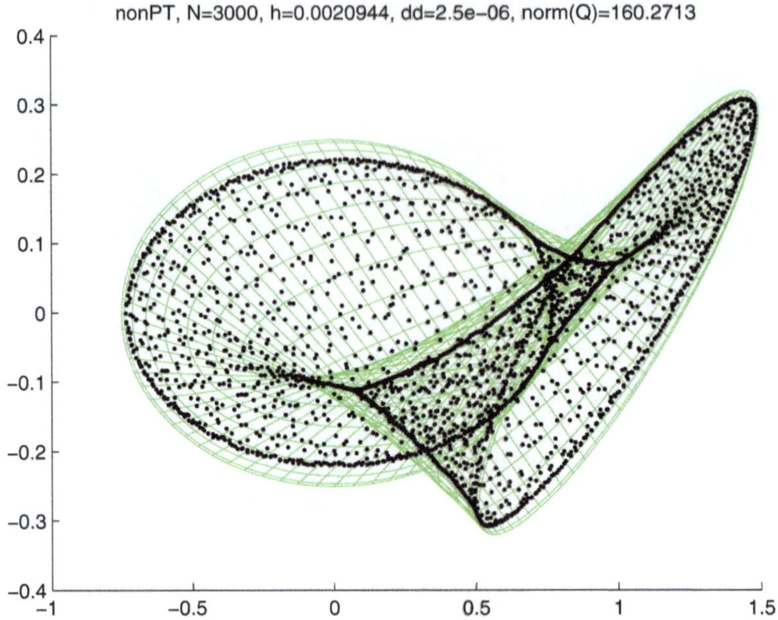

Fig. 20.62 Eigenvalues of $P_\delta = P + \delta Q$ as in (20.5.9) ff., $h = 0.0020944$, $N = 3000$, $\delta = 2.5 \times 10^{-6}$. $\|Q\| = 160.2713$

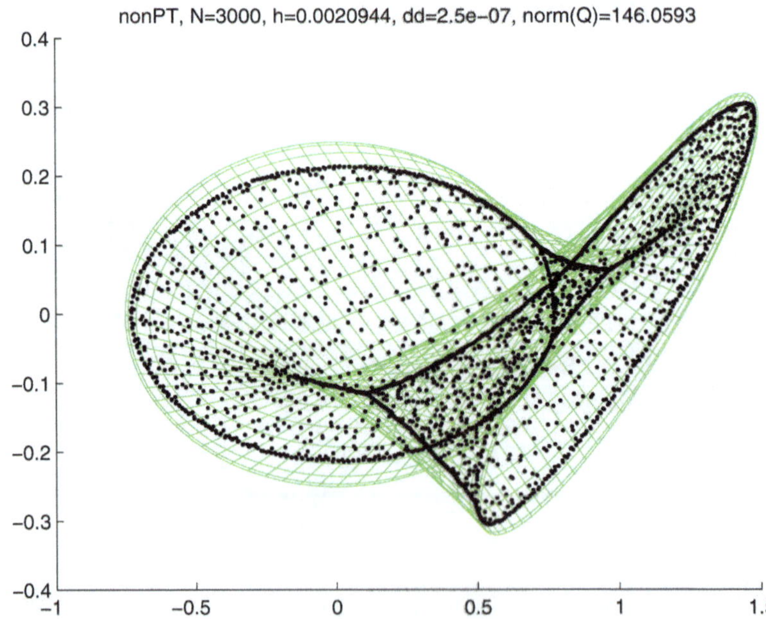

Fig. 20.63 Eigenvalues of $P_\delta = P + \delta Q$ as in (20.5.9) ff., $h = 0.0020944$, $N = 3000$, $\delta = 2.5 \times 10^{-7}$. $\|Q\| = 146.0593$

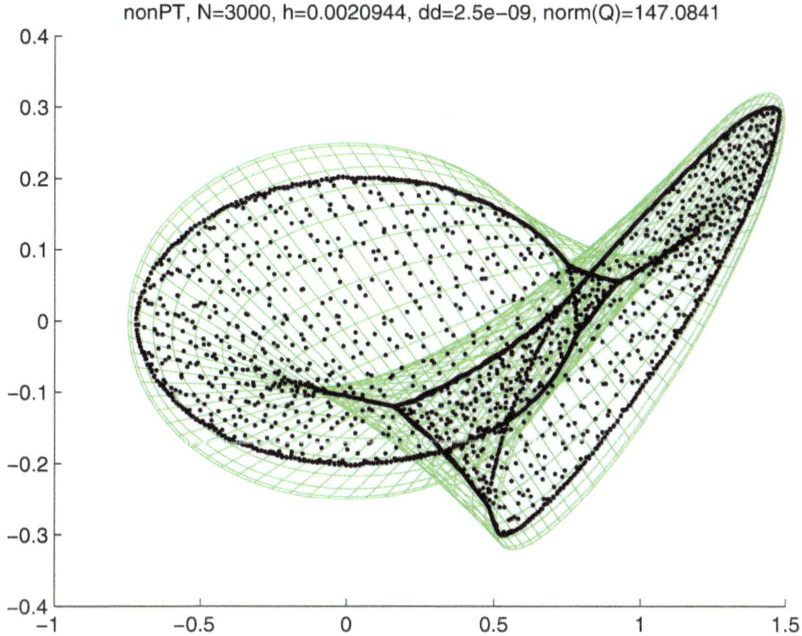

Fig. 20.64 Eigenvalues of $P_\delta = P + \delta Q$ as in (20.5.9) ff., $h = 0.0020944$, $N = 3000$, $\delta = 2.5 \times 10^{-9}$. $\|Q\| = 147.0841$

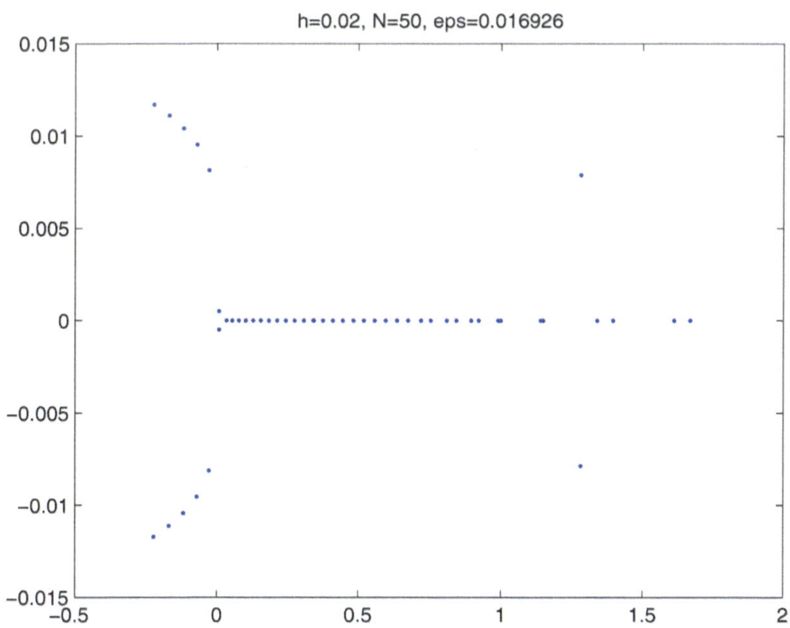

Fig. 20.65 Spectrum of P_ϵ in (20.6.13) with truncations of x and of $h\partial$. $h = 0.02$, $N = 50$, $\epsilon = 0.016926$

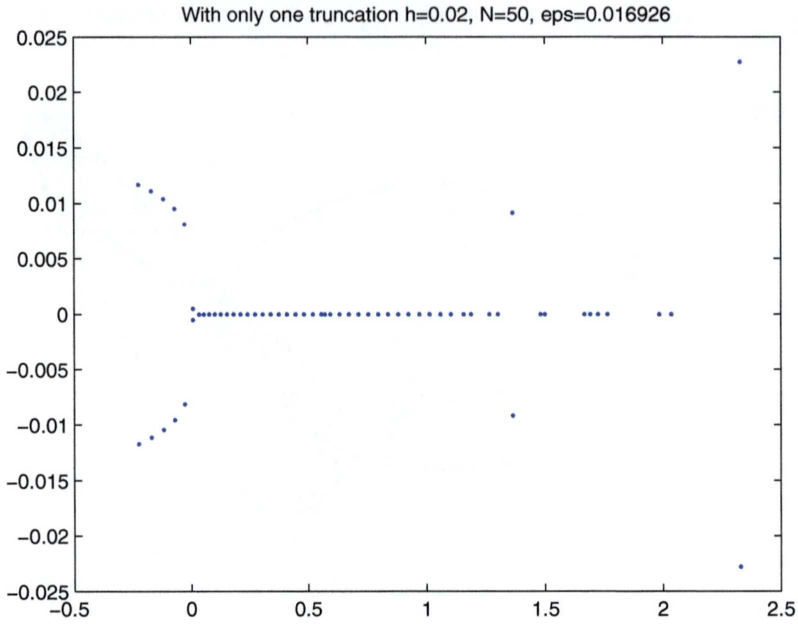

Fig. 20.66 Spectrum of P_ϵ in (20.6.13) with one global truncation. $h = 0.02$, $N = 50$, $\epsilon = 0.016926$

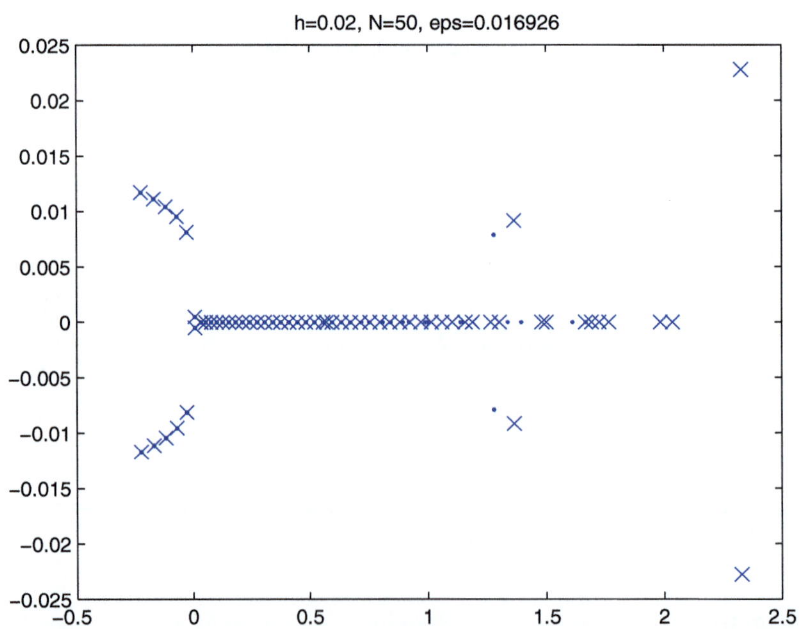

Fig. 20.67 Superposition of the eigenvalues in Figs. 20.65 and 20.66. The former are marked with dots and the latter with cross symbols

Bibliography

1. S. Agmon, On the eigenfunctions and on the eigenvalues of general elliptic boundary value problems. Commun. Pure Appl. Math **15**, 119–147 (1962)
2. S. Agmon, *Lectures on Elliptic Boundary Value Problems,* Van Nostrand Mathematical Studies, No. 2 (D. Van Nostrand Co., Inc., Princeton, 1965)
3. S. Agmon, *Lectures on Exponential Decay of Solutions of Second-Order Elliptic Equations: Bounds on Eigenfunctions of N-Body Schrödinger Operators*. Mathematical Notes, vol. 29 (Princeton University Press/University of Tokyo Press, Princeton/Tokyo, 1982)
4. L. Ahlfors, *Complex analysis. An Introduction to the Theory of Analytic Functions of One Complex Variable,* 3rd edn. International Series in Pure and Applied Mathematics (McGraw-Hill Book Co., New York, 1978)
5. A. Aleman, J. Viola, Singular-value decomposition of solution operators to model evolution equations. Int. Math. Res. Not. IMRN **2015**(17), 8275–8288 (2014). https://arxiv.org/abs/1409.1255
6. A. Aleman, J. Viola, On weak and strong solution operators for evolution equations coming from quadratic operators. J. Spectr. Theory **8**(1), 33–121 (2018). https://arxiv.org/abs/1409.1262
7. Y. Almog, The stability of the normal state of superconductors in the presence of electric currents. SIAM J. Math. Anal. **40**(2), 824–850 (2008)
8. V.G. Avakumović, Uber die Eigenfunktionen auf geschlossenen Riemannschen Mannigfaltigkeiten. Math. Z. **65**, 324–344 (1956)
9. M.S. Baouendi, J. Sjöstrand, Régularité analytique pour des opérateurs elliptiques singuliers en un point. Ark. Mat. **14**(1), 9–33 (1976)
10. M.S. Baouendi, J. Sjöstrand, Analytic regularity for the Dirichlet problem in domains with conic singularities. Ann. Sc. Norm. Super. Pisa Cl. Sci. **4**(3), 515–530 (1977)
11. B. Bellis, Subelliptic resolvent estimates for non-self-adjoint semiclassical Schrödinger operators. J. Spectral Theory **9**(1), 171–194 (2019). https://arxiv.org/abs/1609.00436
12. A. Benbernou, N. Boussekkine, N. Mecherout, T. Ramond, J. Sjöstrand, Non-real eigenvalues for $\mathcal{PT}$-symmetric double wells. Lett. Math. Phys. **106**(12), 1817–1835 (2016). http://arxiv.org/abs/1506.01898
13. C.M. Bender, S. Boettcher, P.N. Meisinger, $\mathcal{PT}$-symmetric quantum mechanics. J. Math. Phys. **40**(5), 2201–2229 (1999)

J. Sjöstrand, *Non-Self-Adjoint Differential Operators, Spectral Asymptotics and Random Perturbations*, Pseudo-Differential Operators 14,
https://doi.org/10.1007/978-3-030-10819-9

14. C.M. Bender, P.D. Mannheim, $\mathcal{PT}$ symmetry and necessary and sufficient conditions for the reality of energy eigenvalues. Phys. Lett. A **374**(15–16), 1616–1620 (2010)
15. P. Bleher, R. Mallison Jr., Zeros of sections of exponential sums. Int. Math. Res. Not. **2006**, 1–49 (2006), Art. ID 38937
16. W. Bordeaux Montrieux, Loi de Weyl presque sûre et résolvante pour des opérateurs différentiels non-autoadjoints, thèse, CMLS, Ecole Polytechnique, 2008. https://pastel.archives-ouvertes.fr/pastel-00005367
17. W. Bordeaux Montrieux, Loi de Weyl presque sûre pour un système différentiel en dimension 1. Ann. Henri Poincaré **12**(1), 173–204 (2011)
18. W. Bordeaux Montrieux, Estimation de résolvante et construction de quasimode près du bord du pseudospectre (2013). http://arxiv.org/abs/1301.3102
19. W. Bordeaux Montrieux, J. Sjöstrand, Almost sure Weyl asymptotics for non-self-adjoint elliptic operators on compact manifolds. Ann. Fac. Sci. Toulouse **19**(3–4), 567–587 (2010). http://arxiv.org/abs/0903.2937
20. C. Bordenave, M. Capitaine, Outlier eigenvalues for deformed i.i.d. random matrices. Commun. Pure Appl. Math. **69**(11), 2131–2194 (2016)
21. D. Borisov, P. Exner, Exponential splitting of bound states in a waveguide with a pair of distant windows. J. Phys. A **37**(10), 3411–3428 (2004)
22. L.S. Boulton, Non-self-adjoint harmonic oscillator, compact semigroups and pseudospectra. J. Operator Theory **47**(2), 413–429 (2002)
23. N. Boussekkine, N. Mecherout, $\mathcal{PT}$ -symmetry and Schrödinger operators – the simple well case. Math. Nachr. **289**(1), 13–27 (2016), French version at http://arxiv.org/pdf/1310.7335
24. L. Boutet de Monvel, Hypoelliptic operators with double characteristics and related pseudo-differential operators. Commun. Pure Appl. Math. **27**, 585–639 (1974)
25. N. Burq, M. Zworski, Geometric control in the presence of a black box. J. Am. Math. Soc. **17**(2), 443–471 (2004)
26. E. Caliceti, S. Graffi, J. Sjöstrand, Spectra of $\mathcal{PT}$-symmetric operators and perturbation theory. J. Phys. A: Math. Gen. **38**(1), 185–193 (2005)
27. E. Caliceti, S. Graffi, J. Sjöstrand, $\mathcal{PT}$ symmetric non-selfadjoint operators, diagonalizable and non-diagonalizable, with real discrete spectrum. J. Phys. A: Math. Theor. **40**(33), 10155–10170 (2007)
28. T. Carleman, *Über die asymptotische Verteilung der Eigenwerte partielle Differentialgleichungen,* Berichten der mathematisch-physisch Klasse der Sächsischen Akad. der Wissenschaften zu Leipzig, LXXXVIII Band, Sitsung v. 15. Juni 1936
29. T. Christiansen, Several complex variables and the distribution of resonances in potential scattering. Commun. Math. Phys. **259**(3), 711–728 (2005)
30. T. Christiansen, Several complex variables and the order of growth of the resonance counting function in Euclidean scattering. Int. Math. Res. Not. **2006**, 36 pp. (2006), Art. ID 43160
31. T. Christiansen, P.D. Hislop, The resonance counting function for Schrödinger operators with generic potentials. Math. Res. Lett. **12**(5–6), 821–826 (2005)
32. T.J. Christiansen, M. Zworski, Probabilistic Weyl laws for quantized tori. Commun. Math. Phys. **299**(2), 305–334 (2010). http://arxiv.org/abs/0909.2014
33. E.B. Davies, Semi-classical states for non-self-adjoint Schrödinger operators. Commun. Math. Phys. **200**(1), 35–41 (1999)
34. E.B. Davies, Pseudospectra, the harmonic oscillator and complex resonances. Proc. Roy. Soc. Lond. A **455**, 585–599 (1999)
35. E.B. Davies, Pseudopectra of differential operators. J. Operator Theory **43**, 243–262 (2000)

36. E.B. Davies, Eigenvalues of an elliptic system. Math. Z. **243**(4), 719–743 (2003)
37. E.B. Davies, Semigroup growth bounds. J. Operator Theory **53**(2), 225–249 (2005)
38. E.B. Davies, M. Hager, Perturbations of Jordan matrices. J. Approx. Theory **156**(1), 82–94 (2009)
39. E.B. Davies, A.B.J. Kuijlaars, Spectral asymptotics of the non-self-adjoint harmonic oscillator. J. Lond. Math. Soc. (2) **70**, 420–426 (2004)
40. N. Dencker, J. Sjöstrand, M. Zworski, Pseudospectra of semiclassical *(pseudo-)differential operators,* Commun. Pure Appl. Math. **57**(3), 384–415 (2004)
41. M. Dimassi, J. Sjöstrand, *Spectral Asymptotics in the Semi-Classical Limit*. London Mathematical Society Lecture Note Series, vol. 268 (Cambridge University Press, Cambridge, 1999)
42. E.M. Dyn'kin, An operator calculus based upon the Cauchy-Green formula. Zapiski Nauchn. semin. LOMI **30**, 33–40 (1972), J. Soviet Math. **4**(4), 329–334 (1975)
43. K.J. Engel, R. Nagel, *One-Parameter Semigroups for Linear Evolution Equations*. Graduate Texts in Mathematics, vol. 194 (Springer, New York, 2000)
44. M.V. Fedoryuk, *Asymptotic Analysis. Linear Ordinary Differential Equations* (Springer, Berlin, 1993)
45. T. Fischer, *Existence, Uniqueness, and Minimality of the Jordan Measure Decomposition*. http://arxiv.org/abs/1206.5449
46. I. Gallagher, Th. Gallay, F. Nier, Spectral asymptotics for large skew-symmetric perturbations of the harmonic oscillator. Int. Math. Res. Not. IMRN **2009**(12), 2147–2199 (2009)
47. V.I. Girko, *Theory of Random Determinants*. Mathematics and Its Applications (Kluwer Academic Publishers, Dordrecht, 1990)
48. I.C. Gohberg, M.G. Krein, *Introduction to the Theory of Linear Non-selfadjoint Operators*. Translations of Mathematical Monographs, vol. 18 (AMS, Providence, 1969)
49. D. Grieser, D. Jerison, Asymptotics of the first nodal line of a convex domain. Invent. Math. **125**(2), 197–219 (1996)
50. A. Grigis, Estimations asymptotiques des intervalles d'instabilité pour l'équation de Hill. Ann. Sci. École Norm. Sup. (4) **20**(4), 641–672 (1987)
51. A. Grigis, J. Sjöstrand, *Microlocal Analysis for Differential Operators*. London Mathematical Society Lecture Notes Series, vol. 196 (Cambridge University Press, Cambridge, 1994)
52. A. Guionnet, P.M. Wood, O. Zeitouni, Convergence of the spectral measure of non-normal matrices. Proc. AMS **142**(2), 667–679 (2014)
53. M. Hager, Instabilité spectrale semiclassique d'opérateurs non-autoadjoints, Thesis, 2005. https://tel.archives-ouvertes.fr/tel-00010848/
54. M. Hager, Instabilité spectrale semiclassique pour des opérateurs non-autoadjoints. I. Un modèle. Ann. Fac. Sci. Toulouse Math. **15**(2), 243–280 (2006)
55. M. Hager, Instabilité spectrale semiclassique d'opérateurs non-autoadjoints. II. Ann. Henri Poincaré **7**(6), 1035–1064 (2006)
56. M. Hager, J. Sjöstrand, Eigenvalue asymptotics for randomly perturbed non-selfadjoint operators. Math. Ann. **342**(1), 177–243 (2008). http://arxiv.org/abs/math/0601381
57. B. Helffer, *Spectral Theory and Its Applications*. Cambridge Studies in Advanced Mathematics, vol. 139 (Cambridge University Press, Cambridge, 2013)
58. B. Helffer, F. Nier, Hypoelliptic Estimates and Spectral Theory for Fokker-Planck Operators and Witten Laplacians. Lecture Notes in Mathematics, vol. 1862 (Springer, New York, 2004)
59. B. Helffer, J. Sjöstrand, Multiple wells in the semiclassical limit. I. Commun. Partial Differ. Equ. **9**(4), 337–408 (1984)
60. B. Helffer, J. Sjöstrand, *Résonances en limite semi-classique*. Mém. Soc. Math. France (N.S.), vol. 24–25 (Gauthier-Villars, Paris, 1986), pp. 1–228

61. B. Helffer, J. Sjöstrand, *Equation de Schrödinger avec champ magnétique et équation de Harper.* Springer Lecture Notes in Physics, vol. 345, 118–197 (1989)
62. B. Helffer, J. Sjöstrand, From resolvent bounds to semigroup bounds, published as part III of the published version of Ref. [138] (2010). http://arxiv.org/abs/1001.4171
63. R. Henry, Spectral projections of the complex cubic oscillator. Ann. Henri Poincaré **15**(10), 2025–2043 (2014)
64. R. Henry, Instability for even non-selfadjoint anharmonic oscillators. J. Spectr. Theory **4**(2), 349–364 (2014)
65. F. Hérau, J. Sjöstrand, C. Stolk, Semiclassical analysis for the Kramers-Fokker-Planck equation. Commun. PDE **30**(5–6), 689–760 (2005)
66. M. Hitrik, Eigenfrequencies and expansions for damped wave equations. Methods Appl. Anal. **10**(4), 543–564 (2003)
67. M. Hitrik, Boundary spectral behavior for semiclassical operators in dimension one. Int. Math. Res. Not. **2004**(64), 3417–3438 (2004)
68. M. Hitrik, L. Pravda-Starov, Spectra and semigroup smoothing for non-elliptic quadratic operators. Math. Ann. **344**(4), 801–846 (2009)
69. M. Hitrik, K. Pravda-Starov, Semiclassical hypoelliptic estimates for non-selfadjoint operators with double characteristics. Commun. Partial Differ. Equ. **35**(6), 988–1028 (2010)
70. M. Hitrik, K. Pravda-Starov, Eigenvalues and subelliptic estimates for non-selfadjoint semiclassical operators with double characteristics. Ann. Inst. Fourier (Grenoble) **63**(3), 985–1032 (2013)
71. M. Hitrik, J. Sjöstrand, Rational invariant tori, phase space tunneling, and spectra for non-selfadjoint operators in dimension 2. Ann. Sci. ENS 4 **41**(4), 511–571 (2008)
72. M. Hitrik, J. Sjöstrand, Non-selfadjoint perturbations of selfadjoint operators in 2 dimensions IIIa. One branching point. Canad. J. Math. **60**(3), 572–657 (2008)
73. M. Hitrik, J. Sjöstrand, Rational invariant tori and band edge spectra for non-selfadjoint operators. J. Eur. Math. Soc. **20**(2), 391–457 (2018). http://arxiv.org/abs/1502.06138
74. M. Hitrik, J. Sjöstrand, Two minicourses on analytic microlocal analysis, in *Algebraic and Analytic Microlocal Analysis, AAMA, Evanston, Illinois, USA, 2012 and 2013.* Springer Proceedings in Mathematics & Statistics, vol. 269 (2018), pp. 483–540. http://arxiv.org/abs/1508.00649
75. M. Hitrik, J. Sjöstrand, J. Viola, Resolvent estimates for elliptic quadratic differential operators. Anal. PDE **6**(1), 181–196 (2013)
76. M. Hitrik, K. Pravda-Starov, J. Viola, Short-time asymptotics of the regularizing effect for semigroups generated by quadratic operators. Bull. Sci. Math. **141**(7), 615–675 (2017). https://arxiv.org/abs/1510.01992
77. M. Hitrik, K. Pravda-Starov, J. Viola, From semigroups to subelliptic estimates for quadratic operators. Trans. Am. Math. Soc. **370**(10), 7391–7415 (2018). https://arxiv.org/abs/1510.02072
78. L. Hörmander, Differential equations without solutions. Math. Ann. **140**, 169–173 (1960)
79. L. Hörmander, Differential operators of principal type. Math. Ann. **140**, 124–146 (1960)
80. L. Hörmander, The spectral function of an elliptic operator. Acta Math. **121**, 193–218 (1968)
81. L. Hörmander, Fourier integral operators, lectures at the Nordic summer school of mathematics, unpublished notes, 1969. Available at http://portal.research.lu.se/portal/files/50761067/Hormander_Tjorn69.pdf
82. L. Hörmander, *On the Existence and the Regularity of Solutions of Linear Pseudo-Differential Equations.* Série des Conférences de l'Union Mathématique Internationale, No. 1. Monographie No. 18 de l'Enseignement Mathématique. Secrétariat de l'Enseignement Mathématique, 69 pp. (Université de Genève, Geneva, 1971)

83. L. Hörmander, *The Analysis of Linear Partial Differential Operators.* I–IV, Grundlehren der Mathematischen Wissenschaften, vols. 256, 257, 274, 275 (Springer, Berlin, 1983/1985)
84. L. Hörmander, *An Introduction to Complex Analysis in Several Variables,* 3rd edn. North-Holland Mathematical Library, vol. 7 (North-Holland Publishing Co., Amsterdam, 1990)
85. R.A. Horn, C.R. Johnson, *Matrix Analysis* (Cambridge University Press, Cambridge, 1985)
86. V. Ivrii, *Microlocal Analysis and Precise Spectral Asymptotics.* Springer Monographs in Mathematics (Springer, Berlin, 1998)
87. O. Kallenberg, *Foundations of Modern Probability.* Probability and Its Applications (New York) (Springer, New York, 1997)
88. T. Kato, *Perturbation Theory for Linear Operators.* Die Grundlehren der mathematischen Wissenschaften, Band, vol. 132 (Springer, New York, 1966)
89. V.V. Kučerenko, Asymptotic solutions of equations with complex characteristics (Russian). Mat. Sb. (N.S.) **95**(137), 163–213, 327 (1974)
90. H.H. Kuo, *Gaussian Measures in Banach Spaces.* Lecture Notes in Mathematics, vol. 463 (Springer, Berlin, 1975)
91. B. Lascar, J. Sjöstrand, Equation de Schrödinger et propagation des singularités pour des opérateurs pseudodifférentiels à caractéristiques réelles de multiplicité variable I. Astérisque **95**, 467–523 (1982)
92. G. Lebeau, *Equation des ondes amorties.* Algebraic and Geometric Methods in Mathematical Physics (Kaciveli, 1993), pp. 73–109. Mathematical Physics Studies, vol. 19 (Kluwer Academic Publishers, Dordrecht, 1996)
93. B.Ya. Levin, *Distribution of Zeros of Entire Functions,* English translation (American Mathematical Society, Providence, 1980)
94. B.M. Levitan, Some questions of spectral theory of selfadjoint differential operators. Usp. Mat. Nauk (N.S.) **11**(72), no. 6, 117–144 (1956)
95. L. Lithner, A theorem of the Phragmén-Lindelöf type for second-order elliptic operators. Ark. Mat. **5**, 281–285 (1964)
96. A.S. Markus, *Introduction to the Spectral Theory of Polynomial Operator Pencils,* Translated from the Russian by H.H. McFaden. Translation ed. by B. Silver. With an appendix by M.V. Keldysh. Translations of Mathematical Monographs, vol. 71 (American Mathematical Society, Providence, 1988)
97. A.S. Markus, V.I. Matseev, Asymptotic behavior of the spectrum of close-to-normal operators. Funktsional. Anal. i Prilozhen. **13**(3), 93–94 (1979), Functional Anal. Appl. **13**(3), 233–234 (1979) (1980)
98. J. Martinet, Sur les propriétés spectrales d'opérateurs nonautoadjoints provenant de la mécanique des fluides, Thèse de doctorat, Université de Paris Sud, 2009
99. V.P. Maslov, *Operational Methods.* Translated from the Russian by V. Golo, N. Kulman, G. Voropaeva (Mir Publishers, Moscow, 1976)
100. O. Matte, Correlation asymptotics for non-translation invariant lattice spin systems. Math. Nachr. **281**(5), 721–759 (2008)
101. N. Mecherout, N. Boussekkine, T. Ramond, J. Sjöstrand, *$\mathcal{PT}$-symmetry and Schrödinger operators. The double well case.* Math. Nachr. **289**(7), 854–887 (2016). http://arxiv.org/abs/1502.06102
102. A. Melin, J. Sjöstrand, *Fourier Integral Operators with Complex-Valued Phase Functions.* Fourier Integral Operators and Partial Differential Equations (Colloq. Internat., Univ. Nice, Nice, 1974), pp. 120–223. Lecture Notes in Mathematics, vol. 459 (Springer, Berlin, 1975)
103. A. Melin, J. Sjöstrand, Fourier integral operators with complex phase functions and parametrix for an interior boundary value problem. Commun. Partial Differ. Equ. **1**(4), 313–400 (1976)

104. A. Melin, J. Sjöstrand, Determinants of pseudodifferential operators and complex deformations of phase space. Methods Appl. Anal. **9**(2), 177–238 (2002). https://arxiv.org/abs/math/0111292
105. A. Melin, J. Sjöstrand, Bohr-Sommerfeld quantization condition for non-selfadjoint operators in dimension 2. Astérique **284**, 181–244 (2003). https://arxiv.org/abs/math/0111293
106. A. Menikoff, J. Sjöstrand, On the eigenvalues of a class of hypo-elliptic operators. Math. Ann. **235**, 55–85 (1978)
107. D.V. Nekhaev, A.I. Shafarevich, Semiclassical limit of the spectrum of the Schrödinger operator with complex periodic potential. Mat. Sb. **208**(10), 126–148 (2017)
108. L. Nirenberg, *A proof of the Malgrange preparation theorem*. In *Proceedings of Liverpool Singularities-Symposium, I (1969/70)*, pp. 97–105. Lecture Notes in Mathematics, vol. 192 (Springer, Berlin, 1971)
109. M. Ottobre, G. Pavliotis, K. Pravda-Starov, Exponential return to equilibrium for hypoelliptic quadratic systems. J. Funct. Anal. **262**, 4000–4039 (2012)
110. A. Pfluger, Die Wertverteilung und das Verhalten von Betrag und Argument einer speciellen Klasse analytischen Funktionen I, II. Comment. Math. Helv. **11**, 180–214 (1938), **12**, 25–65 (1939)
111. K. Pravda-Starov, Étude du pseudo-spectre d'opérateurs non auto-adjoints, thesis, 2006. http://tel.archives-ouvertes.fr/docs/00/10/98/95/PDF/manuscrit.pdf
112. K. Pravda-Starov, A complete study of the pseudo-spectrum for the rotated harmonic oscillator. J. Lond. Math. Soc. (2) **73**(3), 745–761 (2006)
113. K. Pravda-Starov, On the pseudospectrum of elliptic quadratic differential operators. Duke Math. J. **145**(2), 249–279 (2008)
114. K. Pravda-Starov, Subelliptic estimates for quadratic differential operators. Am. J. Math. **133**, 39–89 (2011)
115. P. Redparth, Spectral properties of non-self-adjoint operators in the semiclassical regime. J. Differ. Equ. **177**(2), 307–330 (2001)
116. M. Reed, B. Simon, *Methods of Modern Mathematical Physics. I. Functional Analysis,* 2nd edn. (Academic, New York, 1980), *Methods of Modern Mathematical Physics. II. Fourier Analysis, Self-Adjointness* (Academic, New York, 1975), *Methods of Modern Mathematical Physics. IV. Analysis of Operators* (Academic Press, New York, 1978)
117. F. Riesz, B.-Sz. Nagy, *Leçons d'Analyse Fonctionnelle,* Quatrième édition. Académie des Sciences de Hongrie, Gauthier-Villars (Editeur-Imprimeur-Libraire/Akadémiai Kiadó, Paris/Budapest, 1965)
118. G. Rivière, Delocalization of slowly damped eigenmodes on Anosov manifolds. Commun. Math. Phys. **316**(2), 555–593 (2012)
119. D. Robert, *Autour de l'Approximation Semi-classique*. Progress in Mathematics, vol. 68 (Birkhäuser, Boston, 1987)
120. O. Rouby, Bohr-Sommerfeld quantization conditions for non-selfadjoint perturbations of selfadjoint operators in dimension one. Int. Math. Res. Not. IMRN **2018**(7), 2156–2207 (2018). http://arxiv.org/abs/1511.06237
121. O. Rouby, S. Vũ Ngọc, J. Sjöstrand, Analytic Bergman operators in the semiclassical limit (2018). https://arxiv.org/abs/1808.00199
122. E. Schenk, Systèmes quantiques ouverts et méthodes semi-classiques, thèse novembre, 2009. https://www.ipht.fr/Docspht/articles/t09/266/public/thesis_schenck.pdf
123. R.T. Seeley, Complex powers of an elliptic operator, in *Proceedings of Symposium on Singular Integrals*, vol. 10, pp. 288–307 (American Mathematical Society, Providence, 1967)

124. R. Seeley, A simple example of spectral pathology for differential operators. Commun. Partial Differ. Equ. **11**(6), 595–598 (1986)
125. B. Shiffman, S. Zelditch, S. Zrebiec, Overcrowding and hole probabilities for random zeros on complex manifolds. Ind. Univ. Math. J. **57**(5), 1977–1997 (2008)
126. K.C. Shin, On the reality of the eigenvalues for a class of $\mathcal{PT}$-symmetric oscillators. Commun. Math. Phys. **229**(3), 543–564 (2002)
127. A.A. Shkalikov, Spectral portraits of the Orr-Sommerfeld operator at large Reynolds numbers (Russian). Sovrem. Mat. Fundam. Napravl. **3**, 89–112 (2003), Translation in J. Math. Sci. (N. Y.) **124**(6), 5417–5441 (2004)
128. A.A. Shkalikov, Eigenvalue asymptotics of perturbed self-adjoint operators. Methods Funct. Anal. Topol. **18**(1), 79–89 (2012)
129. B. Simon, Semiclassical analysis of low lying eigenvalues. I. Nondegenerate minima: asymptotic expansions. Ann. Inst. H. Poincaré Sect. A (N.S.) **38**(3), 295–308 (1983), Erratum in ibid **40**(2), 224 (1984)
130. B. Simon, Semiclassical analysis of low lying eigenvalues. II. Tunneling. Ann. Math. (2) **120**(1), 89–118 (1984)
131. J. Sjöstrand, Parametrices for pseudodifferential operators with multiple characteristics. Ark. Mat. **12**(1), 85–130 (1974)
132. J. Sjöstrand, *Singularités analytiques microlocales*. Astérisque 95 (Société Mathématique de France, Paris, 1982), pp. 1–166
133. J. Sjöstrand, *Puits multiples, (d'après des travaux avec B. Helffer),* Sém. Goulaouic-Meyer-Schwartz, 1983–1984, Exposé No. 7, École Polytech., Palaiseau, 1984. http://www.numdam.org/item?id=SEDP_1983-1984____A7_0
134. J. Sjöstrand, Asymptotic distribution of eigenfrequencies for damped wave equations. Publ. RIMS Kyoto Univ. **36**(5), 573–611 (2000)
135. J. Sjöstrand, Resonances for bottles and trace formulae. Math. Nachr. **221**, 95–149 (2001)
136. J. Sjöstrand, *Lectures on Resonances* (2007). http://sjostrand.perso.math.cnrs.fr/Coursgbg.pdf
137. J. Sjöstrand, Eigenvalue distribution for non-self-adjoint operators with small multiplicative random perturbations. Ann. Fac. Sci. Toulouse **18**(4), 739–795 (2009). http://arxiv.org/abs/0802.3584
138. J. Sjöstrand, *Spectral properties of non-self-adjoint operators,* Notes d'un minicours à Evian les Bains, 8–12 juin, 2009, Actes des Journées d'é.d.p. d'Évian 2009, published: http://jedp.cedram.org:80/jedp-bin/feuilleter?id=JEDP_2009___ preprint: http://arxiv.org/abs/1002.4844
139. J. Sjöstrand, Eigenvalue distribution for non-self-adjoint operators on compact manifolds with small multiplicative random perturbations. Ann. Fac. Sci. Toulouse **19**(2), 277–301 (2010). http://arxiv.org/abs/0809.4182
140. J. Sjöstrand, *Resolvent Estimates for Non-self-adjoint Operators via Semi-groups*. International Mathematical Series, vol. 13, pp. 359–384 Around the research of Vladimir Maz'ya, III (Springer/Tamara Rozhkovskaya Publisher, Novosibirsk, 2010). http://arxiv.org/abs/0906.0094
141. J. Sjöstrand, Counting zeros of holomorphic functions of exponential growth. J. Pseudodiffer. Operators Appl. **1**(1), 75–100 (2010). http://arxiv.org/abs/0910.0346
142. J. Sjöstrand, Eigenvalue distributions and Weyl laws for semiclassical non-self-adjoint operators in 2 dimensions. Geometric Aspects of Analysis and Mechanics, vols. 345–352. Progress in Mathematics, vol. 292 (Birkhäuser/Springer, New York, 2011)
143. J. Sjöstrand, *PT Symmetry and Weyl Asymptotics*. The Mathematical Legacy of Leon Ehrenpreis, Springer Proceedings in Mathematics, vol. 16, pp. 299–308 (2012). http://arxiv.org/abs/1105.4746

144. J. Sjöstrand, Weyl Law for Semi-classical Resonances with Randomly Perturbed Potentials. Mém. de la SMF **136**, 144 (2014). http://arxiv.org/abs/1111.3549
145. J. Sjöstrand, M. Vogel, Interior eigenvalue density of large bi-diagonal matrices subject to random perturbations, in *Proceedings of the Conference "Microlocal Analysis and Singular Perturbation Theory"*, RIMS, Kyoto University, October 5–9, 2015, RIMS Conference Proceedings series "RIMS Kokyuroku Bessatsu" (2015). http://arxiv.org/abs/1604.05558
146. J. Sjöstrand, M. Vogel, Large bi-diagonal matrices and random perturbations. J. Spectr. Theory **6**(4), 977–1020 (2016). http://arxiv.org/abs/1512.06076
147. J. Sjöstrand, M. Vogel, Interior eigenvalue density of Jordan matrices with random perturbations. In Analysis Meets Geometry: A Tribute to Mikael Passare. Trends in Mathematics, pp. 439–466 (Springer, Cham, 2017). http://arxiv.org/abs/1412.2230
148. J. Sjöstrand, M. Zworski, Elementary linear algebra for advanced spectral problems. Ann. Inst. Fourier **57**(7), 2095–2141 (2007)
149. M. Sodin, Zeros of Gaussian analytic functions. Math. Res. Lett. **7**, 371–381 (2000)
150. M. Sodin, B. Tsirelson, Random complex zeroes. III. Decay of the hole probability. Isr. J. Math. **147**, 371–379 (2005)
151. L.N. Trefethen, Pseudospectra of linear operators. SIAM Rev. **39**(3), 383–406 (1997)
152. L.N. Trefethen, M. Embree, *Spectra and Pseudospectra. The Behavior of Nonnormal Matrices and Operators* (Princeton University Press, Princeton, 2005)
153. J. Viola, Spectral projections and resolvent bounds for partially elliptic quadratic differential operators. J. Pseudo-Differ. Oper. Appl. **4**, 145–221 (2013)
154. J. Viola, The norm of the non-self-adjoint harmonic oscillator semigroup. Integr. Equ. Oper. Theory **85**(4), 513–538 (2016)
155. M. Vogel, The precise shape of the eigenvalue intensity for a class of non-selfadjoint operators under random perturbations. Ann. Henri Poincaré **18**(2), 435–517 (2017). http://arxiv.org/abs/1401.8134
156. M. Vogel, Two point eigenvalue correlation for a class of non-selfadjoint operators under random perturbations. Commun. Math. Phys. **350**(1), 31–78 (2017). http://arxiv.org/abs/1412.0414
157. A. Voros, *Spectre de l'Équation de Schrödinger et Méthode BKW*. Publications Mathématiques d'Orsay 81, vol. 9, 75 pp. (Université de Paris-Sud, Département de Mathématique, Orsay, 1982)
158. G. Weiss, The resolvent growth assumption for semigroups on Hilbert spaces. J. Math. Anal. Appl. **145**, 154–171 (1990)
159. H. Weyl, Das asymptotische Verteilungsgesetz der Eigenwerte linearer partieller Differentialgleichungen (mit einer Anwendung auf die Theorie der Hohlraumstrahlung). Math. Ann. **71**(4), 441–479 (1912)
160. S. Zrebiec, The zeros of flat Gaussian random holomorphic functions on $\mathbb{C}^n$, and hole probability. Mich. Math. J. **55**(2), 269–284 (2007)
161. M. Zworski, A remark on a paper of E. B. Davies: "Semi-classical states for non-self-adjoint Schrödinger operators". Proc. Am. Math. Soc. **129**(10), 2955–2957 (2001)
162. M. Zworski, Numerical linear algebra and solvability of partial differential equations. Commun. Math. Phys. **229**(2), 293–307 (2002)

Symbols

Terms

Item	Page/formula
Bohr–Sommerfeld condition	154
Canonical transformation	102
Classical principal symbol	314
Cyclicity of the trace	173, (8.4.4)
Determinant	174
Eikonal equation	58, (4.1.17)
ϵ-pseudospectrum	13
FBI	196
Fourier transform	332
Green kernel	228
Hamilton field	185
Index	157
Ky Fan	166
Lagrangian manifold	99, 186, 187
Metaplectic	124, 134
Microlocally	195
Multiindex	313
Order function	67
Poisson bracket	31
Poisson kernel	228
Pseudodifferential operator	68
Semi-classical principal symbol	54, 314
Subharmonic	14
Symplectic coordinates	96
Symplectic form	96, 186
Transpose	315
Weyl inequalities	167
Weyl quantization	68

MIX
Papier aus verantwortungsvollen Quellen
Paper from responsible sources
FSC® C105338

If you have any concerns about our products,
you can contact us on
ProductSafety@springernature.com

In case Publisher is established outside the EU,
the EU authorized representative is:
Springer Nature Customer Service Center GmbH
Europaplatz 3, 69115 Heidelberg, Germany

Printed by Libri Plureos GmbH
in Hamburg, Germany